# Lecture Notes in Networks and Systems

Volume 290

The series "Lecture Notes in Networks and Systems" publishes the latest developments in Networks and Systems—quickly, informally and with high quality. Original research reported in proceedings and post-proceedings represents the core of LNNS.

Volumes published in LNNS embrace all aspects and subfields of, as well as new challenges in, Networks and Systems.

The series contains proceedings and edited volumes in systems and networks, spanning the areas of Cyber-Physical Systems, Autonomous Systems, Sensor Networks, Control Systems, Energy Systems, Automotive Systems, Biological Systems, Vehicular Networking and Connected Vehicles, Aerospace Systems, Automation, Manufacturing, Smart Grids, Nonlinear Systems, Power Systems, Robotics, Social Systems, Economic Systems and other. Of particular value to both the contributors and the readership are the short publication timeframe and the world-wide distribution and exposure which enable both a wide and rapid dissemination of research output.

The series covers the theory, applications, and perspectives on the state of the art and future developments relevant to systems and networks, decision making, control, complex processes and related areas, as embedded in the fields of interdisciplinary and applied sciences, engineering, computer science, physics, economics, social, and life sciences, as well as the paradigms and methodologies behind them.

Indexed by SCOPUS, INSPEC, WTI Frankfurt eG, zbMATH, SCImago.

All books published in the series are submitted for consideration in Web of Science.

More information about this series at http://www.springer.com/series/15179

Samiksha Shukla · Aynur Unal ·
Joseph Varghese Kureethara ·
Durgesh Kumar Mishra ·
Dong Seog Han
Editors

# Data Science and Security

## Proceedings of IDSCS 2021

Springer

*Editors*
Samiksha Shukla
Christ University
Bangalore, India

Joseph Varghese Kureethara
Christ University
Bangalore, India

Dong Seog Han
School of Electronics Engineering
Kyungpook National University
Daegu, Korea (Republic of)

Aynur Unal
Digital Monozukuri
Stanford Alumni
Palo Alto, CA, USA

Durgesh Kumar Mishra
Computer Science and Engineering
Sri Aurobindo Institute of Technology
Indore, Madhya Pradesh, India

ISSN 2367-3370 ISSN 2367-3389 (electronic)
Lecture Notes in Networks and Systems
ISBN 978-981-16-4485-6 ISBN 978-981-16-4486-3 (eBook)
https://doi.org/10.1007/978-981-16-4486-3



This Springer imprint is published by the registered company Springer Nature Singapore Pte Ltd.
The registered company address is: 152 Beach Road, #21-01/04 Gateway East, Singapore 189721, Singapore

# Preface

This Volume contains the papers presented at the International Conference on Data Science, Computation, and Security (IDSCS 2021) held on April 16–17, 2021, organized by CHRIST (Deemed to be University), Pune, Lavasa Campus, India. IDSCS 2021 received approximately 170 research submissions from 4 different countries, viz. South Korea, Iraq, United Arab Emirates, and Ukraine.

Technology is the driving force in this era of globalization for the socioeconomic growth and sustained development of any country. The influence of data and security in shaping the process of globalization, particularly in productivity, commercial, and financial spheres, is highly required. It is the phase of revolution that has significant implications for the current and future societal and economic situation of all the countries in the world. Data and security play a fundamental role in enabling people to rely on digital media for a better lifestyle. It is concluded that data science and security is a significant contributor to the success of the current initiative of Digital India.

The international conference deliberated on topics specified within its scope. It focused on exploring the role of technology, data science, computational security, and related applications in enhancing and securing business processes. The conference offered a platform for bringing forward substantial research and literature across the arena of data science and security. It provided an overview of the upcoming technologies. IDSCS 2021 provided a platform for leading experts, academicians, fellow students, researchers, and practitioners to share their perceptions, provide supervision, and address participants' interrogations and concerns. After a rigorous peer review with the help of program committee members consisting of several reviewers across the globe, 53 papers.

The conference was inaugurated by Dr. Bimal Roy, Chairman, National Statistics Commission, and other eminent dignitaries, including Dr. Fr. Abraham V. M., Dr. A. K. Nayak, Dr. Fr. Jose C. C., Dr. Fr. Joseph Varghese, Dr. Fr. Jossy P. George, Dr. Fr. Arun Antony, and Dr. D. K. Mishra.

The conference witnessed keynote addresses from eminent speakers, namely Dr. Aninda Bose, Senior Publishing Editor, Springer Nature; Dr. Arceloni Neusa Volpato, International Affairs Coordinator, Transcultural Practices Master &

Coordinator, Centro Universitário Facvest–UNIFACVEST; Dr. Marta Zurek-Mortkan, Lukasiewicz Research Network, Institute for Sustainable Technologies, Department of Control Systems, Radom, Poland; Dr. Sunil Kumar Vuppala, Director, Data Science, Ericsson, Bangalore Urban, Karnataka, India; Dr. Sudeendra Koushik, Innovation Director at Volvo Group, Head of CampX, Bangalore; Dr. B. S. Dayasagar, Systems Science and Informatics Unit, Indian Statistical Institute-Bangalore Centre, India.

The organizers wish to thank Dr. Aninda Bose, Senior Editor, Springer Nature, New Delhi, India, for his support and guidance. Ms. Divya Meiyazhagan, Springer Nature. The organizing committee wishes to thank EasyChair Conference Management System as it is a wonderful tool for easy organization and compilation of conference documents.

Bangalore, India — Samiksha Shukla
Palo Alto, USA — Aynur Unal
Bangalore, India — Joseph Varghese Kureethara
Indore, India — Durgesh Kumar Mishra
Daegu, Korea (Republic of) — Dong Seog Han

# Contents

# About the Editors

**Dr. Samiksha Shukla** is currently employed as Associate Professor and Head, Data Science Department, CHRIST (Deemed to be University), Pune, Lavasa Campus. Her research interest includes computation security, machine learning, data science, and big data. She has presented and published several research papers in reputed journals and conferences. She has 15 years of academic and research experience and is serving as reviewer for Inderscience Journal, Springer Nature's International Journal of Systems Assurance Engineering and Management (IJSA), and for IEEE and ACM conferences. She is an experienced and focused teacher, who is always committed to promote the education and well-being of students. She is passionate about innovation and good practices in teaching. She is always engaged in continuous learning to broaden knowledge and experience. She is having core expertise at computational security, artificial intelligence, and healthcare-related projects. She is skilled in adopting a pragmatic approach in improvising on solutions and resolving complex research problems. She possesses an integrated set of competencies that encompass areas related to teaching, mentoring, strategic management and establishing Centre of Excellence via industry tie-ups. Having a track record of driving unprecedented research and development project with international collaboration, instrumental in organizing various National and International Events.

**Dr. Aynur Ünal** educated at Stanford University (class of '73), comes from a strong engineering design and manufacturing tradition, majored in Structural Mechanics-Mechanical Engineering-Applied Mechanics and Computational Mathematics from Stanford University. She has taught at Stanford University till mid 80s and established the Acoustics Institute in conjunction with NASA-AMES research fellowships funds. Her work on "New Transform Domains for the Onset of Failures" received the most innovative research awards. Most recently, she is bringing in the social responsibility dimension into her incubation and innovation activities by getting involved in social entrepreneurial projects on street children and ageing care. She is also Strategic Adviser for Institutional Development to IIT

Guwahati. She is a preacher of Open Innovation, and she always encourages students to innovate and helps them with her support.

**Dr. Joseph Varghese Kureethara** is heading the Centre for Research at Christ University. He has over sixteen years of experience in teaching and research at CHRIST (Deemed to be University), Bengaluru, and has published over 100 articles in the fields of Graph Theory, Number Theory, History, Religious Studies and Sports both in English and in Malayalam. He co-edited three books and authored one book. His blog articles, comments, facts, and poems that has earned about 1.5 lakhs total pageviews. He has delivered invited talks in over thirty conferences and workshops. He is the mathematics section editor of Mapana Journal of Sciences and member of the Editorial Board and reviewer of several journals. He has worked as the member of the Board of Studies, Board of Examiners and Management Committee of several institutions. He has supervised 5 Ph.Ds, 12 M.Phils, and supervising 8 Ph.Ds. Current Profession: He is Professor of Mathematics CHRIST (Deemed to be University), Bengaluru.

**Dr. Durgesh Kumar Mishra** has received M.Tech degree in Computer Science from DAVV, Indore, in 1994 and Ph.D. in Computer Engineering in 2008. Presently, he is working as Professor (CSE) and Director Microsoft Innovation Centre at Sri Aurobindo Institute of Technology, Indore, MP, India. He is having around 24 years of teaching experience and more than 6 years of research experience. His research topics are secure multi-party computation, image processing, and cryptography. He has published more than 80 papers in refereed international/national journals and conferences including IEEE and ACM. He is a senior member of IEEE, Computer Society of India, and ACM. He has played very important role in professional society as Chairman. He has been a consultant to industries and government organization like Sales tax and Labor Department of Government of Madhya Pradesh, India.

**Dr. Dong Seog Han** received his B.S. degree in electronic engineering from Kyungpook National University (KNU), Daegu, Korea, in 1987, and his M.S. and Ph.D. degrees in electrical engineering from the Korea Advanced Institute of Science and Technology (KAIST), Daejon, Korea, in 1989 and 1993, respectively. From October 1987 to August 1996, he was with Samsung Electronics, Co. Ltd., where he developed the transmission systems for QAM HDTV and Grand Alliance HDTV receivers. Since September 1996, he has been with the School of Electronics Engineering, KNU as Professor. He worked as courtesy Associate Professor in the Department of Electrical and Computer Engineering, University of Florida in 2004. He is Director of the Center ICT and Automotive Convergence, KNU, and AI Society Chair at the Korean Institute of Communications and Information Sciences. He has published over 122 articles in the fields of communication theory, autonomous vehicles, and AI and holds 42 patents.

# Towards a Knowledge Centric Semantic Approach for Text Summarization

Siddhant Singh[1] and Gerard Deepak[2(✉)]

[1] Department of Computer Science and Engineering, SRM Institute of Science and Technology, Ramapuram, Chennai, India

[2] Department of Computer Science and Engineering, National Institute of Technology, Tiruchirappalli, India

**Abstract.** Text Summarization is one of the important process for extracting important data from a text document. In the proposed method, the useful text or data collected is obtained as abridged form of the document and it is provided to the user as summary. In this current world where there is almost limitless information online, we must understand what data and its context is relevant to our objective for a certain task. As the enormous information is getting collected on the internet, it is a tedious and a challenging task and to go through the accessible information on the Web. This paper proposes a knowledge centric approach for text summarization. The dataset used for training this system is the DUC 2007 which contains manually created summaries, automatic baseline summaries and additional supporting data with results, documents etc. which is combined with TF-IDF algorithm. A domain-based ontology is created in addition Cross Entropy and Normalized Point Wise Mutual Information and ANOVA Normalized Point Wise Mutual Information is calculated based on which the sentences are grouped and eliminated. The proposed approach is superior in terms of performance and recorded F-Measure and False Negative Rate is 88.20% and 0.14 respectively

**Keywords:** ANOVA-NPMI · Cross Entropy · Text summarization · TF-IDF

## 1 Introduction

Text Summarization has exceptionally grown into a vital and significant asset for encouraging and illustrating text or data in the hindmost and fastest way possible. in this very fast paced world what we witness today it is necessary to model an efficient text summarizer. If we discuss about human limits of retrieving a context from an over-sized document, it is really a challenging task. There is abundance of text present as information in the current structure of the Web and there is a necessity of document summarizer which would yield the best-in-class results. The objective of the proposed system is to get feasible and relevant results for the mentioned topic specific to a topic of choice such that robust results can be furnished.

*Motivation:* The As there is limitless information available in the present-day times there is a need for accurate and better performing models for text summarization. Although


S. Shukla et al. (eds.), *Data Science and Security*, Lecture Notes in Networks and Systems 290,
https://doi.org/10.1007/978-981-16-4486-3_1

existing techniques catered towards text summarization there is always a sparsity and loss of useful entities during summarization. A knowledge centric approach is the need of the hour for text summarization to prevent the loss of entities and to enhance the user experience by consuming less time. The Sematic Web, where almost every entity is labelled and annotated, and the need of finding a better approach which is semantically complaint is required.

*Contribution:* A knowledge centric approach has been proposed, where the pre-processing of dataset is done following which TF-IDF algorithm is applied, based on which feature extraction is accomplished. A term-based ontology modelled using existing Domain Ontology and auxiliary knowledge sources. Cross Entropy & Intersection of NPMI and ANOVA NPMI measures is computed between the feature extracted and the terms extracted in ontology model. Hash-Table is generated using the values and average F-Measure of 88.20% has been achieved for DUC 2007 dataset. The sentences are grouped depending upon the values and the process of redundancy is done. Using lexical and grammatical agents the summarised result is obtained.

*Organization:* The assembly of the rest of the paper is as follows. Section 2 consists of relevant research work which is formerly done in the field of text summarization. Section 3 elucidate the proposed architecture. Section 4 describes in detail about the implementation. Section 5 consists of performance evaluation and observed results. Section 6 depicts on the conclusion and references.

## 2 Related Works

Vetriselvi et al. [1] is a baseline approach where the result of summary of information is obtained by using LCIS and FGSS that are used to enhance the important terms. The following approach gives good harmonic mean and this approach is domain and knowledge independent. Moratanch et al. [2] gives a thorough overview of the of text summarization technique, which is known an extractive text summarization, which interprets methods with a less expendable summary of the text document which is cohesive and contains information in larger ratio N. Bhatia et al. [3] investigates in the field of single and multiple documents which is essential and widely accepted or favored work done and discuss about the pragmatic approaches and extractive methods. Youngjoong Ko et al. [4] he created a method for sentence extraction using statistical approach where he used contextual information and he summarized the text. Aker et al. [5] finds the optimum extractive summary up to a certain value with help of a search algorithm. Barzilay et al. [6] proposes a new algorithm where in a text lexical chains are computed in combination with sturdy source of knowledge. It is processed in the order of certain steps which includes segmentation of original text, construction of lexical chains, their identification and the relevant sentences are extracted. Erkan et al. [7] created an architecture for finding the relative importance computing of textual unit for NLP using a graph-based methodology. Goldstein et al. [8] proposes a system which uses techniques which are domain independent based on statistical processing, the selected passages where the diversity is maximized and values which reduces the redundancy. He also proposed a framework that helps in effortless parametrizing for variety of genres and user requirements. Ferreira et al. [9] it is a descriptive paper where assessment of 15 algorithms is done which is both quantitative and qualitative which is used for scoring sentence and is evaluated in the different format of literature dataset such as news, blog, contexts etc. In [10–23] several supporting ontology focused models are depicted and discussed.

## 3 Proposed Architecture

Figure 1 depicts the Architecture of the proposed knowledge centric text summarization system which involves the pre-processing of dataset at first. The pre-processing involves process of Parsing Tokenization, Lemmatization, and Stop Word Removal. Tokenization which is based on blank space and special character as tokenizer which are tokenized based on the blank space and special characters, to obtain individual words. Lemmatization refers to deriving the base term from the morphology of terms, such that the base form of the term is obtained from its inflectional form. Further TF-IDF is used for feature extraction based on the frequency and rarity of terms over a document corpus. The TF-IDF is depicted by Eq. (1).

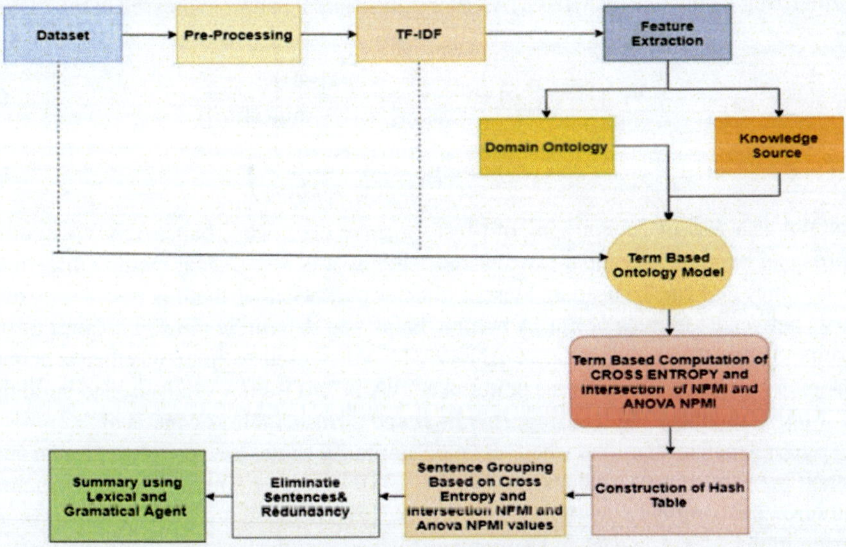


**Fig. 1** Proposed system architecture

$$Tf - idf\,(j,k) = tf\,(j,k) \times \log\left(\frac{l}{(df + 1)}\right) \tag{1}$$

where j is term (word), k (document) and l are known as the count of corpus (set of words).

The feature is extracted based on rarity feature across document and the most frequent words the domain ontology and knowledge sources are integrated. The Domain Ontology is modelled as a static domain ontology based on the terminologies that are extracted from the dataset. The Knowledge source which is integrated into the framework is Wikidata which is knowledge store and incorporates entities web documents which stresses on certain factors like concepts, terminologies, information on certain topic etc. The data is present in the form of property and value. The Wikidata API is used to obtain relevant knowledge source which is relevant to the document and which incorporates the auxiliary knowledge. A term Based Ontology model is built, which is an enriched ontology using the real-world knowledge, the local domain ontology as well as the terms relevant to the dataset. The term-based ontology is built by estimating the Cross Entropy, and intersection of NPMI and ANOVA-NPMI. Cross Entropy is derived from the information theory and computes the bits required to transmit an aggregates event from one distribution to another. Cross Entropy function can be represented in Eq. (2)

$$H(Pp, Qq) = -summinXPp(m) \times \log(Qq(m)) \tag{2}$$

where H (Pp, Qq) is formally known as the Cross Entropy function, Pp(m) is the probability of the event m in P, Q(m) is the probability of event x in Q and log is the base-2 logarithm. To eliminate similar words by sentence duplication extraction a Normalized PMI is used. NPMI finds the similar or repeated occurrence of terms and is processed dynamically over entire dataset after feature extraction. The Normalized PMI is one by normalizing PMI value by MIN–MAX Normalization. It can be expressed in Eq. (3) and (4)

$$X[:s] = x[:a] - \frac{min(x[:a])}{\max(x[:a]) - \min(x[:a])} \tag{3}$$

$$PMI(x, y) = \log\left(\frac{pa(x, y)}{pa(x) \times pb(y)}\right) \tag{4}$$

ANOVA is a way of generalizing of t-test for more than two groups, ANOVA in other words can be used to compare two or more items, it is a statistical method that yields such values that can be used and tested to determine whether there is relevant relation exists between variables or not. A p-value helps you determine the significance of our results. The calculated p-value of ANOVA-NPMI is used to infer whether one must accept or reject a null hypothesis or not. ANOVA-NPMI framework is used to build this term-based ontology model using a p-values and a Hash Table is generated.

Redundancy is identified when semantic similarity values are very close similar such that NPMI values and the values of ANOVA-NPMI which are similar will be eliminated, and further according to the threshold values the sentences are grouped upon their score. Furthermore, a lexical agent and a grammatical agent for the yielding of the summary of the document has been incorporated. Lexical agents will capture lexemes using Wordnet 2.0. A grammatical agent fixes the grammatical mistakes in the summary so that the result is error free and does not contain any grammatical errors that disturbs the structure of the summarized document.

## 4 Implementation

The proposed system architecture was designed and implemented on Windows 10 Operation System. The implementation has been performed using Intel Core i7 9th Gen processor and 16 GB RAM. NLTK, Re, Scikit-learn, Matplotlib which are python libraries which was used for pre-processing data and visualize the values such as F-Measure of baseline model and other approaches. Pythons deep learning API knows as Kera's along with TensorFlow has been utilized for implementation and training process of the Term Based Ontology Model on the dataset of DUC 2007 on which summarization algorithms were run.

**Algorithm 1**: Algorithm of proposed architecture of text summarization

```
Input: Knowledge Source, Domain Ontology and Dataset
Output: Summarized Text
Begin
  1: GET TBV
  2: While (TBV < Model created) {
  3:      import Dataset
  4:      Data. Tokenize ()
  5:      Data. Lemmatize ()
  6:      Data. Arising & Remove Stop Word
  7:      t = words
  8:      d = words. Set ()
  9:      N = Len(d)
  10:     Tf-idf (j, k) = tf (j, k) *log(l/(df+1))
  11:     Extract Frequent terms and Rare terms From Dataset.
  12:     Integrate Domain Ontology and Knowledge source
 13:    Model Created +=1}
 14: Compute Cross Entropy
 15: H = – sum m in X Pp(m) * log (Qq(m))
 16: Compute PMI and normalize it using Min Max Normalization
 17: PMI =log (pa (x, y)/(pa(x)×pb(y)))
 18: NPMI = MIN MAX(PMI)
 20: Compute ANOVA NPMI generate p value.
 21: if (H && intersection (NPMI & ANOVA NPMI) < 0.5) {
     Eliminate ()
     Else
     Keep ()
     }
 22. Construct Hash Table and Sentence Grouping based on values of hash tables.
 23: if (value (< 0.3) {
      Eliminate ()
     Else if (value > 0.5 and value <= 0.6)
     Sa = word.
     Else if (value > 0.7 and value <=0.8)
     Sb = word.
     Else if (value > 0.8 and value <= 0.9)
     Sc = word.
     Else
     Merge () similar sentences}
 24: Use of lexical agent and Grammatical Agent
 25: Return the optimal summarized text.
 End
```

The proposed system algorithm consists of domain ontology, knowledge source and dataset as the input. Here the data set is DUC 2007 which contains auxiliary knowledge sources which is used for implementation and building of the model. It contains summaries, results of various documents. It also contains manually created summaries, automatically created baseline summaries, submitted summaries created by participating groups, tables with results and certain evaluations of it and additional support data. TF-IDF on the dataset is computed and further the feature extraction that gives the most frequent and rarest word. We integrate domain ontology and knowledge source and create a Term Based Ontology ModelThe Cross Entropy values and NPMI and ANOVA-NPMI are computed. if the values are below 0.5 that word is eliminated based on the values, hash tables are made which contains the p values and depending upon those values the words and sentences are grouped in Sa, Sb, Sc and if the values are similar, we will merge. Then lexical agents are used which capture lexemes using WordNet 2.0 and finally the grammatical agent is used to make the summarized text error free and credible.

## 5 Performance Evaluation and Results

The F-Measure, False Negative Rate (FNR), Precision, Recall and Accuracy are used as performance evaluation metric. The F-Measure, Precision, Recall, Accuracy and False Negative Rate (FNR) are used as potential metrics in their standard form. In Fig. 2 the various performance metrices are marked with different color codes on x axis that are Precision, Recall, Accuracy and F-Measure. The Precision of the proposed approach is 90.18. The Recall, accuracy and F-Measure is 86.32, 88.74, 88.20 respectively which as compared to baseline models as we can infer that the IKTWA [1] has precision, recall, accuracy, F-Measure as 80.41, 76.42, 78.21, 78.36. The Sentence Similarity + Textual Entailment has values of precision, recall, accuracy, F-Measure as 78.72, 75.44, 76.87, 77.04. The Sentence Similarity alone has values of precision, recall, accuracy, F-Measure as 59.69, 52.44, 55.78, 55.83, respectively. The other approach that is combination of Fuzzy and Entropy has values of precision, recall, accuracy, F-Measure as 77.41, 72.81, 75.97, 75.03. The Textual Entailment has values of precision, recall, accuracy, F-Measure i.e., 48.32, 45.81, 46.91, 47.03.

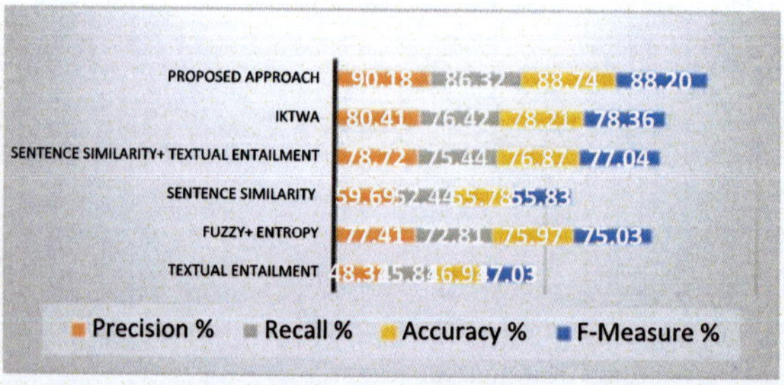


**Fig. 2** Comparison of performance measures of the proposed text summarization

It can be concluded from Fig. 2 that the performance metrices of the proposed system is maximum in comparison of other baseline models because as IKTWA uses the approach of local context information scoring (LCIS) in combination with Fuzzy graph sentence scoring system which brings the performance metric values as mentioned in Fig. 2. The Sentence Similarity is a method which easily allows to compute the semantic similarity between the words. The Textual Entailment is the method where there is relation between words in a series of words, with a certain meaning. The Fuzzy is basically an approximate computing technique whereas the Entropy give the value of informativeness. We can observe that when we combine these methodologies there is improve in precision, recall, accuracy, F-Measure%. But as the results depicts the proposed approach is better in performance of precision, recall, accuracy, F-Measure% as, the proposed system of Text Summarization uses Knowledge Centric approach in combination of Domain Ontology and statistical measure to form a Term Based Ontology Model, thus it results in better performance than other existing system and approaches. The FNR of Textual Entailment, Fuzzy + Entropy, Sentence Similarity, IKTWA and Proposed Approach is 0.54, 0.27, 0.47, 0.24, 0.23, 0.13, respectively. It is evident from the graph that the proposed has less FNR as compared to other baseline model and variations. In the proposed approach a domain-based ontology in combination with knowledge source is used to create a term-based ontology model along with semantic and statistical measures to improve the efficiency of the proposed approach. The IKTWA uses fuzzy graph sentence scoring, local context information scoring and sentence-based weightage to the key terms which therefore results in decrease in the FNR values and makes the proposed approach a better solution.

Figure 3 depicts the No. of Entities vs Accuracy graph, here we know that proposed approach is baselined with Textual Entailment, Fuzzy in combination with Entropy, Sentence Similarity, Sentence Similarity + Textual Entailment, IKTW [1] and Proposed Approach. Analysing the trend of the graph as the no. of entities increase, in general the values of all approaches decrease but at every instant the proposed approach is higher than baseline approaches. When the number of entities is 10 the accuracy, we get is peak accuracy of 91.17. Even the number of entities is increased to 20 the accuracy of the proposed system is 86.32. Comparing the accuracy with baseline models the number of entities increased up to level of 50, the accuracy is 44.12, 73.44, 53.32, 75.29, 76.39,

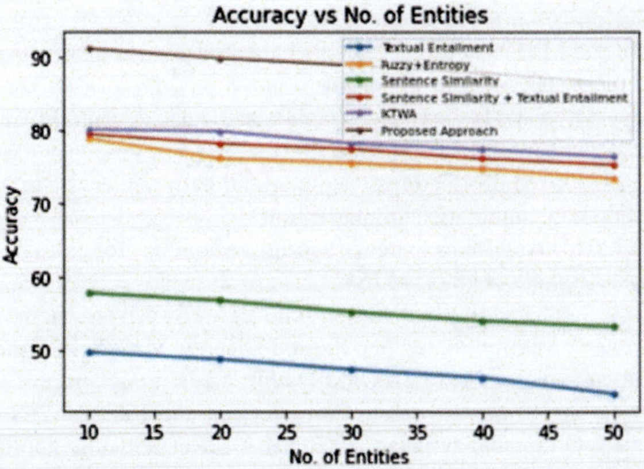


**Fig. 3** Accuracy vs no. of entities of proposed approach and baseline models

86.45, respectively. As the proposed system used the knowledge centric approach along with domain ontology in the implementation the accuracy is higher when compared to other baseline models.

## 6 Conclusion

The Text summarization is one of the key process for productive information gathering the era of Web 3.0 where the contents is in form entities which are labelled and annotated. With the exceptional increase in the data this paper proposes novel strategic approach for Text Summarization using a knowledge centric methodology. The accuracy of the proposed approach is 91.17%, the ascendancy of the proposed when compared to baseline approaches performance is because of the semantically driven modified technique which uses domain ontology, knowledge sources and generic algorithm for optimization. From this it can be concluded that in future the proposed approach of knowledge centric in combination of domain ontology and semantic similarity-based text summarization system can ensure high level performance when compare to baseline approaches.

## References

1. Vetriselvi T, Gopalan N (2020) An improved key term weightage algorithm for text summarization using local context information and fuzzy graph sentence score. J Ambient Intell Humaniz Comput. https://doi.org/10.1007/s12652-020-01856-9
2. Moratanch N, Gopalan C (2017) A survey on extractive text summarization. 1–6. https://doi.org/10.1109/ICCCSP.2017.7944061
3. Bhatia N, Jaiswal A (2016) Automatic text summarization and it's methods - a review. In: 2016 6th international conference - cloud system and big data engineering (confluence), Noida, pp 65–72. https://doi.org/10.1109/CONFLUENCE.2016.7508049
4. Ko Y, Seo J (2008) An effective sentence-extraction technique using contextual information and statistical approaches for text summarization. Pattern Recogn Lett 29(9):1366–1371. https://doi.org/10.1016/j.patrec.2008.02.008. ISSN 0167-8655
5. Aker A, Cohn T, Gaizauskas R (2010) Multi-document summarization using A* search and discriminative learning. In: Proceedings of the 2010 conference on empirical methods in natural language processing
6. Barzilay R, Elhadad M (1999) Using lexical chains for text summarization. In: Advances in automatic text summarization, pp 111–121
7. Erkan G, Radev DR (2004) Lexrank: graph-based lexical centrality as salience in text summarization. J Artif Intell Res 22:457–479
8. Goldstein J et al (2000) Multi-document summarization by sentence extraction. In: NAACL-ANLP 2000 workshop: automatic summarization
9. Ferreira R et al (2013) Assessing sentence scoring techniques for extractive text summarization. Expert Syst Appl 40(14):5755–5764
10. Deepak G, Teja V, Santhanavijayan A (2020) A novel firefly driven scheme for resume parsing and matching based on entity linking paradigm. J Discrete Math Sci Cryptogr 23(1):157–165
11. Deepak G, Santhanavijayan A (2020) OntoBestFit: a best-fit occurrence estimation strategy for RDF driven faceted semantic search. Comput Commun 160:284–298
12. Kumar N, Deepak G, Santhanavijayan A (2020) A novel semantic approach for intelligent response generation using emotion detection incorporating NPMI measure. Procedia Comput Sci 167:571–579

13. Deepak G, Kumar N, Santhanavijayan A (2020) A semantic approach for entity linking by diverse knowledge integration incorporating role-based chunking. Procedia Comput Sci 167:737–746
14. Haribabu S, Kumar PSS, Padhy S, Deepak G, Santhanavijayan A, Kumar N (2019) A novel approach for ontology focused inter-domain personalized search based on semantic set expansion. In: 2019 fifteenth international conference on information processing (ICINPRO), pp 1–5. IEEE, December 2019
15. Deepak G, Kumar N, Bharadwaj GVSY, Santhanavijayan A (2019) OntoQuest: an ontological strategy for automatic question generation for e-assessment using static and dynamic knowledge. In: 2019 fifteenth international conference on information processing (ICINPRO), pp 1–6. IEEE, December 2019
16. Kaushik IS, Deepak G, Santhanavijayan A (2020) QuantQueryEXP: a novel strategic approach for query expansion based on quantum computing principles. J Discrete Math Sci Cryptogr 23(2):573–584
17. Varghese L, Deepak G, Santhanavijayan A (2019) An IoT analytics approach for weather forecasting using Raspberry Pi 3 Model B+. In: 2019 fifteenth international conference on information processing (ICINPRO), pp 1–5. IEEE, December 2019
18. Deepak G, Priyadarshini S (2016) A hybrid framework for social tag recommendation using context driven social information. Int J Soc Comput Cyber-Phys Syst 1(4):312–325
19. Deepak G, Sheeba Priyadarshini J (2018) A hybrid semantic algorithm for web image retrieval incorporating ontology classification and user-driven query expansion. In: Advances in big data and cloud computing. Springer, Singapore, pp 41–49
20. Deepak G, Gulzar Z (2017) ONTOEPDS: enhanced and personalized differential semantic algorithm incorporating ontology driven query enrichment. J Adv Res Dyn Control Syst 9(Specia):567–582
21. Deepak G, Kumar AA, Santhanavijayan A, Prakash N (2019) Design and evaluation of conceptual ontologies for electrochemistry as a domain. In: 2019 IEEE international WIE conference on electrical and computer engineering, pp 1–4
22. Deepak G, Priyadarshini JS (2018) Personalized and enhanced hybridized semantic algorithm for web image retrieval incorporating ontology classification, strategic query expansion, and content-based analysis. Comput Electr Eng 72:14–25
23. Deepak G, Ahmed A, Skanda B (2019) An intelligent inventive system for personalised webpage recommendation based on ontology semantics. Int J Intell Syst Technol Appl 18(1/2):115–132

# Detection of Abnormal Red Blood Cells Using Features Dependent on Morphology and Rotation

Ali Sadam[1(✉)], Hazim G. Daway[1], and Jamela Jouda[2]

[1] College of Science, Department of Physics, Mustansiriyah University, Baghdad, Iraq
{alisadam,hazimdo}@uomustansiriyah.edu.iq

[2] College of Science, Department of Biology, Mustansiriyah University, Baghdad, Iraq
jamela.jouda@oumustansiriyah.idu.iq

**Abstract.** Automatic detection of abnormal Red Blood Cells (RBCs) is an important problem in medical imaging. A new algorithm is proposed in this paper to detect abnormal RBCs, the first step in this algorithm identifies the regions of interest depending on morphology operations and the second step uses the rotate features to verify that the detected object is an abnormal RBCs. Two criteria were used to measure detection accuracy depending on the detection and false alarm rate, from analyzing the results they reach about 86 and 14%.

**Keywords:** Abnormal RBCs · Binary image · RBCs detection · Morphology · Rotation features

## 1 Introduction

Human blood has very important cellular elements such as abnormal red blood cells (RBC), white blood cells (WBC), and platelets. Each of them has a function to keep the body homeostasis. Any defect in these elements could cause dire effects on human health such as Anemia. Anemia is an illness resulting from a lack of RBC function. Anemia reasons are many depend on its type such as hemolytic anemia. Hemolytic anemia is Different abnormalities of the shape and cytoplasm membrane of red blood cells, many of which are hereditarily acquired; make the cells fragile, so that they rupture easily as they go through the capillaries, especially through the spleen. One of the hemolytic anemia is sickle cells. Techniques for diagnosing this type of anemia require close monitoring for abnormal RBCs samples under the microscope, which are dependent on Red blood cell shape usually, this process takes a long time depending on the skills of the expert.

The field of digital image processing includes several branches, including improvement and distinction These fields can be linked with the images captured by optical microscopy for processing and detection [1] usually, the first step with these techniques is to divide the red blood cells according to their desired shape from each other and background.


S. Shukla et al. (eds.), *Data Science and Security*, Lecture Notes in Networks and Systems 290,
https://doi.org/10.1007/978-981-16-4486-3_2

Hence, it is possible to extract the properties from the segmented cells It distinguishes normal red blood cells from abnormal red cells based on their characteristics abnormal RBC is perfect manually via examining blood touches under the microscope. Experienced technicians will make key assessments of this work. Although it is still possible to count cells manually under certain conditions, current hematology research institutions still use automatic blood analyzers for cell counts. The number of different types of RBCs provides a good quantitative record for assessing an individual's health. Since a large number of blood tests are continuously performed on a global scale, a fast and accurate method is indeed required. Visualizing red blood cells from small images can evaluate and discover many diseases, such as sickle cell anemia, and can identify it by examining the red blood cells that cause red blood cell breakdown, sickle cell anemia. All of these refer to sickle-shaped red blood cells, whose shape is different from that of B.Cs.normal R.B.Cs are cells in a humane body that are susceptible to certain diseases, including anemia, and sickle cell loss. It exists in every area of a humane body, including bone marrow and blood. Since the amount of RBC in the blood is usually a sign of a specific disease, counting different types of RBC (called differential testing) plays a noteworthy role in ensuring the well-being of patients at different stages (discovery, treatment, and development). Obviously, the manual counting process is very tedious and slow and depends on the operator's ability and experience. Therefore, the necessity of an automatic differential counting system becomes inevitable. In fact, this kind of morphological investigation only requires images instead of blood tests, and this method is reasonable with minimal effort and a precise remote indication framework. The application of a procedural framework combined with a picture preparation program allows for subjective evaluation and in this way to improves decision-making. Along these lines, programming frameworks that rely on image processing strategies can support hematologists and speed up the surgical process. In the past few years, the use of image processing programs has developed rapidly. These methods help to include cells in the blood and provide data about cell morphology at the same time.

## 2 Literature Survey

There are many previous studies related to the subject of analysis and detection RBCs, The most important of them:

Chadh, Srivastava, Singh et al. [2], proposed the red blood cell count method. The process includes image acquisition, and then a large number of image handling operations, including median filtering, segmentation, edge disclosure, and corrosion.

Elsalamony [3] introduced investigations into three neural networks for all types of anemia and on the geometry of the cell. They proposed an algorithm to detect natural cells, sickle cells, & ellipsoid cells in the microscope. Blood spot using the Hough that takes samples. an algorithm that seeks to highlight the Transform method, The researcher found a high quality in improving the diagnosis of anemia regarding the toward an effective method of diagnosis and natural counting and two popular anemia cells in medical-based blood smear image processing.

Tomari et al. [4] suggest a system for detecting and identifying RBCs from blood sample images, in which an artificial network neural (ANN) classifier is utilized to classify normal & abnormal RBCs. The system includes a combination of three major modules, which are processing & segmentation blocks. Features Extract blocks and classification blocks. In the algorithm in your block, he produced good performance during the completion of the task, and an acceptable error occurred. The result is that low arithmetic resources and high discrimination can be used to distinguish between and normal abnormal. The results he got are very accurate.

Pranati et al. [5] proposed the use of several image processing tools and techniques to determine the correct abnormalities in the natural parameters of RBCs. Draw conclusions from your results, and based on his work, he developed a scale that can determine the abnormal shape of the ball to diagnose the disease.

Di Ruberto, Loddo, Putzu [6] proposed the "edge box" method to detect and measure RBCs and WBCs From microscopic images for blood. He introduced that a complete blood count (CBC) can be completed quickly using a flow cytometer or a laboratory test using a flow cytometer. In this way, the cover of the cells and any abnormalities are described. The result is subjective and error-prone. Instead, using image processing technology, the proposed system is fully automated.

Delgado-Font et al. [7] are recommended to use peripheral blood smear image analysis for automatic differential red blood cell count. In this method, the Chan-Vese active poorness model is used to section the object of avail in the image, and the result of the algorithm is a confusion matrix, which can measure the performance of an algorithm. Each row of the matrix appears the no. of RBCs in a true category, and each column represents the number of red blood cells in a predicted category.

Jameela et al. [8] introduced a method to distinguish between ordinary abnormal RBC and normal RBC by (SVM) is proposed. and in [9] Tao et al. introduced the study of three neural networks for three types of anemia and cell geometry. In your article, the algorithm for disclosure of sickle cells, normal cells, and ellipsoid cells under a microscope is proposed. Using the Hough transform method for blood smear examination, the researchers found that the quality of the collected samples was high in improving the diagnosis of anemia. The algorithm attempts to emphasize an effective diagnostic and normal counting method and two popular anemia cells in medical-based blood smear image processing.

## 3 Proposed Method

The proposed method for detecting red blood cells using two features first features dependent on the binary conversion of R, G, and B compounds at several threshold values, and another feature takes the least rotation area for different objects in the image. The detection algorithm can be divided into two stages, the first to identify the regions of interest (ROI)and the second to make sure that the ROI is a red cell or not.

### 3.1 Determine the ROI

The ROIs is identified in the microscope image is likely to contain red blood cells, If the size of the images is large, it is preferable to reduce their size within the range of 600 ×

800 pixel In order to increase the speed of the program. We can smooth the gray image to reduce some of the high-light areas inside the cell by using:

$$Ic = I(x, y) * kr \tag{1}$$

where I(x, y) is the lightness component given by:

$$I(x, y) = min(r(x, y), g(x, y), b(x, y)) \tag{2}$$

and $*$ denoted the convoluting and kr ones kernel with size $5 \times 5$. Applying the medium filter to the image.

Then the gray images I(x, y) are converted to binary images for several threshold values by using:

$$\left.\begin{array}{r} I(x, y) > th_i \; then \, Ib_i(x, y) = 1 \\ else \quad Ib_i(x, y) = 0 \end{array}\right\} \tag{3}$$

$$th_i = [60\,70\,80\,90\,100] \tag{4}$$

i is the number of the threshold values (i = 5) and $Ib_i(x, y)$ being the binary image at each threshold value as in shown in Fig. 1. In this image, not all areas are important areas with small areas (less than 100 pixels) that can be deleted, The number of regions $Ib_{in}(x, y)$, in the binary image at threefold index i, with number n ($n = 1, 2, 3, \ldots T$), T is the Total number of objects and is determined as the region of interest.

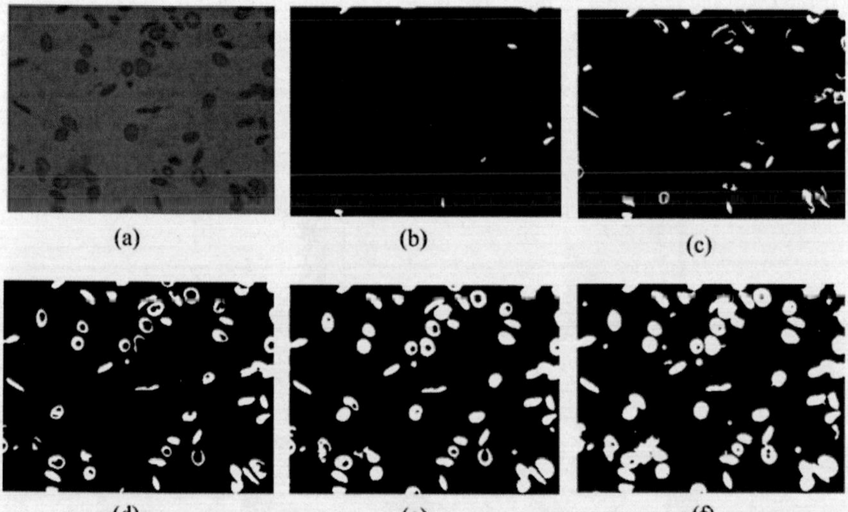

(a) (b) (c) (d) (e) (f)

**Fig. 1** Gray component microscope image in (a), and in (b, c, d, e and f) the binary images conversion of several threshold values (60, 70, 80, 90 and 100) respectively

### 3.2 Features Determination

After identifying the objects in the binary image $Ib_{in}(x, y)$ at the locations x, y Those positions are matched in the original image by:

$$\left.\begin{array}{c}\text{if } Ib_{in}(x, y) = 1 \text{ then} \\ r_{in}(x, y) = r(x, y), g_{in}(x, y) = g(x, y), b_{in}(x, y) = b(x, y)\end{array}\right\} \quad (5)$$

Then he switched to the binary images $Ibd_{in}(x, y)$ again by using:

$$\left.\begin{array}{c}\text{if } 0 > r_{in}, g_{in}, b_{in} > thc \\ Ibd_{in}(x, y) = 1 \\ \text{else } Ibd_{in}(x, y) = 0\end{array}\right\} \quad (6)$$

Each object $Ib_n(x, y)$ is rotated to four angles (10, 20, 30, and 40) Counterclockwise by using:

$$\left.\begin{array}{c}x_j = x\cos(\theta_j) - y\sin(\theta_j) \\ y_j = x\sin(\theta_j) + y\cos(\theta_j)\end{array}\right\} \quad (7)$$

The further away the cell is from the spherical shape, the more abnormal it becomes, often the cutout is rectangular and oval in shape, and there are three cases, whether the shape is a horizontal ellipse or vertical ellipse in both cases, the length is not equal to the width and the shape is rectangular and have a small area, but sometimes it is a diagonal ellipse, so its shape is square, and rotation can be used to distinguish the ellipse shape, so if the length is not equal to the width there is a possibility that the cell is an ellipse. Figure 2 illustrated the normal and abnormal cells, and the abnormal cell is rotated at different angles.

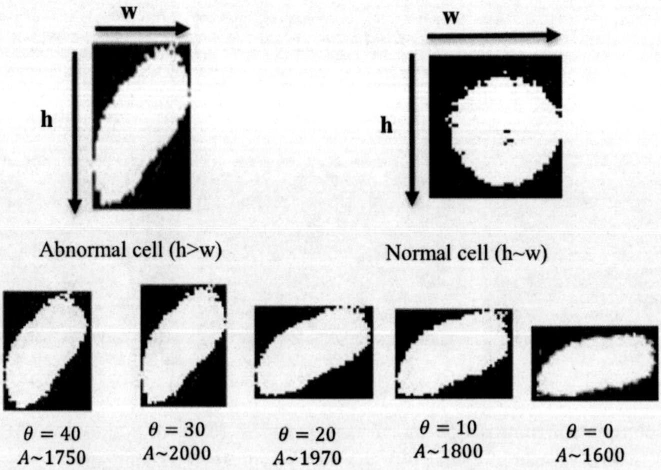


**Fig. 2** Comparison between normal and abnormal and cell, and abnormal image cell rotated by many angles in counterclockwise

If the length is approximately equal to the width, the shape will be circular. Therefore, we can conclude the following feature:

$$\Delta j = |hj - wj| \tag{8}$$

$$Aj = hj \times wj \tag{9}$$

$$\Delta m = \min(\Delta j) \tag{10}$$

$$\left.\begin{array}{c} if\ \Delta m > T1 \\ if\ T2 > A > T3 \\ Then\ (the\ object\ is\ abnormal\ cell) \\ else\ (the\ object\ is\ a\ normal\ cell) \end{array}\right\} \tag{11}$$

in this study, we used (thc = 110, T1 = 7, and T2 = 1000 and T3 = 450, Fig. 3 illustrates a working outline of the proposed algorithm to distinguish normal and abnormal white blood cells.

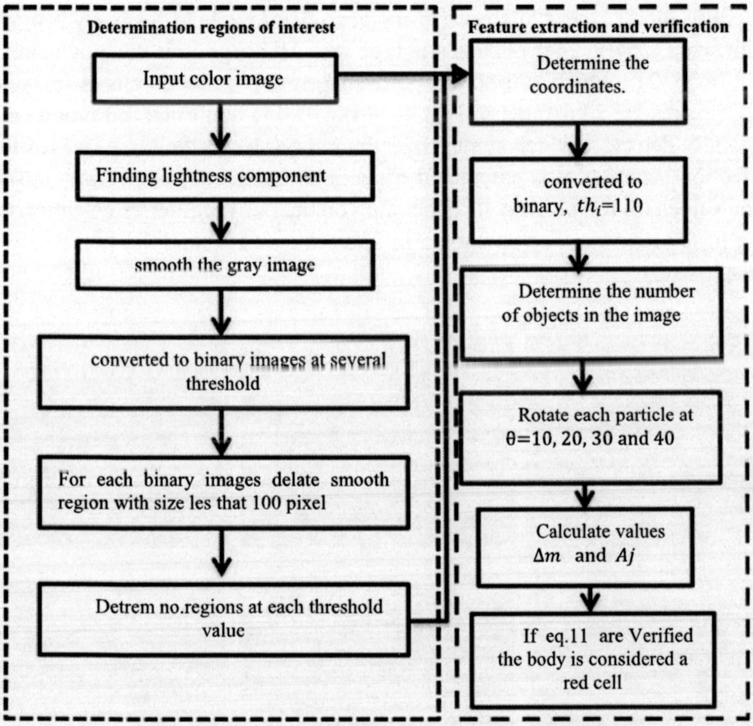


**Fig. 3** Block diagram for the suggested algorithm

## 4 Quality Assessment

The quality is evaluated based on the accuracy measurement between manual counting of abnormal RBC and automatic counting of uncovered cells according to the recommendations for using detection rate and false alarm rate. The detection rate ($\mu_e$) is defined as the ratio between the number of abnormal red blood cells that are correctly detected and the number of abnormal red blood cells determined by:

$$\mu_e = \frac{\text{RBC Corrected auto detection}}{\text{RBC Manual count}} \tag{12}$$

The false alarm rate ($\varepsilon$) is defined as the ratio between the number of RBC objects that have been wrongly identified as abnormal RBC and the number of abnormal RBC which have been determined by the expert. $\varepsilon$ is determined by

$$\varepsilon = \frac{\text{False detection}}{\text{RBC Manual count}} \tag{13}$$

## 5 Experimental Results

In this paper, we propose a new algorithm for detecting abnormal RBCs. This algorithm is applied on blood smear microscope images (40) download form erythrocytes IDB [10] with size (3264 × 2448) pixels and type jpg. All program is done by using Matlab software (R2020a), which is used to detect abnormal red blood cells in images. Some of These images are illustrated in Fig. 4. In Fig. 5, the abnormal red blood cells were automatically detected, which symbolized the red border of the same images in Fig. 4. The detection accuracy was calculated by comparing the auto-detection and manual-detection values for red blood cells depending on the two parameters detection rate ($\mu_e$) and The false alarm rate ($\varepsilon$), as shown in Table 1 as it reached the two values 86 and 14%, respectively.

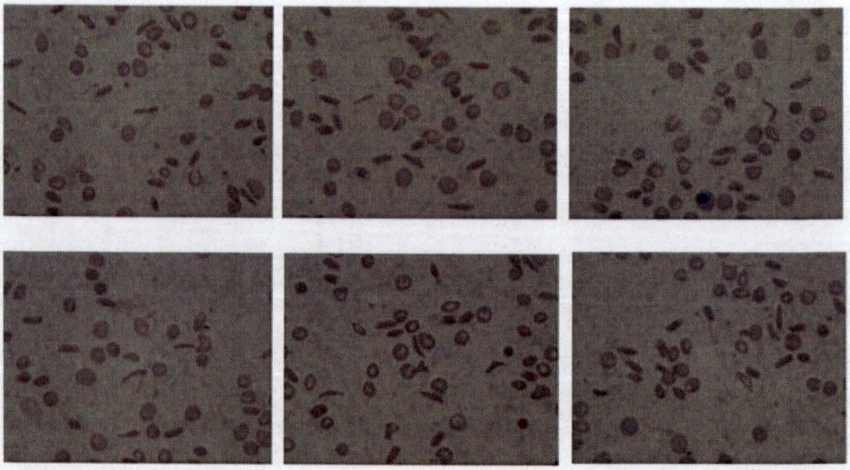

**Fig. 4** Sample of microscope images of blood Smears that were used in this study

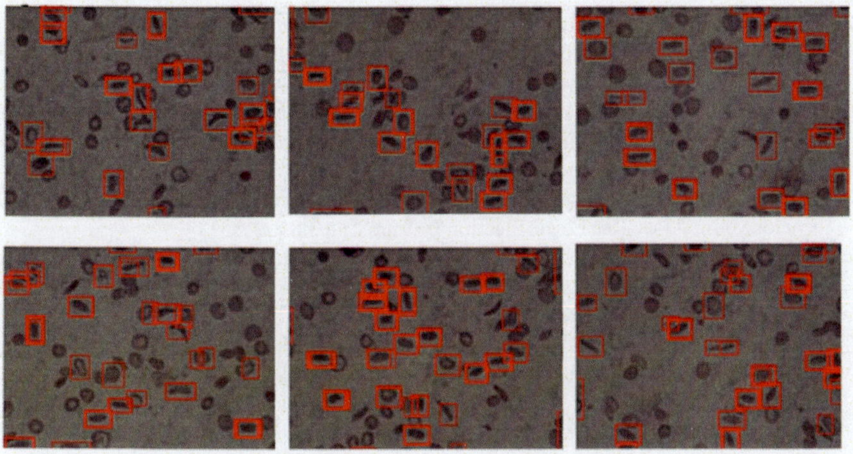

**Fig. 5** Abnormal red cells were discovered in Fig. 4

**Table 1** The quality evaluation algorithm for abnormal red blood cell detection is proposed

| RBC manual count | RBC corrected auto detection | Error in detection | $\mu_e$ | $\varepsilon$ |
|---|---|---|---|---|
| 501 | 433 | 70 | 86% | 14% |

## 6 Conclusion

In this paper, a new algorithm is proposed to detect abnormal red blood cells, dependent on morphology and rotation features, by analyzing the results, good accuracy was obtained $\mu_e = 86\%$ and $\varepsilon = 14\%$, This indicates the great success of the proposed algorithm in distinguishing the abnormal red blood captured by optical microscopy.

## 7 Future Scope

The suggested algorithm that is used to detect abnormal red blood cells can be configured by artificial intelligence techniques, especially in SVM. Circular fitting techniques can also be used to distinguish normal cells.

## References


1. Mohammed MH, Daway HG, Jouda J (2020) Automatic cytoplasm and nucleus detection in the white blood cells depending on hisogram analysis. IOP Conf Ser Mater Sci Eng 871:12071
2. Chadha GK, Srivastava A, Singh A, Gupta R, Singla D (2020) An automated method for counting red blood cells using image processing. Procedia Comput Sci 167:769–778. https://doi.org/10.1016/j.procs.2020.03.408

3. Elsalamony HA (2018) Detection of anaemia disease in human red blood cells using cell signature, neural networks and SVM. Multimed Tools Appl 77:15047–15074. https://doi.org/10.1007/s11042-017-5088-9
4. Tomari R, Zakaria WNW, Jamil MMA, Nor FM, Fuad NFN (2014) Computer aided system for red blood cell classification in blood smear image. Procedia Comput Sci 42:206–213. https://doi.org/10.1016/j.procs.2014.11.053
5. Pandit A, Rangole J (2014) Literature review on object counting using image processing techniques. Int J Adv Res Electr Electron Instrum Eng 3:8509–8512
6. Di Ruberto C, Loddo A, Putzu L (2020) Detection of red and white blood cells from microscopic blood images using a region proposal approach. Comput Biol Med 116:103530. https://doi.org/10.1016/j.compbiomed.2019.103530
7. Delgado-Font W, Escobedo-Nicot M, González-Hidalgo M, Herold-Garcia S, Jaume-i-Capó A, Mir A (2020) Diagnosis support of sickle cell anemia by classifying red blood cell shape in peripheral blood images. Med Biol Eng Comput 58:1265–1284. https://doi.org/10.1007/s11517-019-02085-9
8. Akrimi JA, Suliman A, George LE, Ahmad AR (2015) Classification red blood cells using support vector machine. In: Conference proceedings - 6th international conference on information technology and multimedia UNITEN cultiv. creat. enabling technol. through Internet Things, ICIMU 2014, pp 265–269. https://doi.org/10.1109/ICIMU.2014.7066642
9. Tao L, Asari V (2004) An integrated neighborhood dependent approach for nonlinear enhancement of color images. In: International conference on information technology: coding and computing. ITCC, vol 2, pp 138–139. https://doi.org/10.1109/itcc.2004.1286612
10. http://erythrocytesidb.uib.es/

# A Systematic Review of Challenges and Techniques of Privacy-Preserving Machine Learning

Kapil Tiwari(✉), Samiksha Shukla, and Jossy P. George

Christ University, Bangalore, India
Kapil.tiwari@res.christuniversity.in

**Abstract.** Machine learning (ML) techniques are the backbone of Prediction and Recommendation systems, widely used across banking, medicine, and finance domains. ML technique's effectiveness depends mainly on the amount, distribution, and variety of training data that requires varied participants to contribute data. However, it's challenging to combine data from multiple sources due to privacy and security concerns, competitive advantages, and data sovereignty. Therefore, ML techniques must preserve privacy when they aggregate, train, and eventually serve inferences. This survey establishes the meaning of privacy in ML, classifies current privacy threats, and describes state-of-the-art mitigation techniques named Privacy-Preserving Machine Learning (PPML) techniques. The paper compares existing PPML techniques based on relevant parameters, thereby presenting gaps in the existing literature and proposing probable future research drifts.

**Keywords:** Privacy-Preserving Machine Learning · Privacy threats · Federated learning · SMPC · Differential privacy

## 1 Introduction

Machine Learning (ML) techniques are the backbone of Prediction and Recommendation systems, widely used across banking, medicines, and finance domains. ML techniques such as Deep Neural Networks (DNNs) have been successful in solving complex problems in areas like speech recognition and natural language processing. Today, ML is utilised in finance domain for financial monitoring, algorithmic trading, and customer retention programs. However, they are subject to privacy and security threats such as model invasion or membership inference attack which can lead to expensive rip-offs, and identify theft. Moreover, ML techniques' effectiveness depends largely on the amount, distribution, and variety of data. However, it is challenging to combine data from multiple sources due to privacy and security concerns, competitive advantages, and data sovereignty.

To address this challenge, one needs to ensure privacy while doing Machine Learning or develop Privacy Preserving Machine Learning (PPML). In the recent past, PPML has received substantial interest from the research community. There are a couple of


S. Shukla et al. (eds.), *Data Science and Security*, Lecture Notes in Networks and Systems 290,
https://doi.org/10.1007/978-981-16-4486-3_3

published works summarising privacy and security attacks and techniques to overcome them; however, these studies remain fragmented and inconclusive given the wide range of ML use cases, algorithms, and the process involved. Hence, this paper aims to define privacy and classify privacy threats. Additionally, the paper reviews recent privacy threats, techniques to overcome, and comparative analysis of the published techniques based on crucial parameters. Finally, it talks about future research direction and open questions. This paper converges mainly on privacy-specific vulnerabilities and leaves core security threats such as data poisoning out of its scope. This survey paper concentrates on finding privacy preserving Machine Learning techniques, with an emphasis on the ML methods and their descriptions. Several papers focusing on privacy in Machine Learning and Deep Learning has been chosen for this survey paper. As opposed to the previous survey papers, the kingpin of this paper is to find the most relevant work in this area by applying a well thought filter criteria. Google Scholar queries were performed using "Privacy-Preserving" and "Machine Learning" or "Privacy-Enhancing" and "Machine Learning" or "Privacy-Preserving" and "Deep Learning". Highly cited papers were picked up for deep dive study. However, since this is a new and emerging topic, the filter criteria also picked up impactful and influential papers for study. Overall, research papers were selected to cover all known privacy threats and the relevant technique to overcome them at one place for an effective comparative study. The rest of the paper is organized as follows:

Section 2 focuses on the background and the meaning of privacy in Machine Learning; The current Machine Learning attacks are classified in Sect. 3; Sect. 4 focuses on Privacy-Preserving and Privacy-Enhancing Machine Learning mechanisms; Sect. 5 does a comparative analysis of the current mechanisms, followed by the conclusion.

## 2 Background

### 2.1 What is Privacy in Machine Learning

Privacy has subtle meaning in the context of Machine Learning. Privacy in Machine Learning is about the right to protect training data, model, model parameter, and inference attacks. Table 1 describes the meaning of privacy breaches in Machine Learning. It shows that privacy is said to be breached when an individual's confidentiality is compromised e.g., membership inference from the population or membership inference from the training dataset. In Machine-Learning-as-a-Service (MLaaS) model, the trained model represent the intellectual property of the model owner when the adversary's target is to infer the model while firing multiple queries to the model.

## 3 Classification of Machine Learning Attacks

This section aims to find and categorize prevalent attacks concerning Machine Learning and their impact on privacy. We need to understand the current threats briefly to evaluate the need and means of privacy protection. We have divided the threats into two categories, explicit and implicit attacks. Explicit attacks highlight the vulnerabilities when the actual training data is accessed, exposed, or leaked. In contrast, implicit attacks happen when access to actual training data is not available to the adversary, but he can guess or infer the data with clever techniques.

**Table 1** Categories of machine learning threats

| Category | Threat | Details |
|---|---|---|
| Member inference from population [1] | | The adversary's focus is to find inferences about the population, groups, or classes within the training dataset by using the model |
| | Statistical Disclosure | Using the model, The attacker can learn about a data instance, which was part of the training dataset. Statistical Disclosure Control (SDC) needs to be applied to ensure no information can be inferred about a data instance from the training set by applying the model. However, it is found to be practically unachievable by any model [2] |
| | Model inversion | The values of sensitive attributes used for the training of the model putting to use the model can be induced by an adversary |
| | Inferring class representatives | If the adversary gets access to the model, he can use the model inversion generalization technique to build representatives of the training dataset classes |
| Member inference from training dataset [1] | | The adversary's focus here is to find inference of an individual data instance which was used to train the model |
| | Membership inference | The adversary tries to find out whether a given data point was used to train the model |
| | Property inference | The adversary focuses on finding properties of given data point or subgroup within the training dataset |

(*continued*)

**Table 1** (*continued*)

| Category | Threat | Details |
|---|---|---|
| Model Parameter inference [1] | | When adversary's objective is to infer the model parameter themselves while using the model |
| | Model extraction inference | When adversary's objective is to infer the model parameter themselves while using the model<br>The adversary's objective here is to deduce identical or near-identical model. Usually used with Blackbox access to queryable model available for Machine Learning as Service (MLaaS) |
| | Functionality stealing | The adversary focuses on creating a quick workable model mimicking the actual model |

### 3.1 Explicit Attack

A data breach is a valid event in which sensitive, classified, or otherwise protected data is accessed or disclosed in an unauthorized manner. A data breach is an example of explicit threat to information security and is not limited to the Machine Learning field. Attacker, hackers, virus, malware, or social engineering are the mechanisms through which an adversary can derive a data breach. Information security vulnerability can be exploited to gain access to training data, archived models, or parameters [3, 4]. In a recent attack at Equifax, a security vulnerability with Apache Struts software was exploited to steal user's trade secrets and personal information [5]. Private data is prone to exposure at rest or during transmission without proper encryption. In 2018, Kaspersky lab reported that thousands of Android apps were found to be sending sensitive user data to unauthorized advertiser servers in an unencrypted format [6]. In Machine-Learning-as-a-Service (MLaaS) setting, private data can get exposed through the cloud service; moreover, these services do not transparently deal with the lifecycle of the data obtained from users. These attacks directly compromise an individual's confidentiality and can lead to serious privacy challenges. Since these attacks mostly fall into the category of information security attacks, further discussion on this is beyond this paper's scope [7].

### 3.2 Implicit Attack

We classify implicit attacks into five categories, property inference attacks, model inversion, membership inference, parameter inference, and hyper parameter inference. Table

2 shows a review of the different implicit attacks and their characteristics. As the name suggests, "Access to Model" defines if the adversary needs white-box or black-box access to the model to mount successfully. The black-box approach does not know the architecture or parameters of the target model and relies on query access for inference. Whereas the white-box approach has full access to the target model. The second column gives us a better understanding as to whether the attacker requires access to the output confidence values of the model (the probabilities, logits), or if the predicted labels will be enough to mount successfully.

### 3.2.1 Training Time Attacks

The attack which happens at the time of training of the model are discussed in this section. This is usually the case when adversary is one of the active/passive participants into model training exercise.

#### 3.2.1.1 Poison Attack

During the training of the model, if an adversary collaborates a carefully crafted training to training algorithm, the model output can be unexpected. This is said to be poisoning of the model. These attacks look to damage the model output or the model output is directly controlled by the adversary. The poison attack is not limited to training data poisoning but can be exploited further by algorithm poisoning and model poisoning.

### 3.2.2 Inference Time Attacks

These attacks happen at the time of inference/testing of the model post completion of training. The adversary has the ability to query the model and receive results, it is in this phase where an adversary can exploit security and privacy vulnerability to infer training data model, model parameters or membership information of an individual. These attacks are further categorized below.

#### 3.2.2.1 Model Inversion

When an adversary attempts to infer a sensitive attribute of the given data instance from a published model, it is called a model inversion attack. One such attack was uncovered by Fredrikson et al. [23] by successfully revealing patients' sensitive information such as age, height, and genomic metadata by just having black-box API access to the medicine dosage prediction model. Later, Fredrikson et al. [24] used the same analogy on neural network-based learning, where he proved to have learnt sensitive attribute of training dataset by feeding carefully crafted input to the target model using black-box access.

#### 3.2.2.2 Membership Inference

The Membership Inference Attack (MIA) is the method of learning whether or not a sample belongs to the training dataset of a trained ML model. More specifically, determining if a given data point has membership in the training dataset. This attack is vital from a confidentiality perspective and hence a more significant privacy threat with ML model. Shokri et al. [8] described MIA for the first time and proved how an adversary can get a probability vector for a given data point by querying the target model in black-box access. Adversary later uses the probability vector to deduce the given data point

**Table 2** Classification of known machine leaning attacks

| Attack | Access to model | Access to output | Membership inference | Model inversion | Parameter inference | Property inference |
|---|---|---|---|---|---|---|
| Membership Inference [8] | Blackbox | Logits | ✓ | | | |
| Membership Privacy [9] | Blackbox | Logits | ✓ | | | |
| ML-Leaks [10] | Blackbox | Logits | ✓ | | | |
| The Natural Auditors [11] | Blackbox | Logits | ✓ | | | |
| LOGAN [12] | Both | Logits | ✓ | | | |
| Data Provenance [13] | Blackbox | Logits | ✓ | | | |
| Privacy Risk in ML [14] | Whitebox | Logits | ✓ | ✓ | | |
| Fredrikson et al. [15] | Blackbox | Logits | | ✓ | | |
| MIA with Confidence Values [15] | Both | Logits | | ✓ | | |
| Adversarial NN Inversion [16] | Blackbox | Logits | | ✓ | | |
| Update Leaks [17] | Blackbox | Logits | | ✓ | | |
| Collaborative Inference MIA [18] | Both | Logits | | ✓ | | |
| The Secret Sharer [19] | Blackbox | Logits | | | | ✓ |
| Property Inference of FCNNs [20] | Whitebox | Logits | | | | ✓ |
| Hacking Smart Machine with [21] | Whitebox | Logits | | | | ✓ |
| Cache Telepathy | Blackbox | Logits | | | ✓ | |
| Stealing Hyperparameter [22] | Blackbox | Logits | | | ✓ | |

(*continued*)

**Table 2** (*continued*)

| Attack | Access to model | Access to output | Membership inference | Model inversion | Parameter inference | Property inference |
|---|---|---|---|---|---|---|
| Stealing ML model [20] | Blackbox | Label | | | √ | |

membership in the training dataset. The basic principle here is to create a shadow model trained with the labelled dataset. Effective use of noisy real data or a model inversion attack can help secure the labelled dataset. The shadow models are used later on to train the attack model, which is capable of differentiating the membership of a data point in the training set. Finally, the adversary queries the target model for each given data point and compares it with the attack model in terms of a probability vector; if the probability vector's value is high, membership chances are also high. The critical premise here is that the probability vector value will be low if the model has not seen the data point before; but will be higher if the model has seen it earlier during the training phase; hence membership is proved. MIA can be associated with memorization of data heavily used by Deep Neural Networks (DNN) or over-fitting of the model [25, 26]. DNN often tend to over-fit to the training dataset because of the memorization techniques used [27]. Even Generative Adversarial Networks (GANs) can be susceptive to MIA, as pointed out by Hayes et al. [12].

#### 3.2.2.3 Model Stealing

ML models are the end-results of rigorous Machine Learning activities such as data aggregation, Extract-Transform-Load (ETL), algorithm selection, hyperparameter tuning, and training; therefore, the trained models are portrayed as intellectual property of their owner. A privacy breach is occurred if the model is extracted or compromised. The model can reveal a lot about the training data set in a Deep Neural Network (DNN) based setting where the model carries much memory of the training data.

#### 3.2.2.4 Property Inference

Property Inference attacks work by inferring patterns of information, specific to the target model. A memorization attack aimed at finding sensitive patterns in a target model's training data is an example of the forenamed attacks [19]. Hidden Markov Models (HMM) and Support Vector Machines (SVM) are the targets of Property Inference attacks [28].

## 4 Privacy-Preserving Mechanisms

This section reviews state-of-the-art Privacy-Preserving mechanisms for Machine Learning. Figure 1, classifies the Privacy-Preserving mechanisms into three broad categories split across a typical Machine Learning workflow. The first category describes Privacy-Preserving data collection or aggregation methods. The second category covers devising

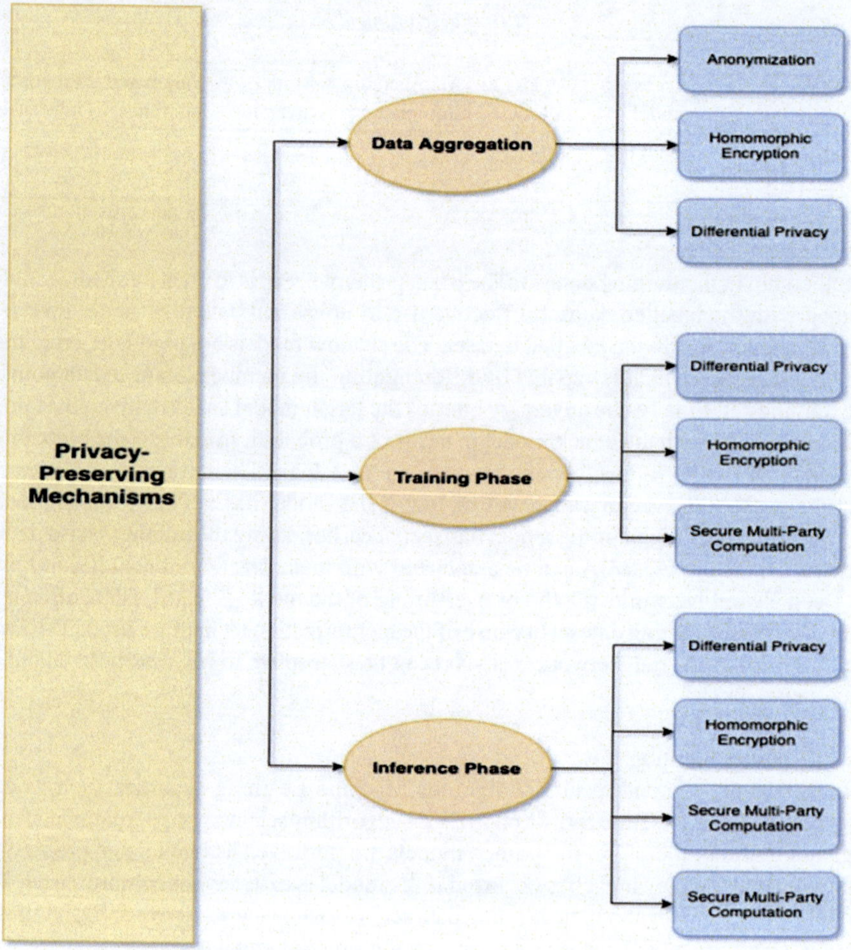


**Fig. 1** Privacy-Preserving machine learning techniques

techniques to preserve privacy during the crucial phase of model training. These mechanisms ensure that the model and training data remain confidential and secure during the training phase of ML. The third and final category intends to cover Privacy-Preserving mechanisms at the test-time inference phase of ML by protecting the model user's privacy [29].

## 4.1 Data Aggregation

This section focuses on ensuring privacy while collecting or aggregating the data during the Machine Learning process. Arguably, preserving privacy when sharing data is the most crucial technique. The techniques here broadly utilize methods such as perturbation, anonymization, and encryption to preserve privacy. The methods can further be classified into two subgroups based on the context-awareness during data sharing. Context-free privacy methods, such as differential privacy, are ignorant of the particular context or the overall objective of why the data is shared, On the contrary, context-aware privacy methods can perform better in achieving privacy because they are cognizant of the context and purpose of data usage, such as information-theoretic privacy [30].

### 4.1.1 Anonymization

These methods use the naivest anonymization techniques where the personally identifiable information such as name, address, and identifiers is removed from the dataset before sharing it with the collaboration engine. However, this is less efficient and does an inferior job in protecting privacy since breaking anonymization is possible using intelligent de-anonymization techniques. One such attack is presented by Narayanan et al. where Netflix prize dataset was de-anonymized [31] using a tiny known bit of information about a subscriber, the whole subscriber data was obtained from Netflix prize dataset. Anonymization is not enough if auxiliary information from other data sources is present. For example, Netflix, in 2007, as part of a competition to find out if anyone can beat their collaborative filtering algorithm, made public a dataset of their user ratings. Even though no personally identifying information was released researchers were able to recover 99% of personal information that was removed from the dataset [32]. In this incident researchers were able to gather the data removed by using auxiliary information from IMDB.

### 4.1.2 Homomorphic Encryption

The homomorphic techniques enable the data to be encrypted first and then shared for collaborative ML to preserve privacy. The techniques use semantic security [32] or probabilistic encryption as a mechanism, which though theoretically breakable, is currently infeasible.

### 4.1.3 Differential Privacy

Privacy at the data aggregation level can also be achieved by using the perturbation technique, one of the implementations is called Differential Privacy (DP). DP intends to present ways to maximize the correctness of queries from databases while minimizing the odds of recognizing its records. DP techniques focus on including random noise into the mix (data, algorithm, output etc.), resulting in noisy and imprecise entries received by any adversary, making it difficult for them to breach privacy. In Local DP (LDP) mode, rather than using a centralized DP server, participants add differentially private randomization to their data before sharing. LDP was inspired by an assumption that the data collector is untrusted and applicable to data aggregation PPML techniques. The LDP-based solutions are widely adopted to solve privacy woes. For example, Google enabled web browser developers to collect usage statistics data privately by using a RAPPOR [33]. Majority of the work around DP [34, 35] has been introduced for various applications. For GAN-based system, Triastcyn & Faltings presented a technique that can provide differential privacy for newly generated synthetic data that still has the real data's statistical properties [36]. Another implementation called Pufferfish on DP was presented by Kifer et al. [37]. The Pufferfish framework proved to be handy for precise applications, for example, Census data release.

### 4.2 Training Phase

This section categorizes Privacy-Preserving techniques during training time in Machine Learning. Ensuring privacy during training time can be achieved by using either differential privacy or encryption. If the training happens on encrypted data itself, privacy can be preserved. The two commonly used methods for achieving the same are Homomorphic Encryption during training and Secure Multi-Party Computation (SMPC). In Table 3 following abbreviations have been used: GM used for Generative Model, AE used for Auto Encoder, LIR used for Linear Regression, LOR used for Logistic Regression, LM used for Linear Means, FLD used for Fisher's Linear Discriminant, NB for Naive Bayes used for RF for Random Forest, IM for Image Classification, FL for Federated Learning.

In this survey, the literature of private training has been divided into three categories of methods that employ: 1) Differential Privacy (DP), 2) Homomorphic Encryption (HE) and 3) Secure Multi-Party Computation (SMPC). Table 3 shows this categorization for the literature that will be discussed in this section.

#### 4.2.1 Differential Privacy

This section focuses on differential privacy methods used during the training phase of ML. Figure 2 shows the application of differential privacy to different phases of ML, here we have taken the example of a Deep Learning ML framework where perturbation techniques of differential privacy are applied at different phases. In short, it can be applied at five places, starting from data input, in the loss function, during gradient update, to the parameters obtained from the model's training, and finally to labels [41].

Input perturbation is as good as applying noise at the data aggregation level before training. To include noise at objective function and output level, one can utilize objective function and output perturbation, respectively. Chaudhuri et al. [53] prove that objective perturbation outperforms output perturbation in the context of the trade-off between learning performance and privacy [53]. Recently, objective perturbation was benchmarked real-world high dimensional data by Iyengar et al. [54] which is general and more practical. However, sometimes non-convexity of objective function plays a more significant role in deciding objective function perturbation effectiveness. Phan et al. [55] proposed to use the approximate convex polynomial function in place of non-convex function before applying objective function perturbation; however, this solution limits the ability of deep neural network due to approximation limits. To address the limitation above, researchers proposed applying perturbation gradient and called this technique gradient perturbation. Despite its wide usage, the technique introduced a side effect of bounded gradient norms, which diluted the effectiveness. Another perturbation approach was perturbating model parameters during distributed Machine Learning to achieve privacy [56]. However, scalability issues are a limitation with the implementation as epsilon needed to be in proportion to the model's size. Analyzing the exhaustion of the privacy budget is a crucial area of consideration with differential privacy [36]. Here, the approach by Abadi et al. [35] was improved by Bu et al. [56] who applied the Gaussian

**Table 3** Categorization of distinguished Privacy-Preserving mechanisms for training

| Implementations | Work | Datasets | DP | HE | SMPC |
|---|---|---|---|---|---|
| DPSGD [35] | IM | MNIST, CIFAR-10 | * | | |
| DP LSTM [34] | LanguageModel w/LSTMs | Reddit Posts | * | | |
| DP GAN [38] | Data Generation w/GAN | MNIST, MIMIC-III | * | | |
| DP GM [39] | Data Generation w/GM | MNIST, CDR, TRANSIT | * | | |
| DP AE [40] | Behaviour Prediction w/AE | Health Social Network Data | * | | |
| PATE [41] | IM | MNIST, SVHN | * | | |
| Scalable Learning w/PATE [42] | IM | MNIST, SVHN, Adult, Glyph | * | | |
| Distributed DP [43] | Classification | eICU, TCGA | * | | |
| DP FL [44] | IM | MNIST | * | | |
| Private Collaborative NN [45] | IM | MNIST | * | | * |
| Secure Aggregation for ML [46] | FL | – | | | * |
| QUOTIENT [47] | Classification | MNIST, Thyroid, Credit | | | * |
| SecureNN [48] | Classification | MNIST | | | * |
| ABY3 [49] | LIR, LOR, NN | MNIST | | | * |
| Trident [50] | LIR, LOR, NN | MNIST, Boston Housing | | | * |
| SecureML [51] | LIR, LOR, NN | MNIST, Gisette,Arcene | | * | * |
| CryptoDL [52] | IM | MNIST, CIFAR-10 | | * | |

Differential Privacy (GDP) notion to Deep Learning, initially proposed by Dong et al. [36]. The key differentiating feature with GDP is Adam Optimizer's use for analyzing exhaustion of privacy budget without developing advanced techniques such as moments accountant proposed by Abadi et al. [35]. Essentially, differential privacy introduces a loss of utility to Machine Learning because of the incorporation of noise and clipping. Bagdasaryan et al. [57] has explained that this loss in utility is diverse across different subgroups of the population of mixed sizes [57]. They prove empirically that the loss of accuracy is more for sub-groups with less representation.

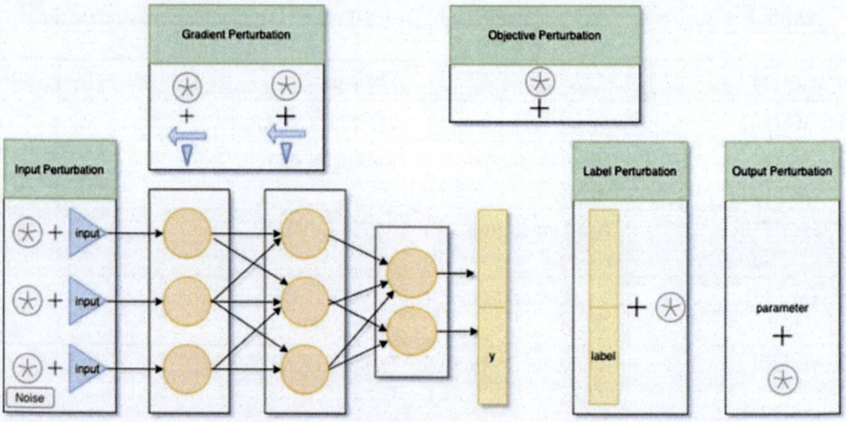


**Fig. 2** Differential privacy application to various phases of deep learning

### 4.2.2 Homomorphic Encryption

Homomorphic encryption [58] enables training to happen over encrypted data. Multiple parties can encrypt the data and send it over to a centralized server that runs a Machine Learning algorithm over this encrypted data itself as if it is not encrypted and then communicate the encrypted result to parties who can decrypt the result. This technique utilizes advanced cryptography and requires a lot of number crunching. Therefore, it is notably computation-intensive task, making it a less favourable choice for production usage [59, 60]. Significantly, few works have utilized homomorphic encryption for the Machine Learning model's private training [52, 61]. Some of the notable works include Somewhat HE (SHE) by Graepel et al. (for liner mean training) and Fisher's Linear Discriminate (FLD) classifiers. The notable limitation with HE algorithms is their ability to deal with non-linear functions.

Graepel et al. avoid complex algorithms such as a Neural Network and suggest division-free algorithms and simple classifiers. Hesamifard et al. [52] experimented with using HE for Deep Learning. They proposed methods for the approximation of Neural Network activation functions (ReLU, Sigmoid, and Tanh) with low-degree polynomials to design effective homomorphic encryption schemes. Afterward, they used the approximate polynomial functions to train Convolution Neural Networks (CNN) and eventually implemented CNN over encrypted data and tested with MNIST optical character recognition tasks. The Convolution Neural Networks were then trained using the approximated polynomial functions and the performance of the models were measured after the eventual implementation of CNN over encrypted data.

### 4.2.3 Secure Multi-party Computation

Secure Multi-Party Computation (SMPC) is a subdiscipline within cryptography. It aims to solve privacy challenges with collaborative computing involving multiple honest/semi-honest/adversary parties [62]. No party in this network has access to more than one encrypted part of the whole dataset in joint computation; hence parties are oblivious of the data held by other parties. SMPC ensures the secrecy of data owners as long as at least one party in the collaborative computing network is trusted by them. SMPC implementations consist of simple functions that can easily be computed, such as secure sum, garbled circuits. However, supporting complex computing functions comes with a restrictive computational cost [60, 63].

In Machine Learning, SMPC based ML protocols enhance privacy by off-loading computation/training on a separate set of non-colluding servers, whereas different parties contribute data for training linear, logistic regression, and neural network models and inference [47, 51]. One of the shortcomings with SMPC is the necessity of all the parties and servers to be online at all times; this results in significant communication overhead [47]. Mohassel et al. [51] introduced Secure ML to privately train ML algorithms (such as linear and logistic regression and neural network) using stochastic gradient descent-based method and secret sharing in multi-party computation setting. Recently, Mohassel et al. [49] extended their research and proposed a framework named ABY3 for secure three-party training of DNNs having mostly honest parties. Agrawal et al. [47] suggest QUOTIENT, they enhanced privacy by proposing a combination of clusterized DNN training algorithm and customized secure two-party protocol to work on it. Its novice work towards designing an optimized algorithm and a secure computation protocol together, instead of applying the encryption on an existing algorithm in the traditional way.

### 4.3 Inference Phase

As presented in Table 4, ensuring privacy during inference is an emerging research area since not much work has been done in this direction until. Inference privacy focuses on ensuring privacy for a system which offers Machine-Learning-as-a-Service (MLaaS) or Inference-as-a-Service. The MLaaS systems provide inference using query or APIs and do not need to be actively trained with newer data sets.

There are some glaring differences between the literature on training and inference privacy. Most of the training literature suggests differential privacy is a defacto choice during training, whereas encryption methods such as homomorphic encryption and SMPC are preferred during inference. One of the reasons is that the computation cost and complexities with encryption methods make them difficult for long-running training jobs [64]. On the contrary, inference jobs are relatively quicker, and adopting encryption here is much simpler because the model is trained already. Adding perturbation techniques comes at the cost of accuracy; requires precise execution; hence adopting it during inference is less favorable.

Below we discuss the literature of each category.

#### 4.3.1 Differential Privacy

As discussed above, due to accuracy and performance concerns with differential privacy in the inference phase, the adoption of differential privacy for inference is significantly less, especially with pre-trained networks, MLaaS, or Inference-as-a-Service setting Wang et al. [71] suggest that random data nullification and random noise addition can preserve privacy during inference. The developed framework is named Arden. Arden

**Table 4** Classification of distinguished Privacy-Preserving mechanisms for inference

| Implementations | Work | Datasets | DP | HE | SMPC |
|---|---|---|---|---|---|
| Cloak [65] | IM | CIFAR-100, CelebA | * | | |
| EPIC [60] | IM | CIFAR-10, MIT, Caltech | | | * |
| DeepSecure [66] | Classification | MNIST, UCI-HAR | | | * |
| XONN [59] | IM | MNIST, CIFAR-10 | | | * |
| Chameleon [48] | Classification | MNIST, Credit Approval | | | * |
| CRYPTFLOW | Classification | MNIST,CIFAR, ImageNet | | | * |
| MiniONN [67] | IM | MNIST, CIFAR-10 | | * | * |
| GAZELLE [64] | IM | MNIST, CIFAR-10 | | * | * |
| DELPHI [68] | IM | CIFAR-10, CIFAR-100 | | * | * |
| Cryptonets [69] | IM | MNIST | | * | |
| Private Classification | IM | MNIST | | * | |
| TAPAS [70] | IM | MNIST, Faces, Cancer | | * | |
| Cheetah | IM | MNIST, Imagenet | | * | |
| ARDEN [71] | IM | MNIST, CIFAR-10, SVHN | * | | |

proposes partitioning the DNN across edge devices and the cloud servers, where complex inference are carried out at the cloud server and simple transformations are done at edge devices such as mobile. Random data nullification and random noise addition make different inference queries indistinguishable, and hence inference user's privacy is preserved.

### 4.3.2 Homomorphic Encryption

In 2016, Dowlin et al. presented CryptoNets [69], a mechanism that encrypted a trained neural network; operated on an encrypted input; produced encrypted output that the calling user later decrypted. To enhance the system's performance, CryptoNets enables the use of Single Instruction Multiple Data (SIMD) operations. Even then, the latency remains high for single queries. Morris et al. [72] proposed low-degree polynomials to approximate the ReLu non-linear activation function and implement a normalization layer before the activation function, resulting in higher accuracy theoretically. However, this did not prove higher accuracy empirically. To counter latency issues with prior HE techniques during private inference, Juvekar et al. suggest GAZELLE [64]. GAZELLE implemented an innovative blend of the traditional two-party computation techniques and homomorphic encryption. Using the low-latency HE library and fast algorithms based on homomorphic linear algebra kernels, GAZELLE proved superior to CryptoNets in terms of latency since it is three orders of magnitude faster. Sanyal et al. used binarized neural networks to make HE inference method quicker and claimed privacy of model and data, unlike CryptoNets, which only preserve data privacy.

### 4.3.3 Secure Multi-party Computation

Liu et al. suggest MiniONN [67] to solve Privacy-Preserving prediction or inference. It applies additively homomorphic encryption (AHE) in a pre-processing phase, as opposed to GAZELLE, which employs AHE to fasten linear algebra directly. MiniONN proves a notable performance gain in comparison to CryptoNets, without losing accuracy. Nevertheless, currently, it is limited to a two-party computation scheme only. Riazi et al. introduce Chameleon [73], a two-party computation framework for secure inference that performs 4.2 times better in terms of latency over MiniONN. Chameleon employs vector dot product of signed fixed-point numbers to gain efficiency in classification prediction based upon large matrix multiplications. Speeding up the computation is a trend with SMPC based private inference solution since the accuracy loss is negligible.

# 5 Privacy-Enhancing Execution Models and Environments

Privacy can also be preserved if the environment and execution enhance privacy, though they are not Privacy-Preserving by themselves. This section briefly discusses privacy-enhancing techniques such as trusted execution environments, split learning, and federated learning. However, these privacy-enhancing techniques need to be coupled with Privacy-Preserving measures discussed earlier. Figure 3 shows the division of privacy-enhancing mechanisms into three categories namely, federated learning, split learning, and Trusted Execution Environment (TEE).

## 5.1 Federated Learning

Federated Learning (FL) is a collaborative Machine Learning setting. FL's core idea is input data does not leave the client's environment; rather different clients collaboratively train the model with their data under the supervision of a central server. FL is an example

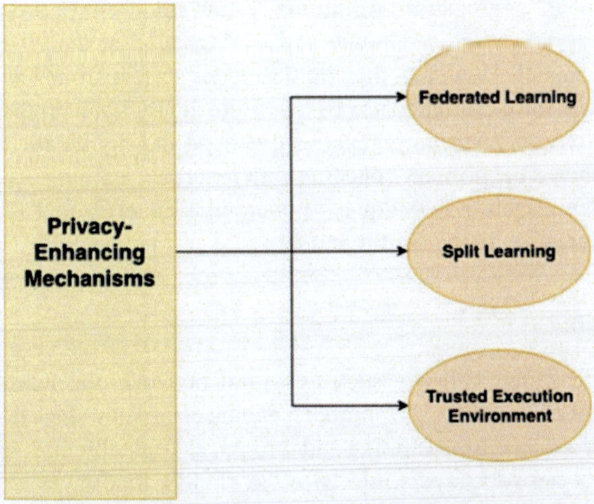


**Fig. 3** Privacy-enhancing machine learning techniques

of distributed? learning securely, thereby addressing privacy concerns of centralized Machine Learning [74].

Federated learning is executed in the following 6 phases:

1. Problem identification: Identify the problem to be solved using federated learning.
2. Client instrumentation: The participating parties or clients are chosen and instructed to facilitate required data and training needs. Typically, parties need to maintain configuration data on which server to communicate to and the core data to be contributed from their side.
3. Simulation prototyping (optional): A simulation is done to test different architectures and hyperparameter tuning by a deployment engineer.
4. Federated model training: Formal training jobs are launched consisting of multiple federated training tasks across participating clients. Here, individual parties train the intermediate model obtained from the federated server by feeding their data, producing a trained model with a newer model parameter.
5. Model evaluation: All the participating clients contribute their training model parameter to a centralized server, where parameter aggregation is performed, and a new intermediate model is obtained. Step 4 and 5 are iterated multiple times to arrive at the final trained model.
6. Deployment: Once the final model is decided, it is subjected to deployment after due quality checks such as quality insurance, live A/B testing, and a staged roll-out.

In the recent past, there is a trend of using FL along with SMPC and differential privacy [46, 74, 75]. Bonawitz et al. used a federated learning setup to implement secure aggregation to privately consolidate local Machine Learning outputs on edge devices to arrive at the trained global model. Secure aggregation is a typical SMPC solution where multiple parties collaborate to find a secure sum without revealing individual party's input data even to the aggregator. This setting enhances privacy in distributed learning scenarios since the data does not leave the client boundary, and a virtually reliable third party performs secure aggregation. Additionally, few studies have proposed not to trust the centralized server, rather, randomize data for training at the client end itself and named it shuffle model. However, the shuffle model suffers from low accuracy. Cheu et al. and Balle et al. have offered to use DP in secure aggregation protocols in the shuffle model [76, 77]. However, the communication overhead and significant time consumption during training remain the primary concerns with federated learning. Recently [78] work has proposed few mitigation techniques to overcome the error that may incur and the issues of communication in the shuffle model.

## 5.2 Split Learning

In a distributed learning scenario, such as neural network, the learning can be split between client and server [30]. The approach enhances privacy since none of the parties get the other party's data during training and inference. Typically, split learning is shaped by splitting neural network layers into two; the client's final layer is called cut layer. Each client computes the neural network until the cut layer and passes the intermediate output data to the server, where the rest of the computation is performed. The forward pass and backpropagation also happen between cut layer and server.

### 5.3 Trusted Execution Environment

Trusted execution environments, aka secure enclaves, enables collaborating parties to move part/whole of the training process or inference process into the cloud-based trusted environment. Additional security measures are adopted to verify or attest to the code running in these trusted environments, thereby providing confidentiality and integrity during execution.

Mo et al. have suggested a framework to restrict the attack surface upon DNN by using edge device's Trusted Execution Environment (TEE) in association with model partitioning [79]. Many organizations have brought into play various forms, including Intel's SGX-enabled CPUs [80–83].

These models assure security during execution. However, the end-user still needs to send their data to an enclave that might be running in a secure remote server. Hence this raw data is still prone to unauthorized access, which can lead to privacy loss [84, 85].

## 6 Comparative Analysis

This section focuses on the comparative analysis of some of the distinguished Privacy-Preserving machines. Table 5 summarizes the details. Some of the parameters over which the techniques are compared are privacy, accuracy, efficiency (time to train and infer), suitable algorithms, a suitable number of participants, and communication overhead.

DPSGD [35] did adequate privacy accounting for functional output of centrally held data but lacked accuracy. It is trained and tested with DBs, e.g., MNIST and CIFR-10. Whereas DP LSTM [34] demonstrated high privacy using DP, however, resulted in high computation cost; the study can be accompanied with federated learning so that sensitive data remains at the client-side only. PATE [41] Private Aggregation of Teacher Ensembles, or PATE, is utilized on simple classification tasks on smaller DBs such as MNIST. Distributed DP [43] does collaborative Deep Learning with various parties using distributed Stochastic Gradient Descent. It uses Selective Stochastic Gradient Descent (Selective SGD or SSGD) protocol. However, it does not address the global parameter server's problem knowing the contributing parties's identity, additionally having a communication overhead when model parameters are shared between global parameter server and client. ABY3 [49] (three-party, most of them honest party) and SecureML are training DNN on encrypted data. Apart from lesser accuracy than vanilla DNN, both the approaches consume significant time in training; in a two-party setting, a DNN with three layers takes 80 h to train MNIST dataset. Moreover, in a wide-area-network (WAN) setting, the same task takes close to 4277 h. QUOTIENT [47] succeeded in getting 50 times faster DNN training in WAN condition over ABY3 and SecureML by customizing ML model and secure protocol jointly. However, it is still slow for CNN training, where the communication load is still high. SMPC protocol for quick training of CNN is a good area for further research. CryptoDL [52] used HE friendly, low degree polynomial as activation function in CNN perform better than Cryptonets [69] in terms of

**Table 5** Comparative analysis of privacy-preserving techniques

| Techniques | Efficiency | Computation cost | No. of parties | Model | Accuracy | Communication cost |
|---|---|---|---|---|---|---|
| Cloak [65] | High performance | One single training super computer | NA | DNN, CNN, semi supervised algo | High | Yes |
| DPSGD [35] | High | NA | Single party | DNN | Low | No |
| DP LSTM [34] | – | High | – | DNN, CNN | – | – |
| PATE [41] | – | High | Multiple | DNN,CNN | Low with SVHN | Yes |
| Distributed DP [43] | – | Low (no use of encryption) | Multiple | MLP, CNN | – | Yes |
| ABY3 [49] | Low | High | Two-party | DNN, | Low | High |
| SecureML [51] | Low | High | Two-party | DNN, regression, logistic regression | Low | High |
| QUOTIENT [47] | High, Low for CNN | Low, High (for CNN) | Two-party | DNN, CNN | 6% > SecureML | Low, high for CNN |
| CryptoDL [52] | High | High | Multiple | DNN, CNN | High | – |
| MiniONN [67] | Higher than CryptoNET | Low | Two-party | DNN, CNN | Higher than SecureML | – |
| Chameleon [48] | 4.2 × higher than MiniONN | Low | Two-party | NN | – | – |
| GAZELLE [64] | Better than Chameleon | High | Two-party | NN | – | High |
| EPIC [60] | 34 × higher than GAZELLE | Low | Multiple | SVM | 7% > GAZELLE | 50x < GAZELLE |

accuracy, communication overhead, and runtime; however, it needs to be further tested against large datasets such as CIFAR-100 or ImageNet. Moreover, it is computationally complex to train CNN on encrypted data and requires expensive hardware such as GPUs. MiniONN [67] only support two-party, and the security model remains passive. Chameleon [48], also support two-party and/but has a restrictive assumption about the semi-honest third party on their training involvement; it produces performance 4.2 times better than MiniONN [67]. DeepSecure [66] and XONN [59] are SMPC based solutions using the garbled circuit to produce secure inference but limited to the two-party setting. Moreover, XONN [59] only supports NN with binary weights. GAZELLE [64] has better efficiency than Chameleon [48], and/but uses additive homomorphic encryption along with two-party SMPC using the garbled circuit. EPIC [60] produced a secure classification, a non-NN base image classification using SVM in SMPC setting, producing 34× times more efficiency, 7% more accuracy, and 50× times lesser communication cost than GAZELLE [64].

Beyond two-parties SMPC based Privacy-Preserving, Machine Learning solutions are limited; usual challenges are complexity, computation cost, and communication overhead. Most collaborative learning considers data is horizontally partitioned; however, very little work has been done when data is vertically partitioned.

## 7 Conclusion

The immense growth in data volume and easy availability of inexpensive computing power has given rise to the evolution and adoption of Machine Learning. Privacy-preserving Machine Learning techniques are required to enable multiple participants to contribute their data without any privacy concerns. This paper aims to present a thorough and methodical summary of Privacy-Preserving Machine Learning techniques across various phases of the Machine Learning process. It is observed that inference phase is an opportunity for future research compared to the aggregation and training phase. Very little attention has been given to vertically partitioned data during the training phase. Additionally, a secure-multi-party-based Privacy-Preserving Machine Learning solution needs to go beyond two-party settings without incurring more computational and communication costs.

## References

1. Xue M, Yuan C, Wu H, Zhang Y, Liu W (2020) Machine learning security: threats, countermeasures, and evaluations. IEEE Access 8:74720–74742. https://doi.org/10.1109/ACCESS.2020.2987435
2. Du W, Han YS, Chen S (2004) Privacy-preserving multivariate statistical analysis: linear regression and classification. In: SIAM proceedings series, pp 222–233. https://doi.org/10.1137/1.9781611972740.21
3. Lipp M et al (2020) Meltdown: reading kernel memory from user space. Commun ACM. https://doi.org/10.1145/3357033
4. Kocher P et al (2020) Spectre attacks: exploiting speculative execution. Commun ACM. https://doi.org/10.1145/3399742

5. Opinion | Chinese Hacking Is Alarming. So Are Data Brokers. - The New York Times. https://www.nytimes.com/2020/02/10/opinion/equifax-breach-china-hacking.html. Accessed 13 Mar 2021
6. Leaking Ads—Is User Data Truly Secure? https://www.slideshare.net/cisoplatform7/leaking-adsis-user-data-truly-secure. Accessed 24 Mar 2021
7. Opinion | FaceApp Shows We Care About Privacy but Don't Understand It - The New York Times. https://www.nytimes.com/2019/07/18/opinion/faceapp-privacy.html. Accessed 24 Mar 2021
8. Shokri R, Stronati M, Song C, Shmatikov V (2017) Membership inference attacks against machine learning models. https://doi.org/10.1109/SP.2017.41
9. Long Y, Bindschaedler V, Gunter CA (2017) Towards measuring membership privacy. arXiv
10. Singh S, Sikka HD (2020) Benchmarking differentially private residual networks for medical imagery. arXiv. https://doi.org/10.31219/osf.io/v2ms6
11. Song C, Shmatikov V (2018) The natural auditor: how to tell if someone used your words to train their model. arXiv Preprint arXiv:1811.00513
12. Hayes J, Melis L, Danezis G, De Cristofaro E (2017) LOGAN: evaluating privacy leakage of generative models using generative adversarial networks. arXiv
13. Song C, Shmatikov V (2019) Auditing data provenance in text-generation models. https://doi.org/10.1145/3292500.3330885
14. Yeom S, Giacomelli I, Fredrikson M, Jha S (2017) Privacy risk in machine learning: analyzing the connection to overfitting. http://arxiv.org/abs/1709.01604
15. Fredrikson M, Jha S, Ristenpart T (2015) Model inversion attacks that exploit confidence information and basic countermeasures. https://doi.org/10.1145/2810103.2813677
16. Yang Z, Chang EC, Liang Z (2019) Adversarial neural network inversion via auxiliary knowledge alignment. arXiv
17. Salem A, Bhattacharya A, Backes M, Fritz M, Zhang Y (2020) Updates-leak: data set inference and reconstruction attacks in online learning
18. He Z, Zhang T, Lee RB (2019) Model inversion attacks against collaborative inference. https://doi.org/10.1145/3359789.3359824
19. Carlini N, Liu C, Kos J, Erlingsson Ú, Song D (2018) The secret sharer: measuring unintended neural network memorization & extracting secrets. arXiv
20. Tramèr F, Zhang F, Juels A, Reiter MK, Ristenpart T (2016) Stealing machine learning models via prediction APIs
21. Ateniese G, Mancini LV, Spognardi A, Villani A, Vitali D, Felici G (2015) Hacking smart machines with smarter ones: how to extract meaningful data from machine learning classifiers. Int J Secur Netw. https://doi.org/10.1504/IJSN.2015.071829
22. Wang B, Gong NZ (2018) Stealing hyperparameters in machine learning. https://doi.org/10.1109/SP.2018.00038
23. Fredrikson M, Lantz E, Jha S, Lin S, Page D, Ristenpart T (2014) Privacy in pharmacogenetics: an end-to-end case study of personalized warfarin dosing
24. Fredrikson M, Jha S, Ristenpart T (2015) Model inversion attacks that exploit confidence information and basic countermeasures. In: Proceedings of the ACM conference on computer and communications security, October 2015, pp 1322–1333. https://doi.org/10.1145/2810103.2813677
25. Truex S, Liu L, Gursoy ME, Yu L, Wei W (2018) Demystifying membership inference attacks in machine learning as a service. arXiv. https://doi.org/10.1109/tsc.2019.2897554
26. Sablayrolles A, Douze M, Ollivier Y, Schmid C, Jegou H (2019) White-box vs black-box: Bayes optimal strategies for membership inference
27. Arplt D et al (2017) A closer look at memorization in deep networks

28. Yang R (2020) Survey on privacy-preserving machine learning protocols. Lecture notes in computer science (including subseries lecture notes in artificial intelligence and lecture notes in bioinformatics), vol 12486. LNCS, pp 417–425. https://doi.org/10.1007/978-3-030-62223-7_36
29. Mireshghallah F, Taram M, Vepakomma P, Singh A, Raskar R, Esmaeilzadeh H (2020) Privacy in deep learning: a survey. arXiv, April 2020. http://arxiv.org/abs/2004.12254. Accessed 11 Mar 2021
30. Gupta O, Raskar R (2018) Distributed learning of deep neural network over multiple agents. J Netw Comput Appl. https://doi.org/10.1016/j.jnca.2018.05.003
31. Narayanan A, Shmatikov V (2008) Robust de-anonymization of large sparse datasets. https://doi.org/10.1109/SP.2008.33
32. Goldwasser S, Micali S (1984) Probabilistic encryption. J Comput Syst Sci. https://doi.org/10.1016/0022-0000(84)90070-9
33. Erlingsson Ú, Pihur V, Korolova A (2014) RAPPOR: randomized aggregatable privacy-preserving ordinal response. https://doi.org/10.1145/2660267.2660348
34. Brendan McMahan H, Ramage D, Talwar K, Zhang L (2017) Learning differentially private recurrent language models. arXiv
35. Abadi M et al (2016) Deep learning with differential privacy. https://doi.org/10.1145/2976749.2978318
36. Triastcyn B, Faltings A (2018) Generating differentially private datasets using GANs
37. Kifer D, Machanavajjhala A (2014) Pufferfish: a framework for mathematical privacy definitions. ACM Trans Database Syst. https://doi.org/10.1145/2514689
38. Xie L, Lin K, Wang S, Wang F, Zhou J (2018) Differentially private generative adversarial network. arXiv
39. Acs G, Melis L, Castelluccia C, De Cristofaro E (2019) Differentially private mixture of generative neural networks. IEEE Trans Knowl Data Eng. https://doi.org/10.1109/TKDE.2018.2855136
40. Phan NH, Wang Y, Wu X, Dou D (2016) Differential privacy preservation for deep auto-encoders: an application of human behavior prediction
41. Papernot N, Goodfellow I, Abadi M, Talwar K, Erlingsson Ú (2017) Semi-supervised knowledge transfer for deep learning from private training data
42. Papernot N, Song S, Mironov I, Raghunathan A, Talwar K, Erlingsson Ú (2018) Scalable private learning with pate. arXiv
43. Beaulieu-Jones BK, Finlayson SG, Yuan W, Wu ZS (2018) Privacy-preserving distributed deep learning for clinical data. arXiv
44. Geyer RC, Klein T, Nabi M (2017) Differentially private federated learning: a client level perspective. arXiv
45. Chase M, Gilad-Bachrach R, Laine K, Lauter K, Rindal P (2017) Private collaborative neural network learning. IACR Cryptology ePrint Archive
46. Bonawitz K et al (2017) Practical secure aggregation for privacy-preserving machine learning. In: Proceedings of the ACM conference on computer and communications security, October 2017, pp 1175–1191. https://doi.org/10.1145/3133956.3133982
47. Agrawal N, Kusner MJ, Shamsabadi AS, Gascón A (2019) QUOTIENT: two-party secure neural network training and prediction. https://doi.org/10.1145/3319535.3339819
48. Wagh S, Gupta D, Chandran N (2019) SecureNN: 3-party secure computation for neural network training. Proc Priv Enhancing Technol 2019(3):26–49. https://doi.org/10.2478/popets-2019-0035
49. Mohassel P, Rindal P (2018) ABY3: a mixed protocol framework for machine learning. https://doi.org/10.1145/3243734.3243760
50. Rachuri R, Suresh A (2019) Trident: efficient 4PC framework for privacy preserving machine learning. arXiv. https://doi.org/10.14722/ndss.2020.23005

51. Mohassel P, Zhang Y (2017) SecureML: a system for scalable privacy-preserving machine learning. In: Proceedings - IEEE symposium on security and privacy, pp 19–38. https://doi.org/10.1109/SP.2017.12
52. Hesamifard E, Takabi H, Ghasemi M (2017) CryptoDL: deep neural networks over encrypted data. arXiv, 14 November 2017
53. Chaudhuri K, Monteleoni C, Sarwate AD (2011) Differentially private empirical risk minimization. J Mach Learn Res 12:1069–1109
54. Iyengar R, Near JP, Song D, Thakkar O, Thakurta A, Wang L (2019) Towards practical differentially private convex optimization. https://doi.org/10.1109/SP.2019.00001
55. Phan N, Wu X, Hu H, Dou D (2017) Adaptive Laplace mechanism: differential privacy preservation in deep learning. https://doi.org/10.1109/ICDM.2017.48
56. Treiber A, Weinert C, Schneider T, Kersting K (2020) CryptoSPN : expanding PPML beyond neural networks ∗. In: ACM CCS 2020, pp 9–14
57. Bagdasaryan E, Shmatikov V (2019) Differential privacy has disparate impact on model accuracy. arXiv
58. Gentry C (2009) Fully homomorphic encryption using ideal lattices. https://doi.org/10.1145/1536414.1536440
59. Sadegh Riazi M, Samragh M, Lauter K, Chen H, Koushanfar F, Laine K (2019) XONN: XNOR-based oblivious deep neural network inference
60. Makri E, Rotaru D, Smart NP, Vercauteren F (2019) EPIC: efficient private image classification (or: learning from the masters). https://doi.org/10.1007/978-3-030-12612-4_24
61. Graepel T, Lauter K, Naehrig M (2013) ML confidential: machine learning on encrypted data. https://doi.org/10.1007/978-3-642-37682-5_1
62. Shukla S, Sadashivappa G (2014) Secure multi-party computation protocol using asymmetric encryption. https://doi.org/10.1109/IndiaCom.2014.6828069
63. Shukla S, Sadashivappa G (2014) A distributed randomization framework for privacy preservation in big data. https://doi.org/10.1109/CSIBIG.2014.7056940
64. Juvekar C, Vaikuntanathan V, Chandrakasan A (2018) GAZELLE: a low latency framework for secure neural network inference
65. Mireshghallah F, Taram M, Jalali A, Elthakeb AT, Tullsen D, Esmaeilzadeh H (2020) A principled approach to learning stochastic representations for privacy in deep neural inference. arXiv
66. Rouhani BD, Riazi MS, Koushanfar F (2017) DeepSecure: scalable provably-secure deep learning. arXiv. https://doi.org/10.1109/dac.2018.8465894
67. Liu J, Juuti M, Lu Y, Asokan N (2017) Oblivious neural network predictions via MiniONN transformations. https://doi.org/10.1145/3133956.3134056
68. Mishra P, Lehmkuhl R, Srinivasan A, Zheng W, Popa RA (2020) DELPHI: a cryptographic inference system for neural networks. https://doi.org/10.1145/3411501.3419418
69. Dowlin N, Gilad-Bachrach R, Laine K, Lauter K, Naehrig M, Wernsing J (2016) CryptoNets: applying neural networks to encrypted data with high throughput and accuracy
70. Sanyal A, Kusner MJ, Gascón A, Kanade V (2018) TAPAS: tricks to accelerate (encrypted) prediction as a service
71. Wang J, Zhu X, Zhang J, Cao B, Bao W, Yu PS (2018) Not just privacy: improving performance of private deep learning in mobile cloud. https://doi.org/10.1145/3219819.3220106
72. Phan NH, Wu X, Dou D (2017) Preserving differential privacy in convolutional deep belief networks. Mach Learn. https://doi.org/10.1007/s10994-017-5656-2
73. Sadegh Riazi M, Songhori EM, Weinert C, Schneider T, Tkachenko O, Koushanfar F (2018) Chameleon: a hybrid secure computation framework for machine learning applications. In: ASIACCS 2018 - proceedings of the 2018 ACM Asia conference on computer and communications security, May 2018, pp 707–721. https://doi.org/10.1145/3196494.3196522

74. Kairouz P et al (2019) Advances and open problems in federated learning. arXiv, p 16, December 2019. http://arxiv.org/abs/1912.04977. Accessed 27 Jan 2021
75. Hitaj B, Ateniese G, Perez-Cruz F (2017) Deep models under the GAN: information leakage from collaborative deep learning. https://doi.org/10.1145/3133956.3134012
76. Cheu A, Smith A, Ullman J, Zeber D, Zhilyaev M (2019) Distributed differential privacy via shuffling. https://doi.org/10.1007/978-3-030-17653-2_13
77. Balle B, Bell J, Gascón A, Nissim K (2019) The privacy blanket of the shuffle model. https://doi.org/10.1007/978-3-030-26951-7_22
78. Ghazi B, Pagh R, Velingker A (2019) Scalable and differentially private distributed aggregation in the shuffled model. arXiv
79. Mo F et al (2020) DarkneTZ: towards model privacy at the edge using trusted execution environments. https://doi.org/10.1145/3386901.3388946
80. Costan V, Devadas S (2016) Intel SGX explained. IACR Cryptol. ePrint Arch.
81. Narra KG, Lin Z, Wang Y, Balasubramaniam K, Annavaram M (2019) Privacy-preserving inference in machine learning services using trusted execution environments. arXiv
82. Hashemi H, Wang Y, Annavaram M (2020) DarKnight: a data privacy scheme for training and inference of deep neural networks. arXiv
83. Tramèr F, Boneh D (2018) Slalom: fast, verifiable and private execution of neural networks in trusted hardware. arXiv
84. Canella C et al (2019) Fallout: leaking data on meltdown-resistant CPUs. https://doi.org/10.1145/3319535.3363219
85. Taram M, Venkat A, Tullsen D (2020) Packet chasing: spying on network packets over a cache side-channel. https://doi.org/10.1109/ISCA45697.2020.00065

# Deep Learning Methods for Intrusion Detection System

Yash Agrawal(✉), Tushar Bhosale, Hrishikesh Chavan, and Deepak Kshirsagar

Department of Computer Engineering and IT, College of Engineering Pune, Pune, India
{agrawalym17.comp,ddk.comp}@coep.ac.in

**Abstract.** With ever increasing use and the number of users on the internet, data has become vulnerable to attacks. One such attack is Denial of Service (DoS) attack. This attack is meant to temporarily or indefinitely make unavailable a machine or network resources thereby making the system inaccessible. In this paper, an intrusion detection system is built using Deep Learning approaches Deep neural network and Convolutional Neural Network to detect DoS attacks. CICIDS2017 dataset is used to train the model and test the performance of the model. The experimental trials show that the proposed model outperforms all the previously implemented models.

**Keyword:** Denial of Service attack · Intrusion Detection System · Convolutional Neural Network · Deep Neural Network

## 1 Introduction

A denial of service (DoS) attack is a malicious cyber threat in which a user is unable to access a system or a network resource. The services may be temporarily or indefinitely unavailable. It is a condition where either the web services are flooded to an extent that they don't respond or they simply crash. A DoS attack is carried out by either flooding or by a crash attack. Flooding is relatively more common as compared to a crash attack. In flooding, a network server is sent a large number of request traffic until a point where the server is unable to respond. Eventually the server stops working. ICMP flood and SYN flood [1] are two such variations of flooding. In an ICMP flood, the targeted system is flooded with echo-requests which make the system unavailable to respond to normal traffic. In a SYN flood, the server is overwhelmed with requests to begin the three-way handshaking which is never completed, thereby exhausting the limit of maximum number of open ports. A crash attack happens when the intruder exploits the loopholes in the system, eventually causing the system to crash.

DoS attacks may be carried out by peers in a same business, hackers or hacktivits. An intent of the same may be to promote a social/political cause or for some monetary benefits as in case of organizations across some business. As a DoS attack can be easily carried out from any location, it becomes a tedious task to find the source of the attack. Thus Intrusion Detection Systems (IDSs) [2] become an important tool against network attacks. An IDS tracks network traffic for malicious activity detection. The scope of an


S. Shukla et al. (eds.), *Data Science and Security*, Lecture Notes in Networks and Systems 290,
https://doi.org/10.1007/978-981-16-4486-3_4

IDS may vary from one computer to a larger network. Also, it is worth noting that with advancement in technology there has also been advancement in the nature of DoS attacks being made. There has been quite a lot of variation in the attacks that were carried earlier to those that are being carried out now. In contrast to this, adequate research and progress hasn't been made to identify and overcome these attacks. Lack of testing and validation dataset may be one of the reasons due to which the intrusion detection approaches are not able to significantly able to perform well. It has been observed that all research work and developments until now have been done using datasets which are out of date and unreliable. Some lack traffic diversity, some have not incorporated the newer known attacks while some do not reflect the current trends.

In this paper, we have proposed a deep learning based DoS attack detection system. The training and testing of the proposed system is done using the CICIDS2017 dataset. CICIDS2017 dataset contains benign and most upto date common real world attack [3]. Our Current work implements binary classification using deep neural network and 1D convolutional neural network. This binary classifier labels the packet as benign or an attack packet. The experimental trials show that the proposed model outperforms all the previously implemented systems.

The paper is organized as follows: Sect. 2 describes the related work. The proposed deep learning based intrusion detection system is described in Sect. 3. Section 4 describes the system implementation. Results and conclusion form the Sect. 5 and 6 respectively.

## 2 Related Work

In the study [4], an IDS model is developed by using Deep Learning—Deep Multilayer Perceptron. The model is tested using the CICIDS2017 dataset. The dataset is reduced using Recursive Feature elimination. The model gave a test accuracy of 91% whereas the test loss was 0.14.

In [5] work, layer configuration and feature selection methods are proposed and performance of each method is compared with accuracy. Feature selection allows them to reduce some irrelevant features whereas in layer configuration they found that more layers only extend learning time and did not provide high accuracy due to overfitting. The model achieved the accuracy of 98.5% on NSL-KDD dataset.

In the paper [6], deep learning based IDS model is developed using Deep Neural Network. Feature extraction with information gain is done to reduce input dimensions and increase accuracy. The model achieved better performance than Decision tress, Naive Bayes, K-Nearest Neighbour, Random forest, Support Vector Machine. Accuracy of 99.37 and 87.74% is achieved on NSL KDD dataset for binary classification on validation and test data set respectively.

The paper [7] proposed a deep learning based IDS which consists of feature selection algorithm based on a denoising auto encoder. The classifier was based on multilayer perceptron. Performance of model was evaluated using UNSW-NB dataset. After feature selection 12 out of 202 features are selected. After classification using a multilayer perceptron with 2 hidden layers, the model achieves accuracy of 98.80%.

The paper [8] proposed a hybrid IDS using deep neural networks with 5 hidden layers. The neural network architecture and optimal parameters of network are selected by performing a number of trials. Average accuracy of the deep neural network for binary classification on CICIDS2017 dataset 94.5%.

The paper [9], IDS is developed for handling huge dataset and identifying different attack types. Results shows that deep neural network based system is reliable and efficient. NSL KDD dataset is used. Non-numerical values are converted into numerical values as a part of preprocessing. The accuracy for binary classification was 97%.

## 3 Proposed Deep Learning Based Intrusion Detection System

To detect and classify DoS attacks using deep learning approach, Intrusion Detection system is designed. Classification of network traffic samples is done by the model. If an attack is detected, an alert is generated. The whole process of DoS attack detection is illustrated in Fig. 1.

The core of this intrusion detection system consists of model and a deep learning algorithm. CICIDS 2017 dataset is used. Deep learning algorithms like deep neural network and convolution neural network are used for classification. Hyperparameters are optimized by trial and error method. While training, the dataset is divided into 2 parts. The first part, which comprises 80% of the dataset is used for training whereas the second part, which comprises 20% of the dataset is used for testing. Various experiments are performed while training and tuning to obtain the best result and the final model is developed for DNN and CNN for those parameters. The whole process of classification, training, tuning and developing of models is discussed below.

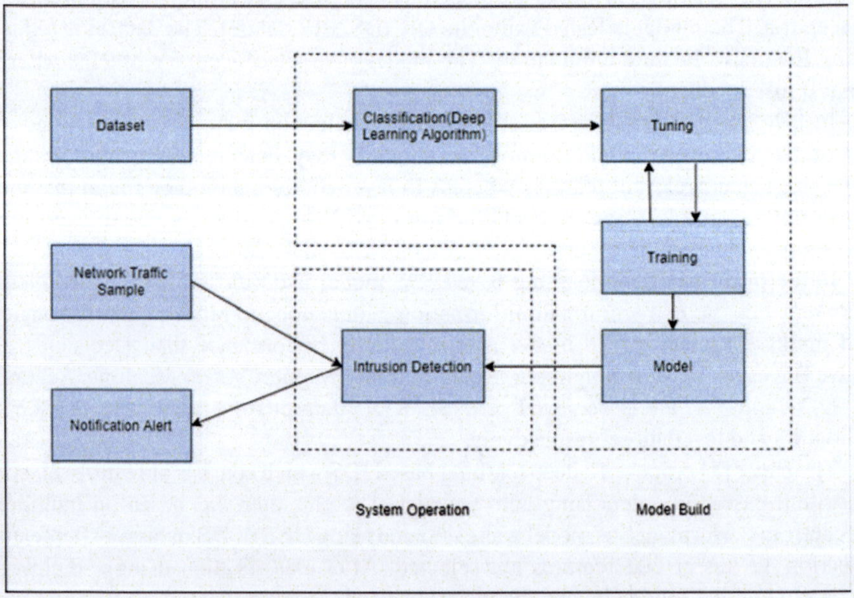


**Fig. 1** Proposed intrusion detection system

Then while in use, collected network samples are given as input to the model. Intrusion detection system checks for attack and if attack is detected, system classifies it and sends attack notification.

# 4 System Implementation

The proposed system is built on a device with 64 bit OS which had Intel Core i5-8250U CPU with 8 GB RAM. The implementation of model is done using Keras API in python3. Manipulation of data is done using Numpy and Pandas.

## 4.1 Dataset

The intrusion detection system was trained and tested on CICIDS2017 dataset [3]. CICIDS2017 dataset contains benign and most upto date common real world attack. It includes various types of DoS and DDoS attacks as mentioned in the Table 1. The size of the dataset after cleaning is 213 MB which has a total of 692,692 instances. Table 1 shows the number of instances of each label in dataset. The dataset has 77 features as mentioned in the work [10].

## 4.2 Data Preprocessing

CICIDS 2017 dataset consist of noise in the form of duplicate features, missing and null values. The data pre-processing is performed as mentioned in the work [11] on CICIDS 2017 dataset.

## 4.3 Finding Optimal Parameters in DNN

Performance of DNN depends on optimal parameters. To identify the ideal parameter for the DNNs, trials with a variable number of hidden units and learning rate were performed. Using CICIDS2017 dataset, the model consists on an input layer having 77 neurons. The hidden layer contains 128, 256 and 512 units. The output layer has 1 neuron in case of binary classification. It classifies the packet as either benign or DoS attack.

**Table 1** Occurrences of each label in CICIDS2017 dataset

| Sr. no | Class label | Number of instances |
|---|---|---|
| 1 | Benign | 440,031 |
| 2 | DoS Slowloris | 5796 |
| 3 | DoS Slowhttptest | 5499 |
| 4 | DoS Hulk | 231,073 |
| 5 | DoS Goldeneye | 10,293 |

**Table 2** Proposed deep neural network architecture

| Layers | Output shape | Activation function | Parameters |
|---|---|---|---|
| Input | (None, 77) | – | 0 |
| Fully connected | (None, 128) | Relu | 9984 |
| Dropout (0.01) | (None, 128) | – | 0 |
| Fully connected | (None, 512) | Relu | 66,084 |
| Dropout (0.01) | (None, 512) | – | 0 |
| Fully connected | (None, 256) | Relu | 131,328 |
| Dropout (0.01) | (None, 256) | – | 0 |
| Fully connected | 1 | Sigmoid | – |

Various layers in deep Neural Network are connected using fully connected layers. To avoid overfitting, dropout layer is also added. Three trials of experiment with one hidden layer were carried out with 128, 256 and 512 hidden units respectively. The model gave maximum accuracy for 128 hidden units.To further build the model, 128 hidden units were kept in the first hidden layer and experiments were carried out with more hidden layers. Deep Neural Network is trained for 50 epochs in all experiments. The results saturated for a further increase in the number of epochs. Three trials of experiments for 50 epochs with learning rate 0.01, 0.05 and 0.001 were run in order to find an optimal learning rate. Table 2 describes the architecture of the proposed DNN model.

### 4.4 Finding Optimal Parameters in CNN

Tuning of parameters is important for CNN also. Proposed solution implement the 1D Convolution neural network. The optimal parameter determination was done for CICIDS 2017 dataset. To identify the ideal parameter for the CNNs, trials with a variable number of hidden units, filter size, kernel size and learning rate were carried out. For CICIDS 2017 Dataset, input layer size is (77,1).

Output layer is a fully connected layer similar to DNN and contains 1 neuron in binary classification. It classifies the packet as either benign or DoS attack. Various experiments carried out for different filters and kernel size. Convolutional Neural Network is trained for 50 epochs in all experiments. Three trials of experiments for 50 epochs with learning rate 0.01, 0.05 and 0.001 were run in order to find an optimal learning rate. Table 3 describes the architecture of the proposed CNN model. Proposed convolutional neural network architecture.

### 4.5 Classification

In both deep neural network and convolutional neural network the last layer is a fully connected layer. As our proposed neural networks are for binary classification we use a

**Table 3** Proposed convolutional neural network architecture

| Type | Output shape | Activation function | Parameters |
|---|---|---|---|
| Input | (None, 77, 1) | – | 0 |
| Conv1D | (None, 77, 32) | Relu | 128 |
| Maxpooling1D | (None, 38, 32) | – | 0 |
| Conv1D | (None, 38, 64) | Relu | 6208 |
| Maxpooling1D | (None, 19, 64) | – | 0 |
| Flatten | (None, 1216) | – | 0 |
| Fully connected | (None, 512) | Relu | 623,104 |
| Dropout (0.01) | (None, 512) | – | 0 |
| Fully connected | 1 | Sigmoid | – |

**Table 4** Hyperparameters for DNN and CNN

| Model parameters | Value |
|---|---|
| Learning rate | 0.001 |
| Batch size | 128 |
| Optimizer | Adam |
| Epochs | 50 |

sigmoid loss function. The sigmoid loss function is defined using binary cross entropy [12]. Mathematically it is defined as:

$$loss(p, t) = (-1/N) \sum_{n=1}^{\infty} [(t_i) \log(p_i) + (1 - t_i) \log(1 - p_i)] \quad (1)$$

where p is a vector containing predicted probabilities for test dataset, t is a vector containing true class labels. The loss of Binary cross entropy is minimized using Adam as an optimizer.

## 5 Results

CICIDS2017 dataset was used to evaluate the performance of the DNN and CNN. The training dataset had 554,151 instances whereas the test dataset had 138,541 instances. Feature scaling was done by subtracting the mean and then scaling to unit variance. Table 4 represents the hyperparameter values for DNN and CNN respectively.

The proposed DNN and CNN model are trained and tested on CICIDS2017 dataset.

**Table 5** Test results of DNN and CNN for binary classification

| Architecture | Accuracy | Precision | Recall | F-score |
|---|---|---|---|---|
| DNN | 0.9967 | 0.9891 | 0.9981 | 0.9942 |
| CNN | 0.9976 | 0.9922 | 0.9990 | 0.9961 |

**Table 6** Performance analysis comparison against existing work

| Work | Dataset | Classifier | Accuracy |
|---|---|---|---|
| [4] | CICIDS2017 | DNN | 0.9100 |
| [5] | NSL-KDD | DNN | 0.9890 |
| Proposed 1 | CICIDS2017 | DNN | 0.9967 |
| Proposed 2 | CICIDS2017 | CNN | 0.9976 |

Table 5 shows that CNN produces a higher accuracy of 99.76% as compared to that produced by DNN which is 99.67%. Hence we can conclude that a 1D convolutional neural network (CNN) is better than deep neural network (DNN) for DoS attack detection.

The implemented system is further compared with traditional systems as shown in Table 6.

From the above Table 6 shown that the proposed model performs better than the other traditional models [4] and [5].

## 6 Conclusion

We have proposed an intrusion detection system using deep learning for DoS attack detection. The proposed solutions were based on deep neural network and 1D convolutional neural network. Various models with different parameters were trained for a comparative analysis. Both CNN and DNN have a good accuracy for binary classification when trained and tested on CICIDS 2017 dataset. The solutions were evaluated based on accuracy metrics. Relatively CNN performed better than DNN. Our proposed model performs better in comparison to other previously implemented classifiers. Our Future work includes Multi label classification of DoS attacks and using Feature selection using deep learning based algorithms to further enhance the performance.

## References

1. Wankhede S, Kshirsagar D (2018) DoS attack detection using machine learning and neural network. In: 2018 fourth international conference on computing communication control and automation (ICCUBEA), Pune, India, pp 1–5. https://doi.org/10.1109/ICCUBEA.2018.8697702

2. Brahmkstri K, Thomas D, Sawant ST, Jadhav A, Kshirsagar DD (2014) Ontology based multi-agent intrusion detection system for web service attacks using self learning. In: Networks and communications (NetCom 2013). Springer, Cham, pp 265–274
3. Sharafaldin I, Lashkari AH, Ghorbani A (2018) Toward generating a new intrusion detection dataset and intrusion traffic characterization, pp 108–116. https://doi.org/10.5220/0006639801080116
4. Ustebay S, Turgut Z, Aydin MA (2018) Intrusion detection system with recursive feature elimination by using random forest and deep learning classier. In: 2018 international congress on big data, deep learning and fighting cyber terrorism (IBIGDELFT), Ankara, Turkey, pp 71–76. https://doi.org/10.1109/IBIGDELFT.2018.8625318
5. Woo J, Song J, Choi Y (2019) Performance enhancement of deep neural network using feature selection and preprocessing for intrusion detection. In: 2019 international conference on artificial intelligence in information and communication (ICAIIC), Okinawa, Japan, pp 415–417. https://doi.org/10.1109/ICAIIC.2019.8668995
6. Kasongo SM, Sun Y (2019) A deep learning method with filter based feature engineering for wireless intrusion detection system. IEEE Access 7:38597–38607. https://doi.org/10.1109/ACCESS.2019.2905633
7. Zhang H, Wu CQ, Gao S, Wang Z, Xu Y, Liu Y (2018) An effective deep learning based scheme for network intrusion detection. In: 2018 24th International Conference on Pattern Recognition (ICPR), Beijing, China, pp 682–687. https://doi.org/10.1109/ICPR.2018.8546162
8. Vinayakumar R, Alazab M, Soman KP, Poornachandran P, Al-Nemrat A, Venkatraman S (2019) Deep learning approach for intelligent intrusion detection system. IEEE Access 7:41525–41550. https://doi.org/10.1109/ACCESS.2019.2895334
9. Potluri S, Diedrich C (2016) Accelerated deep neural networks for enhanced intrusion detection system. In: 2016 IEEE 21st International Conference on Emerging Technologies and Factory Automation (ETFA), Berlin, Germany, pp 1–8. https://doi.org/10.1109/ETFA.2016.7733515
10. Kshirsagar D, Kumar S (2020) Identifying reduced features based on IG-threshold for DoS attack detection using PART. In: International conference on distributed computing and internet technology. Springer, Cham, pp 411–419
11. Kshirsagar D, Kumar S (2021) A feature reduction based reflected and exploited DDoS attacks detection system. J Ambient Intell Humaniz Comput 1–13
12. Zhou Y, Wang X, Zhang M, Zhu J, Zheng R, Wu Q (2019) MPCE: a maximum probability based cross entropy loss function for neural network classification. IEEE Access 7:146331–146341. https://doi.org/10.1109/ACCESS.2019.2946264

# Adaptive Neuro Fuzzy Approach for Assessment of Learner's Domain Knowledge

Varsha P. Desai[1(✉)], Kavita S. Oza[2], and Rajanish K. Kamat[2]

[1] V.P. Institute of Management Studies and Research, Sangli, India
[2] Computer Science Department, Shivaji University, Kolhapur, India
rkk_eln@unishivaji.ac.in

**Abstract.** E-learning lends a significant contribution to the education system. Learner's interests, skills, domain knowledge, learning behavior, and learning styles are essential for developing a personalized e-learning system. Student centric learning process improves learners' satisfaction, leading to improving the education system's outcome. Due to the advancement of ICT technologies variety of e-study material is available through various online sources. It is a challenge for today's e-learners to select the best e-learning material which fulfills their learning needs. This paper explores an intelligent, personalized learning approach to assess learner's domain knowledge. The result of the said assessment helps to recommend a learning path for the learner.

**Keywords:** ANFIS · Domain knowledge · Personalized learning · E-learning

## 1 Introduction

E-learning provides alternative and innovative strategies as compared to traditional classroom teaching learning process. Adaptive e-learning focuses on the need and expectations of the individual learner to provide the best learning experience. It facilitates a personalized learning path according to learning object to improve learner's learning performance [1]. An interactive, customized e-learning system helps to improve learner's learning interest, confidence, and skill. Intelligent strategies for personalized e-learning systems are developed by considering personalized aspects like resources, guidance, communication, learning activities, and interactive user interface [2]. Personalized learning system motivates learner to learn from their own pace as per individual learning needs. Learning objective, contents, source of study material, tools, content sequence are varies from learner to learner.

Adaptive Neuro-Fuzzy System (ANFIS) is a hybrid learning approach that constructs if–then rules from the training dataset. Due to the high capability of ANFIS, it is possible to refine the rules generated by human experts [3]. The nonlinear and structured knowledge representation of ANFIS is the primary advantage over the classical linear approach in adaptive filtering [4]. It is one of the best function approximation tools implements using the Sugeno style fuzzy system. It is a fusion technology that combines capabilities of fuzzy logic and artificial neural network. Fuzzy logic handles


S. Shukla et al. (eds.), *Data Science and Security*, Lecture Notes in Networks and Systems 290,
https://doi.org/10.1007/978-981-16-4486-3_5

imprecision and uncertainty while neural network gives it sense of adaptability [5]. This intelligent fuzzy modeling system accepts information from the training dataset to compute membership functions to generate the best fuzzy inference system from the given input–output dataset.

The present paper propose an intelligent technique for domain knowledge prediction using Artificial Neuro-Fuzzy System for providing the best learning experience for individual learners. The paper is organized in different section. After introduction, the literature survey correlates the significance of domain knowledge of a learner with e-learning resources and learning objects and thereby implies the importance of the ANFIS approach. Thereafter the paper discuss dataset preparation, ANFIS model formation and validation and present result and conclusion.

## 2 Literature Review

It is a challenging task to make the e-learning process more comfortable and exciting for learners. Learners' learning needs, cognitive features identification is essential for developing a personalized learning system [6]. An intelligent e-learning system can be created using k-means, hierarchical clustering, self-organizing map, Apriori approach, classification methods to recommend learning path for each learner as per their learning expectations [7]. An adaptive e-learning system motivates learners to improve their learning capacities. The educational data mining approach for predicting learner performance, learning style, and knowledge help develop an effective personalized e-learning system [8]. An intelligent web teacher system can be created using semantic web technologies as per an individual's ontology to provide a better learning experience [9]. Agent-based personalized e-learning environment reduces learners' efforts to search for suitable content as per their learning behavior and interest. It monitors learning activities and communication and recommends learning resources as per domain-specific requirements [10]. Learning style identification is crucial for advising suitable learning tool for the learner. It affects the learning process and learner's achievements. Learning styles are different from learner to learner. So predicting learning style is a significant contest while developing an interactive learning system. Fuzzy based learning style identification model provides a competitive approach to predict the rate of learning style [11].

According to the literature review, providing e-learning content as per learner's interest, skill, domain knowledge level is a challenging task while developing personalized e-learning system. Domain knowledge prediction is a significant aspect for the recommendation of e-material, learning objects, and learner's learning path for interactive, personalized e-learning systems. This further promotes the application of ANFIS approach for the e-learning domain. During research students domain knowledge data is collected from online portal. Data is preprocess to remove noise and anomalies. Further rules are designed from domain experts and ANFIS model is implemented for prediction. The subsequent details are presented in the following sections.

## 3 Dataset Preparation and ANFIS Model Development

Online tests are constructed for 550 students on the three levels like primary, intermediate, and advanced, to create a domain knowledge dataset. Domain expert knowledge is used to designed rules from each test score to a categorized student into Unknown (UK), Partially Known (PK), and Completely Known (CK). Artificial Neuro-Fuzzy System (ANFIS) model is trained and developed to predict learner's domain knowledge level.

In order to develop the ANFIS model, 400 records are considered for the training dataset, and 100 records are considered testing dataset, whereas 50 records are considered for validation. ANFIS model trained using Linear Gaussian membership function with the hybrid optimization method. Gaussian membership function gives higher degree of accuracy with minimum computational competency. Hybrid optimization method combines least square method and backpropagation to tune the FIS parameters. The grid partition method is implemented to generate FIS with a maximum number of possible rules that combine all input and output combinations. Stooping condition is specify using error tolerance to 0 means training will stopped when number of training epoch is reached.

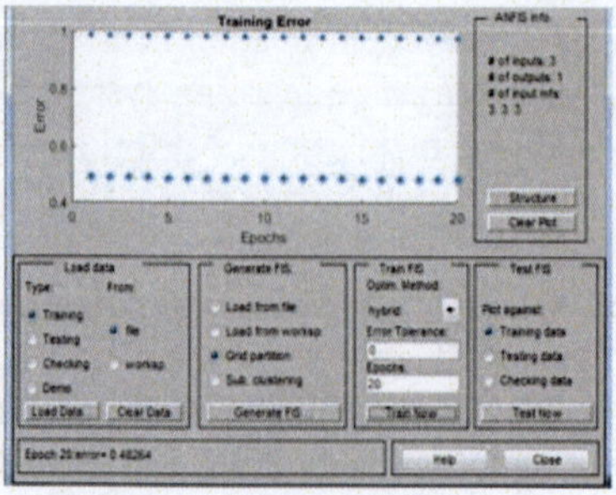


**Fig. 1** Training ANFIS model

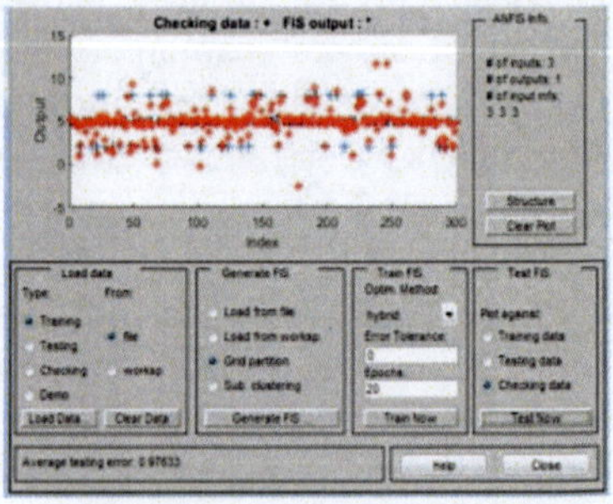


**Fig. 2** Testing ANFIS model

Figure 1 shows ANFIS model training details, where plot shows training error indicated with blue stars and checking error indicated with blue dots for each training epoch. The model trained with 20 epoch gives the best result with RMSE (Root Mean Square Error) is 0.48264.

ANFIS model is trained with three inputs parameters as a tests score with three membership functions to classify learner into specific class as UK, PK, CK. Figure 2 shows plot with output values of testing dataset indicated as blue + and output of trained ANFIS with corresponding testing input values indicated as red *'s Here testing ANFIS model shows that actual output values are correlated with expected output values reveals better result. The optimization back propagation neural network algorithm is used with the least square method for predicting learner's domain knowledge level.

**Table 1** ANFIS structure details

| Sr.No | Particular | Values |
|---|---|---|
| 1. | Number of nodes | 78 |
| 2. | Number of linear parameters | 108 |
| 3. | Number of nonlinear parameters: | 18 |
| 4. | Total number of parameters | 126 |
| 5. | Number of training data pairs | 400 |
| 6. | Number of checking data pairs | 100 |
| 7. | Number of Epoch | 20 |
| 8. | Number of fuzzy rules | 27 |
| 9 | Minimal training RMSE | 0.48264 |
| 10. | Minimal checking RMSE | 0.97633 |

Table 1 shows that the structure of ANFIS model. It is developed with 78 nodes, 108 linear parameters, 18 non-linear parameters, 126 total number of parameters, 400 training data pairs, and 100 checking data pairs with 27 fuzzy rules. ANFIS model's components trained with 20 epoch gives training RMSE 0.48264 and testing RMSE 0.97633. Here RSME is low, indicating ANFIS model provides prediction with high accuracy.

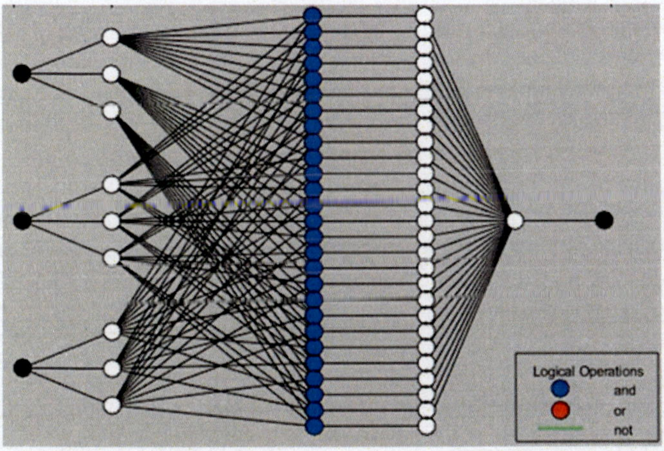


**Fig. 3** ANFIS model for domain knowledge prediction

Figure 3 shows ANFIS structure for domain knowledge prediction. ANFIS model is developed that accepts three inputs, 27 rules are automated with AND (Blue Dots), OR (Red Dots NOT (Green Dots) conditions to classify learner into one of three classes as UK, PK, CK.

Table 2 shows the sample result of domain knowledge prediction using ANFIS Model.

**Table 2** Sample result of ANFIS model

| Sr.No | StudentID | Test-I | Test-II | Test-III | Result | Knowledge Level |
|---|---|---|---|---|---|---|
| 1 | 167896 | 8 | 7 | 8 | 7.159152 | CK |
| 2 | 167897 | 5 | 4 | 5 | 5.060029 | PK |
| 3 | 167899 | 2 | 4 | 5 | 1.070752 | UK |
| 4 | 167900 | 5 | 3 | 5 | 5.298094 | PK |
| 5 | 167901 | 6 | 3 | 5 | 5.096235 | PK |
| 6 | 679021 | 3 | 1 | 2 | 1.753862 | UK |
| 7 | 679033 | 5 | 2 | 2 | 2.002198 | UK |
| 8 | 679044 | 4 | 3 | 5 | 4.630681 | PK |
| 9 | 679055 | 7 | 5 | 8 | 7.159152 | CK |
| 10 | 672962 | 5 | 1 | 2 | 2.013625 | UK |

Above Table 2 depicts that students test score for three different test conducted for domain knowledge prediction, result and learners classification into specific class UK, PK and CK.

## 4 ANFIS Model: Testing and Validation

Artificial neural network algorithm is applied to validate the result and performance of proposed ANFIS model. Performance of model is evaluated through ROC (Reduced Operating Curve). Confusion matrix is used to describe performance of classifier with actual and predicted output.

Figure 4 shows the experimental result of the neural network. As per our dataset three inputs are given for the model, and to obtained good result model was tested by increasing hidden layers one by one. The two layer feed-forward neural network model with 10 hidden neurons and sigmoid activation function gives best result for the proposed ANFIS model. The network is trained using a scaled conjugate gradient backpropagation algorithm (traincg) with a cross-entropy performance of 0.000481 and gradient 0.000704 at 24 epoch.

Training statistics of ANFIS model are depicted using performance and training state plot. In Matlab, plotperform() function is used to plot error vs. epoch for training, validation and test performance of model. To reduce error, number of epochs are increased and training stops as when we obtained best validation performance with less validation error.

Figure 5 shows validation perofrmanc of ANFIS model. Best validation performance for training, validation and testing data indicated with blue, green and red lines respectively with error 0.029684 at 24 epoch.

In matlab, plottrainstate() function is used to plot trainning state values to validate the result of ANFIS model with gredient and epoch. Figure 6 depicts the plot for ANFIS

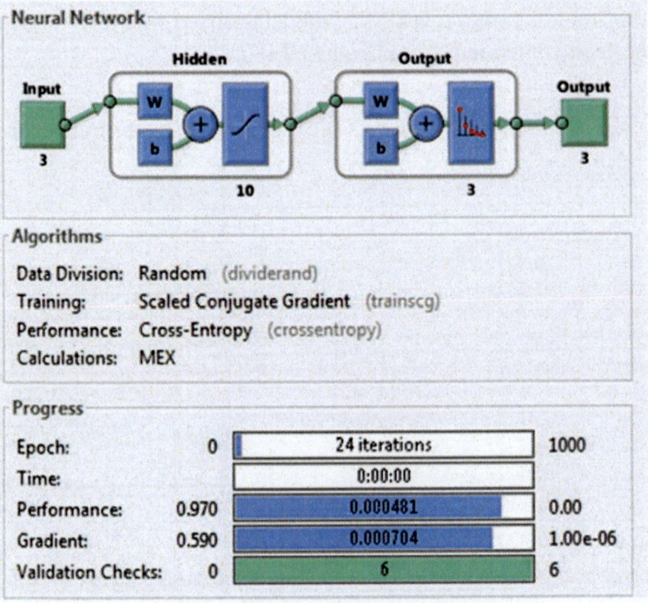


**Fig. 4** Experimental result of ANFIS

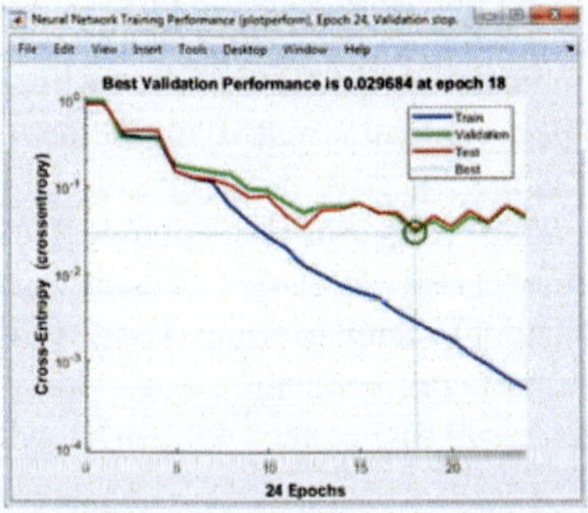


**Fig. 5** Validation performance of ANFIS model

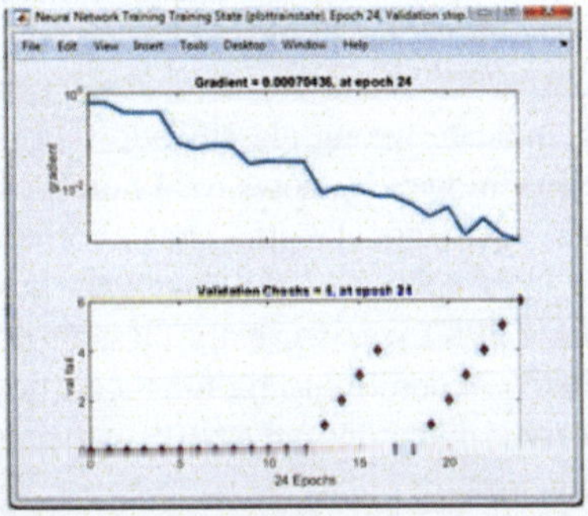


**Fig. 6** ANFIS model trainning state performance

model training state performance. Its shows points at which validation fails during each epoch. Training is stopped at validation check 6 with gradient 0.000070436 at epoch 24. The gradient is less indicates that ANFIS model gives a better result.

The result of network simulation and training shown using ROC and confusion matrix. Receiver Operator Characteristics Curve (ROC) for training, validating, and testing dataset measures the performance of ANFIS model.

ROC cover is formed by plotting true positive rate (TPR) against false positive rate (FPR). In Fig. 7 red line indicate overlapping of ROC curve for three classes (UK, PK,

CK) at the top and left corner reveals that model trainned with 100% true positive rate and data is perfectly classified by trainned ANFIS model.

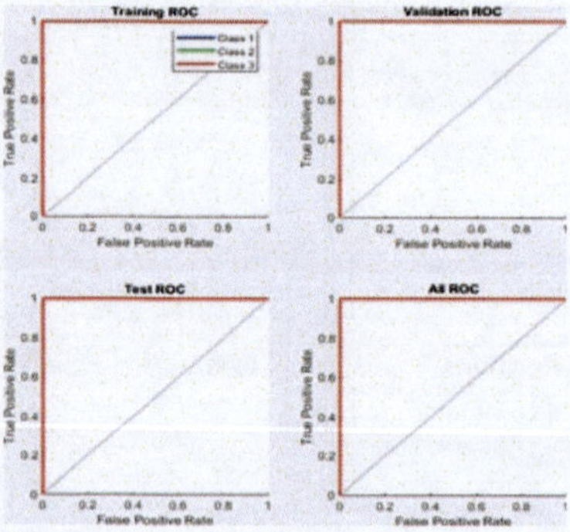


**Fig. 7** ROC for ANFIS model performance measurement

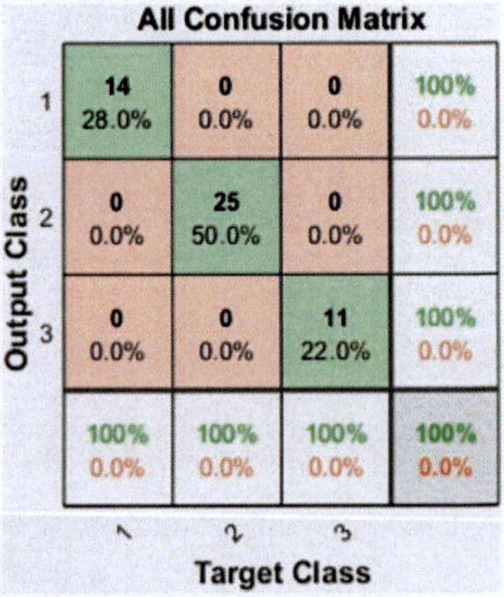


**Fig. 8** Confusion matrix for validation of ANFIS model

Confusion marix is used to validate the accuracy of classification of data into three classes. It shows total number of false positive, false negative, true positive and true negative enties amoung the dataset. Green cell indicate correct classification while pink cell indicate wrong classification. Figure 8 depicts confusion matrix for validation of ANFIS model. It shows total number of samples in respective target class and output class. All diagonal entries shows correct classification. Validation result of model reveals that 14 samples from the class-1 (CK) was correctly classify as class-1. 25 samples from class-2 (PK) was correctly classify as class-2 and 11 samples from class-3 (CK) as class-3. According to training statistics, result of network simulation and training it is observed that proposed ANFIS model provide best result for learner's domain knowledge prediction.

## 5 Conclusion

Domain knowledge prediction helps to develop an interactive, personalized e-learning system. Adaptive Neuro-Fuzzy System designed for learner knowledge prediction combines best features of ANN and Fuzzy. Trained ANFIS model automated 27 rules for prediction with root mean square error is 0.97633. Accuracy of ANFIS model checked using the scaled conjugate backpropagation algorithm with a cross-entropy performance of 0.000481 and gradient 0.000704 at 24 epoch. According to experimentation, proposed ANFIS model gives 100% accuracy in prediction. Results of domain knowledge prediction can be further used for deciding the learning path of the individual learners.

## References

1. Kulaglic S, Mujacic S, Serdarevic IK, Kasapovic S (2013) Influence of learning styles on improving efficiency of adaptive educational hypermedia systems. In: Proceeding of international conference, University of Tuzla, October 2013
2. Duo S, Ying ZC (2012) Personalized e- learning system based on intelligent agent. In: International proceeding of international conference. Elsevier, pp 1899–1902
3. Jang J-SR (1993) ANFIS adaptive-network-based fuzzy inference system. IEEE Trans Syst Man Cybern 23(3):665–685
4. Haykin SS (1991) Adaptive filter theory, 2nd edn. Prentice Hall, Englewood Cliffs
5. Buragohain M (2008) Adaptive network based fuzzy inference system (ANFIS) as a tool for system identification with special emphasis on training data minimization. Thesis, Department of Electronics and Communication Engineering Indian Institute of Technology Guwahati, July 2008
6. Senthil Kumaran V, Sankar A (2013) Study of personalization in e-learning. Int Rev Comput Softw (IRECOS) 8(5):1209–1217. e-ISSN: 2533-1728
7. Markowska-Kaczmar U, Kwasnicka H (2010) Intelligent techniques in personalization of learning in e-learning systems. In: Computational intelligence for technology enhanced learning. Springer, Heidelberg, pp 1–23
8. RamyaSree P (2019) Personalized e-learning system based on user's performance and knowledge: an adaptive technique. Int J Recent Technol Eng (IJRTE) 8(4):8695–8899. ISSN: 2277-3878
9. Gaeta M, Miranda S (2013) An approach to personalized e-learning. Syst Cybern Inform 11(1):15–21. ISSN: 1690-4524
10. John Martin A, Dominic M (2019) Adaptation using machine learning for personalized eLearning environment based on students preference. Int J Innov Technol Exploring Eng (IJITEE) 8(10):4064–4069. ISSN: 2278-3075
11. Alian M, Shaout A (2016) Predicting learners styles based on fuzzy model. Educ Inf Technol 22(5):2217–2234. 0639-016-9543-4

# Comparison of Full Training and Transfer Learning in Deep Learning for Image Classification

Sibu Cyriac[1(✉)], Nidhin Raju[1], and Sivakumar Ramaswamy[2]

[1] Centre for Digital Innovation, CHRIST (Deemed To Be University), Bangalore, India
sibu.cyriac@christuniversity.in

[2] Department of Computer Science, CHRIST (Deemed To Be University), Bangalore, India

**Abstract.** The deep learning algorithms on a small dataset are often not efficient for image classification problems. Make use of the features learned by a model trained on large similar dataset and saved for future reference is a method to solve this problem. In this work, we present a comparison of full training and transfer learning for image classification using Deep Learning. Three different deep learning architectures namely MobileNetV2, InceptionV3 and VGG16 were used for this experiment. Transfer learning showed higher accuracy and less loss than full-training. According to transfer learning results, MobileNetV2 model achieved 98.96%, InceptionV3 model achieved 98.44% and VGG16 model achieved 97.405 as highest test accuracies. The full-trained models did not achieve as much accuracy as that of transfer learning models on the same dataset. The accuracies achieved by full-training for MobileNetV2, InceptionV3 and VGG16 are 79.08%, 73.44% and 75.62% respectively.

**Keywords:** Deep learning · Transfer learning · Full training · Image classification

## 1 Introduction

The deep learning algorithms need long time and large dataset to train the various weights and parameters. At the same time, the availability of public datasets for different research areas are limited. Data augmentation is a way of expanding the datasets for training the deep learning models. But it requires more computational time and storage. The augmented images can be always good as much as original images. As we are applying some transformation operations such as zooming, rotation and so on in generating augmented images, sometimes we need to compromise on its quality. Transfer learning is another method in deep learning that can solve the issue of limited dataset. The transfer learning technique is used to identify and adapt common features from previous tasks to new tasks. The features of new dataset should have some similarity with pre-trained model's dataset to take its advantages in transfer learning.

The Convolutional Neural Networks (CNN) models like VGG16 [1, 2], InceptionV3 [3, 4] and MobileNetV2 [5] were trained using the ImageNet dataset on their original


S. Shukla et al. (eds.), *Data Science and Security*, Lecture Notes in Networks and Systems 290,
https://doi.org/10.1007/978-981-16-4486-3_6

research work. According to Pedro [6], the compilation of a large dataset were costly for a specific image classification task. In the case of a smaller data collection, reliability and high performance could be achieved using the transfer learning. Deep learning models like MobileNetV2, InceptionV3 and VGG16 trained on ImageNet datasets learn to recognize images and later layers building on top of these models could learn more complicated structure. The last layer of every models was used to classify the images into various categories. In this proposed work, we used transfer learning on a small dataset so that the training is faster. As we used the small dataset, the chance of the overfitting of the model was high, but we could reduce it through data augmentation and dropouts.

In this proposed work, we focused on classifying cat and dog images from 3 different pre-trained networks by using transfer learning. A pre-trained model is a saved network trained on a large dataset, usually in a large-scale image classification task. The key aspect behind transfer learning is that the network is trained on a large and simliar dataset. Features learned by these saved networks contributes to train a complex deep learning model on smaller dataset. This is achieved by freezing the initial layers and train only a few layers with this small dataset. A pre-trained model is customized in two ways: feature extraction and fine-tuning. Feature extraction uses the features learned from a previous network to inherit meaningful attributes from new dataset [6]. Adding a new classifier to the saved model will adapt the feature maps of this model for classifying the new dataset. The entire model does not have to retrain. The base convolution network already includes useful features for the classification of images. However, the final classification part of the saved model would be for the specific classification work and then to the designated classes for which the model is to be trained. In fine-tuning, train the added classifier layers and the last layers of the saved model together. This required to unfreeze a few of the top layers of the saved model and helps to fine-tune the saved model to make it more appropriate for any particular task.

The rest of the paper is planned as follows: the literature review is presented in Sect. 2. The full training and transfer learning approaches are described in Sect. 3. Section 4 is dedicated to present and analyze the results from each experiment. The last section (Sect. 5) concludes the work with a discussion of our significant findings.

## 2 Literature Review

Transfer learning is a deep learning method used to perform a new task based on the transmission of knowledge from an already learned task. In computer vision, the object recognition using transfer learning is improving day by day. The training of a new classifier with a limited number of samples would significantly increase the risk of the new data being improperly generalized [7, 8].

A pre-trained deep learning model is a model that is typically trained on a huge dataset for the purpose of classification tasks. In CNN, feature extraction is done by convolutional base and the image classification is done by classifier. In transfer learning, original classifier is replaced by a new classifier and fine-tune based on any of the following 3 categories [6]:

1. Train the entire model: Use the saved model architecture to train our dataset. A huge dataset is required as we need to train the model from beginning
2. Training and freezing layers: make a few layers as frozen to prevent overfitting when the dataset is small and the number of parameters are more.
3. Freeze the convolutional base: It is used if the computing capacity is limited, our dataset size is insufficient or our pre-trained network solves a similar task.

It is recommended to choose a small learning rate measure since a higher rate raise the danger of loss of prior learnings while using a pre-trained network on CNN. ImageNet will be useful if the task is to classify cat and dog since it contains cat and dog images. W. Rawat and Z. Wang used only a small amount of data from the original dataset for the image classification. For fully connected layers and Global average pooling an average of 85% validation accuracy was achieved without overfitting [9]. Huh et al. used the AlexNet model for their experiment. The model was pre-trained on ImageNet and used for image classification, object recognition, action identification, human pose approximation, image segmentation, image captioning etc. [10].

Sajja Tulasi Krishna and Hemantha Kumar Kalluri conducted a survey based on CNN and its architectures like LeNet, AlexNet, GoogleNet, VGG16, VGG19, Resnet50. The transfer learning by pre-trained CNN architecture was used and tested on ImageNet datasets. The increased minimum batch size also could boost device efficiency per iteration [11]. Kornblith et al. made a comparison of 12 different datasets using 16 classification networks. They realized that ImageNet highest accuracy was tightly coupled with transfer accuracy. Fine-tuning provided better results than logistic regression with larger datasets [12].

Shiddieqy et al. developed an automated deep transfer learning architecture using an acute version of the Inception model to detect Covid19 from chest x-ray images. Training and testing analysis was done using accuracy, f-measure, sensitivity, specificity, and kappa statistics [13]. In [14], T. Agarwal and H. Mittal conducted a comparative study for image classification using VGG16, MobileNet, ResNet50 and InceptionV3 models. They evaluated models in terms of accuracy on different datasets called cat and dog for binary classification and plant seedlings dataset for multiclass classification. The VGG16 model was accurate, but it required more computational power than other models. Shiddieqy et al. [15] implemented a CNN for classification from cat and dog dataset. They have used two CNN structures; one with two layers and the other with five layers. The best validation accuracy obtained was 78%, with a learning rate of 0.001 for a batch size of 64. They concluded that using a deeper model might give better accuracy in prediction.

Cengil et al. used cat and dog images from Kaggle to do binary classification using a CNN model. All images were resized to $64 \times 64$. The accuracy rate remained constant after 5000 iterations. They achieved 83% accuracy on test set [16]. In [17], Jajodia et al. created a dataset with 10,000 images of cat and dog from the Kaggle repository. They utilized 4000 images of cat and dog for training and 1000 each for validation and testing with respective classes. The used image resolution was $64 \times 64$. The different data augmentation techniques such as zoom range, rescaling and shear range used to extend the dataset size. The model was trained for 15 epochs and obtained the best test accuracy as 90.10%. The image segmentation approach was applied in the cat and dog dataset

[18]. Due to the poor segmentation result, the authors did not achieve any improvement in accuracy after training the models. The highest accuracy achieved was 71.47% from the Support Vector Machine (SVM) classifier. In second approach, a trainable model could achieve 94% accuracy from SVM classifier. A pre-trained ResNet-18 model in ImageNet provided 93.76% accuracy on cat and dog dataset [19].

## 3 Method

For the VGG16 architecture, we used an image of 224 × 224px as the input and returned a 1000 size vector with the probability of each class. The VGG16 model was pre-trained on ImageNet with 13 convolutional layers, 5 pooling layers and 3 fully-connected layers. It contained multiple 3 × 3 filters with 1PX as a stride at each convolution layer. For the classification, the softmax layer was the last layer used. The Rectified Linear Unit (ReLU) was the activation function used in every convolution block [1, 2].

We considered an image of 299 × 299px as the input and returned a 1000 size vector with the probability of each class in the InceptionV3 model. This model contained 312 layers and 10 blocks in total. This model included 3 inception blocks, 13 convolutional layers and 2 pooling layers for training. It contained multiple 3 × 3 filters with 2PX as a stride at each convolution layer. In the last layer, the number of output nodes was same as the number of categories in the dataset. The softmax layer used as classification layer and ReLU used as activation function in every convolution block [3, 4].

The MobileNetV2 network that we used for this experiment considered input images of 224 × 224px and returned a 1280 size vector with the probability of each class. It contained 16 blocks of layers with 3 × 3 filters with 1PX as a stride at each convolution layer. The only thing that differentiates MobileNet from other CNNs was the use of a depth-wise convolutional split, which separates the convolution into a 3 × 3-depth and a 1 × 1-pointwise convolutions respectively. As same as other two models, softmax used for classification and ReLU used for activation function [5].

Most of the current deep learning neural networks are trained on large datasets. In small datasets, deep convolutional neural networks were uncommon because every model provided lower accuracy and higher loss value during training and testing [6]. In this paper, we suggest a modified deep learning neural network and use this model on a small size dataset for cat and dog classification. The purpose of this work is to show that a properly modified deep pre trained model on the ImageNet can be used for image classification with a good accuracy and low loss.

We trained the models using the cat and dog image dataset from Kaggle, which contains 2000 training images, 800 validation images and 200 testing images. The transfer learning [20] and full training [9] concepts were applied in this proposed work.

### 3.1 Transfer Learning Approach

In this experiment, we considered models MobileNetV2, InceptionV3 and VGG16 as base models for training. The MobileNetV2 [21] and InceptionV3 [22] models were developed at Google and VGG16 was developed by the Visual Graphics Group (VGG) at Oxford [23]. The models were pre-trained on a huge dataset named ImageNet which

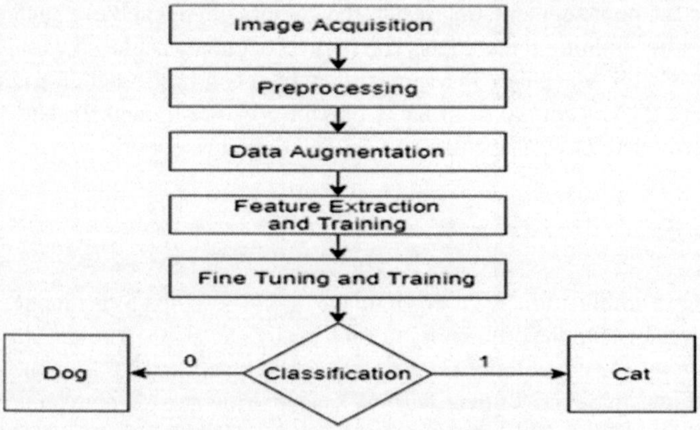


**Fig. 1** Flowchart of proposed work

consists of 1000 classes and 1.4 M images. This knowledge base allows classifying cat and dog from a particular data set. The flow of our proposed model is in Fig. 1.

### 3.1.1 Preprocessing

In the preprocessing step, a total of 3000 cat and dog images were acquired for the experiments. The pre-trained models required pixel value in the range [−1, 1], but the pixel values of acquired images were between [0–255]. Therefore, every image was rescaled as per the requirements of models in the preprocessing phase.

### 3.1.2 Data Augmentation

By applying RandomFlip and RandomRotation, another 3150 augmented images were generated, which helped to avoid overfitting during training. The RandomFlip method is used to randomly flip each image vertically and horizontally. In case of RandomRotation, each image would be randomly rotated by a fixed degree. In this experiment, we used horizontal flipping and 20° for random rotation for generating augmented images.

### 3.1.3 Feature Extraction

For the first step of the feature extraction process, we picked a layer of the model that could be useful to do this task. The last classification layer of the base models was not much useful. So, instead of the last classification layer, we added a classifier to the top that was fully connected. The classifier's weights were only modified in training time, and the pre-trained network kept frozen. If a network does not have classification layers at the top by specifying include_top = False, it is perfect for extracting features. In the next step, froze the convolutional base and added a classifier on top of the model. The dataset was used to train only on the top-level classifier for the first 10 epochs. The rest of all the layers remained frozen. It would avoid upgrading of weights and parameters of the unfreeze layers of the model when training is going on. These models have several layers, so it can freeze all layers by assigning the flag of the whole model as False. For the first 10 epochs, Adam optimizer was used with a learning rate of 0.0001. The Adam optimizer is much faster than other optimizers and its default hyper-parameters usually work well. Binary-cross entropy loss is used in compilation since there are only two classes.

### 3.1.4 Fine-Tuning

We trained only top-level classifier of the model in the feature extraction process. During training, the pre-trained model's parameters and weights didn't get modified. The addition of a new classifier along with the pre-trained model's weights and parameters is one method of enhancing the performance. During training, weights from basic feature maps to update individually associated with the data set would be determined. It can only be attempted just after classifier has been trained with the pre-trained network as untrained. Suppose the random classifier is placed on top of the base model and try to execute complete layers together. In that case, the gradient updates' magnitude is too high due to the classifier's random weights. The re-training by fine-tuning with new dataset to the base model's last few layers helps to adjust weights and its specifically learned features. It is always better to fine-tune only a few top layers rather than the whole layers of base models. The top layer is more specialized in most convolutional networks. The initial layers quickly learn fundamental and basic features that are familiar to nearly every image type. For the top layers, the dataset features on which the model was trained are increasingly specific. The fine-tuning aims to adapt specialized features to the functioning of the new data set instead of re-learning existing features. To do so, first, make the top layers of the base model as trainable and the remaining layers as un-trainable. It is always better to use a low learning rate since the models are large and require to re-adapt the pre-trained weights. Otherwise, the chance to get overfitting is too high during training. The models were fine-tuned from epochs 10 to 50 with the RMSprop optimizer and learning rate 0.00001.

Finally, we need to assess the performance of the trained models on the dataset using a set of test images. As per the binary classification context, the model would predict 0 for dogs and 1 for cats.

## 3.2 Full Training Approach

In this approach, we used three deep learning models called VGG16, MobileNetV2 and InceptionV3 for classification. All models were trained with RMSprop optimizer and learning rate as 0.00001 and 0.0001 each for 50 epochs. As it was full training, all layers in the models trained with the cat and dog dataset. In this approach, the image acquisition, preprocessing and data augmentation steps were the same as the transfer learning approach. All models were trained from scratch and it required more time to complete its training than transfer learning. Each model's number of layers, blocks, and input image resolution were different. However, all models were Sequential models, which meant all layers were arranged in a sequence way. We used ImageDataGenerator from keras.preprocessing to import each image with its labels into models. The ReLU activation function was added with each layer to avoid passing negative values to the next layers. After creating all the convolution blocks, we added 2 units of dense layers with a softmax classifier at the end. The softmax layer provided output value as 0 and 1, with 0 for dog and 1 for cat. The evaluation of each full-training model was also done in the same way as that of transfer learning.

## 4 Results and Analysis

We utilized the accuracy and loss measurements of the training and the validation data for evaluating the model in these experiments. We checked for over-fitting and under-fitting using these measurements. If the validation accuracy is much less than training accuracy and validation loss is considerably greater when compared with training loss, it is overfitting. If the validation accuracy is much greater than training accuracy and validation loss is considerably less when compared with training loss, it is under-fitting.

For the full training experiments, we initialized the deep learning models MobileNetV2, InceptionV3 and VGG16 with learning rates 0.00001 and 0.0001 for the training of 50 epochs. There was no overfitting after using 3150 augmented images. All models showed better results with a learning rate of 0.0001. It required less computational time during training than using a learning rate of 0.00001. Also, it gave better average training and validation accuracy and lower training and validation loss measures. The results achieved using full-training models are shown in Table 1. The best testing accuracy achieved using full training was 79.08% from MobileNetV2 with a learning rate of 0.0001. For the same learning rate, InceptionV3 and VGG16 models provided 73.44 and 75.62%, respectively. As it was a full training, insufficient dataset is the reason behind poor training results for all the deep learning models. Training results for a learning rate of 0.00001 are poorer than a learning rate of 0.0001 for all the models since the learning rate controls how quickly a deep learning model adapted to a classification problem. A model with a larger learning rate can quickly learn features with fewer training epochs.

**Table 1** Full-training results on various models using 2000 training images, 800 validation images 200 testing images

| Model | Learning rate of RMSprop | Epochs | Training accuracy (%) | Validation accuracy (%) | Test accuracy (%) |
|---|---|---|---|---|---|
| MobieNetV2 | 0.00001 | 50 | 57.65 | 59.33 | 60.94 |
| MobieNetV2 | 0.00001 | 100 | 60.92 | 58.51 | 61.94 |
| MobieNetV2 | 0.0001 | 50 | 78.77 | 76.02 | 79.08 |
| InceptionV3 | 0.00001 | 50 | 67.34 | 65.93 | 65.63 |
| InceptionV3 | 0.0001 | 50 | 79.21 | 72.71 | 73.44 |
| VGG16 | 0.00001 | 50 | 67.44 | 65.75 | 74.48 |
| VGG16 | 0.0001 | 50 | 75.57 | 70.42 | 75.62 |

**Table 2** Transfer learning results on various models using Learning Rate for Adam as0.0001 and RMSprop as 0.00001, 2000 training images, 800 validation images 200 testing images.

| Model | Epochs | Number of trained layers | Training accuracy (%) | Validation accuracy (%) | Test accuracy (%) |
|---|---|---|---|---|---|
| MobieNetV2 | 20 | 6 blocks [64 layers] | 96.4 | 97.99 | 97.35 |
| MobieNetV2 | 50 | 6 blocks [64 layers] | 98.16 | 98.28 | 98.44 |
| MobieNetV2 | 50 | 5 layers | 95.64 | 98.11 | 98.96 |
| MobieNetV2 | 50 | 10 layers | 96.29 | 98.23 | 98.96 |
| InceptionV3 | 50 | 2 blocks [64 layers] | 97.76 | 97.65 | 97.40 |
| InceptionV3 | 100 | 2 blocks [64 layers] | 98.81 | 97.7 | 97.92 |
| InceptionV3 | 50 | 3 blocks [84 layers] | 97.86 | 97.61 | 98.44 |
| VGG16 | 50 | 1 block [4 layers] | 97.85 | 96.78 | 97.40 |

Table 2 gives the details of various experiments done with transfer learning with different learning rates, epochs and number blocks or layers. For all transfer learning experiments, the first ten epochs were trained using Adam optimizer and learning rate as 0.0001. Then changed the optimizer to RMSprop and reduced the learning rate to 0.00001 to avoid overfitting due to re-adapting pre-trained weights. The data augmentation technique also was applied and used 3150 augmented images with each experiment to avoid overfitting. All the models provided more than 95% training and validation accuracy and 97% testing accuracy. The best testing accuracy obtained using transfer learning was 98.96%. Here also MobileNetV2 model gave the highest accuracy. However, all other models also gave accuracy that was almost near to accuracy by MobileNetV2 model. All models were pre-trained on the ImageNet dataset. ImageNet already contained similar classes of data as cat and dog. The utilization of similar weights is the reason behind achieving better performance from transfer learning.

To the best of our knowledge, no deep learning models from previous research works have generated more than 95% test accuracy on cat and dog dataset. It was pretty slow to get the result from full-training models as it required training all layers of the models. Thus these full-training models do not achieve an excellent performance; we decided to utilize the transfer learning technique in image classification for obtaining better performance. Improved results were obtained from transfer learning than full-training in image classification. When compared in terms of required computational time for training, the transfer learning experiments showed better performance. It completed the training with each model in a minimum time and lower training and validation loss value than full training.

Some of the cats and dogs face were not clearly visible and blend in with their bodies, made it difficult to extract features. A few of the cats and dogs eyes were closed. Also, some objects like walls, plants, etc., around cats and dogs, made it even harder to recognize cats and dogs. Some other images contained cats and dogs with human beings that covered a small portion of the images. Overall, the presence of noises that made it challenging to identify cats and dogs. Removing noises in future work must help models to make it easy for classification.

## 5 Conclusion

Our proposed research aimed to show how deep learning models like MobileNetV2, InceptionV3, and VGG16 can be used on a minimal size dataset with a total of 3000 images, including 2000 training data, 800 validation data, and 200 testing data, providing excellent performance. All of the models were pre-trained on ImageNet. All of these transfer learning models performed well by giving good accuracy results and low losses. In the case of full-training, the results were not much promising when compared with transfer learning results. The data augmentation techniques were utilized with all of the above-described scenarios to overcome overfitting issues. Overall, the experiments validate the fact that the deep learning models could fit in tiny datasets with proper modifications of their architecture.

## References


1. Song H, Mao H, Dally WJ (2015) Deep compression: Compressing deep neural networks with pruning, trained quantization and huffman coding. arXiv preprint arXiv:1510.00149
2. Rezende E, Ruppert G, Carvalho T, Theophilo A, Ramos F, de Geus P (2018) Malicious software classification using VGG16 deep neural network's bottleneck features. Adv Intell Syst Comput 738:51–59. https://doi.org/10.1007/978-3-319-77028-4_9
3. Szegedy C, Vanhoucke V, Ioffe S, Shlens J (2016) Rethinking the Inception Architecture for Computer Vision
4. Chollet F (2016) Xception: deep learning with depthwise separable convolutions
5. Sandler M, Howard A, Zhu M, Zhmoginov A, Chen L-C (2018) MobileNetV2: inverted residuals and linear bottlenecks
6. Marcelino P (2020) Transfer learning from pre-trained models. Towards data science. https://towardsdatascience.com/transfer-learning-from-pre-trained-models-f2393f124751, Accessed 02 Dec 2020
7. Weiss K, Khoshgoftaar TM, Wang DD (2016) A survey of transfer learning. J Big Data 3(1):1–40. https://doi.org/10.1186/s40537-016-0043-6
8. Tan C, Sun F, Kong T, Zhang W, Yang C, Liu C (2018) A survey on deep transfer learning
9. Rawat Waseem, Wang Zenghui (2017) Deep convolutional neural networks for image classification: a comprehensive review. Neural Comput 29(9):2352–2449. https://doi.org/10.1162/neco_a_00990
10. Huh M, Agrawal P, Efros AA (2016) What makes ImageNet good for transfer learning?
11. Tulasi KS, Kalluri HK (2019) Deep learning and transfer learning approaches for image classification. Int J Recent Technol Eng (IJRTE) 7(5S4): 427–432
12. Kornblith S, Shlens J, Le QV (2019) Google Brain. Do Better ImageNet Models Transfer Better?

13. Narayan Das N, Kumar N, Kaur M, Kumar V, Singh D (2020) Automated deep transfer learning-based approach for detection of COVID-19 infection in chest X-rays. IRBM. https://doi.org/10.1016/j.irbm.2020.07.001
14. Agarwal T, Mittal H (2019) Performance comparison of deep neural networks on image datasets. In: 2019 Twelfth International Conference on Contemporary Computing (IC3), Noida, India, pp 1–6. https://doi.org/10.1109/IC3.2019.8844924
15. Shiddieqy HA, Hariadi FI, Adiono T (2017) Implementation of deep-learning based image classification on single board computer. In: 2017 International Symposium Electronics Smart Devices, ISESD 2017, vol. 2018-Janua, pp 133–137. https://doi.org/10.1109/ISESD.2017.8253319
16. CengilE, Çinar A, Güler Z (2017) A GPU-based convolutional neural network approach for image classification. In: IDAP 2017 - International Artificial Intelligence Data Processing Symposium. https://doi.org/10.1109/IDAP.2017.8090194
17. Jajodia T, Garg P (2019) Image classification-cat and dog images. Int Res J Eng Technol, 570–572. www.irjet.net
18. Suryawanshi S, Jogdande V, Mane A (2020) Animal classification using deep learning. Int J Eng Appl Sci Technol 04(11):305–307. https://doi.org/10.33564/ijeast.2020.v04i11.055
19. Kim B, Kim H, Kim K, Kim S, Kim J (2019) Learning not to learn: training deep neural networks with biased data. In: Proceedings IEEE Computer Society Conference Computer Vision Pattern Recognition, vol 2019-June, pp 9004–9012. https://doi.org/10.1109/CVPR.2019.00922
20. Transfer learning and fine-tuning TensorFlow Core. https://www.tensorflow.org/tutorials/images/transfer_learning, Accessed 02 Dec 2020
21. Review: MobileNetV2—Light Weight Model (Image Classification)|by Sik-Ho Tsang|Towards Data Science. https://towardsdatascience.com/review-mobilenetv2-light-weight-model-image-classification-8febb490e61c, Accessed 18 Dec 2020
22. Inceptionv3 - Wikipedia. https://en.wikipedia.org/wiki/Inceptionv3, Accessed 18 Dec 2020
23. VGG16 - Convolutional Network for Classification and Detection. https://neurohive.io/en/popular-networks/vgg16/, Accessed 18 Dec 2020

# Physical Unclonable Function and OAuth 2.0 Based Secure Authentication Scheme for Internet of Medical Things

Vivin Krishnan(✉) and Sreeja Cherillath Sukumaran

Department of Computer Science, CHRIST (Deemed To Be University), Hosur Road, Bangalore 560029, Karnataka, India
vivin.krishnan@res.christuniversity.in,
sreeja.cs@christuniversity.in

**Abstract.** With ubiquitous computing and penetration of high-speed data networks, the Internet of Medical Things (IoMT) has found widespread application. Digital healthcare helps medical professionals monitor patients and provide services remotely. With the increased adoption of IoMT comes an increased risk profile. Private and confidential medical data is gathered across various IoMT devices and transmitted to medical servers. Privacy breach or unauthorized access to personal medical data has far-reaching consequences. However, heterogeneity, limited computational resources, and lack of standardization in authentication schemes prevent a robust IoMT security framework. This paper introduces a secure lightweight authentication and authorization scheme. The use of the Physical Unclonable Function (PUF) reduces pressure on computational resources and establishes the authenticity of the IoMT. The use of OAuth 2.0 open standard for authorization allows interoperability between different vendors. The resilience of the model to impersonation and replay attacks is analyzed.

**Keywords:** IoMT · Lightweight authentication · PUF · OAuth 2.0

## 1 Introduction

Internet of Medical Things (IoMT) has found rapid and wide acceptance in the healthcare sector. It helps healthcare professionals remotely monitor patients and gather health metrics. IoMT devices monitor health vitals and inform medical professionals in emergencies. Using IoMT, healthcare services can remotely control insulin or drug pumps to release the proper dosages. Doctors can use implantable devices to gather data for analysis. Swallowable camera capsules allow viewing and visualizing the gastrointestinal tract helps in non-invasive procedures.

With the proliferation in adoption, the risk of security exposure also grows. Implementing robust security protocols for IoMT is usually constricted by the lack of computational resources. Computationally intensive tasks need to be limited to preserve onboard power supply. Even with these restrictions, the data gathered by the IoT should be secure at rest and in transit. Device and user authentication should be established


S. Shukla et al. (eds.), *Data Science and Security*, Lecture Notes in Networks and Systems 290,
https://doi.org/10.1007/978-981-16-4486-3_7

between the IoMT and the server before accepting or transmitting data. However, conventional encryption protocols are computationally exhausting and not directly suited for IoMT. This brings in the need for lightweight authentication mechanisms [1]. Physically Unclonable Function (PUF) is a way to achieve this.

This article describes a methodology that utilizes PUF modules for lightweight authentication and a modification on OAuth 2.0 [2] protocol for authorization. The rest of this article is organized as follows. Section 2 covers prevailing work on authentication systems based on PUF and OAuth 2.0. Section 3 covers the proposed methodology. Section 4 provides an analysis of the attacks prevented by the novel scheme. The conclusion is covered in Sect. 5.

## 2 IoMT and Security

While IoMT helps keep track of the health-related data and can ensure faster access to medical services, the security risks involved are significant. Major security concerns around IoMT devices revolve around: confidentiality, integrity, availability, authentication, authorization, privacy, and non-repudiation.

### 2.1 IoMT Architecture

IoMT devices do not have a standard architecture that different vendors implement. Generic representations describe IoTs with three-layers. This model has been further enhanced by adding a Processing and Business layer to represent the complexities of IoT integration [3]. In [4], the authors propose a four-layer architecture for IoMT based on IEEE P2413.1 Reference architecture for Smart City [5].

### 2.2 Security Attacks

In [6–10], authors list the major attacks to which IoT devices are susceptible. The articles cover in detail the attacks based on the security layers and network protocols. Authors in [11] present security issues that are specific to implantable IoMT devices. The authors note the attacks that can be launched against specific implantable devices.

### 2.3 Authentication

Authentication refers to verifying and confirming the identity of a user object or process. IoMT devices deal with private and confidential medical data, and unauthorized access to information has a severe impact.

In [12, 13], the authors cover various authentication schemes available to the IoT. The heterogeneity of IoT devices is a significant factor behind the absence of standardized authentication schemes.

### 2.4 PUF

PUF depends on natural physical variations in semiconductor fabrication that produce responses specific to the inputs. For a given input, the PUF generates a particular response. The input is called a challenge. The output is called a response. The pair of challenge and response (CRP) results from the natural variations in fabrication, and it is impossible to clone a PUF. The unique variations ensure that, given the same input, different PUF produces different results. PUF is implemented mainly in integrated circuits, and this prevents strain on the IoMT computational resources [14].

### 2.5 OAuth 2.0

OAuth 2.0 is an open authorization standard. It is a Bearer token-based mechanism which allows resource access to third party without revealing the owner's credentials [15] (Fig. 1).

```
+--------+                               +---------------+
|        |--(A)- Authorization Request ->|   Resource    |
|        |                               |     Owner     |
|        |<-(B)-- Authorization Grant ---|               |
|        |                               +---------------+
|        |
|        |                               +---------------+
|        |--(C)-- Authorization Grant -->| Authorization |
| Client |                               |     Server    |
|        |<-(D)----- Access Token -------|               |
|        |                               +---------------+
|        |
|        |                               +---------------+
|        |--(E)----- Access Token ------>|    Resource   |
|        |                               |     Server    |
|        |<-(F)--- Protected Resource ---|               |
+--------+                               +---------------+
```

**Fig. 1** OAuth 2.0 protocol flow [15]

## 3 Literature Review

PUF based authentication designs have been researched in the past. In [16], the author analyses the application of PUF in IoT security. Attacks on PUF, including Machine Learning and mathematical attacks are also outlined. Further, an authentication system based on PUF is proposed. The authors in [17] model a practical IoT device authentication framework using PUF. In this model, the authors use Elliptic Curve Cryptography(ECC) algorithm to generate public–private key pairs. The authors in [18] adopt a PUF model-masking approach. In this approach, the CRP does not contain the response stored directly. Instead, the CRP contains the challenge and the challenge encrypted (with the response as the key). This prevents an attacker from guessing the response model considering the challenge is public.

In [19, 20], authors propose a PUF based system to perform lightweight device authentication for IoMT devices. In this literature, the server initiates the connection to the IoMT devices and the authentication. In [21] the authors describe an IoT authentication protocol that uses PUF. In article [22] authors propose a mutual authentication system for IoT based on PUF.

In [23], the authors propose an OAuth 2.0 based authentication scheme for IoT that utilizes the user's proxy agent. The authors' analysis shows the model resistant to impersonation and replay attacks. IoT-Cloud-based secure health service using OAuth 2.0 is discussed in [24]. Article [25] describes a Smart City application that is built on OAuth 2.0 with new integrations. An OAuth-based for secure services in IoT is described in [26]. Article [27] proposes an approach where smartphones of users can act as Authorization servers. This removes the IoMT network's reliance on third-party Authorization servers. The author further proposes a modification to the protocol, where the token issuance happens when authorization server registers with client. In [28], the authors propose an Authorization framework as a service based on interoperable OAuth.

Authors in [29–31] assess the OAuth 2.0 protocol. In [29], the authors evaluate the OAuth 2.0 and find that if the authorization grant is not limited to one-time use and does not have an expiry, the Authorization server is prone to replay attacks. Validation of client authenticity is necessary to prevent impersonation attacks. In [30], the authors cover countermeasures to the standard OAuth 2.0 attacks. In [31], a formal security analysis of OAuth 2.0 rather than specific implementations is performed. The analysis showed the vulnerability of OAuth 2.0 and that it can be strengthened through fixes. These fixes were accepted for a new Request for Comment (RFC) [32].

# 4 Proposed Model

This section describes a novel approach involving PUF modules for authentication with authorization performed using OAuth 2.0 protocol. The proposed model has three different phases - enrolment phase, authentication phase, and resource transmission phase.

## 4.1 Proposed Architecture

In the proposed architecture, the participating entities are the patient or user, the IoMT device, the Medical server, and the Authorization Server (Fig. 2).

Connections from the user to the medical server are made through the IoMT. The IoMT consists primarily of a PUF module, an authentication module, and a module that transmits data to the Medical server. The transmission module also contains the implementation to contact the Authorization Server.

The IoMT contacts the medical server to perform both user and device authentication. The IoMT uses the PUF module to perform device authentication. The IoMT device cannot transmit data to the medical server without first authenticating the user and the device successfully. The IoMT device without the user is considered stolen and should not authenticate or transmit data.

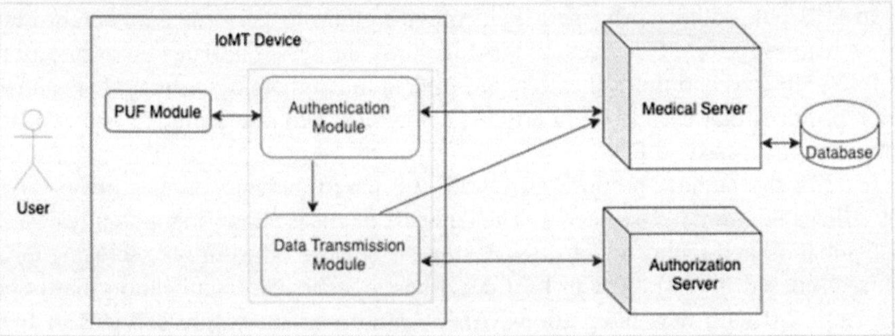


**Fig. 2** Proposed architecture

The Medical server maintains the medical data collected by the various sensor devices. The IoMT devices send the data to the Medical server for aggregation and storage. Depending on the implementation, the Medical server can entirely delegate the OAuth flow to an Authorization server or act as both the Resource and Authorization server.

### 4.2 Algorithm

In the proposed model, the flow proceeds in three phases:

i. Enrolment phase
ii. Authentication phase
iii. Data transmission phase

All communication between the entities in the proposed model is over a secure channel. After successful authentication, the intelligence to invoke the Authorization Server and obtain the Bearer token is embedded in the IoMT device (Table 1).

### 4.3 Enrolment Phase

The enrolment phase proceeds in three steps:

1) User registers with the Medical Server directly. This is a one-time process. This is done by the user directly or by an administrator.
2) User enrolment through the device.
3) Device enrolment.

**Table 1** Notations used in the proposed model

| Notation | Description |
|---|---|
| $D_{iomt}$ | IoMT device |
| $U_i$ | User |
| $U_{identifier}$ | User identifier |
| $M_{server}$ | Medical derver |
| $C_i$ | Challenge |
| $R_i$ | Response |
| {C, R} | Challenge-Response Pair (CRP) |
| $C_S$,$R_S$ | A subset of challenge and response |
| k,r,n | Random numbers |
| t | timestamp |
| h() | Hash function |
| $\oplus$ | XOR function |
| \|\| | Concatenation operation |

| Algorithm: User Enrolment through Device | |
|---|---|
| Device | Server |
| Compute<br>H = h(username‖password)<br>Rn = H $\oplus$ n, where n is a random number<br>LC= {Rn,timestamp $t_{device}$}<br>Send LC to $M_{server}$ through Device $D_{iomt}$ over a secure channel | |
| | Check $t_{server}$-$t_{device}$ > $t_{delta}$. If so,reject.<br>Compute n = Rn $\oplus$ h(username‖password)<br>Generate user identifier $U_{identifier}$<br>Compute R = h(n) $\oplus$ $U_{identifier}$<br>Send R to IoMT Device |
| Receive R<br>Compute $U_{identifier}$ = R $\oplus$ h(n)<br>User $U_i$ notes down the $U_{identifier}$.<br>Proceed with Device authentication. | |

User $U_i$ connects to the Medical Server $M_{server}$ through the Device $D_{iomt}$ over the secure channel in the user enrolment phase. User $U_i$ sends {h(username ‖ password) xor n, current timestamp}, where is a random number to $M_{server}$. If the server receives the request later than a threshold (based on the timestamp), the request is rejected. Server computes $n = Rn \oplus$ h(username‖password) and validates the credentials. If credentials match, server generates a response payload R = {h(n) xor $U_{identifier}$} where $U_{identifier}$ is a unique id assigned to the user. Device $D_{iomt}$ receives computes the $U_{identifier}$ as $U_{identifier} = R \oplus h(n)$.

If n matches, $U_{identifier}$ is recorded by the user and is not stored in the device $D_{iomt}$. The device $D_{iomt}$ proceeds with Device authentication.

| Algorithm: Device Enrolment | |
|---|---|
| **Device** | **Server** |
| Prompt User $U_i$ for the unique identifier.<br>Compute P = { $U_{identifier} \oplus t_{device}$} where $t_{device}$ is the device timestamp<br>Send P to $M_{server}$ over a secure channel. | |
| | $M_{server}$ checks $t_{server}$-$t_{device}$ > $t_{delta}$. If so, reject.<br>Generate set of challenges<br>C = {$C_1$,$C_2$,$C_3$…$C_n$,random string}<br>Send C to IoMT device $D_{iomt}$ |
| Receive challenge set C.<br>Invoke PUF module. Obtain response for the challenge<br>for i=0 to n-1, do:<br>$R_i$=PUF($C_i$)<br>Send R = {h($R_1$),h($R_2$),…h($R_n$), random string} to the server. | |
| | Compute unique identifier $D_{id}$ for $D_{iomt}$<br>Send {$D_{id}$, $D_{limit}$ } to the device over a secure channel, where $D_{limit}$ is a delimiter. |
| Receive {$D_{id}$ ,$D_{limit}$}<br>Store $D_{id}$ and $D_{limit}$ to memory. | |
| | Store {CRP,$D_{id}$,h(username,password), h($U_{identifier}$),$D_{limit}$}securely in the database. |

In the device enrolment phase, the IoMT Device $D_{iomt}$ prompts $U_i$ to enter the unique user identifier. $D_{iomt}$ computes {$U_{identifier}$ xor timestamp} and sends the request payload to $M_{server}$ over a secure channel. If the request does not reach the server within a defined time (based on the request timestamp), the server rejects the request. After verifying the $U_{identifier}$, $M_{server}$ generates a set of challenges $C = \{C_1, C_2, C_3, \ldots C_n\}$ and sends to $D_{iomt}$. Server chooses the number of challenges depending on $D_{iomt}$'s computational capability. $D_{iomt}$ passes challenge set C to the PUF. The PUF module generates responses for each challenge.

$$R_i = PUF(C_i) \tag{1}$$

Device computes the response set $R = \{h(R_1), h(R_2), h(R_3), \ldots h(R_n)\}$ is generated and the challenge-response pair (CRP) is sent to the server. The Medical server chooses a set of delimiters to be used during the data transmission phase. $M_{server}$ creates a unique device identifier $D_{id}$ and sends it to $D_{iomt}$. $D_{iomt}$ stores the identifier in memory. The Medical server stores {CRP,$D_{id}$,h(username,password),h($U_{identifier}$)} securely in the database. This completes the enrolment process.

## 4.4 Authentication Phase

Algorithm: Authentication Phase

| Device | Server |
|---|---|
| Gather User credentials and $U_{identifier}$.<br>Compute $H_{cred}$= h(username‖password)<br>Send {$H_{cred}$,$D_{id}$,h($U_{identifier}$),$t_{device}$} to $M_{server}$. ,<br>where $t_{device}$ is the device timestamp | |
| | Check $t_{server}$-$t_{device}$ > $t_{delta}$. If so, reject.<br>Randomly choose a subset of challenges $C_s$ from the CRPs stored for $D_{id}$<br>Generate a random number K.<br>Sends {$C_s$,K⊕$D_{id}$} to $D_{iomt.}$ over secure channel |
| Retrieve K.<br>Pass $C_s$ to PUF module.<br>Compute response set<br>$R_s$ = {h($R_a$),h($R_d$),h($R_m$),random string)} ,<br>where a, d , m are examples of random challenges from the original set.<br>Send {$R_s$, R⊕K,$t_{device}$} to the server, where R is a random number generated by $D_{iomt}$ | |
| | Check $t_{server}$-$t_{device}$ > $t_{delta}$. If so, reject.<br>Validate $R_s$ against the database<br>Calculate $H_{response}$ = h($U_{identifier}$,$D_{id}$,R,K).<br>Send {$H_{response}$, $t_{server}$ }over a secure channel. |
| Calculate H'= h($U_{identifier}$,$D_{id}$,R,K).<br>If (H' != H), break the flow<br>Else proceed to contact<br>Authorization Server for OAuth Token. | |

Authentication is initiated when the IoMT device contacts the Medical server to transfer data over a secure channel. $D_{iomt}$ obtains user credentials and $U_{identier}$ from the user. $D_{iomt}$ creates a request payload {h(username‖password),$D_{id}$, h($U_{identifier}$), timestamp for $M_{server}$ and transmits securely. The Medical server challenges the IoMT device with a random subset of challenges $C_S$, stored against its Device identifier. The server generates a random number K and sends {$R_S$, K xor $D_{id}$, $t_{server}$} to the IoMT device $D_{iomt}$. The device retrieves K and computes a set $R_s$ of responses for the specific challenge set using the PUF module. The device generates a random number R and sends {$R_S$, R xor K, t} to the server, where $t_{device}$ is the device time. $M_{server}$ verifies freshness of the request and rejects if it has exceeded an interval. This is to ensure freshness in the authentication requests. After validating the response set Rs, the server responds with {h( $U_{identifier}$, $D_{id}$, R, K), server timestamp}. IoMT validates the hash and timestamp to ensure the response comes from the server and is fresh. If the hash H from $M_{server}$ does not match the device calculated hash H', $D_{iomt}$ breaks the authentication flow. If it matches, it proceeds to contact the authorization server for the OAuth token. This completes the authentication phase.

### 4.5 Data Transmission Phase

| Algorithm: Data Transmission |
|---|
| Begin:<br><br>Device $D_{iomt}$ obtains OAuth token T from Authorization server $A_s$.<br><br>while (Token T is not expired)<br>do<br>i) Create String $S=\{t_{current},h(D_{id},R,K)\}$,where $t_{current}$ is the current timestamp,R & K are random numbers from authentication.<br>ii) Place the String S into the Token T, with pre-agreed delimiters $D_{limit}$.<br>iii) Send data to Medical Server $M_{server}$.<br><br>End |

Data transmission sends data to a protected endpoint in the Medical server. OAuth 2.0 protocol performs authorization of the IoMT device. IoMT device $D_{iomt}$ contacts Authorization server As and obtains an OAuth Bearer Token T. T has a fixed expiry and the scopes indicating the authorization level. Once $D_{iomt}$ has received the Bearer token T, $D_{iomt}$ computes $S = \{t_{current}, h(D_{id}, R, K)\}$,where $t_{current}$ is the current timestamp at the client.

The string S is inserted into the Bearer Token wrapped with delimiters agreed in the enrolment phase. The request is transmitted to the Medical server with the modified Bearer token. An attacker may try to impersonate the IoMT device by eavesdropping on the Bearer Token. Mserver will reject the request as the expected randomness will not be found in the incoming token. Similarly, if an attacker replays the request, the timestamp will not be fresh, and the server will reject the request.

## 5 Analysis of the Proposed Scheme

The proposed scheme uses a PUF based lightweight authentication and uses OAuth 2.0 protocol with measures added to ensure freshness. The attacks prevented with the proposed model are:

### 5.1 Replay Attacks

During the authentication phase and data transmission phase, the timestamp is added to the transmission between IoMT and server and validated at either end. This is to ensure the message is not being replayed.

### 5.2 Impersonation Attacks

An attacker cannot clone the PUF module. The challenge set from the server is chosen at random, and the client device has to answer all the challenges. This works to ensure the authenticity of the communicating device.

### 5.3 Eavesdropping Attacks

The IoMT device, Medical server, and Authorization server always communicate over secure channels. This prevents eavesdropping attacks.

### 5.4 Stolen Device

IoMT device to Medical server authentication requires the user to provide credentials and the user's unique identifier. These are not stored in the IoMT device. This prevents an attacker who has stolen the IoMT device from authenticating with Medical Server successfully.

## 6 Conclusion

This article studies the existing authentication schemes that exist for IoMT. It proposes a secure authentication scheme for IoMT devices and medical servers, based on PUF for authentication and OAuth for authorization. The PUF module-based authentication allows lightweight authentication flow between the IoMT devices and the Medical servers. The reliance on the PUF hardware module relieves the IoMT resources of computational strain. The authorization of resource access is achieved using the OAuth 2.0 protocol. The use of the OAuth 2.0 open standard allows for interoperability between various authorization vendor systems. The model offers additional security by randomizing the Bearer Token to prevent replay and impersonation attacks.

## References

1. Ferrag MA, Maglaras LA, Janicke H, Jiang J, Shu L (2017) Authentication protocols for internet of things: a comprehensive survey. Secur Commun Netw 2017:1–41. https://doi.org/10.1155/2017/6562953
2. Hardt D (2012) The OAuth 2.0 Authorization Framework. Accessed 19 Jan 2021, https://tools.ietf.org/html/rfc6749
3. Wu M, Lu TJ, Ling FY, Sun J, Du HY (2010) Research on the architecture of Internet of Things. In: 2010 3rd International Conference on Advanced Computer Theory and Engineering(ICACTE), Chengdu, China, August 2010, pp V5–484–V5–487
4. (2021) IoMT amid COVID-19 pandemic: application, architecture, technology, and security. J Netw Comput Appl 174:102886. Accessed 10 Jan 2021
5. P2413.1 - Standard for a Reference Architecture for Smart City (RASC). https://standards.ieee.org/project/2413_1.html, Accessed 19 Jan 2021
6. Adat V, Gupta BB (2018) Security in Internet of things: issues, challenges, taxonomy, and architecture. Telecommun Syst 67(3):423–441. https://doi.org/10.1007/s11235-017-0345-9

7. Makhdoom I, Abolhasan M, Lipman J, Liu RP, Ni W (2019) Anatomy of threats to the internet of things. IEEE Commun Sur Tutor 21(2):1636–1675. https://doi.org/10.1109/comst.2018.2874978
8. Burhan M, Rehman R, Khan B, Kim B-S (2018) IoT elements, layered architectures and security issues: a comprehensive survey. Sensors 18(9):2796. https://doi.org/10.3390/s18092796
9. Koutras D, Stergiopoulos G, Dasaklis T, Kotzanikolaou P, Glynos D, Douligeris C (2020) Security in IoMT communications: a survey. Sensors 20(17):4828. https://doi.org/10.3390/s20174828
10. Somasundaram R, Thirugnanam M (2020) Review of security challenges in healthcare internet of things. Wirel Netw. https://doi.org/10.1007/s11276-020-02340-0
11. Hassija V, Chamola V, Bajpai BC, Naren SZ (2021) Security issues in implantable medical devices: FACT or fiction? Sustain Cities Soc 66:102552. https://doi.org/10.1016/j.scs.2020.102552
12. El-hajj M, Fadlallah A, Chamoun M, Serhrouchni A (2019) A survey of internet of things (IoT) authentication schemes. Sensors 19(5):1141. https://doi.org/10.3390/s19051141
13. Roy KS, Kalita HK (2017) A survey on authentication schemes in IoT.In: 2017 International Conference on Information Technology (ICIT). https://doi.org/10.1109/icit.2017.56
14. Babaei A, Schiele G (2019) Physical unclonable functions in the internet of things: state of the art and open challenges. Sensors 19(14):3208. https://doi.org/10.3390/s19143208
15. Hardt D (2012) The OAuth 2.0 Authorization Framework. Accessed 19 Jan 2021, https://tools.ietf.org/html/rfc6749#section-1.2
16. Mukhopadhyay D (2016) PUFs as promising tools for security in internet of things. IEEE Des Test 33(3):103–115. https://doi.org/10.1109/mdat.2016.2544845
17. Wallrabenstein JR (2016) Practical and secure IoT device authentication using physical unclonable functions. In: 2016 IEEE 4th International Conference on Future Internet of Things and Cloud (FiCloud). https://doi.org/10.1109/ficloud.2016.22
18. Barbareschi M, Bagnasco P, Mazzeo A (2015) Authenticating IoT devices with physically unclonable functions models. In: 2015 10th International Conference on P2P, Parallel, Grid, Cloud and Internet Computing (3PGCIC). https://doi.org/10.1109/3pgcic.2015.117
19. Yanambaka VP, Mohanty SP, Kougianos E, Puthal D (2019) PMsec: physical unclonable function-based robust and lightweight authentication in the internet of medical things. IEEE Trans Consum Electron 65(3):388–397. https://doi.org/10.1109/tce.2019.2926192
20. Joshi AM, Jain P, Mohanty SP (2020) Secure-iGLU: a secure device for noninvasive glucose measurement and automatic insulin delivery in IoMT framework. In: 2020 IEEE Computer Society Annual Symposium on VLSI (ISVLSI). https://doi.org/10.1109/isvlsi49217.2020.00-17
21. Zhao M, Yao X, Liu H, Ning H (2016) Physical unclonable function based authentication protocol for unit IoT and ubiquitous IoT. In: 2016 International Conference on Identification, Information and Knowledge in the Internet of Things (IIKI). https://doi.org/10.1109/iiki.2016.85
22. Aman MN, Chua KC, Sikdar B (2017) Mutual authentication in IoT systems using physical unclonable functions. IEEE Internet Things J 4(5):1327–1340. https://doi.org/10.1109/jiot.2017.2703088
23. Khan J, et al (2018) An authentication technique based on Oauth 2.0 protocol for internet of things (IoT) network. In: 2018 15th International Computer Conference on Wavelet Active Media Technology and Information Processing (ICCWAMTIP). https://doi.org/10.1109/iccwamtip.2018.8632587
24. Solapurkar P (2016) Building secure healthcare services using OAuth 2.0 and JSON web token in IOT cloud scenario. In: 2016 2nd International Conference on Contemporary Computing and Informatics (IC3I). https://doi.org/10.1109/ic3i.2016.7917942

25. Sucasas V et al (2018) A privacy-enhanced OAuth 2.0 based protocol for smart city mobile applications. Comput Secur 74:258–274. https://doi.org/10.1016/j.cose.2018.01.014
26. Cirani S, Picone M, Gonizzi P, Veltri L, Ferrari G (2015) IoT-OAS: an OAuth-based authorization service architecture for secure services in IoT scenarios. IEEE Sens J 15(2):1224–1234. https://doi.org/10.1109/jsen.2014.2361406
27. Jung SW, Jung S (2017) Personal OAuth authorization server and push OAuth for Internet of Things. Int J Distrib Sens Netw 13(6):155014771771262. https://doi.org/10.1177/1550147717712627
28. Oh S-R, Kim Y-G (2020) AFaaS: authorization framework as a service for Internet of Things based on interoperable OAuth. Int J Distrib Sens Netw 16(2):155014772090638. https://doi.org/10.1177/1550147720906388
29. Yang F, Manoharan S (2013) A security analysis of the OAuth protocol. In: 2013 IEEE Pacific Rim Conference on Communications, Computers and Signal Processing (PACRIM). https://doi.org/10.1109/pacrim.2013.6625487
30. Tysowski P (2016) OAuth standard for user authorization of cloud services. In: Encyclopedia of Cloud Computing, pp 406–416. https://doi.org/10.1002/9781118821930.ch34
31. Fett D, Kuesters R, Schmitz G (2016) A Comprehensive Formal Security Analysis of OAuth 2.0
32. Jones M, Bradley J, Sakimura N (2016) OAuth 2.0 Mix-Up Mitigation – draft-ietf-oauth-mix-up-mitigation-01. IETF. https://tools.ietf.org/html/draft-ietf-oauth-mix-up-mitigation-01

# Sensitivity Analysis of a Multilayer Perceptron Network for Cervical Cancer Risk Classification

Emmanuella A. W. Budu(✉), V. Lakshmi Narasimhan, and Zablon A. Mbero

Department of Computer Science, University of Botswana, Gaborone, Botswana

**Abstract.** In recent times, deep learning and machine learning algorithms are being applied to aid in the diagnosis of cervical cancer to facilitate early diagnosis and reduce mortality rates. This paper concerns sensitivity analysis on an existing cervical cancer risk classification algorithm with respect to the number of epochs, number of neurons in the input layer (NIN), and number of neurons in the hidden layer (NNIHL). Sensitivity analysis is used to analyse the performance of the cervical cancer classification algorithm based on a Multilayer Perceptron Network when changes are made to the setup or architecture of the algorithm. Experimental results reveal that that the algorithm yields a high accuracy when it is trained at 500 epochs with eight input neurons and 100 or 300 or 500 neurons in the hidden layer. We also analyse the execution time of the algorithm under the varied parameters and discover that higher values for NIN, NNIHL and number of epochs all yield longer execution times. These results can aid in the successful application of deep learning for cervical cancer risk prediction.

**Keywords:** Cervical cancer · Multilayer perceptron neural network · Sensitivity analysis

## 1 Introduction

The advancement of technology has introduced the use of deep learning and machine learning in the detection of diseases such as cervical cancer. This is accomplished through the application of these algorithms on datasets obtained from patients' medical history. However, this approach has been met with some difficulties regarding the successful implementation of deep learning and machine learning based prediction and diagnosis systems in clinical settings. A major challenge to date, is that the successful application of deep learning requires years of practice to effectively select optimal regularization, hyper-parameters, and network architecture [1]. Investigating how deep learning can be optimised for greater precision in cervical cancer risk classification, can facilitate early diagnosis, and reduce mortality rates.

In this work, we attempt to conduct sensitivity analysis on an existing deep learning based cervical cancer risk classification algorithm by varying number of epochs, number of neurons in the hidden layer and number of neurons in the input layer.

The rest of the paper is organised as follows, Sect. 2 reviews the related literature, Sect. 3 discusses the methodology, Sect. 4 discusses the results of the experiments and Sect. 5 details the conclusion and provides pointers for future work in this arena.


S. Shukla et al. (eds.), *Data Science and Security*, Lecture Notes in Networks and Systems 290,
https://doi.org/10.1007/978-981-16-4486-3_8

## 2 Related Work

Various algorithms have been proposed by researchers for detecting cervical cancer in clinical settings, for example Alyafeai and Ghouti [2] proposed a deep learning framework, consisting of pre-trained deep learning algorithms to detect cervical cancer using images from the cervix known as cervigrams. Their approach yielded good results, outperforming existing classifiers in terms of accuracy and speed.

In [3], data mining techniques were employed using several machine learning algorithms to analyse cervical cancer risk factors, and predict cervical cancer. Algorithms such as Random Forest, Naïve Bayes, and Simple Logistic Regression were applied to a cervical cancer dataset, yielding accuracies as high as 96.40%.

Rayavarapu and Krishna [4] combined a Voting classifier with a Deep Neural Network to classify cervical cancer from medical history. The Voting Classifier used combined the predictions of several supervised learning algorithms to give labels to new and unseen data. The neural network was trained with ten input neurons, twenty-two hidden layer neurons and a single output layer neuron. The Relu and tanh activation functions were used to activate the hidden layer, with batch size and epoch sizes confined to 10 and 500, respectively. The DNN model achieved a high accuracy of 95% on the Biopsy target variable.

In [5] and [6], algorithms were proposed for cervical cancer prediction based on a combination of other machine learning algorithms to reduce bias and improve performance. The models used were successful in predicting cervical cancer, obtaining accuracies above 85%.

In [7], a pathological model based on the Multilayer Perceptron (MLP) algorithm was used to detect cervical cancer and analyse risk factors. The MLP architecture composed of the input layer, three hidden layers with 100 neurons each and the output layer. The model successfully classified the data with high accuracies ranging between 92-97% across the four target variables.

The literatures reviewed reveal that authors did not conduct extensive analysis on the parameter values under which their proposed algorithms yield a maximum performance. In addition, Artificial Neural Networks were not utilized extensively in the proposed algorithms for cervical cancer risk prediction. The aim of this paper is to conduct analysis on the network parameters under which an MLPN yields the highest performance for cervical cancer risk classification. Thus, yielding insight on how best to structure an MLPN network for cervical cancer classification.

## 3 Methodology

### 3.1 Dataset

The dataset used for this study is obtained from the University of California Irvine machine learning repository. It contains medical records of patients who visited the gynecology facility at Hospital Universitario de Caracas in Venezuela between 2012 and 2013 [8]. There are 858 records in total, with a total of 36 features, 32 independent variables and 4 target variables. The target variables indicate the medical test conducted, these are Biopsy, Citology, Schiller and Hinselmann, together they present a form of examination for cervical cancer [9]. The target variables contain values 0 or 1, where 0 indicates no risk of cervical cancer and 1 indicates a high risk of cervical cancer.

### 3.2 Sensitivity Analysis

Sensitivity analyses are experiments used to determine how sensitive a model is to changes in the values of its parameters. It is used to assess the performance of models, specifically to find the relation between the model's inputs and the output value [10]. The rationale behind using sensitivity analysis in this study is based on the hypothesis that modifying the structure of a deep learning algorithm yields an optimal structure that will maximize performance.

In this experiment sensitivity analysis is used to analyse the performance of the cervical cancer classification algorithms when changes are made to the setup or architecture of the algorithm. Possible parameters for conducting sensitivity analysis include activation functions, number of neurons in the input layer, number of hidden layers, number of epochs and number of neurons in the hidden layer. For this study, we will analyse the algorithm by altering the number of epochs in the ANN architectures, the number of neurons in the input layer (NIN), and the number of neurons in the hidden layer (NNIHL) only.

### 3.3 Algorithm

Algorithm MLPN illustrated below as given by [7] and used in this study predicts the risk of a patient developing cervical cancer using the Multilayer Perceptron Neural Network.

**Algorithm MLPN**

**1.** Pre-process the data by imputing missing values with the most frequent value
**2.** Set the target variable as Biopsy, Schiller, Citology and Hinselmann
**3.** Select *Age, Number of Sexual Partners, Hormonal Contraceptives, First Sexual Intercourse* and *Num. of Pregnancies* as the feature set class
**4.** Divide the data set into training and testing set with the ratio 0.75:0.25
**5.** Feed the data into a Multilayer Perceptron network with:
i) 3 hidden layers each with 100 neurons
ii) maximum number of iterations set at 500
iii) optimizer solver as SDG
iv) initial learning rate of 0.01.
**6.** Obtain the importance index of each attribute in the data using a random forest algorithm on the data
**7.** Calculate and analyse the detection accuracy of the model

# 4 Results and Discussions

In this study, we implemented algorithm MLPN_1 based on algorithm MLPN. We include PCA as a feature selection step to reduce the number of input features, an addition to the sequence of the original algorithm. NIN is varied from 8, 16, to 32. In addition, the number of epochs is also varied from 500, 700, 900, 1100, and 1300. Further, the NNIHL is varied from 100, 200, 300, 400, and to 500. The algorithm is executed four times with respect to each of the four target variables found in the dataset. The experimental results are illustrated in the subsequent sections.

## 4.1 Number of Inputs and Accuracy

The results shown in Tables 1, 2, 3 and 4 and illustrated in Fig. 1 show that the accuracy decreases as NIN is increased across all the target variables, indicating that the number of neurons in the input layer has an effect on the accuracy of the algorithm.

In Fig. 2, the execution time of the algorithm for all four target variables increases as NIN is increased. This is because a large number of neurons in the input layer will require a longer processing time.

**Table 1** Experiment one: MLPN_1 on the Biopsy target variable

| NIN | Accuracy (%) | Execution time (s) |
|---|---|---|
| 8 | 95.34 | 1.343 |
| 16 | 92.09 | 19.538 |
| 32 | 93.02 | 21.448 |

**Table 2** Experiment one: MLPN_1 on the Citology target variable

| NIN | Accuracy (%) | Execution time (s) |
|---|---|---|
| 8 | 97.2 | 1.806 |
| 16 | 95.81 | 19.366 |
| 32 | 94.88 | 19.037 |

**Table 3** Experiment one: MLPN_1 on the Schiller target variable

| Number of inputs | Accuracy (%) | Execution time (s) |
|---|---|---|
| 8 | 93.48 | 1.428 |
| 16 | 93.95 | 18.4 |
| 32 | 91.16 | 19.759 |

**Table 4** Experiment one: MLPN_1 on the Hinselmann target variable

| Number of inputs | Accuracy (%) | Execution time (s) |
|---|---|---|
| 8 | 97.2 | 1.6 |
| 16 | 95.81 | 20.623 |
| 32 | 93.95 | 19.892 |

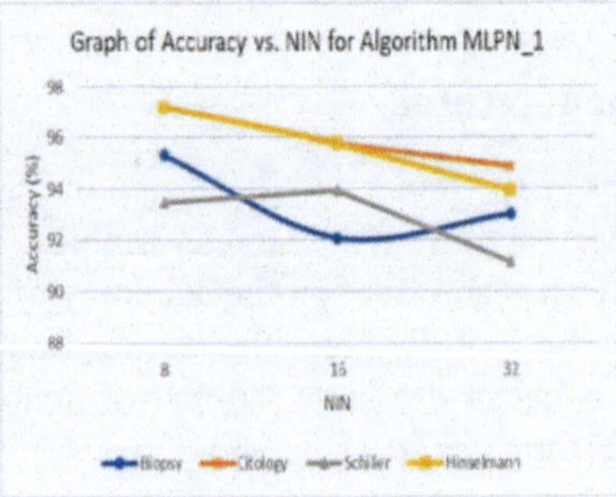


**Fig. 1** Accuracy with NIN values

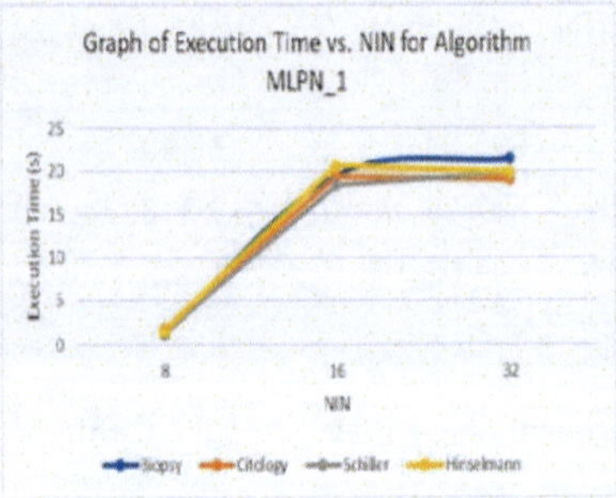


**Fig. 2** Execution time vs. NIN

## 4.2 Number of Epochs and Accuracy

Based on the NIN that yields the highest accuracies, we tested the revised algorithm with a varied number of epochs. As seen in Tables 5, 6, 7 and 8, and illustrated in Fig. 3, the accuracy decreases as the number of epochs are increased. This indicates that a high number of epochs may not always yield higher accuracies.

In Fig. 4, the execution time for all target variables also shows an increase as the number of epochs are increased, this is with the exception of the Schiller variable. The execution time of the Schiller variable first decreases then increases as the number of epochs are increased. A high number of epochs would require more time to run all iterations needed to make classifications on the dataset.

**Table 5** Experiment two: MLPN_1 on the Biopsy target variable

| Number of epochs | Accuracy (%) | Execution time (s) |
|---|---|---|
| 500 | 96.12 | 1.445 |
| 700 | 93.02 | 26.417 |
| 900 | 93.02 | 34.351 |
| 1100 | 91.86 | 41.023 |
| 1300 | 93.8 | 49.09 |

**Table 6** Experiment two: MLPN_1 on the Citology target variable

| Number of epochs | Accuracy (%) | Execution time (s) |
|---|---|---|
| 500 | 95.34 | 1.441 |
| 700 | 93.8 | 31.792 |
| 900 | 94.57 | 42.696 |
| 1100 | 94.57 | 51.103 |
| 1300 | 95.35 | 62.425 |

**Table 7** Experiment two: MLPN_1 on the Schiller target variable

| Number of epochs | Accuracy (%) | Execution time (s) |
|---|---|---|
| 500 | 94.19 | 43.839 |
| 700 | 92.25 | 27.523 |
| 900 | 92.25 | 31.752 |
| 1100 | 91.47 | 40.957 |
| 1300 | 90.7 | 51.767 |

**Table 8** Experiment two: MLPN_1 on the Hinselmann target variable

| Number of epochs | Accuracy (%) | Execution time (s) |
|---|---|---|
| 500 | 97.29 | 1.448 |
| 700 | 94.96 | 27.737 |
| 900 | 95.74 | 37.311 |
| 1100 | 96.12 | 43.771 |
| 1300 | 94.57 | 52.433 |

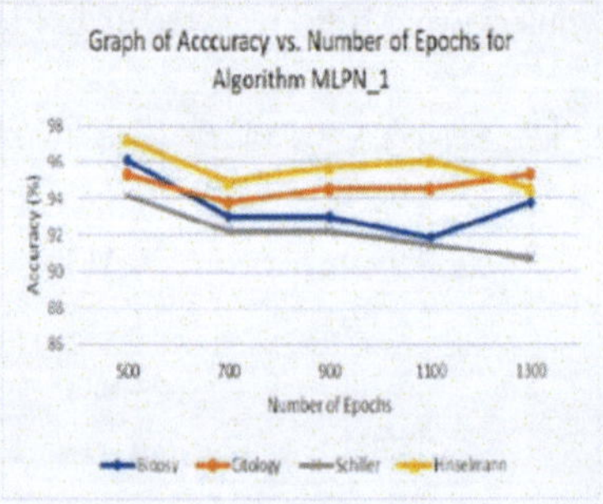


**Fig. 3** Accuracy and number of epochs

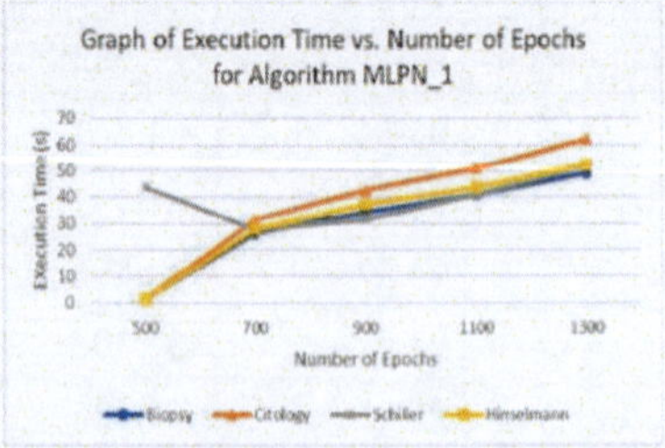


**Fig. 4** Execution time and number of epochs

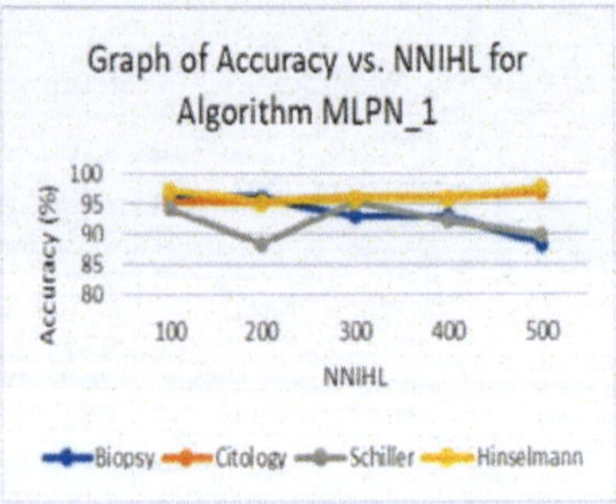


**Fig. 5** Accuracy and NNIHL

### 4.3 NNIHL and Accuracy

In the third experiment, we varied the NNIHL using the values from the previous experiments that yield the highest accuracy. Illustrations in Fig. 5 show that an increase in NNIHL does not always guarantee an increase in accuracy across the four target variables. For the Schiller and Biopsy target variables, the accuracy decreases as the NNIHL value is increased, but high accuracies are recorded at NNIHL values of 300 and 100 respectively.

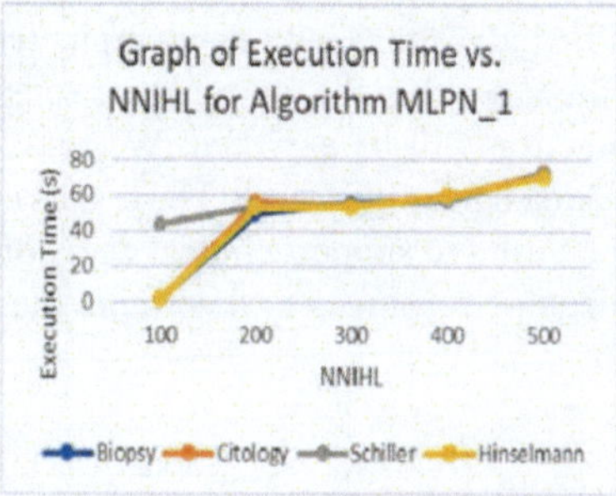


**Fig. 6** Execution time and NNIHL

**Table 9** Comparison table

| Target variable | Highest accuracy from MLPN (%) | Highest accuracy from MLPN_1 (%) |
|---|---|---|
| Biopsy | 96.2 | 96.12 |
| Citology | 95.8 | 96.89 |
| Schiller | 94.4 | 95.34 |
| Hinselmann | 97.2 | 97.67 |

In Fig. 6, the execution time increases significantly as the number of neurons in the hidden layer is increased, this is because the network size increases as more neurons are added, thus leading to a longer processing time.

### 4.4 Performance Comparison

We examined the results obtained from our experiments on algorithm MLPN_1 and the results obtained from the original author's execution of the algorithm. Table 9 gives a comparison of these results.

From the table, it is evident that version MLPN_1 of the algorithm with new network parameters performs just as well as the original version of the algorithm in determining the risk of cervical cancer. In three of the four target variables, algorithm MLPN_1 yields a higher accuracy than the original version of the algorithm.

## 5 Conclusions

The successful application of deep learning algorithms to determine the risk of cervical cancer in patients can assist health professionals in early diagnosis and expediate treatment for patients. In this study, sensitivity analysis was applied to an existing cervical cancer risk classification algorithm to determine the parameters under which the algorithm yields maximum accuracy. Results reveal that a Multilayer Perceptron Network with eight neurons in the input layer, 500 epochs and 100, 300 or 500 neurons in the hidden layer will yield high accuracies. When compared to the original version of the algorithm, the modified algorithm with new parameters yields similar accuracy as the original version of the algorithm. In future, other sensitivity parameters such as the choice of activation function will be explored. We also aim to explore alternative evaluation metrics such as Area Under the Curve. The results confirm that the techniques implemented can possibly be used to support clinical decision support systems that can be used in areas with inadequate resources to determine the risk of cervical cancer in patients.

## References


1. Gandomi A, Haider M (2015) Beyond the hype: big data concepts, methods, and analytics. Int J Inf Manag 35(2):137–144. https://doi.org/10.1016/j.ijinfomgt.2014.10.007
2. Alyafeai Z, Ghouti L (2020) A fully-automated deep learning pipeline for cervical cancer classification. Expert Syst Appl 141:112951. https://doi.org/10.1016/j.eswa.2019.112951
3. Razali N, Mostafa SA, Mustapha A, Wahab MHA, Ibrahim NA (2020) Risk factors of cervical cancer using classification in data mining. J Phys: Conf Ser 1529:022102. https://doi.org/10.1088/1742-6596/1529/2/022102
4. Rayavarapu K, Krishna KKV (2018) Prediction of cervical cancer using voting and DNN classifiers. In: 2018 International Conference on Current Trends towards Converging Technologies (ICCTCT), Coimbatore, pp 1–5. https://doi.org/10.1109/ICCTCT.2018.8551176
5. Nithya B, Ilango V (2020) Machine learning aided fused feature selection based classification framework for diagnosing cervical cancer. In: 2020 Fourth International Conference on Computing Methodologies and Communication (ICCMC), Erode, India, pp 61–66. https://doi.org/10.1109/ICCMC48092.2020.ICCMC-00011
6. Ahishakiye E, Wario R, Mwangi W, Taremwa D (2020) Prediction of cervical cancer basing on risk factors using ensemble learning. In: IST-Africa 2020 Conference Proceedings, p 13
7. Yang W, Gou X, Xu T, Yi X, Jiang M (2019) Cervical cancer risk prediction model and analysis of risk factors based on machine learning. In: Proceedings of the 2019 11th International Conference on Bioinformatics and Biomedical Technology - ICBBT'19, Stockholm, Sweden, pp 50–54. https://doi.org/10.1145/3340074.3340078
8. Fernandes K, Chicco D, Cardoso JS, Fernandes J (2018) Supervised deep learning embeddings for the prediction of cervical cancer diagnosis. PeerJ Comput Sci 4:1–20. https://doi.org/10.7717/peerj-cs.154
9. Geetha R, Sivasubramanian S, Kaliappan M, Vimal S, Annamalai S (2019) Cervical cancer identification with synthetic minority oversampling technique and PCA analysis using random forest classifier. J Med Syst 43(9):286. https://doi.org/10.1007/s10916-019-1402-6
10. Nourani V, Fard MS (2012) Sensitivity analysis of the artificial neural network outputs in simulation of the evaporation process at different climatologic regimes. Adv Eng Softw 47(1):127–146. https://doi.org/10.1016/j.advengsoft.2011.12.014

# Data Encryption and Decryption Techniques Using Line Graphs

Sanjana Theresa(✉) and Joseph Varghese Kureethara

Christ University, Bangalore, India
sanjana.joseph@res.christuniversity.in, frjoseph@christuniversity.in

**Abstract.** Secure data transfer has become a critical aspect of research in cryptography. Highly effective encryption techniques can be designed using graphs in order to ensure secure transmission of data. The proposed algorithm in this paper uses line graphs along with adjacency matrix and matrix properties to encrypt and decrypt data securely in order to arrive at a ciphertext using a shared-key.

**Keywords:** Plaintext · Ciphertext · Line graph · Adjacency matrix · Encryption · Decryption · Cryptography

## 1 Introduction

In this paper, a graph G(V,E) is under consideration, where V is the set of vertices and E is a set of edges. A walk originating from one vertex to another without repeated vertices is a path. A cycle is obtained when the path starts and ends with the same vertex. A cycle graph is one in which the cycle obtained comprises all vertices in the graph. In this study, weighted graphs are also considered in which corresponding weights are assigned to each edge in graph G. The adjacency matrix of a weighted graph can be used to store the weights of the edges. The line graph L(G) of G is the graph whose vertices correspond to the edges of G and two vertices of L(G) are adjacent if and only if the corresponding edges of G are adjacent.

Cryptography involves the study of methodologies to securely transfer data by converting it to an unreadable format called ciphertext. Encryption is the process of converting plaintext to ciphertext while decryption is the process of converting ciphertext back to original plaintext. Cryptography algorithms are majorly of two types: Symmetric-key cryptography (where the sender and receiver shares the same key to encrypt and decrypt data) and publickey or Asymmetric cryptography (where the sender and receiver uses two different keys to encrypt and decrypt data known as public key and private key respectively).

In this study, an algorithm using line graphs along with adjacency matrix and matrix properties is proposed to encrypt and decrypt data followed by an illustrative example using plaintext, 'Crypto'.


S. Shukla et al. (eds.), *Data Science and Security*, Lecture Notes in Networks and Systems 290,
https://doi.org/10.1007/978-981-16-4486-3_9

## 2 Related Work

Hemminger and Beineke [3] did an excellent study on line graphs and line digraphs. In 2012, Yamuna, Gogia, Sikka, and Khan [5] proposed an encryption algorithm where nodes are in a Hamiltonian path and uses adjacency matrix as an additional parameter. In 2014, Etaiwi [2] proposed an encryption algorithm using the concepts of cycle graph, complete graph and minimum spanning tree. In 2015, Yamuna and Karthika [6] used bipartite graphs for secure data transfer. In 2019, Vinothkumar and Balaji [4] used complete graph and matrix properties to decrypt the received message. In the same year, Deepa, Maheswari, and Balaji [1] used the concept of graph labeling and non-singular square matrices for secure data transfer.

## 3 Proposed Algorithm

The first step in this algorithm is to represent data as vertices in a graph. Each character will correspond to a vertex while all adjacent characters in the plaintext will be represented as adjacent vertices in the graph. Keep adding vertices to form a path graph G and then identify the nearest prime number (P) greater than or equal to the values listed in the encryption chart (E). Each vertex is named using the above mentioned prime numbers (P). e.g., the character 'A' corresponds to the value 1 in the encryption chart and the nearest prime number greater than or equal to 1 is 2. Similarly, for 'B' it is 2, 'C' it is 3, 'D' it is 5, and so on (Table 1).

### 3.1 Encryption Algorithm

Assign weights to each edge by taking the difference of the nodal values. Compute the adjacency matrix A of the graph G. Replace the principal diagonal elements of matrix A with the first nodal value of the graph G. After that, multiply A by a predefined Shared-Key K to form M. Complete the graph G to form a cycle and ensure that the first node in the path is adjacent to all other nodes of G, say the graph N.

Add a path $P_2$ to the node that corresponds to the second character in graph N. Calculate the difference (P - E) between the value corresponding to the character in the encryption chart (E) and the nodal values which are primes for each node (P). Pendant edges for the corresponding nodes in graph N are added according to the values (P - E) calculated above except for the node that corresponds to the second character. For this node, add the corresponding pendant edges at the end node of path $P_2$. Finally, construct the line graph L of the new graph N. The ciphertext contains the graph L and Matrix M line-by-line in a linear format.

**Table 1** Encryption chart (Italicised values indicate prime numbers)

| Number | Character | Number | Character | Number | Character |
|---|---|---|---|---|---|
| 0 | SPACE | 31 | e | 62 | ] |
| 1 | A | 32 | f | 63 | ^ |
| *2* | B | 33 | g | 64 | _ |
| *3* | C | 34 | h | 65 | ‘ |
| 4 | D | 35 | i | 66 | { |
| *5* | E | 36 | j | *67* | \| |
| 6 | F | *37* | k | 68 | ~ |
| *7* | G | 38 | l | 69 | ! |
| 8 | H | 39 | m | 70 | “ |
| 9 | I | 40 | n | *71* | # |
| 10 | J | *41* | o | 72 | $ |
| *11* | K | 42 | p | *73* | % |
| 12 | L | *43* | q | 74 | & |
| *13* | M | 44 | r | 75 | ( |
| 14 | N | 45 | s | 76 | ) |
| 15 | O | 46 | t | 77 | * |
| 16 | P | *47* | u | 78 | + |
| *17* | Q | 48 | v | *79* | , |
| 18 | R | 49 | w | 80 | - |
| *19* | S | 50 | x | 81 | . |
| 20 | T | 51 | y | 82 | / |
| 21 | U | 52 | z | *83* | 0 |
| 22 | V | *53* | : | 84 | 1 |
| *23* | W | 54 | ; | 85 | 2 |
| 24 | X | 55 | < | 86 | 3 |
| 25 | Y | 56 | = | 87 | 4 |
| 26 | Z | 57 | > | 88 | 5 |
| 27 | a | 58 | ? | *89* | 6 |
| 28 | b | *59* | @ | 90 | 7 |
| *29* | c | 60 | [ | 91 | 8 |
| 30 | d | *61* | \ | 92 | 9 |

### 3.2 Decryption Algorithm

Compute $B = MK^{-1}$and from the adjacency matrix B, construct a weighted graph P (regardless of the principal diagonal) and name the nodes as a, b, c and so on. In the graph P, the value of the first node a is taken to be the value of the principal diagonal elements. Compute all the nodal values (NV) by adding the corresponding weights to the previous nodal value (e.g., Node b = node a + ab). Find the inverse of the line graph L, say Q, and name the nodes as A, B, C and so on. In the graph Q, identify a cycle which contains a node adjacent to all other nodes. This particular node is considered to be the first node.

The second node is taken to be the node in the cycle graph where the corresponding pendant edges are at the end node of path $P_2$. Suppose this node is towards the left (or right) of the identified first node, the nodes are taken from the left (or right) until the first node in order is reached. Compute the number of pendant edges (PE) in the above order. Finally, subtract the corresponding number of pendant edges (PE) from the nodal values (NV) to obtain the original plaintext by using the encryption chart.

## 4 Encryption and Decryption of the Plaintext - 'Crypto'

As an illustrative example of the above proposed algorithm, the encryption and decryption process of plaintext 'Crypto' is explained in this section (Table 2).

### 4.1 Encryption

**Table 2** Original plaintext for encryption

| Plaintext | Encryption chart values | Primes |
|---|---|---|
| C | 3 | 3 |
| r | 44 | 47 |
| y | 51 | 53 |
| p | 42 | 43 |
| t | 46 | 47 |
| o | 41 | 41 |

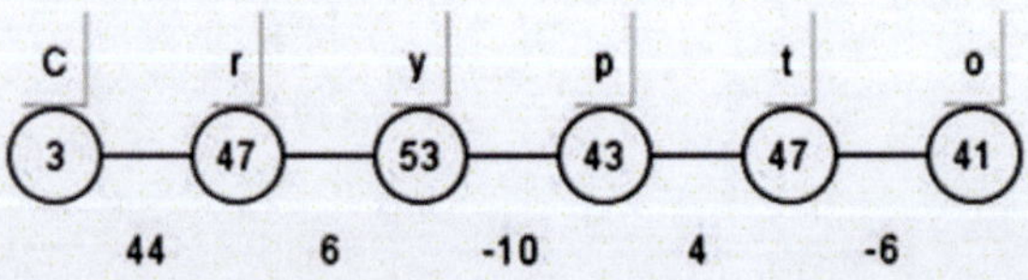


**Fig. 1** Each node represents the primes (P) with corresponding edge weights. This shows a figure consisting of Graph G

$$\text{Edge weight Cr} = \text{Node r - Node C} \tag{1}$$

$$\text{Adjacency matrix, A} = \begin{bmatrix} 0 & 44 & 0 & 0 & 0 & 0 \\ 44 & 0 & 6 & 0 & 0 & 0 \\ 0 & 6 & 0 & -10 & 0 & 0 \\ 0 & 0 & -10 & 0 & 4 & 0 \\ 0 & 0 & 0 & 4 & 0 & -6 \\ 0 & 0 & 0 & 0 & -6 & 0 \end{bmatrix}$$

The prime value is stored corresponding to the first character in the plaintext C which is 3 in the diagonal instead of 0's (Fig. 1).

$$\text{Modified A} = \begin{bmatrix} 3 & 44 & 0 & 0 & 0 & 0 \\ 44 & 3 & 6 & 0 & 0 & 0 \\ 0 & 6 & 3 & -10 & 0 & 0 \\ 0 & 0 & -10 & 3 & 4 & 0 \\ 0 & 0 & 0 & 4 & 3 & -6 \\ 0 & 0 & 0 & 0 & -6 & 3 \end{bmatrix} \text{ and Pre-defined key, K} = \begin{bmatrix} 1&1&1&1&1&1 \\ 0&1&1&1&1&1 \\ 0&0&1&1&1&1 \\ 0&0&0&1&1&1 \\ 0&0&0&0&1&1 \\ 0&0&0&0&0&1 \end{bmatrix}.$$

Multiply Modified A by K to get the matrix M. Therefore,

$$M = A \times K = \begin{bmatrix} 3 & 47 & 47 & 47 & 47 & 47 \\ 44 & 47 & 53 & 53 & 53 & 53 \\ 0 & 6 & 9 & -1 & -1 & -1 \\ 0 & 0 & -10 & -7 & -3 & -3 \\ 0 & 0 & 0 & 4 & 7 & 1 \\ 0 & 0 & 0 & 0 & -6 & -3 \end{bmatrix}$$

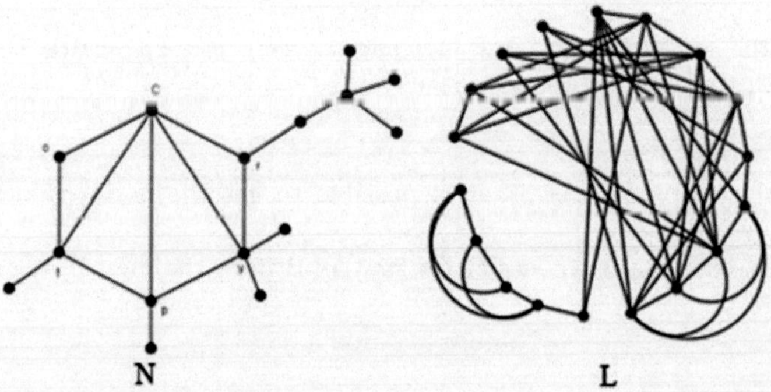


**Fig. 2** Number of pendant edges for each node = P-E. This shows a figure consisting of the new graph N and a line graph L of N

The data is encrypted as L+M, i.e., the graph L (Figs. 2) and the matrix M in row format is sent to the receiver.

Ciphertext: L, 3 47 47 47 47 47 44 47 53 53 53 53 0 6 9 −1 −1 −1 0 0 −10 −7 −3 −3 0 0 0 4 7 1 0 0 0 0 −6 −3

## 4.2 Decryption

Decryption is done using the matrix operations as follows.

$B = M \times K^{-1}$ where

$$K^{-1} = \begin{bmatrix} 1 & -1 & 0 & 0 & 0 & 0 \\ 0 & 1 & -1 & 0 & 0 & 0 \\ 0 & 0 & 1 & -1 & 0 & 0 \\ 0 & 0 & 0 & 1 & -1 & 0 \\ 0 & 0 & 0 & 0 & 1 & -1 \\ 0 & 0 & 0 & 0 & 0 & 1 \end{bmatrix}$$

$$B = \begin{bmatrix} 3 & 47 & 47 & 47 & 47 & 47 \\ 44 & 47 & 53 & 53 & 53 & 53 \\ 0 & 6 & 9 & -1 & -1 & -1 \\ 0 & 0 & -10 & -7 & -3 & -3 \\ 0 & 0 & 0 & 4 & 7 & 1 \\ 0 & 0 & 0 & 0 & -6 & -3 \end{bmatrix} \times \begin{bmatrix} 1 & -1 & 0 & 0 & 0 & 0 \\ 0 & 1 & -1 & 0 & 0 & 0 \\ 0 & 0 & 1 & -1 & 0 & 0 \\ 0 & 0 & 0 & 1 & -1 & 0 \\ 0 & 0 & 0 & 0 & 1 & -1 \\ 0 & 0 & 0 & 0 & 0 & 1 \end{bmatrix} = \begin{bmatrix} 3 & 44 & 0 & 0 & 0 & 0 \\ 44 & 3 & 6 & 0 & 0 & 0 \\ 0 & 6 & 3 & -10 & 0 & 0 \\ 0 & 0 & -10 & 3 & 4 & 0 \\ 0 & 0 & 0 & 4 & 3 & -6 \\ 0 & 0 & 0 & 0 & -6 & 3 \end{bmatrix}$$

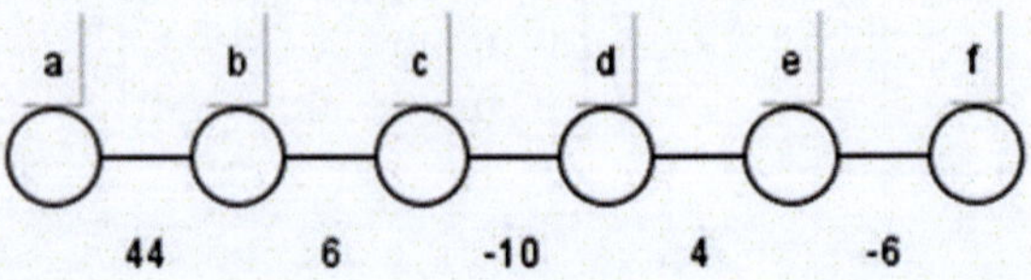


**Fig. 3** Let the nodes of the graph P be named as *a,b,c,* etc. The matrix B (regardless of the principal diagonal) represents the above graph P.

From the graph P (Fig. 3), it is possible to get the nodal values (NV) as follows:

Since the diagonal elements of the matrix B have the digit 3, take the value of node a to be 3.

$$\text{Node b} = \text{Node a} + 44 \tag{2}$$

$$\text{Node c} = \text{Node b} + 66 \tag{3}$$

Thus, the nodal values (NV) are as follows (Table 3):

From the graph Q (Fig. 4), ABCDEFA is a cycle where the node A is adjacent to all the vertices of the above mentioned cycle. Therefore, node A is considered to be the first node. For the node that corresponds to the second character, we take the node in the cycle ABCDEFA where the corresponding pendant edges are at the end node of path $P_2$. In this case, this would be node F and hence we

**Table 3** Nodal values

| Nodes | Values |
|---|---|
| a | 3 |
| b | 47 |
| c | 53 |
| d | 43 |
| e | 47 |
| f | 41 |

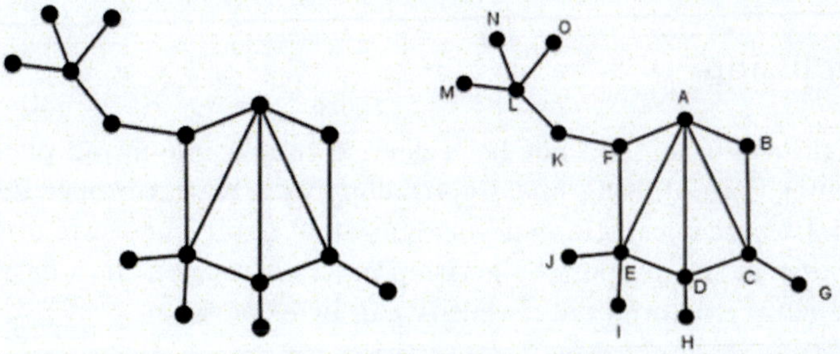


**Fig. 4** Let the nodes of graph Q be named as A, B, C etc. The inverse of line graph L is the above graph Q.

can conclude to choose the nodes from the left of the above cycle. Finally, the order of nodes to be considered is as follows:

Node A, Node F, Node E, Node D, Node C, Node B

Corresponding nodes have the following number of pendant edges (PE) (Table 4). Note that for Node F, the number of pendant edges of node L is considered since it does not have direct pendant edges.

**Table 4** Number of pendant edges (PE)

| Nodes | Number of pendant edges (PE) |
|---|---|
| A | 0 |
| F | 3 |
| E | 2 |
| D | 1 |
| C | 1 |
| B | 0 |

The original text 'Crypto' from the encryption chart is obtained as shown in the Table 5.

**Table 5** Final decrypted plaintext

| Nodal values (NV) | Number of pendant edges (PE) | NV-PE | plaintext |
|---|---|---|---|
| 3 | 0 | 3 | C |
| 47 | 3 | 44 | r |
| 53 | 2 | 51 | y |
| 43 | 1 | 42 | p |
| 47 | 1 | 46 | t |
| 41 | 0 | 41 | o |

## 5 Conclusion

In this work, a new method has been devised (applicable for all plaintext with length greater than 4) using line graphs along with matrix properties to enable efficient and secure encryption and decryption of data. The usage of line graphs and its inverse in this proposed algorithm makes it tougher for a data breach to occur and hence enhanced data security can be achieved.

## References

1. Deepa B, Maheswari V, Balaji V (2019) Creating ciphertext and decipher using graph labeling techniques. Int J Eng Adv Technol 9(1):206–212
2. Etaiwi WMA (2014) Encryption algorithm using graph theory. J Sci Res Rep 3(19):2519–2527
3. Hemminger RL, Beineke LW (1978) Line graphs and line digraphs. In: Lowell WB, Wilson JB (eds) Selected Topics in Graph Theory. Academic Press, New York, pp 271–305
4. Vinothkumar L, Balaji V (2019) Encryption and decryption technique using graph theory. Int J Res Analytical Rev 6(2):192–197
5. Yamuna M, Gogia M, Sikka A, Khan MJH (2012) Encryption using graph theory and linear algebra. Int J Comput Appl. 2(5):102–107
6. Yamuna M, Karthika K (2015) Data transfer using bipartite graphs. Int J Adv Res Sci Eng 4(2):128–131

# Aerial Image Enhanced by Using Dark Channel Prior and Retinex Algorithms Based on HSV Color Space

Hana H. Kareem[1(✉)], Rana T. Saihood[1], and Hazim G. Daway[2]

[1] Department of Physics, College of Education, Mustansiriyah University, Baghdad, Iraq
hanakr2007@uomustansiriyah.edu.iq
[2] Physics Department, Science College, Mustansiriyah University, Baghdad, Iraq
hazimdo@uomustansiriyah.edu.iq

**Abstract.** Enhancing aerial images plays an important role in many remote sensing applications; aerial images are often affected by dust, smoke and fog. In this paper, a new algorithm was proposed to improve aerial images. The proposed algorithm using Dark Channel Prior (DCP) to enhance the color compounds only in HSV space and a Multiscale Retinex (MSR) using to enhance the lightness component in the same space. To know the efficiency of enhancement aerial images, the no-reference quality measures as Natural Image Quality Evaluator (NIQE) and Wavelet Quality Evaluator (WQE) Have been calculated, which are used to see the quality of the proposed method in comparison with DCP, Histogram equalization (HE), Image Entropy and Information Fidelity (IEIF) and Multiscale Retinex with color restoration (MSRCR). Through the analysis of the results, it is found that the proposed method has succeeded in improving aerial images better than other methods.

**Keywords:** Aerial Image enhancement · Dark channel prior · HSV color space · Retinex algorithm

## 1 Introduction

Enhance and restore distorted color images an important area of digital image processing [1–3]. This area includes noise, distortion in lighting, and lack of contrast, This is often the case in aerial images, in which the lack of contrast plays an important role. The atmosphere has been interactd with light because of the forming of the atmosphere according to steady conditions, and it includes a many of particles and tiny suspended particles are called aerosols. The popular problems is aerial images degenerate by these particles. Aerosols are suspended very fine solid or liquid particles. Ithe atmosphere at so low fall velocity, it is generally between 102 and 100 μm in size. Furthermore it, dust, fog, spray, smoke, snow and rain are models of aerosols [4].

Several previous studies have included improving aerial and hazy images. One of the most traditional methods of improving lighting and contrast is histogram equalization [5], this technique is adjusting the intensity of an image to improve contrast with this


S. Shukla et al. (eds.), *Data Science and Security*, Lecture Notes in Networks and Systems 290,
https://doi.org/10.1007/978-981-16-4486-3_10

modification, the intensity can be better distributed on the graph. This allows areas with low local contrast to having higher contrast.

In [6] they suggested single enhanced for hazy images. This method used a white balance, a contrast enhancemt method and reduced artifacts by using a multistate fashion, depending on Laplacian pyramid representation, Their method improved the video clips distorted by haze. Main of the basic and important methods that are considered uncomplicated algorithms are prior dark channel prior, It automatically improves the high haze areas, and the haze-free areas are not distorted [7].

Some algorithms depending on non-local prior [8], This method assumes the colours of a haze free image are good close to by a small hundred singular colours, which form strict clusters in the RGB colours space.

The main observation is that the pixels in a certain group are often non-local, that is, they are spread over the entire plane of the image and are located at various distances from the camera. Each color cluster in the free haze image becomes a line in RGB colours space, This form can get good improvement.

Some hazy image enhancement is entropy dependent [9], in this method the atmospheric light is represented by quad-tree subdivision and the transmission is dependent on an objective function that is comprised of information quality by using weighted least squares optimization.

A common way to enhance contrast and brightness is an algorithm multi-scale retinex algorithm with color restoration [10], it can be used to improve the hazy image, this algorithm relies on the high-pass filter (Gaussian filter) With the Logarithm mapping. But image enhancement often suffers from Hula effected. However, it can be characterized and flown to improve underwater and hazy images.

## 2 Proposed Method

The suggested method is based on combining two techniques for enhancement, The first DCP technique was used to enhance areal images. The second is the Retinex technique, which is used to improve images with low lightness and contrast.

### 2.1 DCP Algorithm

The general model describing dust or fog in a scene is given in relation [7, 11]:

$$I(x) = J(x)tr(x) + Ac(1 - tr(x)) \quad (1)$$

where $I$ is the intensity of the hazy image, $J$ is the radiance in the scene, the true color that wants to $A$ being atmospheric light, and $tr$ is the transmission. One of the most common algorithms for removing blur from images is the DCP, which is based on the assumption Most of the non-sky patches are for the free haze outdoor images, which have very low intensities in at least one of the RGB components.

Thus, DCP for an arbitrary image *Ji* is defined as [7]:

$$J_{dark}(x) = min_{i\in\{r,g,b\}}(min_{y\in\Omega(x)}(Ji(y))) \tag{2}$$

where *Ji* is a color image of J and $\Omega(x)$ being a local patch has x center Depending on the DCP algorithm, the intensity of the dark channel of *Ji* is low and go to be zero where *Ji* is an outdoor haze-free image, except the bright region:

$$J_{dark}(x) \cong 0 \tag{3}$$

Thus the transmission value can be determined by [7]:

$$tr(x) = 1 - w_{miny\in\Omega(x)}\left(min_{i\in\{r,g,b\}}\frac{I^c(y)}{A}\right) \tag{4}$$

The transmission can be refined by using map soft mapping. If the haze is removed, the image appears unnatural, so the value $(0 < w < 1)$, $w$ is fixed at 0.95 and the best value of atmospheric light $A = 0.1$, with patch size $(15 \times 15)$ [7]. Then recover the scene radiance by using [12]:

$$J(x) = \frac{I(x) - A}{max(tr(x), 0.1)} + A \tag{5}$$

### 2.2 HSV Color Space

One of the color spaces that depend on the human vision system Is HSV color space, the image enhanced can be converted using DCP algorithm From basic RGB space to HSV space [13], after that we keep only the color compounds hue and saturation (H and S).

### 2.3 MSR Algorithm

An important algorithm for improving low light images is MSR, This algorithm will be applied to the lighting component only according to the HSV color space by using:

$$Rt(x, y, s) = logI(x, y) - log(F(x, y, s) * I(x, y)) \tag{6}$$

where $Rt(x, y, s)$ the enhanced of the grey image at coordinate $x, y$, is the Gaussian matrix, $I(x, y)$ is the image account for channel i and the symbol * being the convolution. $F(x, y, s)$ Gaussian surrounds function that is studied by [14].

$$F(x, y, s) = Ne^{\frac{-(x^2-y^2)}{s^2}} \tag{7}$$

$N$ is given by:

$$\iint F(x, y, s)dxdy = 1 \tag{8}$$

the final enhancement for a gray component by using SSR is [14].

$$MSR(x, y, w, s_n) = \sum_{n=1}^{N} w_n Rt(x, y, s_n) \tag{9}$$

This treating can affect the hues of the image in a way that the image hues become dark, thus, they proposed an additional processing step [14].

$$R' = MSR.I'(x, y, a) \tag{10}$$

where I' given by:

$$I'(x, y, a) = log\left[1 + a\frac{I(x, y)}{\sum_{i=1}^{3} I(x, y)}\right] \tag{11}$$

The constants $a = 125$ , $(w_1 = w_2 = w_3 = 1/3)$ and $(c_1 = 250, c_2 = 120, c_3 = 80)$ have been employed in this work [14]. Due to the shift of the distribution the value $R'$ can be corrected using:

$$R' = 0.1(R' + 0.9) \tag{12}$$

### 2.4 The Combination of Enhancement Compounds

After the chromatic compounds (He and Se) are enhancing using DCP And improve the lighting (Ve) by using MSR All three compounds are combined Then reverse transform from HSV to RGB is used to obtain the final improvement. This method of optimization made use of the strengths of DCP For color compounds and improve lighting by retinex method. All the aforementioned steps are shown in the diagram in Fig. (1).

## 3 Determine the Quality Assessment

In this study two measures of quality were used, the first is Natural Image Quality Evaluator (NIQE) [15] dependent on color component and the second is Wavelet Quality Evaluator (WQE) [16], which is work on lightness component.

The NIQE is a non-reference metric that measures the amount of contrast in color images, Depending on chromatic details. It can be calculated using a multivariate Gaussian (MVG) model and statistical features. Every part of the image after dividing it into several areas with size (32 × 32) to (160 × 160). This scale depending on the range between the MVG model and the quality-aware feature model and depending on [15]:

$$D = \sqrt{\left((R_1 - R_2)^T\left(\frac{S1 + S2}{2}\right)^{-1}(R_1 - R_2)\right)} \tag{13}$$

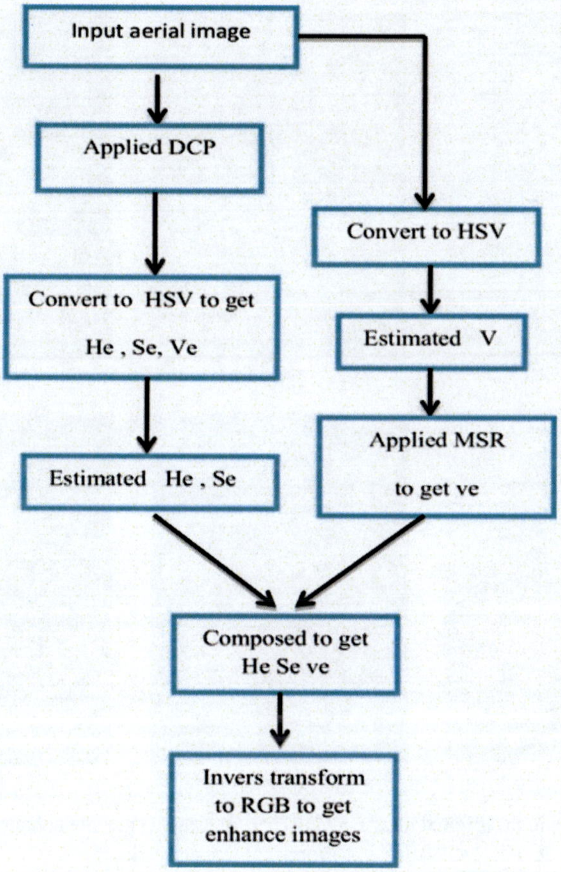


**Fig. 1** A diagram of the suggested algorithm

$R1$, $R2$ being the average vectors and $S1 + S2$ covariance matrix of the natural MVG pattern and the MVG of the distortion image, respectively. The minimum value of NIQE means good quality. The *WQE* depending on HL by using [16]:

$$WQE = 1/std(HL) \tag{14}$$

It indicates the amount of detail in the image, the more detail the higher the quality.

## 4 Result and Discussion

In this study aerial images have been enhanced, Four images were used, which are shown in Figs. (2, a), (3, a), (4, a), and (5, a) two images with size (980 × 1280) and the other two (1280 × 980) all images with JPG type. All enhancements algorithms were done by using MATLAB R2018a. The Figs. (2, 3, 4 and 5) show aerial images enhanced using the methods (proposed, MSRCR, HE, DCP, and IEIF), from a subjective evaluation,

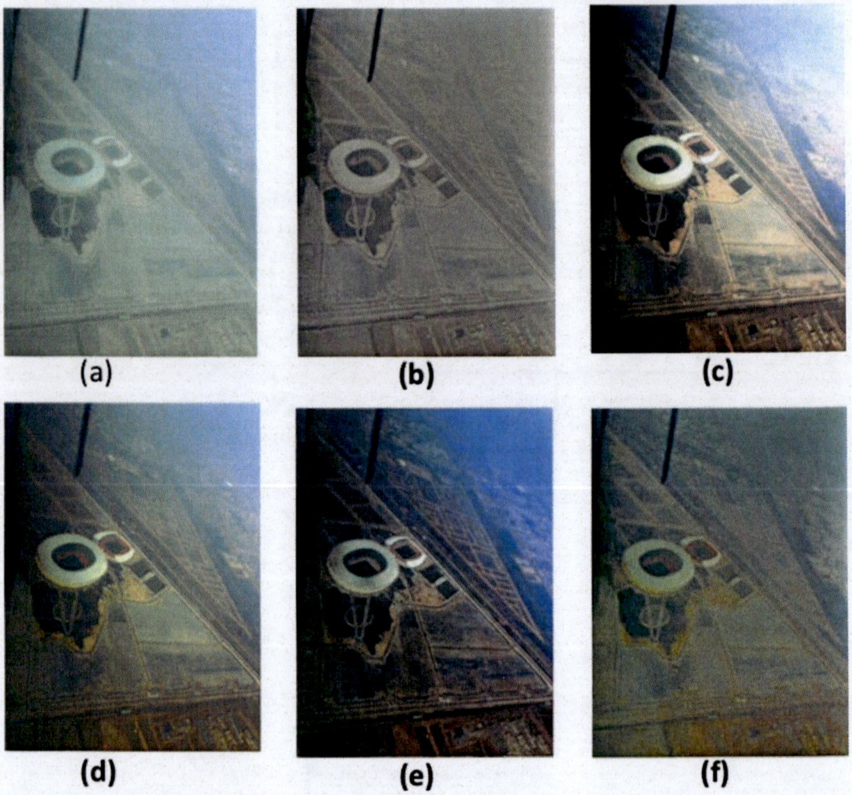


**Fig. 2** First image, in (a) original image, and in (b, c, d, e, and f) that are enhancement by different algorithms (MSRCR, HE, DCP, IEIF and proposed) respectively

we notice that the best way to improve is due to proposed and DCP algorithms. It is identical to the non-reference measures NIQE and WQE in Table 1, where the lowest values of the scale NIQE were mostly due to the proposed method, mostly the highest values of the scale WQE also belong to the proposed method. This indicates that the proposed method has succeeded in improving aerial images and it indicates increased detail because that is increased the Contrast and brightness of these images better than other methods.

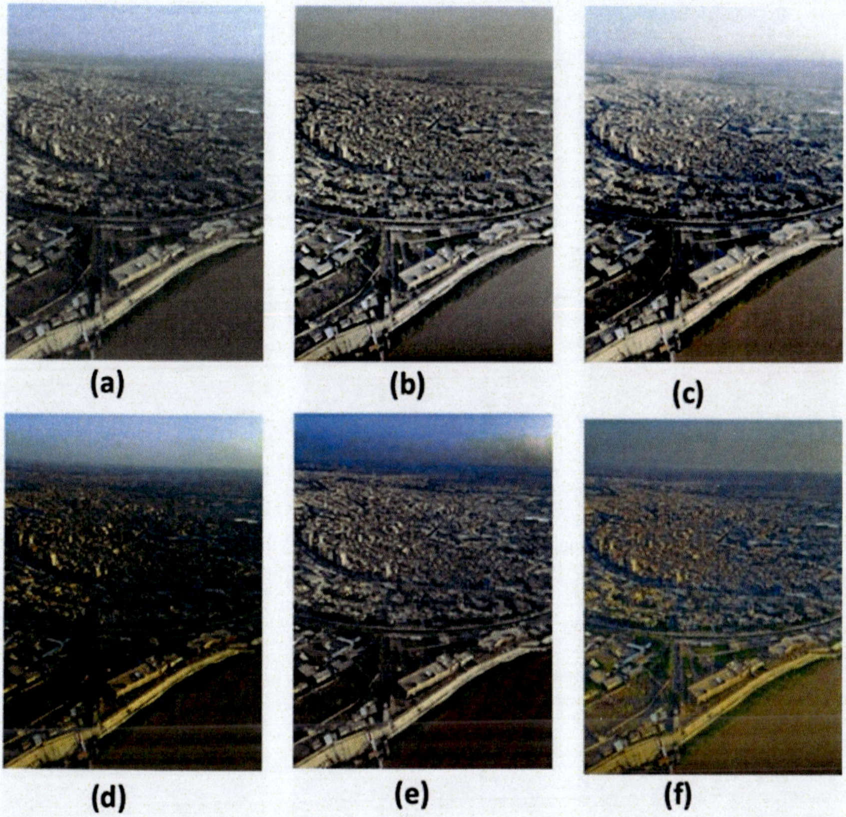


**Fig. 3** Second image, in (a) original image, and in (b, c, d, e, and f) that are enhancement by different algorithms (MSRCR, HE, DCP, IEIF and proposed) respectively

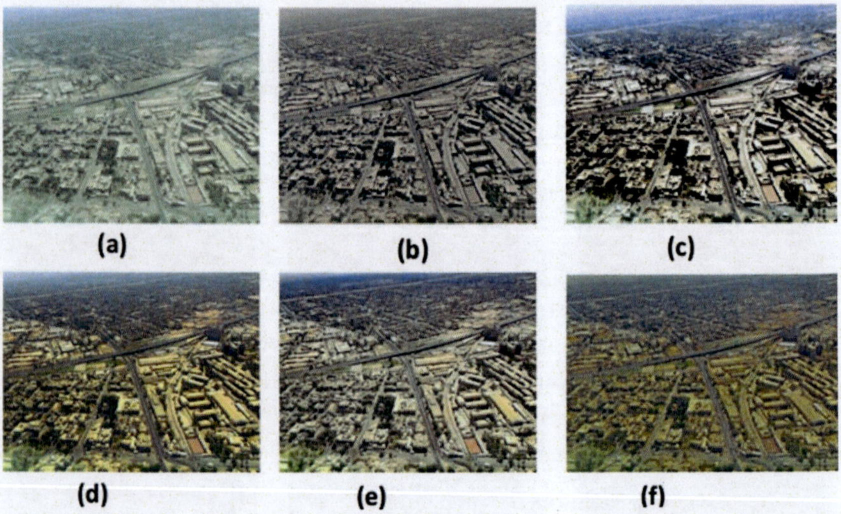


**Fig. 4** Third image, in (a) original image, and in (b, c, d, e, and f) that are enhancement by different algorithms (MSRCR, HE, DCP, IEIF, and proposed) respectively

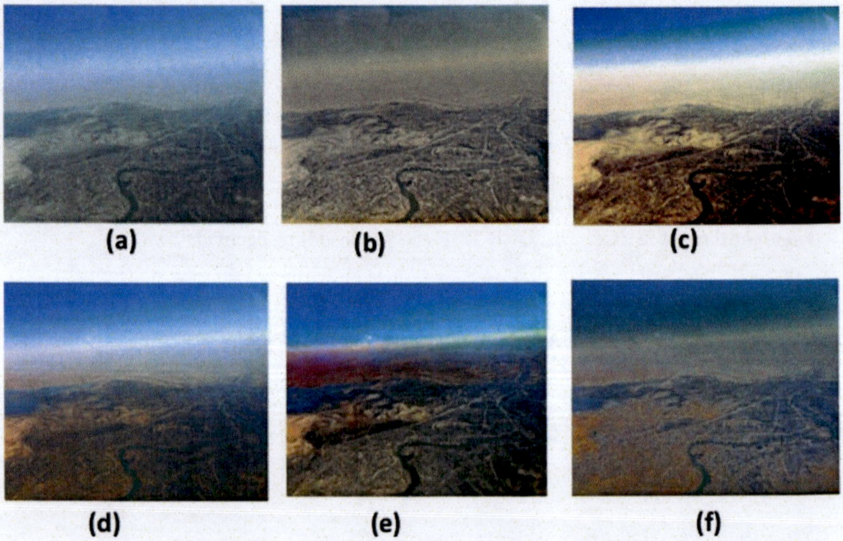


**Fig. 5** Forth image, in (a) original image, and in (b, c, d, e, and f) that are enhancement by different algorithms (MSRCR, HE, DCP, IEIF, and proposed) respectively

**Table 1** The quality metrics for aerial image enhancement

| Second image | | | First image | | |
|---|---|---|---|---|---|
| WQE | NIQE | Methods | WQE | NIQE | Methods |
| 0.07 | 2.29 | Proposed | 0.18 | 3.29 | Proposed |
| 0.04 | 2.62 | MSRCR | 0.13 | 3.88 | MSRCR |
| 0.04 | 2.62 | HE | 0.06 | 4.03 | HE |
| 0.05 | 2.96 | DCP | 0.08 | 4.90 | DCP |
| 0.06 | 2.41 | IEIF | 0.07 | 4.55 | IEIF |
| Fifth image | | | Third image | | |
| WQE | NIQE | Methods | WQE | NIQE | Methods |
| 0.12 | 2.98 | Proposed | 0.06 | 2.79 | Proposed |
| 0.06 | 3.08 | MSRCR | 0.04 | 2.77 | MSRCR |
| 0.06 | 3.05 | HE | 0.03 | 3.05 | HE |
| 0.08 | 3.83 | DCP | 0.03 | 3.16 | DCP |
| 0.07 | 3.40 | IEIF | 0.04 | 3.03 | IEIF |

## 5 Conclusions

In this paper, a new algorithm is proposed to enhance the number of aerial images that suffer from multiple distortions. It was compared with the methods MSRCR, HE, DCP, and IEIF by using no-reference quality metrics NIQE and WQE, by analyzing the results, it can be concluded the proposed method succeeded in improving aerial images better than other methods.

## References


1. Ameer Z , Daway H, Kareem H (2019) Enhancement underwater image using histogram equalization based on color restoration. J Appl Eng Sci 14(2):641–647
2. Mirza NA, Kareem HH, Daway HG (2019) Low lightness enhancement using nonlinear filter based on power function. J Theor Appl Inf Technol 96(1):61–70
3. Ghada Sabah Karam (2018) Blurred ımage restoration with unknown point spread function. Al-Mustansiriyah J Sci 29(1):189. https://doi.org/10.23851/mjs.v29i1.335
4. Daway HG, Mohammed FS, Abdulabbas DA (2016) Aerial ımage enhancement using modified fast visibility restoration based on sigmoid function. Adv Nat Appl Sci 10(11):16–22
5. Gonzalez RC, Wintz P (1987) Digital ımage processing. Addision-Wesley Publishing Company, Boston, pp 275–281
6. Ancuti CO, Ancuti C (2013) Single ımage dehazing by multi-scale fusion. IEEE Trans Image Process 22(8):3271–3282. https://doi.org/10.1109/TIP.2013.2262284
7. He K, Sun J, Tang X (2010) Single image haze removal using dark channel prior. IEEE Trans Pattern Anal Mach Intell 33(12):2341–2353
8. Dana B, Avidan S (2016) Non-local ımage dehazing. İn: Paper Presented at the Proceedings of the IEEE Conference on Computer Vision and Pattern Recognition

9. Park D, Park H, Han DK, Ko H (2014) Single ımage dehazing with ımage entropy and ınformation fidelity. İn: Paper presented at the 2014 IEEE International Conference on Image Processing (ICIP)
10. Jobson D, Rahman Z, Woodell G (1997) Properties and performance of a center/surround retinex. IEEE Trans Image Process 6(3):451–462
11. Tan RT (2008) Visibility in bad weather from a single ımage. In: Paper presented at the 2008 IEEE Conference on Computer Vision and Pattern Recognition
12. Li L, Sang H, Zhou G, Zhao N, Danwen Wu (2017) Instant haze removal from a single image. Infrared Phys Technol 83:156–163
13. Wine S, Horne R (1998) The Color Image Processing Hand Book. International Thomson
14. Jobson DJ, Rahman Z, Woodell GA (1997) A multiscale retinex for bridging the gap between color images and the human observation of scenes. IEEE Trans Image Process 6(7):965–976. https://doi.org/10.1109/83.597272
15. Mittal A, Soundararajan R, Bovik AC (2013) Making a "completely blind" ımage quality analyzer. IEEE Signal Process Lett 20(3):209–212. https://doi.org/10.1109/LSP.2012.2227726
16. Ahmed AAS, Daway HG, Rashid HG (2020) Quality of medical microscope ımage at different lighting condition. In: Paper presented at the IOP Conference Series: Materials Science and Engineering

# CARPM: Comparative Analysis of Routing Protocols in MANET

Vijay Rathi(✉) and Raj Thaneeghaivel

RKDF Institute of Science and Technology, Sarvepalli Radhakrishnan University, Bhopal, Madhya Pradesh, India

**Abstract.** MANET has evolved as a multiple hop based, autonomous and time being nature with wireless connectivity paradigm of network that can handle in constrained environments with constraints like energy, power and bandwidth. Wormhole attack is a highly alarming attack in mobile ad hoc networks., where an attacking node, on receipt of a packet at one location, replays them at another one, which is situated remotely far distance. In this comparative analysis, we deeply understand the working of various routing protocols such as AODV, DSR and ZRP, PA-DSR for its QoS performance, evaluate the performance analysis, considering adverse effects of wormhole attacker nodes. By strategic placement of various wormhole nodes within the network, we determine the performance with respect to packet delivery ratio, throughput, packet loss, average end to end delay and jitter. At last depending upon the simulation we explored the most impacted routing protocol with respect to network metrics.

**Keywords:** Routing protocols · Mobile Ad hoc network (Manet) · Wormhole Metrics Network simulator Throughput Packet loss · Packet delivery ratio · Jitter · Random mobility model

## 1 Introduction

A MANET [1] is the floating network where various mobile nodes are considered as vertices and these vertices are connected to various other vertices through wireless edges or links. All the nodes are omni directional, i.e. every node can send or receive from any connecting nodes which make every node behave in dual nature i.e. either an intermediate node or an end node. As every node has its freedom to transmit and connect to any neighboring node, the nodes are ad hoc and dynamic in nature. because of this dynamic nature, nodes do not have a static infrastructure for data transmission at all. MANETs comprise of nodes which are mobile in nature which gives this network a dynamic infrastructure for data transmission. The nodes can transmit data to other nodes by simply communicating to the neighboring nodes. So the hop count to be calculated from a source node to destination node for the data transmission completely depends on direct links between the nodes.

The MANET Routing Protocols have always been an interesting area of research because of the scope of improvement in the QoS in it. The QoS parameters such as


S. Shukla et al. (eds.), *Data Science and Security*, Lecture Notes in Networks and Systems 290,
https://doi.org/10.1007/978-981-16-4486-3_11

End to End Delay, Throughput, Jitter etc have been playing vital role in evaluating various routing mechanisms proposed for MANET data transmission. But it has been observed that not all parameters are enhanced by any individual routing protocol, and hence there is the need to design novel and efficient routing protocols for MANET. The present research work emphasizes on addressing various issues or research gaps in existing routing protocols and thereby design or propose a novel routing mechanism which overcomes the shortcoming identified in analysis of protocols. The research work also makes in depth for various parameters of QoS and then worked upon with the proposed protocol to enhance the same.

The remaining paper comprises of Related Work in Sect. 2, then Sect. 3 Comprises of Comparative Analysis of Protocols in depth, Sect. 4 emphasizes on identifying Drawbacks of various routing protocols, Sect. 5 Comprises of an overview of the proposed routing protocol.

## 2 Related Work

Number of previous researches compared the routing techniques with presence of wormholes for the work done, and proposed different approaches to safeguard networks. Below are the details for the survey:

Awerbuch et al. [2], ODSBR which is a secure and unicast in nature protocol for routing in MANET is checked against and analyzed against existing AODV considering various forms of attacks such as wormhole, rushing and blackhole attacks. Authors' analysis evaluated that the most vulnerable position in the MANET is the middle or aixs of the network. Evaluation done also determines that in multicast networks also, the highly targeted areas by attackers are the middle or center of the network areas.

Garg et al. [3], Qualnet 4.5 was used in the comparative study of reactive protocols with wormholes considering the mobility and non-mobile environments. The study determined that wormhole attack (threshold) is highly intensive due to the major packet loss. The obtained comparison evaluation determined that AODV is highly susceptible to wormhole attacks considering performance degradation in terms of throughput and also average jitter, mobility, mean EtoE delay, the ration of packet delivery.

Mahajan et al. [1], evaluated in-band and self-contained wormhole analysis based on the various working scenarios like doubtful, unsuccessful, successful, uninteresting and interesting. Evaluation determined that the positions of the attacker nodes grab a significant position in the severity of the wormhole strikes. The observations also determined that end to end delay is adversely affected (increases) with increase in the strength of the wormhole strike.

Sanaei et al. [5], compared AODV with DSR considering a wormhole and without wormhole environments. The study observations were done for the efficieny analysis depending upon ratio of packets delivery, ,throughput and EtoE delay metrics. The research work experimented on 500 m square areas with some 30 nodes having a 52 bit packet size. The Observations evaluated that theAODV is less affected compared to DSR with the wormhole attack.

# 3 Comparative Analysis of Proposed Routing protocol with DSR and AODV and PA-DSR and GC-DSR

In mobile ad hoc networks i.e. MANET [1], the major dependency on data transmission lies on the mobile nodes which are continuously moving i.e. dynamic in topology and therefore routing the packets through this requires additional table to maintain the details of the hop count i.e. distance between a node to another node. Also, there is a need of neighbouring nodes tables which maintain the present neighbor count of the current node n1 and its distance with other neighboring nodes. As the nature of the network is completely dynamic and any node can join or leave the network at any point of time, the data in these tables is updated as soon as the change in network occurs. and hence the very recent information is available in the tables. Let's explore the various types of protocols in MANET.

## 3.1 Proactive Routing Protocols

Proactive routing protocols are usually dependent on routing tables. Every node in the network has to maintain the separate information regarding the neighbor node, ots hop count (distance) etc. (Fig. 1).

As the underlying network is dynamic in nature and the topology changes very suddenly, these tables are updated from time to time to maintain the latest information for routing decisions. Which leads to a limitation of this kind of routing not suitable for large networks as the routing tables at every node then become bulky in nature which may cause performance degradation. Let's explore the type of Proactive Protocols proposed previously:

1. **Destination Sequenced Distance Vector Routing Protocol (DSDV):** DSDV protocol extends the very popular distance vector routing protocol used in wired networks;

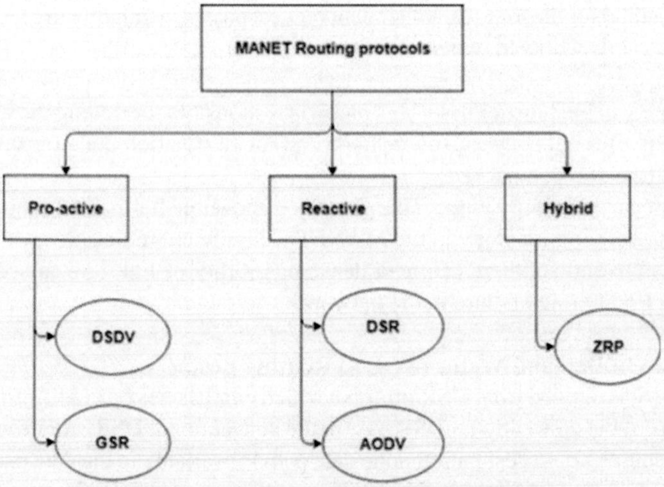


**Fig. 1** Various MANET Routing protocols classification

the working principle is designed on the basis of Bellman-ford routing algorithm. DV protocol contradicts the need of mobile ad-hoc networks because of the identified count-to-infinity problem. So, an extended version named as DSDV i.e. Destination Sequenced Distance Vector Routing Protocol was designed. Destination order or sequence counter is summed with every single routing data in the routing table hosted by every node in the network. Every node will allow the new data entry in the table only if the new entry depicts the higher sequence counter for the route.

2. **Global State Routing (GSR):** GSR basically is derived from or subset of the route protocols used in wired natured networks; the working principle for which is on the basis of Dijkstra's shortest path routing technique. MANETS are not compatible with link state routing as the information if flooded to every node in the network i.e. Global flooding that can cause tremendous traffic of packets in the mobile network. To avoid this challenge, GSR Protocol which is Global State Routing was derived. GSR routing refrains from flooding the linkage status packets straight away into the entire mobile network. On the other hand, GSR mechanism ensures that each and every node in the MANET endorses one list and maintains three tables for three different data entries namely topology table, distance table, next hop table,and a list named as adjacency list.

### 3.2 Reactive Routing Protocols

A special category of on Demand Routing Protocols. In such routing, the route from source to destination is established only if there is the need. So it's quite obvious that the route establishing will be done by significantly flooding the route request over the entire network. So, because of this nature of on demand route establishment, there are two prominent phases for this routing i.e. route establishment and route sustainability.Let's explore the type of Proactive Protocols proposed previously:

1. **Dynamic Source Routing i.e. DSR:** In DSR, source to destination route establishment task occurs only if demanded. This route establishment takes place using flood method for the route establishment request packets straightly in MANET entire network. DSR works in two phases:

    - Route Establishment Phase:This phase is responsible for finding the best suitable (optimal) route between source node and the destination node for the requested data transmission packets.
    - Route Sustainability Phase:This phase is responsible for maintaining the route as the underlying topology in the MANET is highly unpredictable as it is dynamic in nature and so, there can be a dense possibility of link broken scenarios that may lead to route failure in the network.

2. **Ad-Hoc On Demand Vector (AODV) Routing protocol:**
AODV [6] is derived from DSR protocol which emphasizes on eradicating the shortcomings of the conventional DSR. As, in the conventional DSR, after route is established, source not to destination node data path is available in the data packet header Therefore, as and when the network size gets bigger and bigger, the header data i.e. source to destination path gets lengthier which leads in slower transmission rate and

which ultimately slows down the network. So, AODV protocol was designed which maintains the path in different ways. AODV maintains the source to destination route path in the routing table, while on the other hand, DSR maintains this information in the packet (data) header. AODV also comprises two phases similar DSR.

### 3.3 Hybrid Routing Protocols

Hybrid Protocol [8] merges the advantages of both the type of protocols i.e. proactive and reactive protocols. Hybrid protocols are very adaptive and can get compatible as per the source and destination zones or positions in the network. The significant hybrid protocol identified so far is **Zone Routing Protocol (ZRP)**.

The main concept behind ZRP is that the entire network is subdivided into smaller zones and accordingly the source node position and destination node positions are observed. If in the observation, the source node and the destination nodes are found to be in the single zone, data transmission takes place with the proactive routing, otherwise, reactive routing is preferred for data transmission.

## 4 Challenges or Drawbacks in DSR, AODV and ZRP

The various Routing protocols previously proposed have their own set of pros and cons. Various MANET architectures adopt with one of them with the predefined priorities set according to their needs: Below are a few shortcomings for routing protocols observed so far (Table 1):

## 5 Proposed Green Corridor Protocol

The proposed Green Corridor i.e. GC protocol is designed considering the shortcomings of the various previously existing routing protocols. GC routing is the derived protocol

**Table 1** Disadvantages of various Routing Protocols in MANET

| Protocol | Shortcomings |
|---|---|
| DSDV | Dependency on Sequence Number causes delay in updation of tables as sequence number may change frequently |
| DSR | Flooding of route discovery may lead to congestion in the network and may result in an end to end delay increase |
| AODV | Maintaining the source to destination path in the routing table leads to overhead of changing the path frequently on occurrence of link failure in the node network |
| ZRP | ZRP being a derivative of proactive as well as reactive routings, it adopts the challenges of both the routing types and therefore zoning the nodes will not provide an optimal solution. Because nodes n1 and n2 at time t1 can be a part of zone 1, while after some time at time t2 nodes n1 and n2 might belong to different zones, which leads to failure of the routing protocol deployed at time t1 |

from the existing ZRP i.e. Zone routing protocol with an additional feature of managing the broken linkage due to change in network caused because of mobility of nodes in MANET.

Consider a Scenario, where the source to destination packet transmission request is placed in the MANET. At time *t1* when the nodes identified as source *ns* and node *nd* are identified as the nodes belonging to same zone *z1*, the chosen routing protocol is proactive protocol, but if at time *t2* if the nodes *ns* and *nd* due to dynamic in nature gets to different zones *z1* and *z2* , the protocol fails as now they belong to different zones at time *t2*. So the proposed routing protocol addresses this issue and proposes a green corridor mechanism in ZRP where in the initial phase, based on the zones and the nodes position, the routing protocol is chosen, and once the link is established, the zone table is superseded by the link routing protocol chosen. This will lead to the dynamic nature of routing protocol in ensuring the source to destination packet delivery.

### 5.1 Proposed Model Algorithm Pseudo-code

```
Protocol_choosen_flag=false
Route_established_flag=false
If(Transmission Request generated):
        While(Route is being established):
                If(node n1 and node n2 in same zone):
                        Use proactive Routing
                        Protocol_choosen_flag=True
                Else:
                        Use Reactive Routing
                        Protocol_choosen_flag=True
If(Route_established_flag):
        If(Protocol_choosen_flag=True):
                If(node n1 and node n2 change the positions):
                        Use choosen Protocol to proceed without considering
                changed positions.
```

### 5.2 Results and Discussion

Based on various protocols and its evaluation parameters demonstrated by various researchers, the proposed routing protocol seemingly ensures the higher throughput and higher QoS as it makes use of Hybrid Routing Protocol for initial choice of protocol and Green Corridor Mechanism for Route Sustainability. As the Green Corridor Protocol will always focus on maintaining the nodes connectivity, and in case the node changes its position, the protocol remains the same with next available node replacing the disconnected node.Hence the reliability is ensured in proposed protocol.

## 6 Conclusion

As per the comparative analysis done so far, existing routing protocols have their own set of pros and cons, which sometimes override some significant drawbacks of the protocols.

So in this comparative analysis, we have addressed such loopholes and identified the major research area to propose significant work in hybrid protocol, where zonal switch of the nodes while in transmission may lead to link breakage, which has to be tackled. So proposed work emphasized on this issue and hereby propose that will prove to be an optimal and efficient routing protocol.

**Acknowledgements.** I would like to acknowledge everyone who played a role in my Presentation accomplishments. First of all, my parents, who supported me with love and understanding Without you, I could never have reached this current level of success. Secondly, my Co author Raj Thaneeghaivel, who has provided patient advice and guidance throughout the research process. Thank you all for your unwavering support.

## References

1. Mahajan V, Natu M, Sethi A (2008) Analysis of wormhole intrusion attacks in MANET. In: MILCOM 2008–2008 IEEE military communications conference. IEEE
2. Awerbuch B, et al (2004) Mitigating byzantine attacks in ad hoc wireless networks. Department of Computer Science. Johns Hopkins University. Technical report, Version 1, p 16
3. Garg G, Kaushal S, Sharma A (2014) Reactive protocols analysis with wormhole attack in ad-hoc networks. In: 2014 International conference on computing, communication and networking technologies (ICCCNT), pp 1–7. IEEE
4. Sanaei MG, Isnin IF, Bakhtiari M (2013) Performance evaluation of routing protocol on AODV and DSR under wormhole attack. Int J Comput Netw Commun Secur 1
5. Vandana CP, Devaraj AFS (2013) Evaluation of impact of wormhole attack on AODV. Int J Adv Netw Appl 4(4):1652
6. Hu Y-C, Perrig A, Johnson DB (2003) Packet leashes: a defense against wormhole attacks in wireless networks. In: INFOCOM 2003, twenty-second annual joint conference of the IEEE computer and communications, IEEE Societies, vol 3. IEEE.
7. Poovendran R, Lazos L (2007) A graph theoretic framework for preventing the wormhole attack in wireless ad hoc networks. Wireless Netw 13(1):27–59
8. Chiu HS, Lui KS (2006) DelPHI: wormhole detection mechanism for ad hoc wireless networks. In: 2006 1st International symposium on wireless pervasive computing, p 6. IEEE

# Content-Restricted Boltzmann Machines for Diet Recommendation

Vaishali M. Deshmukh[1(✉)] and Samiksha Shukla[2]

[1] New Horizon College of Engineering, Bengaluru, India
vaishali.md@newhorizonindia.edu
[2] Christ University, Bangalore, India
samiksha.shukla@christuniversity.in

**Abstract.** Nowadays, society is leading towards an unhealthy and inactive and lifestyle. Recent studies show the rapid growth of people suffering from diseases caused due to unhealthy lifestyles and diet. Considering this, recognizing the right type and amount of food to eat with a suitable exercise set is essential to obtain good health. The proposed work develops a framework to recommend the proper diet plans for thyroid patients, and medical experts validate results. The experiments' results illustrate that the proposed Content-Restricted Boltzmann Machines (Content-RBM) produces more relevant recommendations with content-based information.

**Keywords:** Recommender system · Restricted Boltzmann Machines · Diet and exercise plans · Content-based features · Thyroid disease

## 1 Introduction

The key to following a healthy lifestyle corresponds to practicing healthy habits. It is particularly essential for people suffering from some minor or significant diseases to follow a healthy lifestyle. Digital health motivations and research investigations offer several recommendation systems to patients for the improvement of their health. Usually, people ignore the intake of nutritional food in the right amount. A customized diet plan consists of the necessary dietary elements like calcium, proteins, and vitamins needed to fulfill an individuals' nutritional requirements. People need to find the right diet plan for themselves which meets their dietary requirements and tastes.

Extensive research works have been investigated in the field of healthcare recommendation systems. Various studies have designed and developed either physical workout-based or diet and nutrition-based recommendation systems. Very few studies are analyzed with consideration of user preferences and health conditions together. The current research works have designed diet recommendation systems for diabetic and cancer patients. The diet recommendation framework for thyroid patients is, however, under-researched. Several existing works proposed different food recommendation systems mentioned as follows:


S. Shukla et al. (eds.), *Data Science and Security*, Lecture Notes in Networks and Systems 290,
https://doi.org/10.1007/978-981-16-4486-3_12

- Content-based methods generate recommendations by segregating the ingredients from recipes. It computes the scores using ingredient lists of positively rated recipes.
- Collaborative filtering methods recommend the recipes based on user similarity and filtered the foods as per nutritional needs [1].
- In context-aware approaches, context data such as climate conditions and food availability are considered.
- Health-aware approaches incorporate health-related data into the recommendation for improving nutritional habits [2].

Prominent research work is investigated in the food recommendation domain for diabetic patients. It is also crucial to recognize that thyroid is common and the most prominent disease among women. It affects the human body with many adverse effects, including metabolism [3]. A suitable diet plan may provide significant health improvement for thyroid patients. The proposed model recommends the diet and plans based on rating information and preference scores.

The recipe data are collected using Yummly's API and scraped from allrecipes.com, which contains ingredients for each recipe. The proposed work constructs tables for recipe-foods and food-nutrient values. Multiple resources are extracted for nutrition data, such as Kaggle's Open Food Facts, Allrecipes, and Indian Food Composition table (IFCT 2017). The constructed dataset provides the nutrition detail for each type of food, such as proteins, carbohydrates, iodine, selenium, fats, and sugar. Along with this information, the system collects the data from thyroid blogs and forums to get user-profiles and food items suitable for thyroid patients.

## 2 Related Work

Food recommendation systems are considered a promising solution to facilitate the patients' food intake and health conditions [6]. Various authors have proposed diverse methods to achieve the customized and efficient food recommender systems. This paper highlights some of these recent research works in this section.

A personalized diet recommender system was implemented using the artificial bee colony method in [7] to find required daily nutrition. The authors proposed a framework that efficiently used a rule-based fuzzy ontology model to generate nutritional-based recommendations. A genetic algorithm was used to recommend relevant food lists. The framework relied on the Google fit Application Programming Interface for information regarding the user's energy requirements and daily activities. The system also needed the patient's past medical records to generate personalized diet recommendations for users.

Authors in [8] used Clustering analysis to generate a nutrition recommender system for diabetic patients. They proposed that food and nutrition is an essential key factor for healthy living. Toledo et al. [9] computed daily nutritional needs based on user's physical information. The study needs to present a process that handles healthy and preference management simultaneously. This current work uses nutrient data and user information together. It gives a solution for the long-term to make sure that patients with thyroid diseases are protected.

The authors investigated a new method to generate a healthy diet using a predictive modeling algorithm [10]. They implemented a model to propose healthy food habits.

The system identified appropriate food patterns for users based on the needed amount of macronutrients and the required calories burned. Though, this research work was useful for predicting healthy diets for patients. However, this model achieved a minimal designing solution as per the patient's needs. The current model addresses this problem using a content-RBM model with accurate precision that identifies the patient nutritional requirements.

A nutrition assistance framework was developed by Leipold et al. [11] to provide feedback based on patient's dietary behavior. The model accommodated the behavioral eating change using a self-monitoring and tracking approach. An automated nutrition recommendation system could offer significant benefits to human nutritionists to generate customized diet plans. It also faces several limitations ranging from usability, efficiency, and efficacy to satisfaction. The results were notwithstanding and required to integrate contextual and social information and enhance the received input data accuracy. The system can use daily feedback to find the desired effects in the long term using mobile platform applications.

Agapito G. [12] implemented the Diet Organizer System to create a user profile using a real-time questionnaire compiled by medical professionals. The system referred to as DIETOS can recommend specific foods in the same category, have similar health grades, and provide nutritional-related suggestions for some health issues. Sahoo et al. [4] used Restricted Boltzmann Machine algorithm and Convolutional Neural Network to generate a health recommender system. The framework designed an intelligent model to predict the physical health condition of patients and their social activities.

Alian et al. [13] implemented the ontological model based on American-Indian user geographical status, cultural and behavioral patterns. A nutrient-based recommendation system was discussed in [14] for children, and diet recommendations were created based on user information.

We have analyzed various existing research papers and identified few lacunas in the previously developed systems. In the proposed method, an efficient recommendation system is implemented based on patient information. The system uses a medical dataset to identify the food items based on the patient's thyroid disease information and preferences. Other content-based features are considered age, gender, weight, calories, and required nutritional information like protein, fat, sodium, fiber, and cholesterol.

## 3 Methodology

The research work proposes a diet and workout recommender framework to combine content-based parameters with conditional RBM. The hybrid model consists of four main steps, which are as below:

- First, we gathered thyroid patients' information, food preferences, food and their corresponding nutritional facts, exercises with their intensities and durations.
- Next, the framework constructs a dataset with patients' physiological and medical data like age, gender, height, weight, food preferences, thyroid profile (T3, TT4, and TSH), and activity level.
- The proposed model generates a user-food rating preference matrix using the vector space model as a content-based filtering method.

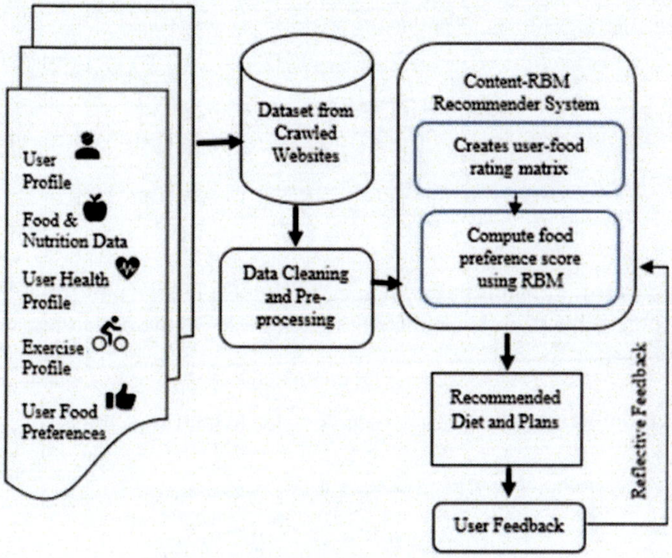


**Fig. 1** Proposed recommender system

- It uses RBM as a collaborative filtering appraoch to deal with missing ratings of the user-food rating matrix. It recommends the diet plans based on recommendation scores of highly rated food items, preferences, and needed nutritional values.

Figure 1 shows a novel health-aware recommendation framework that suggests healthy and tasty (based on preferences) diet plans. A meal's healthiness can be decided by its amount of nutritional values specified for thyroid patients. Its tastiness can be determined by the average ratings provided by the total users. However, to achieve this, the model retrieves the nutrition content and ratings of thyroid-related foods and recipes.

Proposed Content-RBM model integrates content information and conditional RBM together. The system has $N$ users and $R$ recipes, and the ratings are integers between 1 and 5, or 0 if not rated. An RBM has $K$ units of $v$ as a visible layer and $L$ binary units for hidden layer $h$. Both hidden and visible units are fully connected without interconnection within each layer. Hidden and visible units have $a_m$ and $b_n$ as biases, respectively. RBM uses contrastive divergence for the learning process [4]. The proposed algorithm uses the content-based vector space model to construct a user-recipe matrix and RBM to generate the missing ratings. It follows the following steps:

**Proposed Algorithm**

- Obtain all recipes and user profiles from scrapped websites and represent every recipe and user as unique.
- Extract all ratings provided by the user for every recipe.
- Construct a preference-based food matrix on the recipe vectors.

- Obtain the missing ratings, if any using RBM.

$$p(h_n = 1|v) = \sigma\left(b_n + \sum_{m=1}^{K} w_{mn} v_m\right) \tag{1}$$

$$v_m = a_m + \sum_{n=1}^{L} w_{mn} h_n \tag{2}$$

The conditional probability is computed in (1) of a hidden unit for a given visible unit. The visible unit's value is estimated in (2) and uses as the preference score to rank foods for the users.

- The cosine similarity metric $S_{a,b}$ is computed between user *a* and *b*.

$$S_{a,b} = \frac{\sum_i (a_i \times b_i)}{\sqrt{\sum_i (a_i)^2} + \sqrt{\sum_i (b_i)^2}} \tag{3}$$

- Predicted rating score $R_{a,f}$ is computed by averaging the similar users' ratings for food *f*.

$$R_{a,f} = \sum_{a,b\ \varepsilon\ U} S_{a,b} \times r_{b,f} \tag{4}$$

- The system uses the highest predicted ratings to recommend diet plans for a user.

## 4 Results and Discussions

Experimental evaluation analyses the proposed model's performance by calculating mean absolute error and root mean squared error in the test set. The proposed model is compared with other collaborative filtering baseline approaches and achieved lower error rates. The current research study evaluates the quality of food lists generated against the baseline recipe substitution models [5]. Rating prediction is a prevalent task that is commonly tackled by collaborative filtering algorithms. The framework implemented collaborative filtering algorithms, namely KNN, SVD (Singular Value Decomposition), RBM, and the proposed hybrid content-RBM, to predict users' ratings on unseen recipes.

Table 1 shows the results of implemented models using Root Mean Squared Error (RMSE). It is estimated by comparing the actual rating and the rating predicted by the model for every user-food-item pair and its known label. Figure 2 shows the performance evaluation of the Content-RBM over the other collaborative baselines methods. The system converts the ratings 1 to 5 into a binary problem as relevant food-items and not relevant food-items. Any actual rating above 3.5 is considered as a relevant food-item and below 3.5 is not a relevant food-item.

Recommendation systems are mainly interested in suggesting top-N recommendations to the user. The proposed model computes the recall metric for the first N food-items instead of considering all the recipes. Recall metric at *k* is estimated as the ratio

**Table 1** Performance of recommendation models on test set

| Models/Error metrics | *RMSE* |
|---|---|
| KNN | 0.815 |
| SVD | 0.859 |
| RBM | 0.804 |
| **Content-RBM** | **0.722** |

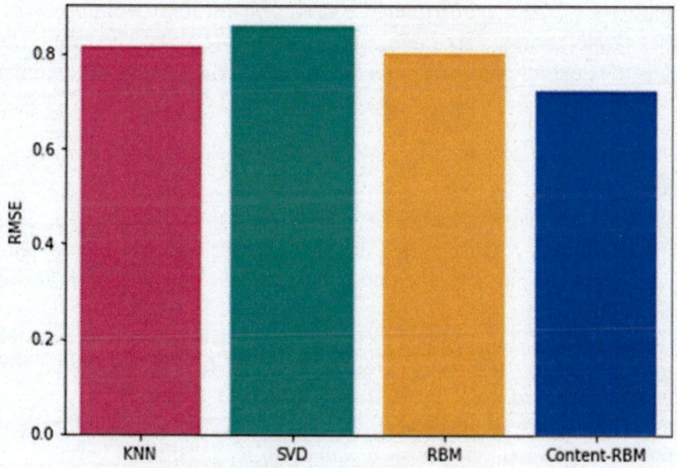


**Fig. 2** Performance of recommender models

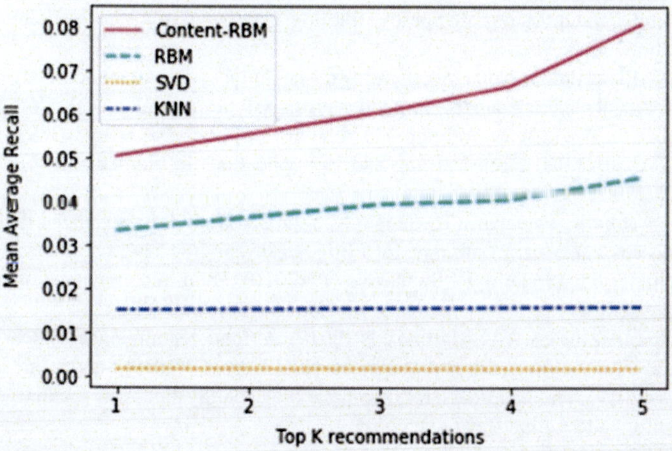


**Fig. 3** Comparision of the recommendation models based on mean average recall metric for top-k recommendations

of relevant food-items identified in Top-k recommendations. The proposed framework computes the recall metric for each user's list of recommendations and averages overall predictions for Top-k recommendations as depicted in Fig. 3.

## 5 Conclusion

The main innovation of this approach is to integrate preferences of recipes with individuals' health. Most diet recommender systems or apps recommend meals based on users' comments and ratings. However, this work incorporates nutritional constraints of thyroid disorders and suggests the best suitable diet plans based on consumed calories and food choices. In this paper, the proposed content-RBM recommender system uses implicit feedback data to predict a new user's ratings and protect data privacy. The results on a real dataset presents its superior performance and significant improvements over the prevalent popularity (KNN) and matrix factorization method (SVD). The future improvement for this work is to consider cultural characteristics and context data (availability, climate) of food.

## References

1. Xie J, Wang Q (2019) A personalized diet and exercise recommender system for type- 1 diabetes self-management: an in silico study. Smart Heal 13:100069. https://doi.org/10.1016/j.smhl.2019.100069
2. Trattner C, Elsweiler D (1892) Food recommender systems: important contributions, challenges and future research directions. In: Maxwell JC (ed) A Treatise on Electricity and Magnetism, 2017, 3rd ed, vol 2, pp 68–73. Clarendon, Oxford
3. Hackney A, McMurray R, Judelson DA, Harrell J (2004) Relationship between caloric intake, body composition, and physical activity to leptin, thyroid hormones, and cortisol in adolescents. Jpn J Physiol 53:475–479. https://doi.org/10.2170/JPhysiol.53.475
4. Sahoo A, Pradhan C, Barik DR, Dubey H (2019) DeepReco: deep learning-based health recommender system using collaborative filtering. Computation 7:25. https://doi.org/10.3390/computation7020025
5. Trattner C, Elsweiler D (2017) Investigating the healthiness of internet-sourced recipes: implications for meal planning and recommender systems. In: Proceedings of the 26th International Conference on World Wide Web, pp 489–498. https://doi.org/10.1145/3038912.3052573
6. Chensi C et al (2018) Deep learning and its applications in biomedicine. Gen Proteomics Bioinf 16:17–32
7. Raut M, Prabhu K, Fatehpuria R, Bangar S, Sahu S (2018) A personalized diet recommendation system using fuzzy ontology. Int J Eng Sci Invent 7(3):51–55
8. Maiyaporn P, Phathrajarin P, Suphakant P (2010) Food recommendationsystem using clustering analysis for diabetic patients. Int Conf Inf Sci Appl 6(2):5–14
9. Toledo RY, Alzahrani AA, Martinez L (2019) A food recommendersystem considering nutritional information and user preferences. IEEEAccess 7(2019):96695–96711
10. Jaiswal V (2019) A new approach for recommending healthy diet using predictive data mining algorithm. Int J Res Anal Rev 6(2):58–65
11. Leipold N, Lurz M, Bohm M (2018) Nutrilize a personalized nutritionrecommender system: an enable study. Health Rec Syst 3(4):4–10
12. Agapito G (2016) DIETOS: a recommender system for adaptive diet monitoring and personalized food suggestion. In: IEEE 12th International Conference on Wireless and Mobile Computing, Networking and Communications (WiMob), New York, NY, pp 1–8. https://doi.org/10.1109/WiMOB.2016.7763190

13. Alian S, Li J, Pandey V (2018) A personalized recommendation system to support diabetes self-management for American Indians. IEEE Access 6:73041–73051. https://doi.org/10.1109/ACCESS.2018.2882138
14. Banerjee A, Nigar N (2019) Nourishment recommendation framework for children using machine learning and matching algorithm. In: 2019 International Conference on Computer Communication and Informatics (ICCCI), Coimbatore, Tamil Nadu, India, pp 1–6

# PIREN: Prediction of Intermediary Readers' Emotion from News-Articles

Rashi Anubhi Srivastava[1] and Gerard Deepak[2](✉)

[1] Department of Electrical Engineering, Central University of Karnataka, Kalaburagi, India
[2] Department of Computer Science and Engineering, National Institute of Technology, Tiruchirappalli, India

**Abstract.** Stimuli like text narratives from news articles and editorials trigger numerous emotions as a response in the readers. As seen in previous works of news documents classification, the attention has been centralized more towards writer's perspective rather than how a particular article affects its readers. This work focuses on a reader's stance, one forms after reading a certain content. Ontology driven knowledge base is used for semantic matching to further calculate the coherence of the words to distinct intermediary emotions. On reception of a new document, frequency of terms are calculated which are then matched with ontologies and hence classification is done using deep learning-based classifier. A series of experiments are taken up on these news documents, and hence an inference for the proposed method is marked out for it being much reliable than other existing systems for emotion detection of a text document.

**Keywords:** Deep learning · Emotion prediction · Ontology · Natural Language Processing · WordNet

## 1 Introduction

Prediction of human emotions plays a very crucial role in the developing world. A huge stream of documents and news stories are generated on web on a daily basis. A massive amount of content is created with rapid rise in internet population; predicated to reviews, original compositions and opinions built up in people regarding the ongoing issues. Taking into account, the realm of Natural Language Processing (NLP), researchers have come forward with considerable number of insightful experimentations on semantic and syntactic level processing of text. The conveyance of such tools into being calls for solutions to the challenge of dealing with humanistic and social aspect of texts such as sentiment, belief, attitude, emotions, etc.

**Motivation:** Considering all the previous works, most of them have centered around text mining based on writer's perspective. And the ones based on text categorization adhering to reader's perspective are mostly News-Headlines based classification which doesn't take the whole content into account, the results of which could be misleading. Moreover, the performance of the existing models and the clustering algorithms used is


S. Shukla et al. (eds.), *Data Science and Security*, Lecture Notes in Networks and Systems 290,
https://doi.org/10.1007/978-981-16-4486-3_13

seen to be quite unstable. And for some of the models the community structure keeps on deviating for every restart of the framework. Therefore, a new algorithm is proposed to enhance the model stability as well as the various clusters formed for the news articles and contents generated.

**Contribution:** The following hypothesis constitutes the major contributions for this framework:

1 Term frequency calculation has been incorporated for Phase-I classification of the documents.
2 Community formulation is based on emotion ontologies and SentiWordNet.
3 The encompassment of concept similarity and inclusion of entropy functions ensures the stability of the proposed framework.
4 Similarity matching has been taken up on three different lines, i.e. TF-IDF calculation for ontology matching, calculation of NPMI measure for obtaining semantic-similarity in relevance to the ontology and finally the entropy calculation for each term.
5 The integration of deep-learning based classifier and semantic-matching are novel concepts that have helped in performance escalation of this work being presented.

**Organization:** This paper is organized in the following format - Sect. 2 gives a detailed description of relevant works done in this field. Section 3 consists of the complete framework architecture/technical approach. The implementation of the method put forth is described in Sect. 4. In Sect. 5, performance of the given algorithm is discussed. The paper has been concluded in Sect. 6.

## 2 Literature Survey and Related Works

Ohana and Tierney et al. [1] proposed a method to apply SentiWordNet in order to obtain a data-set of features and performed sentiment classification on film reviews based on this polarity data-set. Naresh Kumar and Gerard Deepak et al. [2] used linguistic semantic approach to detect emotions integrating NPMI and NAVA words. Jose Antonio, Mar Canovas-Gracia, Rafael Gracia in their study et al. [3] highlighted ontology-driven aspect-based approach for opinion analysis on infectious diseases. Dang, N.C., De la Prieta M.N., Moreno-García, et al. [4] in their research showed a comparative study sentiment prediction and evaluated the results given by deep learning-based models. Mohsen, Hesham, Amira et al. [5] used the SentiWordNet sentiment lexicon along with TF-IDF for estimating emotion polarities. Bhowmick Plaban in his study et al. [6] put forth a framework for multi-label emotion classification based on reader's perspective driven by semantic similarity and RAKEL technique. Nipun, Shashikant, Priyank performed their research of sentiment identification based on maximum entropy function et al. [7]. Machine Learning approach was incorporated built on entropy classification framework. Umang, Akanshu, Srikanth, Puneet et al. [8] proposed a multi-class classification model for emotion detection using LSTM based Deep Learning model. Advantages of semantic and sentiment-based embeddings has been exploited in their research. Sykora, Martin and Jackson et al. [9] devised a framework wherein the problem of emotion detection is taken up around encoded semantic knowledge based on an emotion ontology. Classification for eight basic emotions is taken into account. Franck, Sander brought in the idea of context-based ontology driven emotion prediction in their research et al. [10]. Formation of four emotion communities based on the highest ratings that are present in the news articles to construct final prediction result is taken up in the works of Ramya R S et al. [11]. Another approach of Opinion Network Community (ONC) was carried out in the works of X. Li et al. [12]. In [13–22] several ontology focused approaches in support of the proposed system are addressed (Fig. 1).

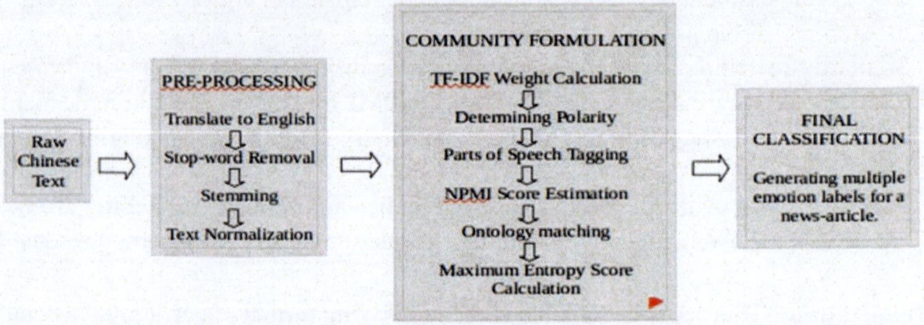


**Fig. 1** Proposed architecture

## 3 Proposed Methodology

A novel approach has been proposed in this paper that predicts intermediary emotions evoked in reader's predicated to the given news articles and documents. The working of this system has been distributed into three phases namely, Pre-processing, Community Formulation and Final Classification. Generally, the emotion induced by a news article is a combination of base emotions, hence this study falls under the category of multi-label classification. The proposed framework also incorporates translation tools to first convert the news article received in any language to English. Since, in the initial phase the given text is categorized under positive or negative label, two different set of keywords are formed based on which this categorization is made. The final classification is then derived conditioned to the categorization made earlier by processing it in a deep-learning classifier. To obtain the sentiment polarity represented by an entity, SentiWordNet has been used. Semantic concepts are then derived on the lines of emotion ontology to make the final prediction. Matching of semantic concepts are done around three tracks. Firstly, TF-IDF is calculated which is used to measure the similarity in accordance with the ontology, followed by the estimation of NPMI measure to implement similarity matching and in the end entropy function is calculated for each term to be fed to the classifier for the final output.

The pre-processed text corpus obtained from phase I, is then given for community formulation based on polarities acquired. Each news article passes through distinct stages in this phase. A lexicon is formed comprising each word present in the given dataset. TF-IDF weighting is calculated for each term present in this lexicon. These weights are provided to each term so as to measure its significance to a given news-article or the whole document. The objective of incorporating TF-IDF is to estimate the occurrence of a term across a specific sentence and then compare this occurrence past the whole document to assign weights which falls in accordance to the importance of the term in prediction of emotion. After calculating TF-IDF, SentiWordNet sentiment lexicon is utilized to draw out the sentimental classification of the weighted terms by adding polarity $P_s$, for the text being either negative, positive, or neutral. These scores reflect polarity biases of $-1$, 0 and 1; for negative, neutral and positive respectively. SentiWordNet biasing is followed by categorizing the terms into parts of speech which is derived from the WordNet lexicon. Every token in the document is tagged with the parts of speech it belongs to, so that the biasing scores are applied correctly. Stanford Part of Speech Tagger is used for this framework. Overall score $S_c$, is estimated for each part of speech and the ratio of scores is calculated along-with; relative to the number of terms as shown by Eq. 1.

$$S_c = S\frac{r_i}{R}C \tag{1}$$

where, $C$ is a constant and $r_i$ gives the term position w.r.t. the total number of terms, R in the news article. For the purpose of multi-label classification, collocation and association between words are found using Pointwise Mutual Information (PMI) and Normalized Pointwise Mutual Information (NPMI). PMI score serves the pivotal role for generating responses which later helps in identifying emotions using the Eq. 2.

$$pmi(a; b) = log\frac{p(a, b)}{p(a)p(b)} = log\frac{p(a/b)}{p(a)} = log\frac{p(b/a)}{p(b)} \tag{2}$$

This measure being relative demands to be normalized; for which NPMI is used. NPMI is a normalized PMI measure. The semantic relation between two words are estimated using NPMI measure on the lines of their co-occurrence, denoted by the Eq. 3.

$$npmi(a; b) = \frac{pmi(a; b)}{h(a, b)} \tag{3}$$

where, h(a,b) is the joint self-information which is estimated as $-log_2 p\ (X = a,\ Y = b)$.

On the basis of these NPMI scores, sentences are appropriately organized into particular sets and thereafter classification of these sets are done according to given emotions. The emotions taken into consideration in the proposed methodology are; emo = {happy, anger, sad, sorrow, resentment, surprise, fear and disgust}. An emotion vector is generated for each sentence while the context computation for these sentences is also taken into account. The sentiment tags given to the terms are now checked if they are emotion-denoting words referring to the emotional categories from emotion ontology. Emotion ontology has been derived from WordNet Affect Database. This lexicon has over 900 synsets (noun, verb, adverb synonym terms) representing concepts which forms a hierarchical set under emotional categories. Each word present in this lexicon is annotated with the emotion they evoke. The weighted terms generated as a processed result of SentiWordNet polarity bias having their co-occurrence measure are now matched with the concepts of this emotion ontology where ontology concept is formed as a triplet of subject-verb-object, E *(sub, vb, ob)* along with emotion tags. Semantic similarity is calculated between these ontology concepts and the terms present in the news article. Ontology matching is made with all the ontologies present in the ontology base, and then the emotion with highest similarity score is assigned. Once, similarity matching and assignment of emotions is done using ontology, maximum entropy function is employed given by Eq. 4, to combine the various chunks of these subjective texts.

$$H = \sum P(V)log(P(V))\forall V \tag{4}$$

so as to calculate the probability of a certain feature emotion class possessing a definite context. The probability that the given news article belongs to a specific feature category is estimated by Eq. 5.

$$P(cs_i \vee doc) = \frac{1}{Z(doc)} exp\left(\sum_j \lambda_j f_j(doc, c_i)\right) \tag{5}$$

where, $P(cs_i/doc)$ is the probability, that the particular feature class occurs for a given news-article. Linguistic Feature selection, $L_s$ for each document is done here. These extracted features are given as input to deep learning model built up on LSTM subsequently for multi-class classification. LSTM forms an extended part of RNN which is favorable for time-dependent processing and filters out the redundant data while the process.

## 4 Implementation

The proposed methodology of PIREN has been implemented using Python on Anaconda IDE. For ontology construction Web Protege and OntoCollab are used. Semantic relatedness is computed for each emotion based on the constructed ontology. All the formulae used have been implemented explicitly in Python to estimate the final predictions for each instance in the algorithm put forth. The emotion prediction algorithm has been depicted in Table 1 which takes a news document as an input and gives out a set of emotions as output corresponding to the news-article. The algorithm put-forth takes into account a high accuracy achieving deep learning model based on LSTM which forms a part of RNN. Calculation of TF-IDF and NPMI co-occurrence score along with maximum entropy calculation is incorporated in the algorithm. Furthermore, semantic match-making is done based on ontology concepts.

## 5 Results and Performance Evaluation

This experiment was taken up on a collection of News documents, The Yanghui Rao's corpus from the society channel of Sina. These documents are in Chinese Language and comprise of 4000 news articles which needs to be converted to English first. To evaluate the performance of this multi-label classification model, different metrics have

**Table 1.** Algorithm for Prediction of intermediary emotions based on reader's perspective

**Input:** A News-Article $N_A$
**Output:** A set of emotions $E_K$ belonging to the document.
***begin***
**Step 1:** Initialize per-processing techniques on $N_A$
**Step 2:** **for each** term in $N_A$
    Calculate TF-IDF
    Add polarity bias $P_s$ to each term based on SentiNET lexicon
    Tag each word with certain parts of speech
    Estimate overall score $S_C$ for each term tagged with particular parts of speech
    **end for**
**Step 3:** **for each** word $W_R$ having score $S_c$
    Compute NPMI co-occurrence measure to estimate semantic relation between two words
    Form a set *ST* of sentences marked with some emotion, *emo*
    **end for**
**Step 4:** Form ontology concept E (sub, vb, obj) for ontology matching
**Step 5:** Compute maximum entropy function
**Step 6:** Perform linguistic feature selection for each document and extract features $L_S$
**Step 7:** Predict an emotion set $E_K$, forming a multi-label classification output.
***end***

been used. Metrics of Precision, Recall, Accuracy, F-Measure, and False Negative Rate. (FNR) has been used in their standard formulations. The prediction of emotions made by the proposed framework is evaluated for 8 intermediary emotions, emo = {Happy, Sad, Anger, Resentment, Sorrow, Surprise, Fear, Disgust}. Average of these metrics has been considered as the foremost evaluation criteria. The average precision for all the emotions achieved by the proposed model is 92.78%. This framework yields an average recall of 90.21%, average accuracy of 91.87%, average F-measure of 91.1% and an average FNR rate of 0.1. The values achieved for all these evaluation metrics are quite significant and remarkably appreciable. Table 2 gives the complete computation details carried out while experimentation for the proposed architecture in line with reader's emotions prediction.

To demarcate contrast, of the proposed system with other existing models, we have taken into account EPNDR [11] and ONC [12] as baseline models. It was seen that the proposed framework is more efficient than the other two. The higher accuracy and precision achieved by the proposed framework is subjected to the usage of semantic similarity measurement along with deep learning based multi-label classifier. The approach of community formation based on distance in ONC results in a system imbalance, which gives out new estimation on each restart. Similarly, in EPNDR, calculation of just the term frequency and then prediction of emotion based on probability leads to

**Table 2** Performance measures of the proposed approach

| Predicted Emotions | Precision % | Recall % | Accuracy % | F – Measure % | FNR |
|---|---|---|---|---|---|
| Happy | 95.27 | 93.18 | 94.89 | 93.55 | 0.07 |
| Sad | 96.12 | 94.81 | 95.74 | 95.18 | 0.06 |
| Anger | 96.14 | 94.12 | 95.02 | 95.23 | 0.06 |
| Resentment | 92.14 | 88.37 | 90.14 | 90.33 | 0.12 |
| Sorrow | 94.18 | 90.14 | 91.89 | 92.38 | 0.1 |
| Surprise | 89.33 | 87.14 | 88.69 | 87.76 | 0.13 |
| Fear | 91.69 | 88.14 | 90.17 | 89.62 | 0.12 |
| Disgust | 87.43 | 85.79 | 88.44 | 84.81 | 0.15 |
| **Average** | **92.78** | **90.21** | **91.87** | **91.1** | **0.1** |

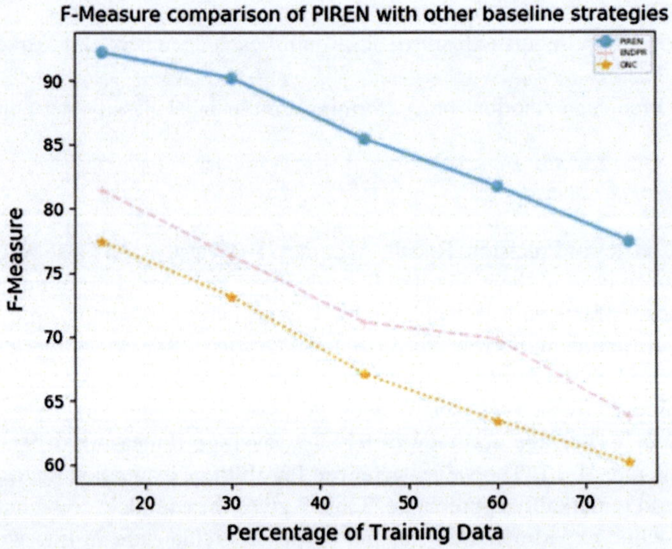


**Fig. 2** F-Measure vs. percentage of training data

biased estimation in some cases. Moreover, both the baseline models give prediction for very few emotions whereas the proposed PIREN framework gives prediction for eight intermediary emotions. Furthermore, calculation of NPMI co-occurrence score adds to better performance given by PIREN.

Figure 2 gives the comparative analysis for all the three models in terms of F-measure. Since, F-measure gives the weighted harmonic mean of recall and precision, thus, greater the F-measure value, the more efficient the model is. It is evident from Fig. 2 that the F-measure score achieved by PIREN is much higher than ONC or EPNDR. Incorporation of maximum entropy calculation helps in enhancing the efficiency of the proposed algorithm. Estimating polarity for each term and then classifying using LSTM deep learning classifier further amplifies the performance. The graphical estimation of F-measure achieved by all the three models in consideration w.r.t. the training data used shows the competence of PIREN. PIREN tends to achieve highest F-measure score for each set of training data evaluated. Of all the data considered, it can be approximated that the best efficiency is shown by the methodology put forth, PIREN. Henceforth, it can be concluded that PIREN is the best-in-class model for prediction of reader's emotions based on news-documents received.

## 6 Conclusion

An ingenious and robust strategy has been put forth through this paper for the prediction of reader's emotion given a News-Document. The proposed algorithm PIREN classifies documents based on deep learning model structured on LSTM framework. PIREN incorporates semantic match-making based on ontology concepts for per-classification emotion tagging. It is an ensemble of multiple techniques which form a sequential procedure for document classification. TF-IDF is calculated at initial stages to find negative–positive polarity for each term using SentiWordNet lexicon. Following polarity assignment, NPMI co-occurrence score is estimated which helps in categorizing each term given in the news-article. Each term is categorized under 8 intermediary emotion labels. Probability is then calculated for each such emotion class followed by final classification using a multi-class classifier. The performance displayed by PIREN is quite efficient and yields satisfactory results for each document given as input. The average precision achieved by the proposed methodology is 92.78% while the average recall value obtained is 90.21%. Furthermore, the average accuracy yielded towards all the eight emotions taken into consideration is 91.87%. The average F-Measure value reaches 91.1% and the FNR rate secured by the proposed system is 0.1. All these estimations help in arriving at an approximation of PIREN being best-in-class model for reader's emotion predictions. The future work upon this system could be probably focused towards increasing the overall relevance. Also, different techniques of semantic similarity or co-occurrence estimation can be used to achieve even better values for evaluation metrics.

## References


1. Ohana B, Tierney B (2009) Sentiment classification of reviews using SentiWordNet. In: 9th IT&T conference, technological university Dublin, Dublin, Ireland, 22–23 October 2009. https://doi.org/10.21427/D77S56
2. Naresh Kumar D, Deepak G, Santhanavijayan A (2020) A novel semantic approach for intelligent response generation using emotion detection incorporating NPMI measure. Procedia ComputSci 167:571–579
3. Antonio J, Canovas-Gracia M, Valencia-Gracia R (2020) Ontology-driven aspect-based sentiment analysis classification: an infodemiological case study regarding infectious diseases in Latin America. Future Gener Comput Syst 112:641–657

4. Dang NC, Moreno-García MN, De la Prieta F (2020) Sentiment analysis based on deep learning: a comparative study. Electronics 9:483
5. Mohsen H (2016) Amira "documents emotions classification model based on TF-IDF weighting measure." World Acad Sci Eng Technol Int J Comput Electr Autom Control Inf Eng 10:252–258
6. Bhowmick P (2009) Reader perspective emotion analysis in text through ensemble based multi-label classification framework. Comput Inf Sci 2. https://doi.org/10.5539/cis.v2n4p64
7. Mehra N, Khandelwal S, Patel P (2002) Sentiment identification using maximum entropy analysis of movie reviews
8. Gupta U, Chatterjee A, Srikanth R, Agrawal P (2017) A sentiment-and-semantics-based approach for emotion detection in textual conversations
9. Sykora M, Jackson T, O'Brien A, Elayan S (2013) Emotive Ontology: extracting fine-grained emotions from terse, informal messages. Int J Comput Sci Inf Syst 8:106–118
10. Berthelon F, Sander P (2013) Emotion ontology for context awareness. In: Coginfocom 2013 - 4th IEEE conference on cognitive infocommunicaitons, December 2013, Budapest, Hungary. ffhal-00908543f
11. Ramya, RS, Madhura K, Sejal Venugopal KR, Iyengar SS, Patnaik LM (2020) EPNDR: Emotion prediction for news documents based on readers' perspectives. Int J Sci Technol Res 9(01):531–539
12. Li X, Peng Q, Sun Z, Chai L, Wang Y (2017) Predicting social emotions from readers' perspective. IEEE Trans Affect Comput 10(2):255–264
13. Deepak G, Teja V, Santhanavijayan A (2020) A novel firefly driven scheme for resume parsing and matching based on entity linking paradigm. J Discrete Math Sci Crypt 23(1):157–165
14. Deepak G, Santhanavijayan A (2020) Onto best fit: a best-fit occurrence estimation strategy for RDF driven faceted semantic search. Comput Commun 160:284–298
15. Kumar N, Deepak G, Santhanavijayan A (2020) A novel semantic approach for intelligent response generation using emotion detection incorporating NPMI measure. Procedia Comput Sci 167:571–579
16. Deepak G, Kumar N, Santhanavijayan A (2020) A semantic approach for entity linking by diverse knowledge integration incorporating role-based chunking. Procedia Comput Sci 167:737–746
17. Haribabu S, Kumar PSS, Padhy S, Deepak G, Santhanavijayan A, Kumar N (2019) A novel approach for ontology focused inter-domain personalized search based on semantic set expansion. In 2019 fifteenth international conference on information processing (ICINPRO), December 2019, pp 1–5. IEEE
18. Deepak G, Kumar N, Bharadwaj GVSY, Santhanavijayan A (2019) OntoQuest: an ontological strategy for automatic question generation for e-assessment using static and dynamic knowledge. In 2019 fifteenth international conference on information processing (ICINPRO), pp 1–6. IEEE
19. Kaushik IS, Deepak G, Santhanavijayan A (2020) QuantQueryEXP: a novel strategic approach for query expansion based on quantum computing principles. J Disc Math Sci Crypt 23(2):573–584
20. Varghese L, Deepak G, Santhanavijayan A (2019) An IoT analytics approach for weather forecasting using Raspberry Pi 3 Model B+. In: 2019 fifteenth international conference on information processing (ICINPRO), December 2019, pp 1–5. IEEE
21. Deepak G, Priyadarshini S (2016) A hybrid framework for social tag recommendation using context driven social information. Int J Soc Comput Cyber-Phys Syst 1(4):312–325
22. Deepak G, Priyadarshini JS (2018) A hybrid semantic algorithm for web image retrieval incorporating ontology classification and user-driven query expansion. In: Advances in big data and cloud computing, pp 41–49. Springer, Singapore https://doi.org/10.1007/978-981-10-7200-0_4

# Automated Organic Web Harvesting on Web Data for Analytics

Lija Jacob(✉) and K. T. Thomas

Christ University, Bangalore, India
lija.jacob@christuniversity.in

**Abstract.** Automated Web search and web data extraction has become an inevitable part of research in the area of web mining. The web scraping has immense influence on ecommerce, market research, web indexing and much more. Most of the web information is presented in an unstructured or free format. Web scraping helps every user to retrieve, analyze and use the data suitably according to their requirement. There exist different methodologies for web scraping. Major web scraping tools are rule based systems. In the proposed work, an automated method for web information extraction using Computer Vision is proposed and developed. The proposed automated web scraping method comprises of automated URL extraction virtual extraction of required data and storing the data in a structured format which is useful in market research.

**Keywords:** Organic web harvesting · Data scraping · Automated scraping · Indexing · SERP pages

## 1 Introduction

There is an unbelievable growth of data in the internet over the last five years. The reason behind this data evolution is the exponential growth in the number of internet users in this decade. Public or individuals, industries, and devices have all become data factories that are propelling out incredible quantities of information to the web each day. In every minute of a day, in the internet, we can find data from social media like Facebook, Twitter, emails, mobile device data, data created by IoT, data generating services such as Amazon.

Web scraping is the extraction of data through any mode rather than a program interacting with an API or through a human using a web browser [1, 2]. The task of web scraping can be attained by scripting an automated program that probes a web server to requests data, and further analyzes the data to be extracted for the prerequisite information [3, 4]. In reality, web scraping involves a wide variety of programming techniques and technologies, such as data analysis and information security.

Web harvesting is getting more popular since last decade. There can be manual or automatic web scraping. Manual web scraping involves copying and pasting the web content. It is a highly tedious and repetitive task to be carried out. Manual scraping is recognized as a expensive, error prone and time consuming method of data extraction [5].


S. Shukla et al. (eds.), *Data Science and Security*, Lecture Notes in Networks and Systems 290,
https://doi.org/10.1007/978-981-16-4486-3_14

Manual web scraping is comparatively difficult to analyze especially for non-experts. Moreover, the time required for data analysis is much more than when compared to automatic scraping. As technology is moving faster and the rate of growth of data is growing exponentially, a manual scraping of data can be a failure owing to high cost, error prone and at the same time taking a whole heap of time. Instead, an automatic web scraping can be preferred.

Automatic web scraping has become common in todays' techie world. One of the major scopes of web scraping is Intelligence gathering for content marketers. The intelligent gathering helps business people to formulate a good decision, strategize or plan based on market research data. An expert web scraping tool can help us to find data that is vital and associated to the required business. This helps business men to focus on mining the valuable understandings from the delivered data. It helps users to analyze some trends or common facts, such as the product which is more prevalent among the different age groups and which gender prefers what beauty products and so on. Web scraping helps to gather the data and conduct such an analysis. Brand monitoring including product reviews monitoring, price comparisons or price wars. Stock Market Tracing, Job search, chasing latest trends are all the benefits of web scraping [6, 7].

The different methods of automatic scraping include HTML Parsing, Document Object Model Parsing, Vertical Aggregation, XPath and many other types of web grabbers developed by different organizations based on their on their own requirement. These scrapping tools extracts contents from the HTML pages, XML files etc. Such scraping methods can result in an action of stealing of data and thus it is illegal. Hence such scraping can be represented as an unethical method of grabbing data. The proposed paper proposes a computer vision-based web scraping which can be represented as Optical Web Harvesting which is an Organic mode of web harvesting.

The sections of the proposed paper are arranged as follows. Section 2 gives an outline of currently existing auto scraping methods and visual based web scraping methods. Section 3 contains the details the proposed system of Organic Web Harvesting. Section 4 explains the experiments and results followed by Conclusion in Sect. 5.

## 2 Literature Review

Web scraping is a form of web mining. The ultimate objective of the web scraping process is to extract the required loads of data from related websites and transform it into comprehensible and clear information and represent it in a structured format like spreadsheets, database or a comma-separated values (CSV) file.

In [8], S Sirisuriya has given a comparative study on different web scraping techniques and tools. The most traditional approach is the copy and paste method which is a manual one. In this manual approach, the humans' manual inspection and copy-paste method is required. But the technique is an error-prone, dull and tedious procedure when users need to scrap lots of data. Another method of web scraping includes Text gripping and regular expression approach, Hyper Text Markup Language (HTML) Parsing, Document Object Model (DOM) Parsing, Vertical aggregation platforms, Semantic annotation recognizing. Most of these automatic scraping techniques are found to be very slow for larger sites, difficult to setup and require high CPU and memory usage.

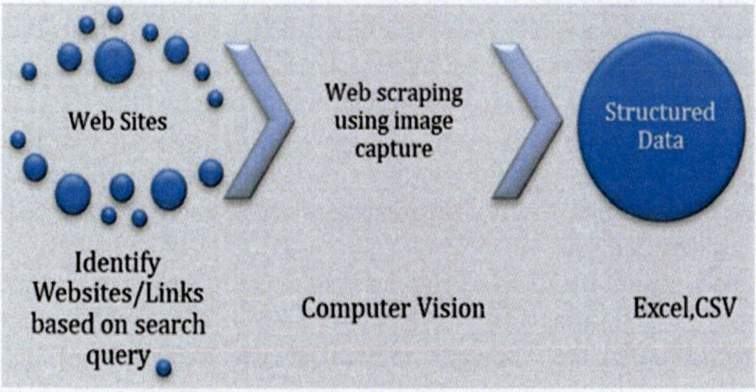


**Fig. 1** Basic process of web scraping

It is not ideal for large projects and some tools which use techniques are not beginner friendly [9].

In [12], the paper has also introduced Computer vision-based web-page analyzers that uses machine learning and computer vision algorithms to identify and extract information from web pages. The algorithm mimics the human vision process and interprets the details in the pages visually similar to what a human being might do. While comparing the code based or browser-based crawlers, the proposed computer vision crawlers present some great advantages. They offer easiness when compared to other web crawlers, that is, even the non-developers can easily learn what content need to be extracted. The paper has shown the algorithms' performance in extracting organized content from sources it has no knowledge about.

The proposed system explains a completely automated computer vision-based system, which uses image processing techniques to identify the links of the websites based on the query; and scrap relevant data from each site to convert unstructured web data to a structured content of data which can be used for analytics.

## 3 Overview of Web Scraping System

The major aim of the web scraping system is to mine information from different websites and transform the extracted unstructured data into comprehensible structure like spreadsheets, database or a comma-separated values (CSV) file. The basic methodology of the web scraping process is shown as follows (Fig. 1).

## 4 The Proposed System

The proposed system of computer vision-based Web Harvesting can be separated into 3 modules-

- Fetch Phase
- Extraction phase
- Transformation phase

The system uses the optical or the visual information of the web pages.

In the fetch phase, a query passed by the user will prompt the display of the SERP pages listing out the list of links of the websites that contains the relevant data to be accessed. In the extraction phase, each of the links displayed in the SERP page is automatically opened, and the required data is extracted. In the transformation phase, the obtained data is transformed to any structured data, which can further be used for storage, presentation or data analysis.

The proposed vision-based web scraping system completely tries to extract the data from a web page in the same way as a human does. This is entirely an automated web harvesting method which has nothing to do with the code of the webpage but can extract the data displayed in the web pages. Usually, a web page consists of multimedia data especially images and text. The foremost and the common visual information of any web page includes mainly its layout. The layout, font and arrangement of data in a web page are the key essentials considered in the proposed system.

**Algorithm**

```
Algorithm CVWebHarvest
  STEP 1: Input Search Query through the user interface
  STEP 2: Open the browser automatically
  STEP 3: Query Automatically passed through the browser
and query is sent to the web server for the results
  STEP 4: The browser returns the SERP pages
  STEP 5: Automated screenshot capture of the SERP page
using Computer Vision procedures
  STEP 6:  Image processing procedures done for the
extraction of the links from the result pages
  STEP 7: Identification of the text data from the links
  STEP 8: Automated link extraction using CV techniques
and virtual keyboard and mouse functions
  STEP 9: Store each link obtained.
  STEP 10: Navigating each link obtained as the result
and open the corresponding web page
  STEP 11: Data extraction from the opened web page and
storing in a data table in the structured format
```

Figure 2 shows the mechanism of how we can turn web scraping problem into an image-based solution. The elaborated process flow of the proposed system is shown in the figure below. The search query is passed through a user interface. The browser is automatically opened and the search query is passed to it. The result of the queries will be displayed as snippets in the web results.

The snippet is mainly a single search result in a set of search results. It usually consists of a heading title, a URL and a description of the page. The content of a snippet matches parts of the search query and we can see the keyword highlighted in the snippet description [10] (Fig. 3).

The Fig. 4 below shows a snippet obtained as part of the search query. From the description and figure of a snippet, it is understood that it contains a URL description and title. The appearance of each component of the snippet is different. There lies a difference in the font and color of the text in each of the component. These differences

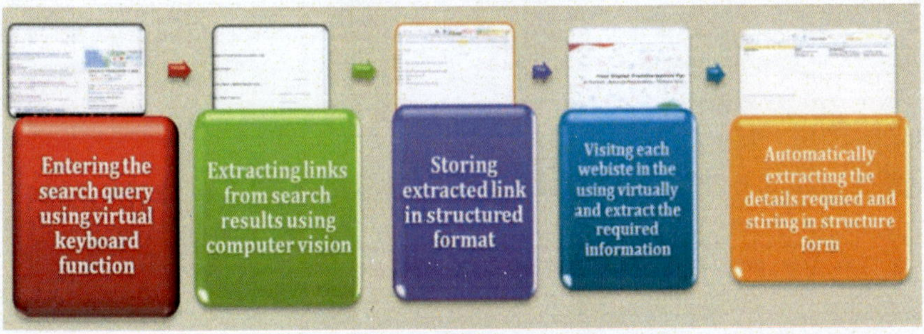


**Fig. 2** Proposed system

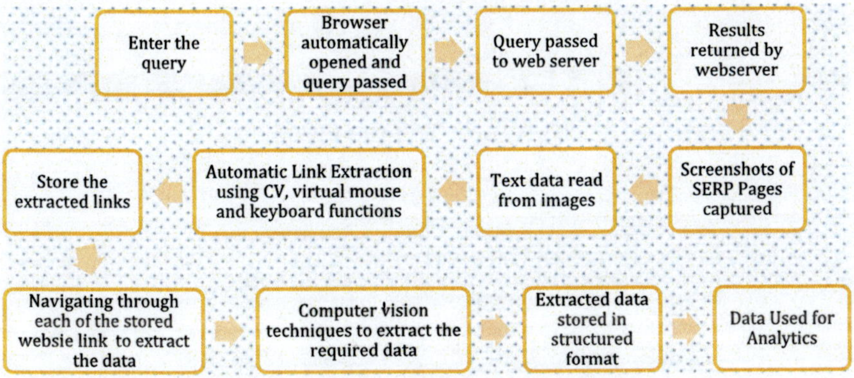


**Fig. 3** The process flow of the proposed system

are considered as the identification factor or feature of interest in the proposed system. It helps to identify the URL and all the URLs can be stored.

**Anatomy of Snippet in SERP Page**

The snippet is obtained as the result of a single search in a set of search results. It mostly consists of a title, a URL and an explanation of the page. The snippet content matches the parts of the search query. The keyword which the user searched will be highlighted in the snippet description. The search engines define and display the best possible snippets based on the search. An organic search can result produce snippets in any of the 3 types.

- Regular Snippet
- Rich Snippet
- Featured Snippet

Figure 5 shows the different types of snippet obtained in SERP pages as part of Organic Search. From the figure it is evident that the snippet components have different styles which can be easily identified.

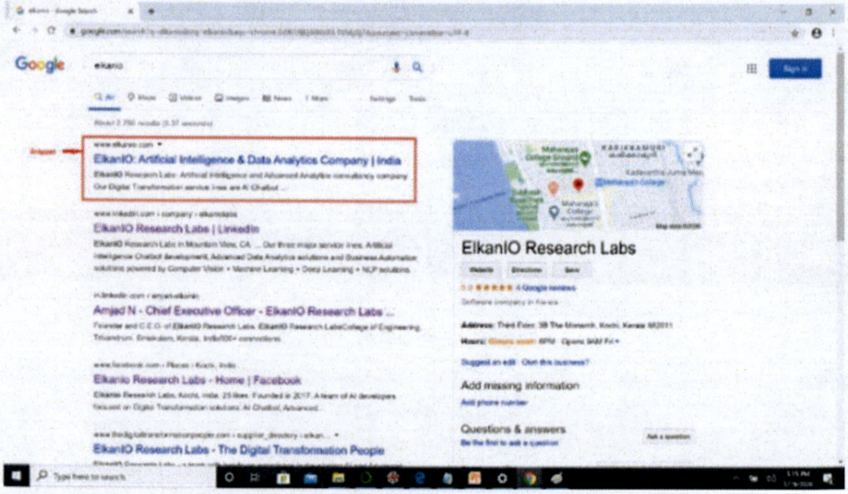


**Fig. 4** Figure showing a snippet

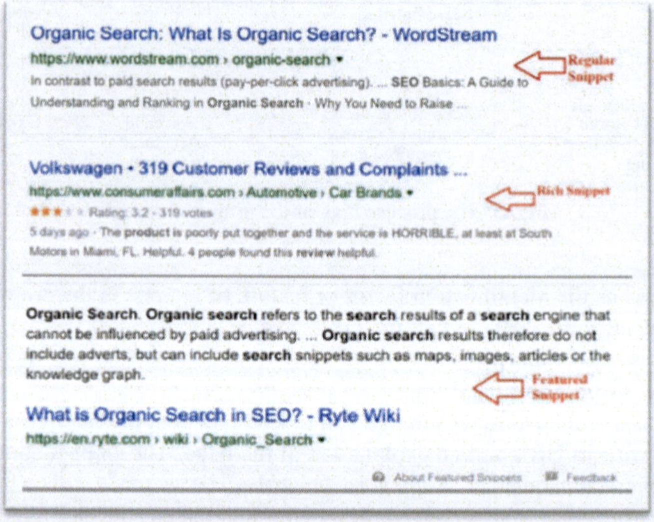


**Fig. 5** Figure depicts the different types of snippets

**Fetch Phase Using Visual Features of the SERP Page**
In the proposed system, a screenshot of each SERP page is captured. Here the visual characteristics of the Search Result page including the font, position are identified. A few image processing steps are performed to identify the URLs of the screen. This includes color space conversions, filtering so that the URL segments are highlighted. Once the preprocessing of the image is done and the image with the URLs is obtained, text localization is performed on the image. The text localization and extraction is performed

with the help of the library called pytesseract. Python-tesseract is an optical character recognition (OCR) tool for python [11]. It will help to recognize and "read" the text embedded in images. Thus pytesseract can be used to extract words from an image. The image_to_data() will convert the image to a data output and the number of detected items can also be identified. The location of detected items is stored. These positions are the actual positions of the URLs in the search page. The browser is loaded with the same query and the mouse cursor passes through the stored positions extracting the same links. PyAutoGUI is used to perform all the automated mouse and keyboard movements such as capture the screenshots, save them to files, and locate images within the screen, virtual mouse clicking on stored coordinates etc. The link is copied during each virtual click and stored in an excel file. Thus, the large amount of data which is obtained as the search results are stored in a structured format in database.

**Extraction Phase**

Once all the URLs are identified extracted and the address stored in a structured format, the websites of each link is visited to extract and store the required data. The data extraction phase focus on identification of the required data records and extract them. The data records are extracted in such a way that it identifies the correct data from the region extracted and at the same time least amount of data item is missed. In the proposed system, the data extraction is performed at various steps such as correctly identifying the URLs for traversing and correctly extracting the required data from the visited web pages. The data extraction phase includes noise filtering, data block grouping by identifying the data boundary. The data blocks are grouped if the distances between two blocks are below a given threshold.

**Transformation Phase-Unstructured to Structured Presentation of Information**

The transformation phase includes data item extraction and data segmentation.

Data Records-The required data from each visited page is automatically extracted. The item extracted is an image and it consists of multiple details. Each data required can be segmented and is stored in required format. The extracted data can be easily used for presentation and for analysis. In the proposed system, the data is used for market research.

## 5 Experiments and Results

The system is developed in Python and OpenCV. OpenCV is used to design a computer vision system to extract information from the real websites for extracting information and performing data analysis. Different libraries including pyautogui, win32api, pytesseract, numpy are used for doing different functions in the proposed system. The Fig. 6 shows the results of the preprocessing of the image.

The proposed system visits each website link stored in the file and extracts the required information using the computer vision techniques. For example: to identify whether a website owns a Chabot or not; or the price of a product and so on. Such data extracted is stored in a structured format which is used for data analytics in business research. Consequently, using a combination of machine learning and computer vision

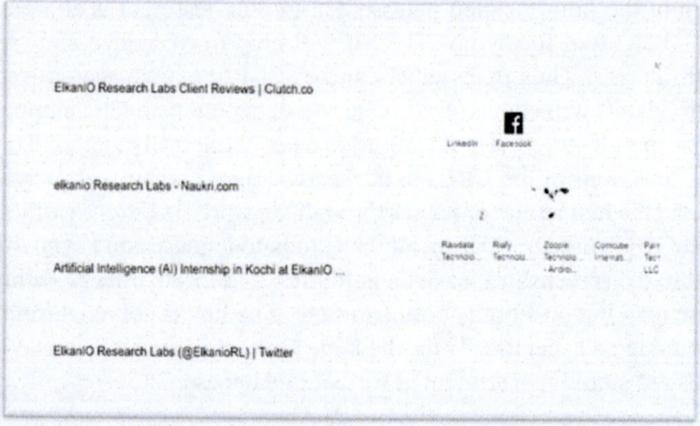


**Fig. 6** Figure showing the result of preprocessing of the image. The bigger texts are the URL links in the webpage

techniques, the proposed organic web harvesting method becomes very similar to what the human visual processing system is doing, when a user is viewing a web page and trying to make intelligence out of it. The computer vision methods identify the relationships and context between the various types of data on the web **pages,** enabling the system to automatically extract the data and store as structured data.

The web data scraping in the proposed system has the following advantages.

a. The system almost mimics human behavior, which means that by using Computer Vision the biological mode of web harvesting is performed. Therefore, the performance is similar to or above the Human-Level Precision.
b. Computer vision-based web scraping is simpler than code-based crawlers because even the non-developers can teach the system what content needs to be extracted.
c. Compared to other types of web scraping tools; the computer vision-based scrawling is more ethical.

But along with this it is noted that the computer vision crawlers do perfectly good for sites that look almost the same but may not perform well in large scale different structured websites.

Figures 6, 7 and 8 explains the different process undergone during the automated web scraping.

**Performance Evaluation**
The performance of the proposed computer vision based webscraping system is based on different accuracy measures:

$$\mathbf{Acc(LE)} = \frac{\mathbf{NLExtracted}}{\mathbf{NLActual}} \tag{1}$$

**where**

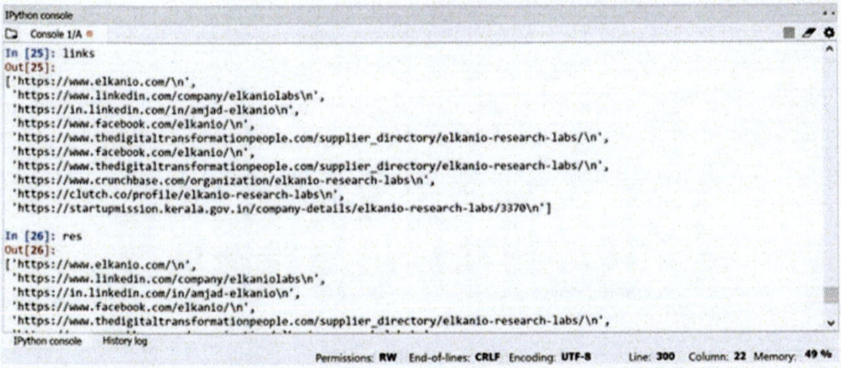


**Fig. 7** Screenshot showing the linked automatically extracted

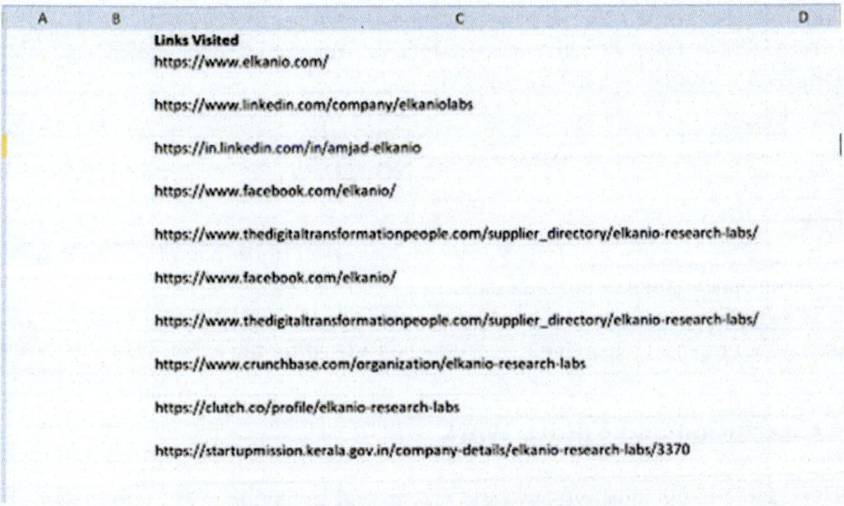


**Fig. 8** Extracted data stored in Excel file

Acc(LE) is the Accuracy of link extraction; NLExtracted = No: of Links correctly extracted from SERP pages; NLActual, = No: of Actual Links in the page().

$$\mathbf{Acc(CW)} = \frac{\mathbf{NWeb}}{\mathbf{NLActual}} \tag{2}$$

where

Acc(CW) is the Accuracy of the correct websites visited; NWeb = No: of websites automatically opened.

$$\mathbf{Acc(DE)} = \frac{\mathbf{NData}}{\mathbf{NOrigData}} \tag{3}$$

where

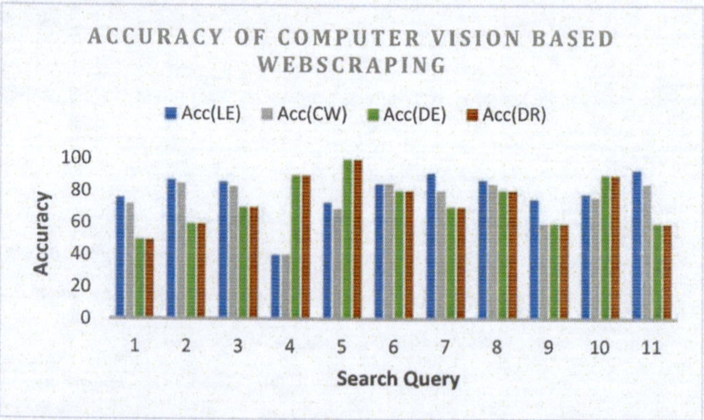


**Fig. 9** Performance evaluation of the proposed Computer Vision based Web scraping

Acc(DE) = Data Effectiveness; NData = Amount of data pulled and stored; NOrigData = Actual data to be extracted.

$$\mathbf{Acc(DR)} = \frac{\mathbf{NCorrectData}}{\mathbf{NOrigData}} \tag{4}$$

where

Acc(DR) = Accuracy of Data Retrieved; NCorrectData = No: of correct data;
NOrigData = Total Number of data.

The graph in Fig. 9 shows the different accuracy obtained using the computer vision based web scraping.11 search queries were considered for the experiments.

## 6 Conclusion and Future Work

Web scraping is the most effective and resourceful technique to get information from websites automatically. The data available in the web in different format like text, image, video can be used for extracting information in different fields of research. These data is in an unstructured format. There needs to be mechanism which may help to scrap data in an ethical manner. Optical web harvesting helps the user to easily extract unstructured data from single or multiple websites into a structured data automatically. The proposed system uses the visual method of scraping data. The main aim of the technique of converting the data from unstructured to structured form is to mine information from the web and aggregate it into new knowledge and dataset which can be used for analytical and research purpose of organizations. The proposed paper has implemented the mechanism to completely automate the web data harvesting in visual mode without any human intervention. The system can be implemented in all major domain of research where analysis and analytics plays a vital role. The system performance can be improved using deep learning algorithms.

## References


1. https://www.webharvy.com/articles/what-is-web-scraping.html
2. https://wscraper.com/what-is-data-harvesting-and-how-to-prevent-it/
3. Ashiwal P, Tandan SR, Tripathi P, Miri R (2016) Web information retrieval using python and beautifulsoup. Int J Res Appl Sci Eng Technol 4(VI). ISSN: 2321–9653
4. Peterson A (2021) BeautifulSoup: Web Scraping with Python
5. https://www.shieldsquare.com/what-are-the-different-scraping-techniques/
6. https://towardsdatascience.com/https-medium-com-hiren787-patel-web-scraping-applications-a6f370d316f4
7. https://www.import.io/post/web-scraping-explained/
8. Sirisurya S (2015) A comparative study on web scraping. In: Proceedings of 8th International Research Conference, KDU
9. https://www.analyticsvidhya.com/blog/2020/04/5-popular-python-libraries-web-scraping
10. https://yoast.com/what-is-a-snippet/
11. https://pypi.org/project/pytesseract/
12. Liu W, Meng X, Meng W (2010) ViDE: a vision-based approach for deep web data extraction. IEEE Trans Knowl Data Eng 22:447–460. https://doi.org/10.1109/TKDE.2009.109

# Convolutional Autoencoder Based Feature Extraction and KNN Classifier for Handwritten MODI Script Character Recognition

Solley Joseph[1](✉) and Jossy George[2]

[1] Science and Commerce for Women, Carmel College of Arts, Nuvem, Goa, India
solley.joseph@res.christuniversity.in
[2] Christ University, Bangalore, India

**Abstract.** Character recognition is the process of identifying and classifying the images of printed or handwritten text and the conversion of that into machine-coded text. Deep learning techniques are efficiently used in the character recognition process. A Convolutional Autoencoder based technique for the character recognition of handwritten MODI script is proposed in this paper. MODI script was used for writing Marathi until the twentieth century. Though at present, Devnagari is taken over as the official script of Marathi, the historical importance of MODI script cannot be overlooked. MODI character recognition will not be an easy feat because of the various complexities of the script. Character recognition-related research of MODI script is in its initial stages. The proposed method is aimed to explore the use of a deep learning-based method for feature extraction and thereby building an efficient character recognition system for isolated handwritten MODI script. At the classification stage, the features extracted from the autoencoder are categorized using KNN classifier. Performance comparison of two different classifiers, such as KNN and SVM, is also carried out in this work.

**Keywords:** Handwritten character recognition · MODI script · CNN · Autoencoder · KNN

## 1 Introduction

The process of identifying and classifying the images of printed or handwritten text and the conversion of that into machine-coded text is called character recognition. It is a branch of Computer Vision and Pattern Recognition [1]. Over the years, researchers in this area have contributed to the development of efficient character recognition for various languages. Compared to most of the foreign scripts, the Indian scripts are generally complex in nature and therefore, the development of a character recognition system for such scripts is a difficult task [2]. MODI script is one of the examples of a complex Indian script [3].

MODI script was widely used in Maharashtra between the twelfth and twentieth centuries as the official script of the state. The usage of the script for official purposes was put to an end because it was hard to typeset and print MODI script. Though at


S. Shukla et al. (eds.), *Data Science and Security*, Lecture Notes in Networks and Systems 290,
https://doi.org/10.1007/978-981-16-4486-3_15

**Fig. 1** The character set of MODI script

present, the script is not used officially, its historical importance cannot be overlooked. Huge collections of historical MODI documents are seen in many libraries across the country and abroad [4].

There are 36 consonants and ten vowels in MODI script, which constitute a total of 46 basic characters of the script. A unique characteristic of MODI manuscripts is that the absence of a demarcation symbol for words or characters in the documents [5]. Word segmentation in MODI manuscripts is a challenging task due to this reason. Figure 1 depicts the various characters of the MODI script.

The introduction of deep neural network-based methods opened up new avenues for doing research in the area of pattern recognition. This work focuses on the implementation of a deep learning-based feature extraction technique in MODI script character recognition. An autoencoder using a Convolutional Neural Network is implemented at the feature extraction stage of the recognition process of MODI script characters. The classification technique used is K-Nearest Neighbor (KNN). The experiment is repeated using a Support Vector Machine (SVM) classifier also, for the performance comparison of these two classifiers.

The remaining part of the paper is structured in four sections. Section 2 of the paper carries a literature review. In Sect. 3, the proposed methodology is described and the Sect. 4 details the experimental study. The conclusion of the work is given in Sect. 5.

## 2 Review of Literature

A survey of the literature shows that only a few research works are reported on MODI script character recognition. Kulkarni et al. [6] performed a review on MODI character recognition systems by listing out the research work done in this field by various researchers and it is concluded that, in comparison with other Indian scripts MODI script character recognition is challenging. Joseph et al. [3] Performed a study on various methods used in MODI character recognition and listed out the various techniques implemented in the feature extraction and character recognition phases. The techniques used at the feature extraction stage include Chain Code Histogram, Zernike Moments, Structure Similarity, Hybrid Techniques, Affine Moment Invariant and Hu's Moments. The classification techniques implemented in MODI script character recognition process are Decision Tree, Euclidian Distance, KNN and SVM [3].

Deep learning based methods are extensively used in numerous pattern recognition problems. These techniques are efficiently implemented at the various phases of the process [7, 8]. Convolutional Neural Network (CNN) based methods show very good results in various pattern recognition tasks and many researchers have experimented with CNN for character recognition tasks [9]. Arabic character recognition using CNN is experimented by Najadat et al. [10] and 94.9% accuracy is achieved. Deep neural network based feature extraction methods are implemented by various researchers. Autoencoder with a Deep Convolutional Neural Network was implemented by Shopon et al. [11] for the handwritten digit recognition of Bangal Script. Implementation of a discriminative autoencoder framework is used for the character recognition of numeral dataset (MNIST data set) by Anupriya et al. [12]. In order to extract relevant representation from the data set, a stacked architecture was built using the basic discriminating autoencoder as a unit. A similar method of Stacked Autoencoder (SAE) has also experimented for Arabic handwritten digit recognition by Mohamed et al. [13].

## 3 Methodology

The proposed method of MODI script character recognition uses a deep learing based feature extraction method. An autoencoder using Convolutional Neural Network is used as the feature extractor in this model. K-Nearest Neighbor classifier is implemented at the classification stage. Data augmentation is used to increase the volume of the training dataset. Data augmentation can be performed in two ways such as offline augmentation and on-the-fly augmentation. In this method, an on-the-fly augmentation method is used. Data augmentation helps to increase the diversity of data and prevents overfitting at the training stage. This way, the data augmentation helps in achieving better results while training the network [14].

The original image is given as the input. The input data was then augmented using on-the-fly data augmentation technique. At the feature extraction stage, the CNN autoencoder performs the extraction of features. In the classification stage, the KNN classifier is applied, and the recognition is performed. The procedure used in our experiment is depicted in Fig. 2.

### 3.1 Feature Extraction

Feature extraction is an important task in the process of character recognition. Various statistical and structural techniques can be used in this task. Deep learning based methods are successfully used in image processing and pattern recognition tasks. In this study, the feature extraction is implemented by means of an autoencoder. An autoencoder is a kind of feedforward neural network employed to learn effective data codings in an unsupervised manner. The three major components of an autoencoder are the encoder, code and decoder. The input image is transformed into a low-dimensional representation which is called a latent representation or code, using the encoder part of the autoencoder. The decoder part then regenerates the input using this latent representation. Thus the autoencoder is able to learn the important representations from the input data, which are

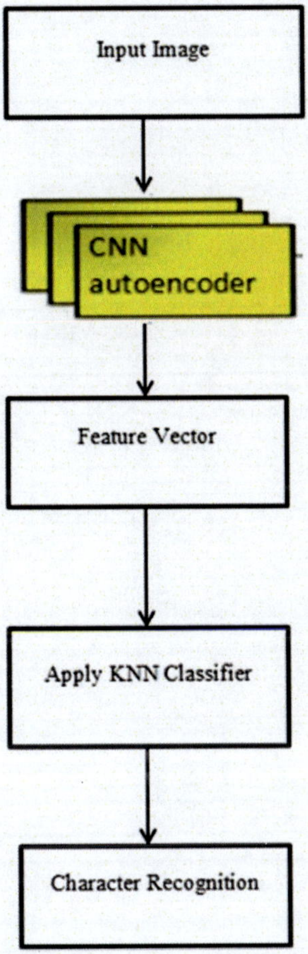


**Fig. 2** The proposed methodology

nothing but the features of the input image. Autoencoder can be constructed using different types of neural networks. In the proposed method, the autoencoder is constructed with Convolutional Neural Network. Figure 3 depicts the architecture of the autoencoder used in the proposed model.

The autoencoder in this system is constructed using three convolutional layers (of 3 * 3 kernel size). ReLU is used as the activation function in all the layers. A pooling layer is used for down-sampling after each convolutional layer (2 by 2 max-pooling). This is followed by a Flatten layer. Thus the features are converted to a single column. A similar architecture of the encoder is used in the decoder as well (as shown in Fig. 3) and all the layers have ReLU activation function. The optimizer used in the architecture is Adam optimizer. The Loss function used in this architecture is Mean Squared Error (MSE) loss. The structure of a basic encoder-decoder is illustrated in Fig. 4.

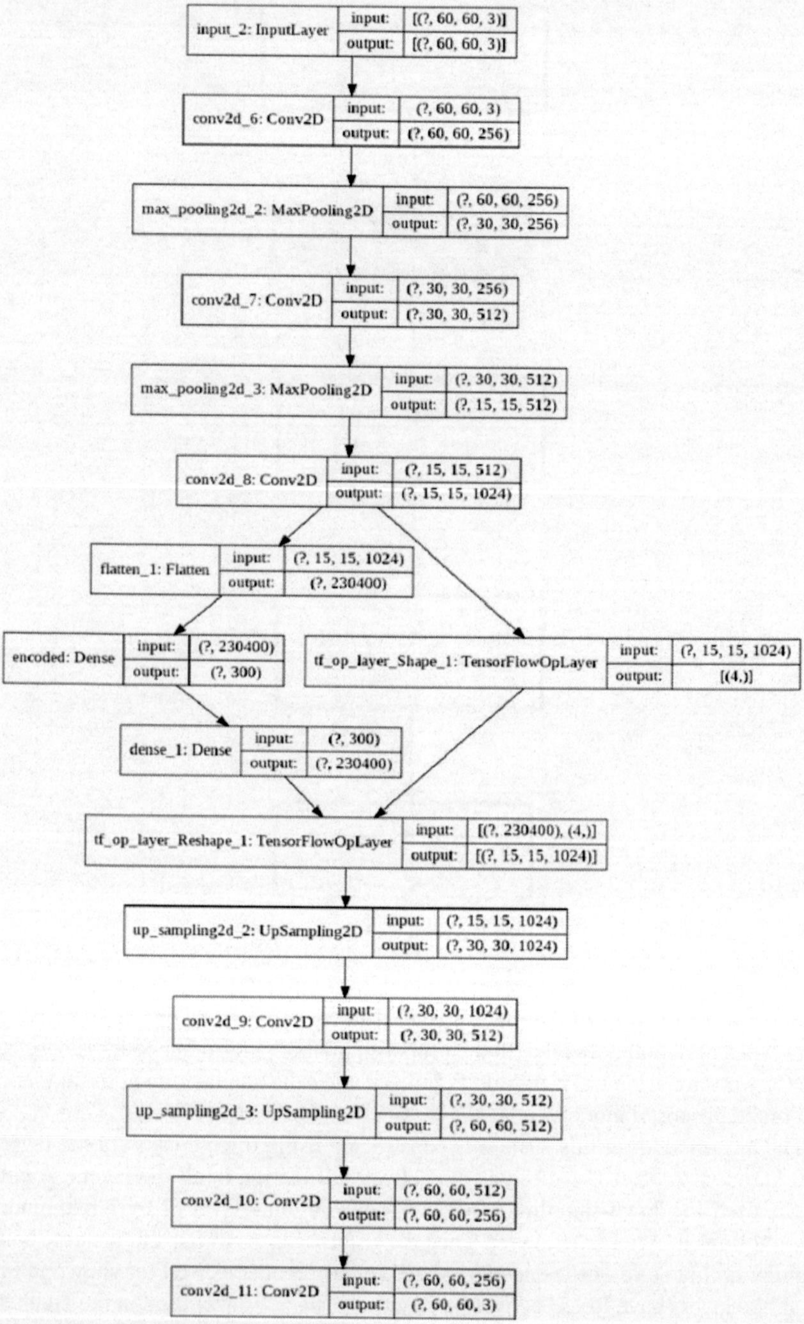


**Fig. 3** CNN autoencoder architecture of the proposed model

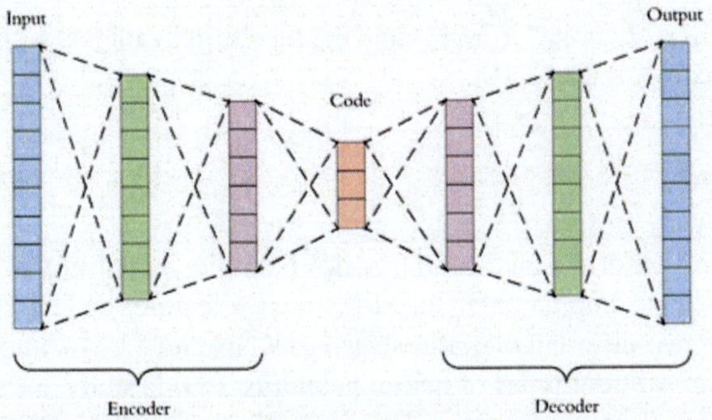


**Fig. 4** The architecture of basic encoder-decoder

## 3.2 Classification

At the classification stage of the character recognition process, the test samples are categorized into various classes. The proposed model uses K-Nearest Neighbors (KNN) classifier for categorizing the test dataset into various classes. The features extracted using the autoencoder are used as the input at this classification stage. KNN technique is easy to apply and robust to the noisy training data. KNN is a supervised machine learning algorithm which can be used for classification as well as regression problems. However, it is more commonly used for classification tasks. KNN is a non-parametric method as it does not make any assumptions on underlying data. The classification task is repeated using SVM classifier for performance comparison. SVM is commonly used for classification problems in machine learning.

# 4 Experimental Results

Isolated MODI script characters written by different persons are used as the experimental data set. The training dataset is comprised of 3220 characters and the test dataset consisted of 1380 characters. The implementation is performed in Python programming environment.

The dataset consists of the grayscale image of isolated MODI characters. These images are size normalized to 60 * 60 pixels. An inbuilt function (ImageDataGenerator of Keras) is used for performing the on-the-fly augmentation. Scaling and rotation techniques are performed on the dataset and the randomly generated dataset is given as the input for the feature extractor (CNN autoencoder). Important representation is learned at the feature extraction stage and a 300-dimensional vector was generated as a feature vector (from the original image of 60 * 60 sizes).

At the classification stage, KNN classifier was used to classify the extracted features and an accuracy of 99.4% is achieved. The experiment was repeated using another classification method. SVM with RBF Kernel was used in the second experiment and the accuracy achieved is 99.3%. A comparison of the performance of KNN and SVM classifiers in combination with CNN autoencoder indicates that the KNN classifier performed better in the classification of MODI script characters. The method using KNN classifier achieved the highest accuracy of 99.4% and the results indicate that KNN classifier gave better accuracy compared to SVM.

## 5 Conclusion

Character recognition of isolated MODI script is implemented in this work. Convolutional autoencoder was implemented for extracting the features and the classification was carried out using two different classifiers such as KNN and SVM. Autoencoders can be constructed using different types of neural networks. In this study, a CNN autoencoder is implemented for feature extraction. The implementation of the CNN autoencoder at the feature extraction level reduced the feature size to 300 (from 3600). Classification of the extracted feature using KNN and SVM fetched recognition accuracy of 99.4 and 99.3%, respectively. The experiment using CNN autoencoder in combination with KNN classifier gave better performance.MODI script character recognition is a challenging task and very limited research work is reported in this area. Pattern recognition of MODI script is tedious in comparison with other Indian scripts due to the various complexities of the script. A huge amount of MODI documents are preserved at libraries and temples in various parts of India and abroad. There is a need for an efficient MODI script character/text recognition system to retrieve the vast amount of knowledge in them. As a future plan, the focus will be on the text recognition and document analysis of MODI manuscripts.

## References

1. Chaudhuri A, Mandaviya K, Badelia P, Ghosh SK (2017) Optical Character Recognition Systems for Different Languages with Soft Computing, vol 352 (2017)
2. Joseph S, George J (2021) Efficient handwritten character recognition of MODI script using wavelet transform and SVD. In: Data Science and Security. Lecture Notes in Networks and Systems. Springer, Singapore, vol 132. https://doi.org/10.1007/978-981-15-5309-7_24
3. Joseph S, George J (2019) Feature extraction and classification techniques of MODI script character recognition. Pertanika J Sci Technol 27(4)1649–1669
4. Joseph S, Datta A, Anto O, Philip S, George J (2021) OCR system framework for MODI scripts using data augmentation and convolutional neural network. In: Data Science and Security. Lecture Notes in Networks and Systems, vol 132, pp 201–209. Springer, Singapore. https://doi.org/10.1007/978-981-15-5309-7_21
5. Joseph S, George JP, Gaikwad S (2020) Character recognition of MODI script using distance classifier algorithms. In: Fong S, Dey N, Joshi A (eds) ICT Anal. Appl. Lecture Notes in Networks and Systems, vol 93. Springer, Singapore. https://doi.org/10.1007/978-981-15-0630-7_11
6. Kulkarni S, Borde P, Manza R, Yannawar P (2015) Review on recent advances in automatic handwritten MODI script recognition. Int J Comput Appl 115(19):975–8887
7. Solley T (2018) A Study of Representation Learning for Handwritten Numeral Recognition of Multilingual Data Set. Lecture Notes Networks and Systems, vol 10, pp 475–482. Springer
8. Maggipinto M, Masiero C, Beghi A, Susto GA (2018) A convolutional autoencoder approach for feature extraction in virtual metrology. Procedia Manuf 17:126–133

9. Joseph S, George J (2020) Handwritten character recognition of MODI script using convolutional neural network based feature extraction method and support vector machine classifier. In: IEEE 5th International Conference on Signal and Image Processing (ICSIP), Nanjing, China, pp 32–36
10. Najadat HM, Alshboul AA, Alabed AF (2019) Arabic handwritten characters recognition using convolutional neural network. In: 2019 10th International Conference on Information and Communication Systems ICICS 2019, pp 147–151, January 2019
11. Shopon M, Mohammed N, Abedin MA (2017) Bangla handwritten digit recognition using autoencoder and deep convolutional neural network. In: IWCI 2016 - 2016 International Workshop on Computational Intelligence, pp 64–68, December 2017
12. Gogna A, Majumdar A (2019) Discriminative autoencoder for feature extraction: application to character recognition. Neural Process Lett 49(3):1723–1735
13. Loey M, El-Sawy A, EL-Bakry H (2017) Deep Learning Autoencoder Approach for Handwritten Arabic Digits Recognition (2017)
14. Joseph S, George J (2021) Data augmentation for handwritten character recognition of MODI script using deep learning method. In: Senjyu T, Mahalle PN, Perumal T, Joshi A (eds) Information and Communication Technology Intelligent Systems ICTIS 2020. Smart Innovation System Technology, vol 196. Springer, Singapore. https://doi.org/10.1007/978-981-15-7062-9_51

Check for updates

# ODFWR: An Ontology Driven Framework for Web Service Recommendation

N. Manoj[1] and Gerard Deepak[2(✉)]

[1] Department of Computer Science and Engineering, SRM Institute of Science and Technology, Ramapuram, Chennai, India

[2] National Institute of Technology, Tiruchirappalli, India

**Abstract.** In recent years, with the spread of the Internet, the use of the web services has grown and diversified. There is always a need for better web service recommender system which is ontology focused and complaint to Web 3.0. In view of this challenge, we present Ontology Driven Framework for web service recommendation (ODFWR) which considers User Query, Web Usage Data of the user, Metadata, Dynamically Modeled Ontology, Real-World Application Stores, Lin Similarity, Bagging and Dynamic User Clicks to recommend web services. The dataset used in ODFWR is the metadata crawled from UDDI and WSDL which uses Lin similarity to aggregate knowledge and bagging to classify it. The ODFWR performance is evaluated and it is compared with baseline models and other variations of the ODFWR and it was found that ODFWR is superior in terms of performance where the precision, recall, accuracy, F-measure and FDR of 96.89, 98.84, 97.18, 97.86% and 0.0311 respectively.

**Keywords:** Bagging · Metadata · Ontology · Semantic Web · Web services

## 1 Introduction

Improvements in Internet technology have strengthened. Several revolutionary innovations, including web services. The web services provide solutions for the deployment of a web application in large collaborative enterprise software environment. The key advantage of Service Oriented Architecture (SOA) is that it supports the composition of the network provider that deploys a complex workflow for enterprise application integration that explicitly leverages web applications on the site. There are also issues that need to be addressed during the development process, considering the benefits of the SOA. To solve these challenges, the key challenge is to construct a workflow during runtime for complex business applications without human involvement to react automatically to dynamic changes in user requests.

**Motivation***:* Most people don't exactly know the proper requirement for getting a proper webservice from the internet because there are thousands of web services and its difficult for traditional web search engine to know the exact need of each and every person using the services. The proposed framework is a step in the direction of how unique the requirements are needed to give the closest recommendation. There is always a need for


S. Shukla et al. (eds.), *Data Science and Security*, Lecture Notes in Networks and Systems 290,
https://doi.org/10.1007/978-981-16-4486-3_16

web service recommender system which is knowledge centric, semantic in nature, user oriented as well as analyzes the metadata of web services. The system which integrates four approaches are rare, the existing system only focus on only one goal, whereas the proposed system combines the user requirements, understand the metadata of the web services, understand the needs of the user and is sematic in nature as it incorporates ontology for recommending web services is essential.

**Contribution:** The following are the propositions in this paper. A framework for recommending web services which incorporates metadata, dynamically models Ontology, which considers user requirement as well as semantically driven has been proposed. Experiments show that a high accuracy can be achieved by using a Web Service Ontology driven framework when compared to some traditional recommender system. The ODFWR system achieves an overall Precision, Recall, Accuracy, F-measure and FDR of 96.89, 98.84, 97.18, 97.86% and 0.03 respectively.

**Organization:** The latter portion of the paper is divided into the following sections. The second section addresses briefly Related Work. The third Section explains the proposed systems architecture. The fourth section consists mainly of the implementation and performance of the technique. The fifth section consists of conclusion drawn from the experiment.

## 2 Related Work

Arul et al. [1] has proposed a single semantic-oriented architecture that combines the corresponding automated web service composition algorithm with a systematic process of updated multi-stage composition and web semantics. Giri et al. [2] have put forth a semantic algorithm where query ontologies are projected by rctaining the connections between ontological entities. By combining collaborative filtering and textual content, Xiong et al. [3] used a hybrid approach for Web service recommendation. Su et al. [4] have put forward a TAP confidence-aware method for accurate personalized prediction of QoS. Li et al. [5] implemented a new QoS-aware web service recommendation model that acknowledges the qualitative characteristics of various services. With a combination of Social Balance Theory and CF, Qi et al. [6] have put forward a new data-sparsity tolerant recommendation approach to Ser-RecSBT + CF. Yin et al. [7] used a shared recommendation strategy consisting of three inventive prediction models based on two approaches, i.e., matrix factorization (MF) and network location awareness neighbor selection. Zhang et al. [8] have proposed a Covering Algorithm constructed on the Quotient Space Granularity on Start inquiry in large-scale scenarios, a scalable technique for accurate Web benefit proposal. A strengthened collective filtering approach has been used by Zou et al. [9] where all similar users and resources are integrally considered into a singleton CF. Xie et al. [10] have recommended an asymmetric relationship strategy of regularized matrix factorization (MF), which consequently combines deviated relationship and topsy-turvy relationship engendering. In [14–21] various supportive ontology models have been discussed.

## 3 Proposed System Architecture

The proposed system consists of 4 phases which finally results in recommendation of useful and relevant web services. The first phase deals with user data pre-processing. The second phase focuses on ontology construction for the web service from UDDI and WSDL repositories. The third phase involves comparing the pre-processed user data and constructed ontology using Lin Similarity [11]. The fourth phase is the classification phase which is achieved by using Bagging sandwiched with Support Vector Machine (SVM) [12] and Random Forest Algorithm [13]. The first phase involves pre-processing of the user requirements addressing both functional and non-functional properties. Further user's click information and web usage data is being tokenized into terms, subjected to lower case conversion, number to text conversion, punctuations removal, lemmatization, stop word removal and word stemming is done. The pre-processed query is extracted and Vocabulary is formulated; further Axiomatization, Logical Induction, Reasoning and Formalization are done in order to yield the constructed ontology.

The second phase is Ontology modeling. where the metadata crawled from UDDI and WSDL repositories. The index terms and descriptions from web service repository are pre-processed by tokenizing it into terms, lower case conversion, number to text conversion, punctuations removal, lemmatization, stop word removal and word stemming. OntoCollab and web protégé is used for construction of ontology. The metadata extracted from UDDI and WSDL repositories serves as potential web service indicators that describes the nature, functionality, behavior, usage, and other essential information about the web services.

The Ontology is dynamically generated by composing the terms extracted from the metadata. WSDL documents describe resources as endpoint network sets, or ports. The abstract concept of endpoints and messages in WSDL is distinguished from their specific deployment of networks or bindings of knowledge formats. This permits abstract meaning to be reused.

The proposed architecture's framework is created as shown in the Fig. 1. Apart from the metadata constituents, the ontology also formalizes information from the web service repository and application stores. OntoCollab integrates the term sets from these heterogeneous sources and is subjected to Axiomatization and inconsistences are checked using the Hermit reasoner. The third phase is to compare the Ontology and the individual words. The similarity function that is used here is called Lin Similarity. This function is used to find the text similarity between two documents.

$$sim_{lin} = \frac{2 * IC(lcs * (c_1, c_2))}{IC(c_1) + IC(c_2)} \tag{1}$$

Lin is the similarity between the quotient two notions of two times the IC of the LCS concepts and the sum of the IC of the two concepts as shown in Eq. 1. The fourth phase involves classification by means of bagging of SVM and Random Forest algorithm.

Bootstrap aggregating, also called bagging, is an algorithm developed by the machine learning ensemble to improve the consistency and accuracy of machine learning algorithms used in statistical classification and regression problems. It eliminates variation as well as helps to avoid over-fitting. While it is generally applicable to methods of the

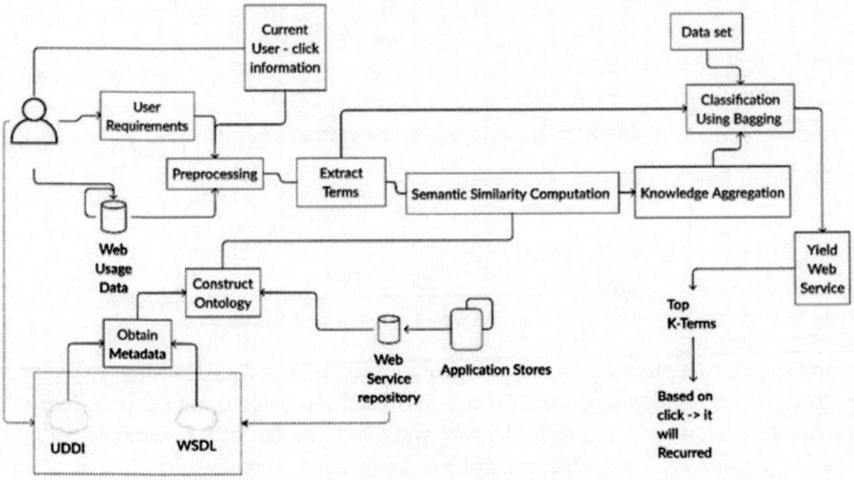


**Fig. 1** Proposed architecture of ODFWR

decision tree, it can be used for any form of structure. A special case of the approach for modeling averages is called as bagging. In the training of samples of input, sparse matrices are only accepted if they are supported by the base estimator.

SVM is a supervised algorithm for machine learning which can be used for classification or regression problems. To convert data, it uses a method called the kernel trick, and then finds an optimum boundary between potential outputs based on these conversions. Simple terms, several incredibly complicated data transformations take place, then works out how to distinguish the data depending on the labels or outputs you have identified. SVM is capable of doing both regression and classification. Non-linear SVM means that a straight line does not have to be the boundary that the algorithm measures. The gain is that, without having to perform complex transformations on your own, you can capture a much more nuanced relationship between your data points. The downside is that planning time is much longer than intensive computing.

The Random Forest is a supervised algorithm for classification. We can see that from its name, which is to create a forest to make it random in some way. The frequency of trees in the forest is directly associated with the results that can be achieved; the higher the number of trees, the more accurate the results.

Random Forest sets up a number of individual decision-making trees in preparation, to the form the ultimate expectation, the expectations from all trees are pooled.

$$Gain\,(P,\,Y) \;=\; Entropy(P) - Entropy\,(P,\,Y) \tag{2}$$

$$na_b = \; w_a C_b - w_{left(b)} C_{left(b)} - w_{right(b)} C_{right(b)} \tag{3}$$

$$fa_a = \frac{\sum_{b:node\ b\ splits\ on\ feature\ a} na_b}{\sum_{k \in all\ nodes} na_k} \tag{4}$$

$$normfa_a = \frac{fa_a}{\sum_{j \in all\ features} fa_b} \tag{5}$$

$$RFfa_a = \frac{\sum_{ball\ tree} normfa_{bb}}{T} \quad (6)$$

$$fa_a = \sum_{b:nodes\ b\ splits\ on\ feature\ a} s_b C_b \quad (7)$$

$$normfa_a = \frac{fa_a}{\sum_{b \in all\ features} fa_b} \quad (8)$$

$$RFfa_a = \frac{\sum_b normfa_{ab}}{\sum_{k \in all\ feature,\ k \in all\ trees} normfa_{bk}} \quad (9)$$

In Eq. (2), Y denotes feature to split on, P represents target and Entropy (P, Y) denotes the entropy that is calculated after the data has been separated into feature Y. In Eq. (3), $na_b$ denoted the *node b*, *w sub(b)* denotes weighted number of test samples coming to *node b*, $C_b$ is denoted by the impurity rate of *node b*, *left(b)* denotes child node from the left split of *node b* and the right *split(b)* denotes the child node of node b from the right split. The value of each function on a decision tree is then determined in the Eq. (4) in which *fa sub(a)* denotes the importance of the feature *a* and $na_b$ denotes the significance of the *node j*. These can at that point be normalized to be esteem between 0 to 1 by separating by the entirety of all include significance values yields Eq. (5).At the level of the Random Forest, over all the trees, the final feature value is his average. In Eq. (6), the addition of the significance value of feature on each and every tree is computed and divided into total number of trees in which $RFfa_a$ denotes the significance of a characteristic measured from each and every tree in the Random Forest model, $RFfa_a$ denotes the significance of the normalizing feature for *a in tree b* and *T* denotes the total number of trees. Spark calculates the significance of a function for each decision tree by adding up the gain, the number of test samples passing through the node provides Eq. (7) in which $fa_b$ denotes the significance of feature a, s b denotes the number of samples reaching *node b*, and $C_b$ denotes the impurity value of *node b*. To begin with, include significance for each tree is normalized in a connection to the trees abdicate Eq. (8) to decide the ultimate highlight noteworthiness at the Arbitrary Timberland level, in which $normfa_a$ indicates the normalized noteworthiness of *feature a* and *fa sub(a)* indicates the importance of highlight a. At that point highlight significance values from each tree are summed normalized in condition Eq. (9) where $RFfa_a$ signifies the significance of include a calculated from all trees within the Irregular Timberland show and $normfa_{ab}$ indicates the normalized include significance for a in tree b.

Once web services are classified using bagging, the top 5% of web service of each class is recommended based on the user clicks on each class or items or web services. The further clicks strategically ordered on the basis of the user clicks by computing the semantic similarity.

## 4 Implementation and Performance Evaluation

In order to facilitate future research and make our studies reproducible, our real-world QoS data set has been published publicly. The website link is https://wsdream.github.io/dataset/wsdream_dataset1.html. The dataset used for the modelling ontology contains

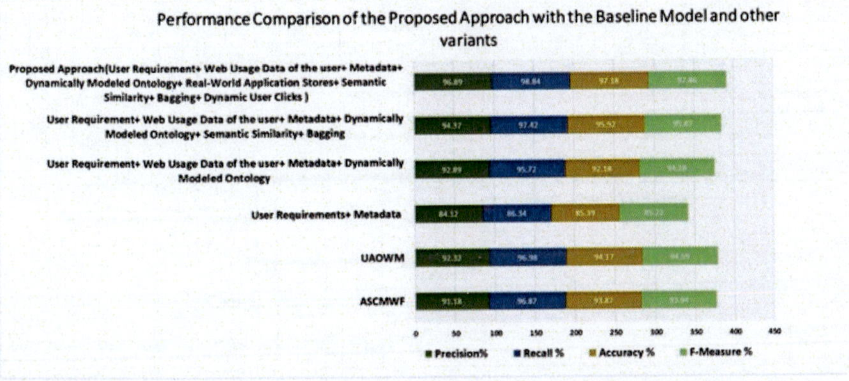


**Fig. 2** Performance comparison of the proposed approach with the baseline model and other variants

WSDL Address, Web service provider name, Country and Description. The Fig. 2 shows the performance evaluation for baseline model namely, ASCMWF [1] and UAOWM [2]. It can be observed that the Precision, Recall, Accuracy, F-Measure and False Discovery Rate (FDR) for ASCMWF is 91.18, 96.87, 93.87, 93.94% and 0.08 respectively and for UAOWM it is 92.32%, 96.98, 94.17, 94.59% and 0.07 respectively.

The Fig. 2 also shows the different variation of proposed model are User Requirements + Metadata, User Requirement + Web Usage Data of the user + Metadata + Dynamically Modeled Ontology, User Requirement + Web Usage Data of the user + Metadata + Dynamically Modeled Ontology + Semantic Similarity + Bagging and Proposed Approach (User Requirement + Web Usage Data of the user + Metadata + Dynamically Modeled Ontology + Real-World Application Stores + Semantic Similarity + Bagging + Dynamic User Clicks). It can be observed that the Precision, Recall, Accuracy, F-Measure and FDR for User Requirements + Metadata it is noted to be 84.12, 86.34, 85.39, 85.22% and 0.15 respectively.

For User Requirement + Web Usage Data of the user + Metadata + Dynamically Modeled Ontology it is 92.89, 95.72, 92.18, 94.28% and 0.07 respectively and for User Requirement + Web Usage Data of the user + Metadata + Dynamically Modeled Ontology + Semantic Similarity + Bagging it is 94.37, 97.42, 95.92, 95.87% and 0.05 respectively and for Proposed Approach(User Requirement + Web Usage Data of the user + Metadata + Dynamically Modeled Ontology + Real-World Application Stores + Semantic Similarity + Bagging + Dynamic User Clicks) it is 96.89, 98.84, 97.18, 97.86% and 0.03 respectively. From the Fig. 2 It can be inferred that the proposed approach is performing significantly better in comparison to the baseline model (Fig. 3).

It is inferable that the performance of the proposed approach is computed using Precision, F-Measure, Recall, Precision, Accuracy, and FDR as the potential measure. Recall is the proportion of ontologies recovered and applicable to the total number of ontologies that are relevant. Precision is characterized as the proportion of the retrieved and significant ontologies to the overall number of recovered ontologies. For precision and recall measures, accuracy is specified as the average (Fig. 4).

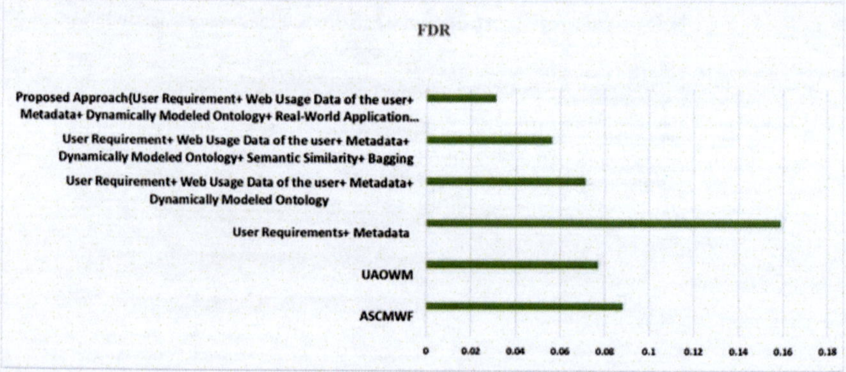


**Fig. 3** Performance comparison of the proposed approach with the baseline model and other variants

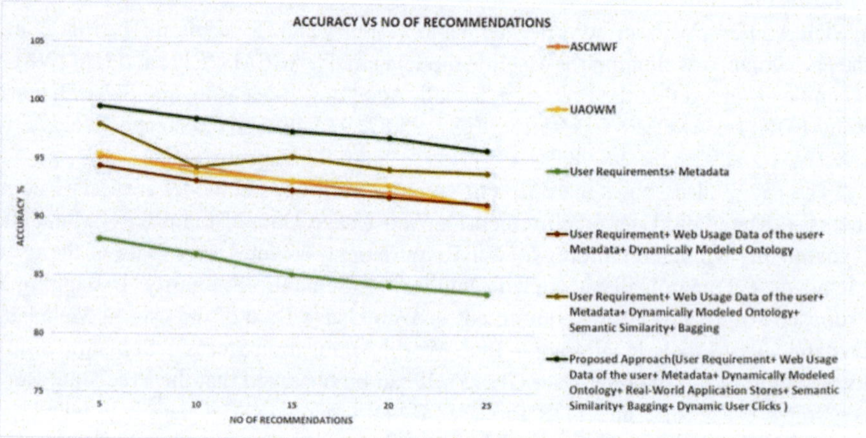


**Fig. 4** Performance comparison of accuracy the proposed approach with the baseline model and other variants and different quantity of recommendation

The performance of the proposed approach is evaluated by comparing it with the ASCMWF and the UAOWA models for web service recommendations. Also, since the proposed approach is an incremental hybridized approach, two incremental aggregations are depicted as variations. The ASCMWF has considered user requirements and web service metadata for Ontology based recommendation of web services but these two aren't enough to provide an accurate recommendation since it lacks variables like user preference, web usage, and current user click – information. From Fig. 2 and 3, it can be inferred that the Proposed ODFWR has better Precision and Recall, 96.89 and 98.84% respective in comparison to ASCMWF which is 91.18 and 96.87%. Another baseline model UAOWA is a conventional method which yield an accuracy of 94.17% in comparison with the proposed ODFWR which has the accuracy of 97.18%. In such a growth in technology accuracy is give more preference. ODFWR considers more data in the process of recommending the web services. The ODFWR is an Ontology based recommender system which not only considers user requirements (functional and non-functional requirements) but also the current user click information and web usage data of the user. Since it is an Ontology based system which also uses knowledge aggregation using the Lin Similarity Measure, helps reducing human error from dataset point of view. The proposed system is much better than the existing approaches as it dynamically models ontology from the metadata which is obtained from the WSDL and UDDI. Also, the dynamically modeled Ontology is further used for knowledge aggregation using the Lin Similarity Measure. Moreover, the user requirements and the web usage data of the user ensures that the web service recommendation is both user centric as well as requirement centric. It implicitly incorporates personalization. The encompassment of Classification of the web services using bagging enhances the relevance and the quality of recommendation. Further, the Click based recommendations ensures that the proposed recommendations are relevant. Moreover, the proposed model is an incremental hybridization of several strategic schemes for web service recommendations which increases the overall accuracy of the recommendation.

## 5 Conclusions

There is always a need for web service recommender system which is knowledge centric, semantic in nature, user oriented as well as considers the metadata of the web services. This paper discusses an effective method for the same. The proposed approach makes accurate predictions as compared to others due to the Ontology-driven context where the Web Service Ontology model is integrated with Knowledge aggregated from Lin similarity. It can be concluded that in future, ensuring techniques combined with Ontology driven frameworks can produce better results for recommending web services. An overall F-Measure of 97.86% has been achieved with a very low FDR of 0.03 which makes the proposed ODFWR the best-in-class approach for recommending web services.

## References


1. Arul U, Prakash S (2019) A unified algorithm to automatic semantic composition using multilevel workflow orchestration. Cluster Comput 22(6):15387–15408
2. Giri GL, Deepak G, Manjula SH, Venugopal KR (2018) OntoYield: a semantic approach for context-based ontology recommendation based on structure preservation. In: Proceedings of International Conference on Computational Intelligence and Data Engineering, pp 265–275. Springer, Singapore
3. Xiong R, Wang J, Zhang N, Ma Y (2018) Deep hybrid collaborative filtering for web service recommendation. Expert Syst Appl 110:191–205
4. Su K, Xiao B, Liu B, Zhang H, Zhang Z (2017) TAP: a personalized trust-aware QoS prediction approach for web service recommendation. Knowl-Based Syst 115:55–65
5. Li S, Wen J, Luo F, Gao M, Zeng J, Dong ZY (2017) A new QoS-aware web service recommendation system based on contextual feature recognition at server-side. IEEE Trans Netw Serv Manage 14(2):332–342
6. Qi L, Zhou Z, Yu J, Liu Q (2017) Data-sparsity tolerant web service recommendation approach based on improved collaborative filtering. IEICE Trans Inf Syst 100(9):2092–2099

7. Yin Y, Aihua S, Min G, Yueshen X, Shuoping W (2016) QoS prediction for Web service recommendation with network location-aware neighbor selection. Int J Softw Eng Knowl Eng 26(04):611–632
8. Zhang YW, Zhou YY, Wang FT, Sun Z, He Q (2018) Service recommendation based on quotient space granularity analysis and covering algorithm on Spark. Knowl-Based Syst 147:25–35
9. Zou G, Jiang M, Niu S, Wu H, Pang S, Gan Y (2018) QoS-aware Web service recommendation with reinforced collaborative filtering. In: International Conference on Service-Oriented Computing, pp 430–445. Springer, Cham
10. Xie Q, Zhao S, Zheng Z, Zhu J, Lyu MR (2016) Asymmetric correlation regularized matrix factorization for web service recommendation. In: 2016 IEEE International Conference on Web Services (ICWS), pp 204–211. IEEE
11. Lin D (1998) An information-theoretic definition of similarity. In: ICML, vol 98, No 1998, pp 296–304
12. Vapnik V, Guyon I, Hastie T (1995) Support vector machines. Mach Learn 20(3):273–297
13. Ho TK (1995) Random decision forests. In: Proceedings of 3rd International Conference on Document Analysis and Recognition, vol 1, pp 278–282. IEEE
14. Deepak G, Teja V, Santhanavijayan A (2020) A novel firefly driven scheme for resume parsing and matching based on entity linking paradigm. J Discrete Math Sci Cryptograph 23(1):157–165
15. Deepak G, Santhanavijayan A (2020) Onto best fit: a best-fit occurrence estimation strategy for RDF driven faceted semantic search. Comput Commun 160:284–298
16. Kumar N, Deepak G, Santhanavijayan A (2020) A novel semantic approach for intelligent response generation using emotion detection incorporating NPMI measure. Procedia Comput Sci 167:571–579
17. Deepak G, Kumar N, Santhanavijayan A (2020) A semantic approach for entity linking by diverse knowledge integration incorporating role-based chunking. Procedia Comput Sci 167:737–746
18. Haribabu S, Kumar PSS, Padhy S, Deepak G, Santhanavijayan A, Kumar N (2019) A novel approach for ontology focused inter-domain personalized search based on semantic set expansion. In: 2019 Fifteenth International Conference on Information Processing (ICINPRO), pp 1–5. IEEE
19. Deepak G, Kumar N, Bharadwaj GVSY, Santhanavijayan A (2019) OntoQuest: an ontological strategy for automatic question generation for e-assessment using static and dynamic knowledge. In: 2019 Fifteenth International Conference on Information Processing (ICINPRO), pp 1–6. IEEE
20. Kaushik IS, Deepak G, Santhanavijayan A (2020) QuantQueryEXP: a novel strategic approach for query expansion based on quantum computing principles. J Discrete Math Sci Cryptograph 23(2):573–584
21. Varghese L, Deepak G, Santhanavijayan A (2019) An IoT analytics approach for weather forecasting using raspberry Pi 3 model B+. In: 2019 Fifteenth International Conference on Information Processing, pp 1–5. IEEE

# Smart Contract Security and Privacy Taxonomy, Tools, and Challenges

Jasvant Mandloi(✉) and Pratosh Bansal

Department of Information Technology, Institute of Engineering and Technology, Devi Ahilya Vishwavidyalaya, Indore, India

**Abstract.** Blockchain smart contract technology implementations are growing rapidly. The resilient and accurate design of smart contracts for these intelligent applications however is a huge challenge because of the complexities involved. Smart Contract modernizes traditional processes of production, technology, and industry. It is self-verifiable, self-executable, and enclosed in Blockchain that removes the demand of trustworthy mediator. The major issue that needs to be addressed to make smart contract successful is security and privacy issues. In the paper, a survey was conducted on available taxonomy for security and privacy concerns and a new taxonomy proposed that could accommodate all potential threats. A detailed review of available security and privacy audit tools has also been done for common smart contract platforms. At last, identified the challenges required to be addressed to make the smart contract more efficient.

**Keywords:** Taxonomy · Mining · Consensus · Timestamp

## 1 Introduction

With the popularization of Blockchain technology, there is an emerging new age of trust-based applications. Blockchain 1.0 is for transactions, particularly on the implementation of Cash-related cryptocurrencies such as money transfers, financial transactions, and electronic paying systems, which involve Bitcoin. Blockchain 2.0 is Blockchain 1.0 extension. It involves the anonymity of the non-native asset Blockchain, smart contracts, and the beginning of tokens and capabilities. Blockchain 3.0 also focuses on Blockchain to incorporate decentralized applications. In version 2.0 and 3.0, One of the most crucial features of Blockchain technology is smart contracts and created a lot of hype over the years. The smart contract is a computer code that can be activated to execute any function if certain predefined conditions are met. A smart contract is a piece of code that is stored on the Blockchain, ensure that the specifications listed are fulfilled to satisfy the user's requirements [1]. The implementation of decentralized apps for various domain fields is an important research sector with the most notable ones being decentralized identity management, land registry management, e-governance mechanisms, etc. Blockchain allow you to preserve a tamper-proof log that is used to determine the control of activities. A decentralised architecture is a back-end technology that uses a peer-to-peer network to link customers and suppliers directly. Additionally, the latest and emerging Blockchain


S. Shukla et al. (eds.), *Data Science and Security*, Lecture Notes in Networks and Systems 290,
https://doi.org/10.1007/978-981-16-4486-3_17

Iteration—Blockchain version 4.0 provides substantial value opportunities. It involves the integration of artificial intelligence (AI) into Blockchain systems on two different technology sides. Artificial Intelligence is based on a predictive analysis that describes uncertainty. It is constantly evolving, and algorithms are supposed to conjecture or presume truth. By comparison, Blockchain has a deterministic hash algorithm that yields the same results while the inputs remain constant [2].

The Blockchain markets are expected to increase to $23.3 billion by 2023 from $1.2 billion in 2018 to 80.2% year-on-year in 2018–23 [3]. Smart contracts play a major role in it and it opens research opportunities for the security and privacy issues in the Smart contract and Blockchain platform involvement in it. Considering all safety enhancements and monitoring tools [4], Blockchain-based smart contract also faces difficulties in coping with various security threats [5]. The number of attacks is continually started to disrupt the usual flow or even to disrupt the network entirely. Attacks that are related to cryptocurrency wallets, Smart contract applications, Secure transactions, Pool Mining, and options are available to exploited opponents using the Blockchain network security breaches. DAO attacks, Multiplayer Games, and King of the Ether Throne due to bugs or errors in the smart contract code, some smart contract-based attacks have occurred [6]. This paper tends to focus on the analysis of attacks based on smart contracts as well as the implications of their exploitation. To analyze it an updated taxonomy for security and privacy is required. We have proposed a new taxonomy that cannot be platform specific. Afterward also analyzed the security tools available for the popular smart contract platforms. The paper is structured as follows; in first section we have done literature review on the work done earlier in this field. The next section proposed a new taxonomy for the Smart contracts. Afterword's listed the audit tools available today to test security and privacy and classified them based on their performance features and shortcomings. At last, highlighted the open challenges still that are not addressed or going to be addressed in the future and finished it with the conclusion of our research work.

## 2 Literature Review

Mense A [7] has surveyed the vulnerabilities in smart contracts and identified them through the literature research and analysis of available information at various platforms. Also Compares the currently available code review tools to define and detect vulnerabilities in smart contracts based on a vulnerability taxonomy. This also demonstrated the best practice to avoid severe vulnerabilities using the example of the DOA attack.

Monika A [8] They carried out a research survey in the attempt to bridging the gap by taking into account all resources, regardless of their source, and deploying and evaluating them. It helps those who would like to study code already deployed, build stable smart contracts, or plan to teach and train. By doing so, they also evaluated 27 tools for the study of Ethereum smart contracts based on availability, maturity, strategies used, and safety issues.

Gupta B [9] has done research work in his Mtech thesis with the aim that his work serves as the reference for the smart contract developers as well as users and guide them. The work highlighted various security vulnerabilities concerning the Ethereum Smart contract platform. Afterword's also listed tools available to audit the security features of the on-chain contract and it helps to secure contracts in an anonymous and trustless environment. Also proposed a taxonomy to classify the security vulnerabilities in the Ethereum smart contract.

Groce A [10] In the paper they have summarized the audit performed by the 23 professional stakeholders on the Ethereum smart contract. They perform audits on the leading Blockchain security platforms using the free open-source tools and licensed also, with this also included professional manpower to perform code analysis. In that, they identified approx. 250 plus faults that can be used to compare the severity and frequency of various types of attack, compare smart contract and non-smart contract faults, and put auto-mated sensing techniques to the test.

Tasca P [11] has performed a comparative analysis of the most commonly accepted Blockchain technologies is carried out in a bottom-up manner. Blockchains have been broken down into their constituent blocks. Each part is divided into main and subcomponents in a hierarchical order. The variations of the subcomponents are then described and compared. To summaries the research and provide a navigation tool for different Blockchain architectural implementations, a taxonomy tree is used.

Demir M [12] Conducted a literature review to recognize vulnerabilities that smart contract programmers and users must avoid, and categorized security bugs based on their existence. This also carried out an analysis of applications that identify these vulnerabilities and provide details on their approach and scope. A major outcome of smart contract evaluation as a platform or mission-critical application support such as access control systems.

Sayeed S [13] They have classified Blockchain exploitation techniques into Four groups depending on the attack target that are targeting consensus mechanisms, bugs in the code of the smart contract, malicious software running on the operating system, and fraudulent users. The key emphasis is on smart contract flaws, evaluating the Seven most relevant attack techniques, and assessing their true effect on smart contract technology. It also recognizes that even after the implementation of the ten most widely used tools for discovering smart contract vulnerabilities all have security bugs, giving a risky false sense of security.

Durieux T [14] Present an empirical evolution on the latest automated analysis tools and used the two datasets. One contains 69 annotated vulnerable smart contracts that contain the accuracy of the analysis tools and the second one is based on the Ethereum smart contract developed using the solidity language. This dataset they have used is part of the Smart bug platform that has a feature to integrate and compare multiple analysis tools available for the analysis of the Ethereum smart contract. Their study suggested that still, tools are generating false-positive results for the vulnerabilities in case of a smaller number of the datasets.

Parizi M.R [15] With a specific aim to make it easier to implement Blockchain smart contracts with safety and privacy, we have to first consider their limitations before widespread implementation. In this paper, they have undertaken a broad empirical assessment of the state-of-the-art static security test methods for smart contracts for Blockchain, the most used.

# 3 Taxonomy for Smart Contract Vulnerabilities

Based on the Information available in academic research, the first attempt to design study and classify smart contract Vulnerabilities is done by Atzei et al. [6]. They classified the flaws into three groups. But the vulnerabilities discussed did not provide a detailed image, as they did not even include common flaws such as authentication, functional visibility, and transaction order dependence. In the series latest work found in the series is done by Gupta B [9] proposed a state of art taxonomy for the Ethereum smart contract that is hierarchical and attempted to cover all the available security vulnerabilities reported. But again, this will be only for the Ethereum smart contract and not applicable to the other platforms.

# 4 Proposed Taxonomy for Blockchain Smart Contracts

The present scenario requirement is a taxonomy that has a broad scope and applicable to almost all the Blockchain smart contract platforms and applications. Earlier taxonomy proposed is platform-specific they are not covering the security threats of all the platforms. We are trying to cover more than one platform and it will help developers and users of Blockchain Smart contracts.

We make an attempt to cover the majority of the security threats that have been identified and are common to most platforms. They may be identified and called under different names on different platforms. Also, we are trying to categorize it in such a way that it will reduce the redundancy and represent proper categorization. Our proposed work will highlight the security threats in a better manner to the security researchers in the field of Blockchain smart contracts and issues related to it.

The taxonomy is systemic and thus evaluation using this taxonomy would provide a good understanding of the main security problems in smart contracts to the security expert. The severity level is marked with high red, medium orange, and low green. The level of intensity was defined in the context of our research, the OWASP risk rating methodology [18].

## 4.1 OWASP Risk Rating Methodology

(See Table 1).

**Table 1** Define Severity levels

| | | | | |
|---|---|---|---|---|
| Effect | High | Medium | High | Critical |
| | Medium | Low | Medium | High |
| | Low | Note | Low | Medium |
| | | Low | Medium | High |
| | | LIKELIHOOD | | |

## 4.2 Proposed Taxonomy

(See Table 2).

**Table 2** A new taxonomy for the blockchain platforms [16–22]

| Platform | Language Used | Intermediate Software | Blockchain |
|---|---|---|---|
| Ethereum | Solidity | EVM | |
| Attacks | Authorization through tx. Origin Unprotected Ether Withdrawal Unprotected self-destruct | Short Address Attack | Bad Randomness |
| | Integer Overflow & Underflow | | Untrustworthy Data Feeds |
| | Floating Point & Precision | | |
| | Solidity Programming Issues | | Transaction Order Dependence |
| | Unchecked Call | Immutable bugs Stack size limit | Timestamp Dependence |
| | Gasless Send | | |
| | Call Stack Limit | | |
| | Assert Violation | | |
| | Requirement Violation | | |
| | Call to the Unknown | | Demystifying Honeypots |
| | Denial of Service | | |
| EOS | C++ | EOS-VM | |
| Attacks | Path Explosion | Fake EOS | DoS Vulnerability |
| | Memory Overlap | Fake Recipient | |
| | | Missing Permission Check | |
| | | Remote Code Execution | |
| NEO | JavaScript / Python, Ruby, C# / VB.Net / F#, Visual-Studio, Java | Neo-VM | |
| Attacks | NEP-5 storage injection vulnerability | Double-spend vulnerability | DoS Vulnerability Fork issue |

## 5 Tools and Methods Used for the Testing

Intelligent contract programming involves an engineering approach different from what you would use. In the present scenario, various tools and methods are available that can be used to perform the security and privacy check on the smart contract before deployment. It is always required to follow the good practice in smart contract development and deployment else it may proceed to a great loss of money and information. This section includes a list of many security & Privacy tools and methods for chronologically evolving smart contracts. It is summarized in the tabular form (Table 3).

**Table 3** Summary of the latest Audit/testing tools used for Blockchain Smart Contract [22–25]

| Tools | Analysis method | Used to test | Vulnerability coverage |
|---|---|---|---|
| MythX | Static Analysis, Dynamic Analysis, Symbolic Execution | Ethereum | Property Verification and Assertions, Byte-code Safety Controls on Authorization Control Flow, ERC Standards |
| Mythril | Symbolic Execution | Ethereum Hedera Quorum | Integer underflows Owners overwrite to Ether withdrawal |
| Slither | Static Analysis | Ethereum | Automated vulnerability detection, Automated optimization detection |
| Neo-Debug tools | Static Analysis | NEO | Debug smart contracts |
| Eosafe | Static Analysis Symbolic Execution | EOS | Symbolic Analysis Framework |
| Manticore | Symbolic Execution | Ethereum | Error Discovery |
| Smart Embed | Similarity checking techniques | Ethereum | Repetitive contract code, clone-related bugs |
| Octopus | Symbolic Execution | EOS, NEO, Ethereum, Bitcoin script | Byte Code, Control Flow |

# 6 Open Challenges

During our research work, we have searched literature related to the tools used to check the security and In the smart contract, there are privacy concerns. Most of the tools found in the literature available are related to the Ethereum smart contract and for the solidity language. Based on the research done it has also come to our knowledge that there is no proper mechanism to perform the testing of the Smart contracts before deploying them on the network. Also, after deploying if found it is not appropriate and has security or privacy flaws then also no mechanism to rolling back to the initial stage. In the present scenario, all the tools are developed based on the symbol execution or static analysis very less work is found in which AI or machine learning is used to develop such a mechanism.

## 6.1 Universal Taxonomy

In the literature, it has been identified that most of the Taxonomy is Ethereum centric or Blockchain-based. To accommodate all the attacks that are common to all the Smart contracts platforms or the specific one requires a universal or standard Taxonomy.

## 6.2 AI-Based Security Tools

Checking smart contracts for bugs has now become important, and some tools have been created to determine a smart contract code's safety level. The execution results of smart contracts are all deterministic as of today and may not be probabilistic. In the literature review, it has been noticed that most of the errors are due to the bugs in the smart contract code or due to the weakness of the programming language used to write the code. It may happen due to human error sometimes and it requires AI-based tools that can guide the user as well as make auto-correction in the code if required.

## 6.3 The Mechanism to Recall Smart Contract

Research tells that many smart contracts that are extensively used are prone to serious malicious attacks that may allow attackers to Steal valuable assets from the parties concerned. Therefore, analysis and automatic recall techniques have to be applied to identify and repair flaws in smart contracts before being implemented.

## 6.4 Auditing Tool that can Support More than One Language

Smart contract development is not restricted to any language or platform now. In the present scenario, all the auditing tools are language-specific or platform-specific, if in the whole system more than one smart contract platform or language is used that can require different tools for the audit. So, a common tool is required that can support more languages and developed to check the common privacy and security flaws in the smart contract.

### 6.5 Strategy for Testing a Smart Contract

Designing and executing consensus scenarios is the largest obstacle in Blockchain and smart contract testing. It is never enough to brainstorm scenes of multiple factors decentralizing across networks, and it is not easy to organize facilities to simulate such contexts. In carrying out a test case, it will be expensive to ensure that the consensus works properly. Despite test automation, we will not efficiently test the Blockchain smart contract. Because of the combinatorial explosion of several inputs and environmental variables in the preconditions of the test case, checking a contract manually is not just too complicated, it is inefficient.

## 7 Future Work and Conclusion

Application development using the Blockchain smart contract is going on and it is one of the trending platforms preferred by the companies. It is in the initial phases and has the potential to accommodate different types of services through it. We proposed in this paper a taxonomy based on three popular smart contract platforms and analyzed tools for security and privacy analysis. Also identified challenges due to the security and privacy issues in the smart contract, it is required to have a standard testing mechanism for the smart contract that can be applicable for all the smart contract platforms. With this to avoid common errors at the time of development of smart contracts, AI can be used and helps to improve the performance. In the future, we will work on it. At last, in the future, we will try to solve the defined problems and try to adapt AI techniques to build tools that will make them more useful for testing and auditing smart contracts.

## References

1. Rosic A (2016) What Are Smart Contracts? [Ultimate Beginner's Guide to Smart Contracts]. Accessed 03 Apr 2020, https://blockgeeks.com/guides/smart-contracts
2. Angelis J, Ribeiro da Silva E (2019) Blockchain adoption: a value driver perspective. Bus Horiz 62(3):307–314. https://doi.org/10.1016/j.bushor.2018.12.001
3. Biosensors Market|Size, Share and Global Market Forecast to 2024|MarketsandMarketsTM (2019). https://www.marketsandmarkets.com/Market-Reports/blockchain-technology-market-90100890.html, Accessed 29 Mar 2020
4. Ma F, et al (2019) EVM ∗ : from offline detection to online reinforcement for ethereum virtual machine. In: SANER 2019 - Proceedings of the 2019 IEEE 26th International Conference on Software Analysis, Evolution, and Reengineering, pp. 554–558. https://doi.org/10.1109/SANER.2019.8668038
5. Lin IC, Liao TC (2017) A survey of blockchain security issues and challenges. Int J NetwSecur 19(5):653–659. https://doi.org/10.6633/IJNS.201709.19(5).01
6. AtzeiNMBB, Cimoli T (2017) A survey of attacks on ethereum smart contracts (SoK), July 2015:164–186. https://doi.org/10.1007/978-3-662-54455-6
7. Mense A, Flatscher M (2018) Security vulnerabilities in ethereum smart contracts. In: ACM International Conference Proceeding Series, pp 375–380. https://doi.org/10.1145/3282373.3282419
8. Di Angelo M, Salzer G (2019) A Survey of Tools for Analyzing Ethereum Smart Contracts
9. Gupta BC (2019) Analysis of ethereum smart contracts - a security perspective

10. Groce A, Feist J, Grieco G, Colburn M (2020) What are the Actual Flaws in Important Smart Contracts (and How Can We Find Them)? Accessed 14 Apr 2020, https://trailofbits.com
11. Tasca P, Tessone CJ (2019) A taxonomy of blockchain technologies: principles of identification and classification. Ledger 4:1–39. https://doi.org/10.5195/ledger.2019.140
12. Demir M, Alalfi M, Turetken O, Ferworn A (2019) Security smells in smart contracts. In: ProceEdings - Companion 19th IEEE International Conference Software Quality Reliability Security QRS-C, pp 442–449. https://doi.org/10.1109/QRS-C.2019.00086
13. Sayeed S, Marco-Gisbert H, Caira T (2020) Smart contract: attacks and protections. IEEE Access 8:24,416–24,427. https://doi.org/10.1109/ACCESS.2020.2970495
14. Durieux T, Ferreira JF, Abreu R, Cruz P (2019) Empirical Review of Automated Analysis Tools on 47,587 Ethereum Smart Contracts. https://doi.org/10.1145/3377811.3380364
15. Parizi RM, Dehghantanha A, Choo K-KR, Singh A (2018) Empirical vulnerability analysis of automated smart contracts security testing on blockchains, no September, pp 103–113. http://arxiv.org/abs/1809.02702
16. NEO Smart Contract Introduction. https://www.apriorit.com/dev-blog/571-neo-nep-5-vulnerabilities, Accessed 09 Sept 2020
17. Peng Z, QihooYC (2018) All roads lead to Rome: Many ways to double spend your cryptocurrency
18. https://owasp.org/www-community/OWASP_Risk_Rating_Methodology
19. Lee S, Kim D, Kim D, Son S, Kim Y (2019) Who Spent My EOS? On the (In)Security of Resource Management of EOS.IO
20. Torres CF, Steichen M (2019) The Art of The Scam: Demystifying Honeypots in Ethereum Smart Contracts. https://etherscan.io/
21. Torres CF, Steichen M (2019) The art of the scam: demystifying honeypots in ethereum smart contracts. https://etherscan.io/
22. GitHub - Relfos/neo-debugger-tools: A set of tools to develop NEO smart contracts. https://github.com/Relfos/neo-debugger-tools#overview, Accessed 20 Sept 2020
23. He N, et al (2020) Security analysis of EOSIO smart contracts. Accessed 20 Sept 2020, http://arxiv.org/abs/2003.06568
24. Mossberg M, et al (2019) Manticore: a user-friendly symbolic execution framework for binaries and smart contracts. In: Proceedings - 2019 34th IEEE/ACM International Conference Automation Software Engineering ASE 2019, pp 1186–1189. Accessed 27 Sept 2020, http://arxiv.org/abs/1907.03890
25. Mossberg M, et al (2019) Manticore: a user-friendly symbolic execution framework for binaries and smart contracts. https://doi.org/10.1109/ASE.2019.00133

Check for updates

# Heteroskedasticity Analysis During Operational Data Processing of Radio Electronic Systems

Maksym Zaliskyi[1](✉), Oleksandr Solomentsev[1], Olga Shcherbyna[1], Ivan Ostroumov[1], Olha Sushchenko[1], Yuliya Averyanova[1], Nataliia Kuzmenko[1], Oleksandr Shmatko[2], Nikolay Ruzhentsev[2], Anatoliy Popov[2], Simeon Zhyla[2], Valerii Volosyuk[2], Olena Havrylenko[2], Vladimir Pavlikov[2], Kostiantyn Dergachov[2], Eduard Tserne[2], Tatyana Nikitina[3], and Borys Kuznetsov[4]

[1] National Aviation University, Huzara av. 1, Kyiv 03058, Ukraine
maximus2812@ukr.net

[2] National Aerospace University H.E. Zhukovsky "Kharkiv Aviation Institute", Chkalov st. 17, Kharkiv 61070, Ukraine

[3] Kharkiv National Automobile and Highway University, Ya. Mudroho st. 25, Kharkiv 61002, Ukraine

[4] State Institution Institute of Technical Problems of Magnetism, National Academy of Sciences of Ukraine, Industrialna st. 19, Kharkiv 61106, Ukraine

**Abstract.** The paper deals with the problem of mathematical models building for diagnostic variables of radio electronic systems in the case of data heteroskedasticity. The heteroskedasticity is associated with event when clusters from single sample are characterized by different values of variance. In operation theory heteroskedasticity occurs in case of equipment technical condition deterioration. The paper describes the method for taking into account heteroskedasticity, which includes the following steps: building and choosing the best regression model; calculation of heteroskedasticity index based on minimization of weighted sum of residuals squares; determining correction coefficients of heteroskedasticity and obtaining the final regression equation. The proposed methods can be used during utilization of condition-based maintenance with diagnostic variables monitoring.

**Keywords:** Statistical data processing · Heteroskedasticity · Segmented regression · Radio electronic system · Technical condition deterioration

## 1 Introduction

Radio electronic systems (RES) in civil aviation are used in the process of providing air navigation services [1]. The efficiency of the intended use of RES is ensured in the operation system [2].

The technical condition of RES during operation can deteriorate [3]. An important component in the tasks of identifying and evaluating the parameters of deterioration of the technical condition of RES is the choice of the most correct mathematical model for describing statistical data on the diagnostic variables and reliability parameters [4].


S. Shukla et al. (eds.), *Data Science and Security*, Lecture Notes in Networks and Systems 290,
https://doi.org/10.1007/978-981-16-4486-3_18

The analysis shows that at the different intervals of quasi-stationarity before the deterioration of the technical condition and after its occurrence, the diagnostic variables and reliability parameters can be characterized by different values of the variance [5]. Therefore, on the process of building mathematical models of these parameters, it is advisable to use the heteroskedasticity analysis.

During heteroskedasticy dependences building, weighted coefficient are calculated for each empirical value, which take into account the instability of the variance for different parts of the statistical sample [6]. Based on the use of these coefficients, analytical dependences can be found that describe the empirical data more accurately (with a smaller value of the weighted sum of squares of deviations between the measured value and the approximated one) [7, 8]. Therefore, the purpose of this paper is the synthesis and analysis of methods for taking into account heteroskedasticity, which can be used in solving problems of detecting deterioration of the technical condition based on statistical processing of data on the diagnostic variables of RES.

## 2 Literature Analysis and Problem Statement

One of the tasks of mathematical models building is to substantiate the best model [9, 10]. The following criteria are usually used alone or in combination as criteria for making this choice: the smallest number of mathematical formula coefficients that are compatible with a given error; the simplest description; reasonable physical substantiation; the minimum sum of the squares of the deviations between the approximated and empirical values; minimum variance [8, 11–13].

Literature analysis in the field of mathematical models building shows that insufficient attention is paid to the heteroskedasticity accounting for empirical data and the use of polygonal regressions to describe them [14, 15]. This leads to less accurate predicted estimates of the studied random variables.

Classical empirical procedures for testing data for heteroskedasticity were proposed by Goldfeld-Quandt and Glaser and demonstrated in [16–21]. However, these procedures have disadvantages: first of all, mentioned tests do not give a specific value to the heteroskedasticity assessment. In addition, uniform approximation functions are most often used to approximate empirical data [6, 8, 13], and insufficient attention is paid to multi-segmented functions.

The problem of building and choosing the best mathematical model can be formed mathematically as follows. Suppose that for a set of two-dimensional statistical data $(t_i;\ y_i)$ there is a certain set of approximation functions $\hat{y}_i = f_j(t_i,\ \vec{c}_{m,j})$, which establishes the correlation between them (where $t_i$ is time moment of parameters measuring, $y_i$ is diagnostic variable, $\vec{c}_{m,j}$ is the vector of $m$ parameters for the approximation function, $j$ is the number of the approximation function). For each approximation function, the standard deviation $\sigma$ between the real values $y_i$ and their estimates $\hat{y}_i$ can be calculated.

Then the choice of the best mathematical model will be made according to the following criterion

$$p = \inf\big(s \in N\,\forall j:\ \sigma(f_s(x_i,\ \vec{c}_{m,s})) \le \sigma(f_j(x_i,\ \vec{c}_{m,j}))\big). \tag{1}$$

## 3 Models of Diagnostic Variable in the Case of Heteroskedasticity

It is known that due to disregard for heteroskedasticity there are the following weaknesses: the least squares method will not allow to construct an estimator with the smallest variance (estimates of linear regression coefficients will not be effective); significance tests will be too high or too low, standard errors will be shifted along with the corresponding confidence intervals.

The problem of detecting heteroskedasticity can be reduced to the task of testing hypotheses [22]. The hypothesis $H_0$ corresponds to the assumption of equality of variances (i.e. $\sigma(y_1) = \sigma(y_2) = \ldots = \sigma(y_n)$), and the alternative $H_1$ is the inequality of variances (i.e. $\sigma(y_1) \neq \sigma(y_2) \neq \ldots \neq \sigma(y_n)$).

Analysis of the processes of the technical condition deterioration for RES shows that in these cases the variance of the diagnostic variables at the different intervals of observation is not constant [23–26]. In general case, after the occurrence of the deterioration, the variance of the statistical data can increase over time [27].

The problem of deterioration detection is especially urgent in case of condition-based maintenance with diagnostic variables monitoring [28, 29].

Let us make statistical data analysis in case of heteroskedasticity. An example of diagnostic variable data in cases of variance increasing over time and the probability density function (PDF) for this diagnostic variable is shown in Fig. 1.

The data in Fig. 1 represent a trend of change of the diagnostic variable for the case of linear deterioration of the technical condition. Such model of deterioration can be described according to the equation:

$$y(t) = Z_0 h(t) + v(t - t_{sw})h(t - t_{sw}) + \vartheta(t), \quad (2)$$

where $Z_0$ is an initial value of diagnostic variable, $h(t)$ is Heaviside step function, $v$ is a deterioration parameter (velocity), $t_{sw}$ is a time moment of deterioration occurrence, $\vartheta(t)$ is a random component of diagnostic variable.

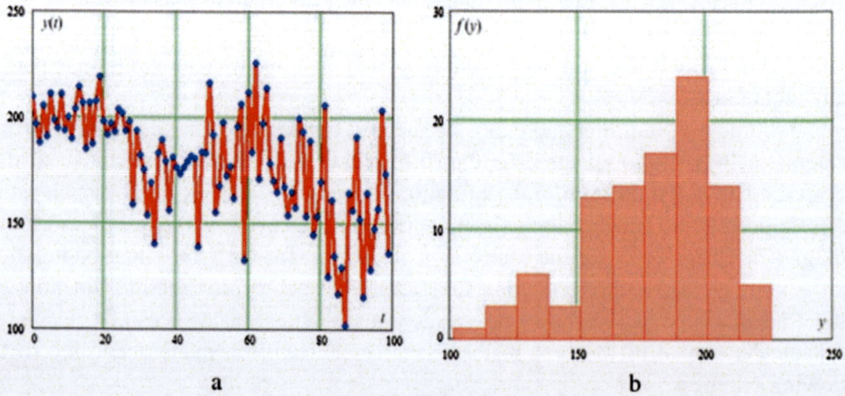


**Fig. 1** Statistical data on RES diagnostic variable in case of variance increasing over time (a) and the probability density function of diagnostic variable for the case of heteroskedasticity (b)

The trend in Fig. 1 was constructed for the following parameters of the general population: $Z_0 = 200$, $t_{sw} = 25$, $v = 0.5$, sample size $N = 100$, before deterioration $\vartheta$ is described by normally distributed random variable with mean $m_1(\vartheta) = 0$ and standard deviation $\sigma(\vartheta) = 10$, after deterioration the standard deviation increases linearly with the velocity $0.2i$, where $i$ is a number of measuring.

The PDF of diagnostic variable (shown in Fig. 1) for the model (2) can be represented as a polygaussian model

$$f(y) = \frac{k-1}{N} \frac{1}{\sigma(\vartheta)\sqrt{2\pi}} e^{-\frac{(y-Z_0)^2}{2\sigma^2(\vartheta)}} + \frac{1}{N-k+1} \sum_{i=k}^{N} \frac{1}{(\sigma(\vartheta)+0.2i)\sqrt{2\pi}} e^{-\frac{(y-Z_0+vi)^2}{2(\sigma(\vartheta)+0.2i)^2}}, \tag{3}$$

where $k$ is the sample number that corresponds to the deterioration occurrence.

## 4 Method for Taking into Account Heteroskedasticity During Analysis of the Diagnostic Variable Trend

Consider examples of using a new method of taking into account heteroskedasticity during data processing on the diagnostic variables. Let assume that the technical condition of the RES deteriorates according to the linear model (2). In this case, let sudden failures do not occur, and the error of the control and measuring equipment is insignificant, so it can be neglected.

For a more detailed analysis of the new method, let perform simulation of the diagnostic variable trend. The initial parameters of the general population during the simulation are: $t_{sw} = 15, N = 30, Z_0 = 220, v = 0.5$, $\vartheta$ is described by normally distributed random variable and before deterioration mean $m_1(\vartheta) = 0$ and standard deviation $\sigma(\vartheta) = 8$, after mean $m_1(\vartheta) = 0$ and standard deviation $\sigma(\vartheta) = 15$.

The simulation results of the diagnostic variable trend are given in Table 1.

During regression models building, the linear, quadratic and piecewise linear regressions were used. The calculations give the following equation for linear and quadratic approximation (according to ordinary least squares method):

$$\hat{y}_1(t_i) = 213.998 + 0.701t_i, \tag{4}$$

$$\hat{y}_2(t_i) = 224.989 - 1.360t_i + 0.066t_i^2. \tag{5}$$

The results of the approximation according to the models (4) and (5) are shown in Fig. 2. The sum of the squares of the residuals for linear regression is $\Sigma_1 = 2542$, and for quadratic $\Sigma_2 = 1942$.

To perform a piecewise linear approximation, consider five options of the values of the abscissa of the switching points $t_{sw} = \{9, 12, 15, 18, 21\}$. For each approximation option, the sums of the squares of the residuals are calculated, which are respectively equal $\Sigma_3(t_{sw}) = \{2218, 2003, 1947, 1896, 2029\}$. To find the optimal value of the switching point, the dependence $\Sigma_3(t_{sw})$ is approximated by the quadratic function according to the ordinary least squares method [30]. The minimum of this parabola

**Table 1** Diagnostic variable trend

| Time $t_i$ | Value $y_i$ | Time $t_i$ | Value $y_i$ | Time $t_i$ | Value $y_i$ |
|---|---|---|---|---|---|
| 1 | 221.44 | 11 | 223.91 | 21 | 227.37 |
| 2 | 225.43 | 12 | 212.09 | 22 | 223.17 |
| 3 | 219.80 | 13 | 205.40 | 23 | 233.27 |
| 4 | 215.01 | 14 | 215.84 | 24 | 242.50 |
| 5 | 219.64 | 15 | 235.57 | 25 | 210.76 |
| 6 | 222.73 | 16 | 215.86 | 26 | 241.22 |
| 7 | 211.81 | 17 | 209.51 | 27 | 235.28 |
| 8 | 230.82 | 18 | 220.09 | 28 | 249.92 |
| 9 | 216.73 | 19 | 220.68 | 29 | 241.29 |
| 10 | 229.57 | 20 | 230.07 | 30 | 238.93 |

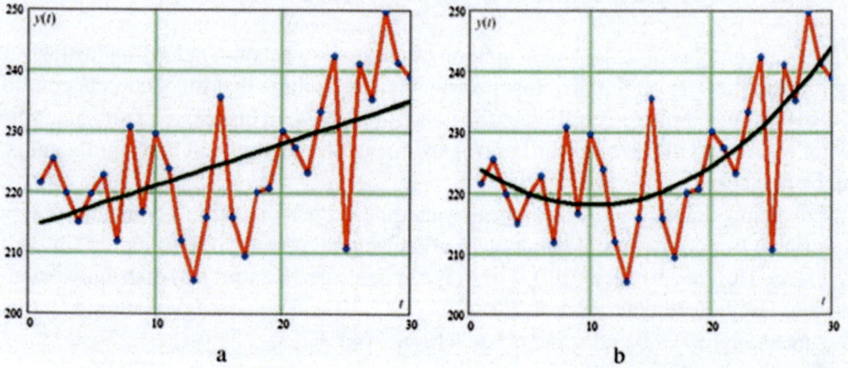


**Fig. 2** Approximation of the diagnostic variable by linear (a) and quadratic (b) regressions

corresponds to the optimal value of the abscissa of the switching point $t_{\text{sw opt}} = 16.451$. As a result, the optimal piecewise linear regression can be calculated

$$\hat{y}_{3\,\text{opt}}(t_i) = 222.077 - 0.277t_i + 2.162(t_i - 16.451)h(t_i - 16.451). \quad (6)$$

The sum of the squares of the residuals for this equation is $\Sigma_3(t_{\text{sw opt}}) = 1880$.

After comparing the values $\Sigma_1$, $\Sigma_2$ and $\Sigma_3$, according to the criterion of the minimum sum of squares of the residues (1), it can be concluded that the best is the optimal piecewise linear regression (6). Therefore, the Eq. (6) will be used as a basic function for calculating the coefficients of heteroskedasticity $W_i$.

The coefficients of heteroskedasticity

$$W_i = \left(\frac{m_1(y)}{\hat{y}(t_i)}\right)^h, \quad (7)$$

where $h$ is a heteroskedasticity index.

**Table 2** Coefficients of heteroskedasticity

| Time $t_i$ | Coefficient $W_i$ | Time $t_i$ | Coefficient $W_i$ | Time $t_i$ | Coefficient $W_i$ |
|---|---|---|---|---|---|
| 1 | 1.044 | 11 | 1.087 | 21 | 0.983 |
| 2 | 1.048 | 12 | 1.091 | 22 | 0.957 |
| 3 | 1.053 | 13 | 1.095 | 23 | 0.933 |
| 4 | 1.057 | 14 | 1.1 | 24 | 0.909 |
| 5 | 1.061 | 15 | 1.104 | 25 | 0.886 |
| 6 | 1.065 | 16 | 1.109 | 26 | 0.864 |
| 7 | 1.069 | 17 | 1.094 | 27 | 0.842 |
| 8 | 1.074 | 18 | 1.065 | 28 | 0.821 |
| 9 | 1.078 | 19 | 1.037 | 29 | 0.801 |
| 10 | 1.082 | 20 | 1.009 | 30 | 0.782 |

The heteroskedasticity coefficients were calculated for the five options of the heteroskedasticity index $h = \{-5;\ -2.5;\ 0;\ 2.5;\ 5\}$. The average value of the diagnostic variable during the observation interval is $m_1(y) = 224.857$. For each approximation option, the weighted sums of the squares of the residuals were calculated, which are respectively equal to $\Sigma_e(h) = \{1945,\ 1905,\ 1880,\ 1870,\ 1873\}$. To find the optimal value of the heteroskedasticity index, the dependence $\Sigma_e(h)$ was approximated by the quadratic function according to the ordinary least squares method. The minimum of this parabola corresponds to the estimated value of the heteroskedasticity index $\hat{h} = 3.164$. For the obtained estimate, the correction coefficients of heteroskedasticity $W_i$ were calculated according to (7), which are given in Table 2.

As a result, the optimal equation with taking into account heteroskedasticity can be obtained

$$\hat{y}_{h\ \mathrm{opt}}(t_i) = 222.088 - 0.278t_i + 2.163(t_i - 16.451)h(t_i - 16.451). \tag{8}$$

The results of the approximation according to the model (8) are shown in Fig. 3. The weighted sum of the squares of the residuals for this equation is $\Sigma_e(\hat{h}) = 1869$. Therefore, as a result, the lowest value of the weighted sum of the squares of the residuals is achieved, so the accuracy of the approximation increases.

A more accurate regression model can be used to predict the values in the trend of the diagnostic variable, and therefore, the veracity of decision-making can be increased in case of using a condition-based maintenance policy with preventive and adaptive threshold.

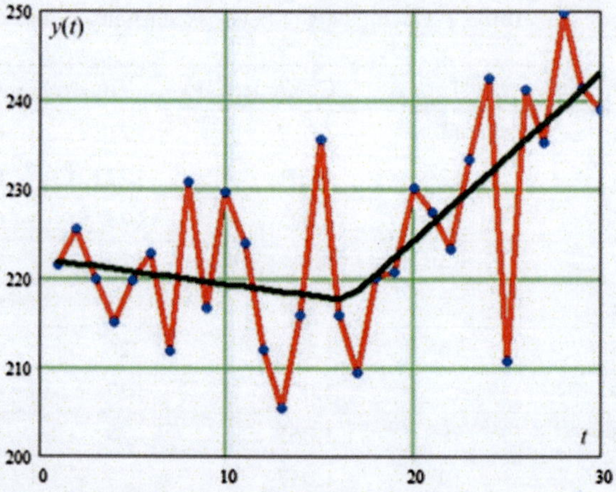


**Fig. 3** Approximation of the diagnostic variable by optimal piecewise linear regression with taking into account data heteroskedasticity

## 5 Conclusion

The analysis showed that during the monitoring of the diagnostic variable in case of deterioration of the technical condition of RES in the trends of their changes, several intervals of quasi-stationarity can be distinguished, each of which is characterized by different value of variance. So, to build more correct mathematical models, it is advisable to use the heteroskedasticity accounting methods.

Comparative analysis showed the advantage of the method of taking into account heteroskedasticity by the criterion of minimum weighted sum of the squares of the residuals. The proposed method allows to build more adequate mathematical models of diagnostic variable trends for RES. The research results can be used during design and improvement of RES operation systems.

## References

1. Kuzmenko NS, Ostroumov IV, Marais K (2018) An accuracy and availability estimation of aircraft positioning by navigational aids. In: IEEE international conference on methods and systems of navigation and motion control (MSNMC), Kyiv, Ukraine, pp 36–40
2. Solomentsev OV, Melkumyan VH, Zaliskyik MY, Asanov MM (2015) UAV operation system designing. In: IEEE 3rd international conference on actual problems of unmanned air vehicles developments (APUAVD), Kyiv, Ukraine, pp 95–98
3. Hryshchenko Y (2016) Reliability problem of ergatic control systems in aviation. In: IEEE 4th international conference on methods and systems of navigation and motion control (MSNMC), Kyiv, Ukraine, pp 126–129
4. Levin BR (1978) Theory of reliability of radio engineering systems. Radio, Moscow. (in Russian)
5. Solomentsev O, Zaliskyi M, Herasymenko T, Petrova Y (2019) Data processing method for deterioration detection during radio equipment operation. In: Microwave theory and techniques in wireless communications (MTTW 2019). Riga, Latvia, pp 1–4

6. Kuzmin VN (2001) The sequential statistical analysis of econometric data under heteroskedasticity. In: Computer data analysis and modeling. robustness and computer intensive methods. Minsk, Byelorussia, pp 37–42
7. Mitropolsky AK (1971) The technique of statistical computing. Nauka, Moscow (in Russian)
8. Himmelblau DM (1970) Process analysis by statistical methods. John Wiley and Sons, New York
9. Kolganova O, et al (2020) Method for improving the efficiency of online communication systems based on adaptive multiscale transformation. In: Advanced computer information technologies (ACIT). Deggendorf, Germany, pp 824–829
10. Huet S, Bouvier A, Poursat M-A, Jolivet E (2004) Statistical tools for nonlinear regression. a practical guide with S-PLUS and R Examples. Springer-Verlag, New York
11. Greene WH (2003) Econometric analysis. Pearson Education Inc., New York
12. Shutko V, Tereshchenko L, Shutko M, Silantieva L, Kolganova, O (2019) Application of spline-fourier transform for radar signal processing. In: IEEE 15th international conference on the experience of designing and application of CAD systems. Polyana, Ukraine, pp 110–113
13. Ezekiel M, Fox KA (1959) Method of correlation and regression analysis linear and curvilinear. John Wiley and Sons, New York
14. Weisberg S (2005) Applied Linear Regression. John Wiley and Sons, New York
15. De Groot MH (1970) Optimal statistical decisions. John Wiley & Sons, New York
16. Breusch TS, Pagan AR (1979) A simple test for heteroscedasticity and random coefficient variation. Econometrica 47(5):987–1007
17. Glejser H (1969) A new tests for heteroscedasticity. J Am Statist Assoc 64:316–323
18. Godfrey LG (2006) Tests for regression models with heteroskedasticity of unknown form. Comput Stat Data Anal 50(10):2715–2733
19. Goldfeld SM, Quandt RE (1965) Some Tests for Homoscedasticity. Journal of American Statist. Assoc. 60:539–547
20. White H (1980) A Heteroskedasticity-consistent covariance matrix estimator and a direct test for heteroskedasticity. Econometrica 48(4):817–838
21. Cheng T-C (2011) Robust diagnostics for the heteroscedastic regression model. Comput Statist Data Anal 55:1845–1866
22. Lehmann EL, Romano JP (2008) Testing statistical hypotheses. Springer, New York
23. Barlow RE, Proschan F (1965) Mathematical theory of reliability. John Wiley and Sons, New York
24. Goncharenko AV (2017) Optimal UAV maintenance periodicity obtained on the multi-optional basis. In: IEEE 4th international conference on actual problems of UAV developments, Kyiv, Ukraine, pp 65–68
25. Zaliskyi M, Petrova Yu, Asanov M, Bekirov E (2019) Statistical data processing during wind generators operation. Int J Electr Electron Eng Telecommun 8(1):33–38
26. Solomentsev O, Zaliskyi M, Herasymenko T, Kozhokhina O, Petrova, Yu (2018) Data processing in case of radio equipment reliability parameters monitoring. In: Advances in wireless and optical communications (RTUWO 2018). Riga, Latvia, pp 219–222
27. Solomentsev O, Zaliskyi M, Herasymenko T, Kozhokhina O, Petrova Yu (2019) Efficiency of operational data processing for radio electronic equipment. Aviation 23(3):71–77
28. Condition-based Maintenance Recommended Practices. SAE standard ARP-6204 (2014)
29. A guidebook to implementing condition based maintenance (CBM) (2018) Using Real-time Data. OSIsoft, LLC, USA
30. Reklaitis GV, Ravindran A, Ragsdell KM (1983) Engineering optimization methods and applications. John Wiley and Sons, New York

# Role of Data Science in the Field of Genomics and Basic Analysis of Raw Genomic Data Using Python

S. Karthikeyan(✉) and Deepa V. Jose

Department of Computer Science, CHRIST (Deemed to be University), Bangalore, Karnataka, India
karthikeyan.s@cs.christuniversity.in,
deepa.v.jose@christuniversity.in

**Abstract.** The application of genomics in identifying the nature and cause of diseases has predominantly increased in this decade. This field of study in life sciences combined with new technologies, revealed an outbreak of certain large amounts of genomic sequences. Analysis of such huge data in an appropriate way will ensure accurate prediction of disease which helps to adopt preventive mechanisms which can ultimately improve the human quality of life. In order to achieve this, efficient comprehensive analysis tools and storage mechanisms for handling the enormous genomic data is essential. This research work gives an insight into the application of data science in genomics with a demonstration using Python.

**Keywords:** Genome · Genetics · Data science · Analysis · Cloud architecture

## 1 Introduction

Genomics is a study of genes in genetic data. A genome is a genetic component of an organism which is made up of Deoxyribonucleic acid (DNA). It has sufficient data to construct that organism with its details; i.e., all the information essential to model that organism including the coding as well as non-coding component of that organism's DNA. A human genome data or size exceeds in billions. Next generation sequencing has been very efficient and plays an important role in reading DNA. Basic biomedical problems are hidden in this data which show how diseases spread and how the mutations happen [1]. There are three steps of analysis in the field of bioinformatics research; the primary analysis, secondary analysis and the tertiary analysis.

In primary analysis, reading of DNA in smaller segments happens. Secondary analysis results in arranging these smaller segments in a manner that makes understanding and discovering the new features effectively. Tertiary analysis focuses on understanding how these segments relate with each other and to create a meaning out of it. Data science concepts such as big data analytics helps in recognizing patterns on large scale data. Correlation exists between these data and such analytics are very helpful in mapping relations among the data. In the past few years genomic data have been growing


S. Shukla et al. (eds.), *Data Science and Security*, Lecture Notes in Networks and Systems 290,
https://doi.org/10.1007/978-981-16-4486-3_19

rapidly and the processing of such pairs of genomic data will yield gigabytes of data in different formats [2]. Analysing and interpreting the results from such data is a major task. Genomic data that is available is not in a user friendly or a readable format. Using python, analysis has been done on the data so that one can find meaningful information from it.

This paper is organised as follows. The recent literature specifying the role of data science in the field of genomics is mentioned in Sect. 2. Section 3, 4 gives an overview of the methodology adopted in this study, the experimental set up used and the discussions and observations of the results achieved followed by the recent findings in this field in Sect. 5.

## 2 Literature Review

Parallel processing is used at a high level to process the data to reduce the cost and time of reading the DNA and a large amount of investment is done for this sequencing technique nowadays. Many disorders have been studied using this technology. The reason for these disorders is present in these data. A simple example would be how cancer cells mutate and its dependence on external factors. Bioinformatics is made up of a combination of all the sciences; especially, which are interrelated with biology and statistical concepts are used for making any inferences from the data [3]. A machine learning perspective combined with bioinformatics is implemented to manage genomic data by the method of parallel computing. Iterative processing is better in terms of performing operations and the problem with the earlier approach was that the tools perform operations in batch-mode.

There are many new methods of machine learning invented such as incremental and multi-view algorithms and these algorithms do help in the input output operations for the iterative process but there are no proper tools to be used for many important bioinformatics problems, such as fast analysis of huge DNA, Ribonucleic acid (RNA), and protein sequence data etc. [4]. One of the major applications of MapReduce is Apache Hadoop1 which is open source. For all these systems to work there is a main node called the master node which controls the operations. Worker nodes perform some of the general operations on the data. Master node takes care of breaking the data into segments and worker nodes take care of analysing and performing the necessary operations on them [5].

Genomics will still remain to be an emerging concept in data science in increasing volume and its presence. This data can be represented in terms of data volume, data velocity, data variety, measurement, mining, modelling, and manipulating [6]. Volume, velocity and variety are known as the 3V framework and measurement, mining, modelling, and manipulating are known as 4M framework. The 3V framework is used for identifying problems which need new measures and tools to manage data [7]. Velocity in genomic data can be understood in two different points: the speed with which the data is generated and the speed with which the data is processed and available for use.

Primary data in the FASTQ format, a text-based format which stores a biological sequence along with its quality scores, which is a combination of sequence and its quality scores are being discarded and the compressed and mapped file formats such as

Binary Alignment Map (BAM), variant cell format (VCF) are preferred for storage [8]. Generating or sequencing of the human genome is done within a day's time today. This is come from months taken to generate the data from famous technologies or undisturbed genomic generation by Human Genome Project.

There are some technologies for which the cost of using them has been reduced considerably such as diagnostic imaging and microarrays therefore resulting data are much quicker to generate [9]. Cloud computing facilitates data storage and major computing operations. These features help in integrating genomic and clinical data. Being a cloud-based architecture, the results can be shared with anyone around the world ensuring security and privacy. A quick example can be the Pan-Cancer Analysis of Whole Genome project where such a technology is being used [10]. Cloud computing has the ability to perform analysis of extremely huge data scales such as a petabyte scale from research centres.

Federated cloud is made up of two services namely public and private. The advantage of this model is that it can be built using different locations by following the rules and regulations of that place. Such a model could help to classify personal data and additional information [11]. Cloud based data managers are made for data from specific regions which support a genome-oriented language known as the GenoMetric Query language (GMQL). The is language inherits the terms and functioning of relational databases and uses it on genomic datasets. Unary operations such as merge, group, sort etc., and binary operations such as join, union etc., are available in GMQL. There also exist domain specific operations such as extraction of regions, properties from respective datasets based on the genome configuration. The GenoMetric Query language system has a modular architecture. The operations under this system are executed by Apache Spark engine. Apache Spark engine has proven to be a very efficient data framework in handling large genomic queries as well as supporting a variety of repository types like Hadoop file system etc. The queries that are executed in the system are supported by Spark. This is again under the support of high-end cloud vendors like Google, Amazon etc. The system includes two interfaces for programming. One is a python-based interface supported by pyGMQL library and second is a R based interface supported by the RGMQL package. The workflows are supported by FireCloud and Galaxy [12].

Genomics aspect is like two sides of a coin. One side consists of generating and sequencing the data where the nucleotides are mapped to the genome and then generate the remaining part. The sequencing of data has a huge variety of smaller components which are used to measure many aspects of genomes. The other side is how these components are interrelated [13].

## 3 Methodology for Analysing Genomic Data Using Python

A basic analysis of genomic data using Python is presented here. The genome file in General Feature Format (GFF3) format can be obtained from websites like Ensemble etc. Python is used as the tool to handle this genomic data. Necessary and basic packages like pandas are also installed in order to perform the analysis. Data at hand needs to be pre-processed well in order to avoid the issues of unlabelled and missing data etc.

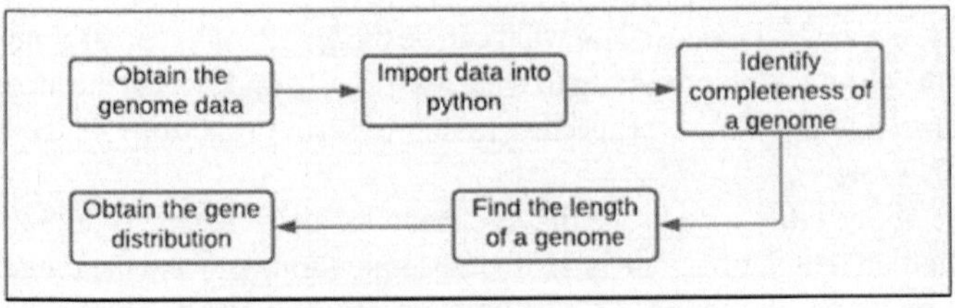


**Fig. 1** Overview of feature analysis

### 3.1 Experimental Setup

The raw data at hand cannot be interpreted without performing an exploratory data analysis on it. Once the data is obtained it is loaded into python, the tool which consists of many useful packages and libraries for handling such data like Pandas. Pandas is an important package as GFF3 format is a tab-delimited file. It helps in flexible and easier analysis for understanding and manipulating the types of data present in the file. The file does not contain headers. Using simple commands this can be handled and can specify names for each column which will be meaningful for further analysis.

Using this pre-processed data, the analysis of the following is done: identify the length, completeness of a genome and gene distribution among chromosomes and an overview of the same is depicted in Fig. 1.

## 4 Results and Discussion

The first point analysed is the completeness of the genome in the genomic data. Genome Reference Consortium (GRCh38) consists of the chromosome information. The data consists of both assembled and unassembled sequences. A sequence ID for each chromosome is present and it covers information of all well assembled sequences. Unassembled sequences turn out to be around 0.4%. All the entries having a sequence ID are extracted and in order to find the fraction of the genome that is incomplete, the sum of the lengths of all entries with a sequence ID must be known. The length of each entry after calculation ranges between 2.5 to 3 billion bases long.

**Table 1** Length of a typical gene

| Attributes | Values |
|---|---|
| count | 4.247000e + 04 |
| mean | 2.583348e + 04 |
| std | 9.683485e + 04 |
| min | 7.000000e + 00 |
| max | 3.304997e + 06 |

Secondly, the analysis focused on what can be the length of a typical gene in genomic data. A length column is added similar to what was done to understand the incompleteness of the genome. Output of the properties (using describe function) of the column are mentioned in Table 1.

From the above table we can see that the mean length of a gene is 25,800 bases long (approx.), median length of a gene is 4130 bases long (approx.) and the range is between 7 to 3.3 million bases long (approx.). The importance of knowing the length of a genome is important in multiple situations.

Example: A human gene with longer length is a part of early life functions whereas genes with smaller length tend to be a part of the entire life. The third analysis is related to the gene distribution among chromosomes. The data file needs to be filtered in order to obtain the unassembled sequences which is done with the help of chromosome list and is represented in Table 2.

**Table 2** Gene distribution

| Sequence ID of chromosomes | Values |
|---|---|
| 1 | 3865 |
| 2 | 2501 |
| 18 | 2314 |
| 16 | 2298 |
| X | 1674 |
| Y | 346 |

Chromosomes based on sequence ID with number of genes shows that the chromosome with sequence ID 1 contains the largest number of genes. The last Chromosome in the list is Y which consists of the least number of genes but it should not be confused with it being the smallest chromosome in the list. Knowing the gene distribution in a genome is very important as it can help understand the effects of certain viruses and relationships among them.

Example: Phages which are also known as bacteriophages refer to a type of virus. They halt bacterial infection by injecting a viral genome. Knowing the gene distribution can help understand the relationship of phages and further allow classification among them.

## 5 Recent Findings in the Field of Genomics

Recent findings in genomic data in terms of volume as well as dimension has created an urge for the scientists to not only store such data but also manipulate this data.

The features of GMQL system will enable sharing of data as well as process of this data in specific GMQL instances which are active on cloud systems. This will help in sending the data and its processing work to whichever regions required. This gives an advantage of avoiding costly transfers and downloads of data as well as enabling the usage of resources that are available in those regions.

## 6 Conclusions

This study has been done in order to get an understanding of genes, its presence in the genomic data and some properties of the components present in the genomic data. In the near future there will be a huge increase in raw genomic data with a scale of almost 20 to 40 times compared to astronomical data. It will also overtake the data of total videos that are uploaded in social sites. Some examples of the current projects that are a proof of genetic data volume increasing enormously are 100 K genomes and 500 K Finnish citizens.

## References

1. A Brief Guide to Genomics (2019) 15 August 2015. https://www.genome.gov/about-genomics/fact-sheets/A-Brief-Guide-to-Genomics. Accessed 2 Dec 2019
2. Quilez Oliete J: A step-by-step guide to DNA sequencing data analysis, Kolabtree Blog, 23 March 2020. https://www.kolabtree.com/blog/a-step-by-step-guide-to-dna-sequencing-data-analysis/. Accessed 9 Apr 2020
3. Zhang X, Li A, Zhang Y, Xiao Y (2012) Validity of cluster technique for genome expression data. In: 2012 24th Chinese control and decision conference (CCDC), Taiyuan, pp 3737–3741. https://doi.org/10.1109/CCDC.2012.6244599
4. Jimenez-Lopez J, Gachomo E, Sharma S, Kotchoni S (2013) Genome sequencing and next-generation sequence data analysis: a comprehensive compilation of bioinformatics tools and databases. Am J Mol Biol 3:115–130. https://doi.org/10.4236/ajmb.2013.32016
5. Leggett RM, Ramirez-Gonzalez RH, Clavijo BJ, Waite D, Davey RP (2013) Sequencing quality assessment tools to enable data-driven informatics for high throughput genomics. Front Genet 4:288. https://doi.org/10.3389/fgene.2013.00288
6. Schatz MC (2015) Biological data sciences in genome research. Cold Spring Harb Lab Press Perspect 25:1417–1422. https://doi.org/10.1101/gr.191684.115.
7. Ceri S, Kaitoua A, Masseroli M, Pinoli P, Venco F, Milano P (2016) Data management for next generation genomic computing. EDBT 485–490. https://doi.org/10.5441/002/edbt.2016.46.
8. Roy S, LaFramboise WA, Nikiforov YE, Nikiforova MN, Routbort MJ, Pfeifer J, Nagarajan R, Carter AB, Pantanowitz L (2016) Next-generation sequencing informatics: challenges and strategies for implementation in a clinical environment. Arch Pathol Lab Med 140(9):958–975. https://doi.org/10.5858/arpa.2015-0507-RA Epub 2016 Feb 22 PMID: 26901284
9. He KY, Ge D, He MM (2017) Big data analytics for genomic medicine. Int J Mol Sci 18:1–18. https://doi.org/10.3390/ijms18020412
10. Molnár-gábor, F, Lueck R, Yakneen S, Korbel JO (2017) Computing patient data in the cloud: practical and legal considerations for genetics and genomics research in Europe and internationally. Genome Med 9:1–12. https://doi.org/10.1186/s13073-017-0449-6
11. Navarro, FCP, Mohsen H, Yan C, Li S, Gu M, Meyerson W (2019) Genomics and data science: an application within an umbrella. Genome Biol 20:1–11. https://doi.org/10.1186/s13059-019-1724-1
12. Ceri S, Pinoli P (2020) Data science for genomic data management: challenges, resources, experiences. SN Comput Sci 1:1–7. https://doi.org/10.1007/s42979-019-0005-0.
13. Kashyap H, Ahmed HA, Hoque N, Roy S, Bhattacharyya DK (2015) Big data analytics in bioinformatics: a machine learning perspective. CoRR abs/1506.05101

# Automatic Detection of Smoke in Videos Relying on Features Analysis Using RGB and HSV Colour Spaces

Raghad H. Mohsin[1], Hazim G. Daway[2(✉)], and Hayfa G. Rashid[1]

[1] Department of Physics, College of Education, Mustansiryah University, Baghdad, Iraq
dr.hayfa@uomustansiriyah.edu.iq

[2] Department of Physics, College of Science, Mustansiryah University, Baghdad, Iraq
hazimdo@uomustansiriyah.edu.iq

**Abstract.** Automatic smoke video detection plays an important role in saving human life and preserving the environment. In this study, digital image processing has been addressed to detect smoke. The suggested algorithm relies on detecting moving objects follows by several features based on the basic colour space RGB and HSV colour space. To find out the efficiency of the proposed method in detection, the proposed method was compared with several other methods by calculating the accuracy, by analyzing the results the proposed method has got a high accuracy of detection nearly approximate to 99.02% for different videos.

**Keywords:** Color space · HSV · Image processing · RGB · Smoke detection

## 1 Introduction

Image processing techniques have entered many fields of application, such as medical, military, industrial, environmental, and other applications [1–3]. In the field of environment and safety, early detection of fire and smoke has become very important in preserving human life and nature, we can now use digital image processing techniques in smoke and fire detection [4]. The smoke is represented one of the main signals of fire, the detection of fire in a video is especially viable in industrial monitoring and surveillance used to monitor buildings and environment as a part of an early warning technique that reports preferably start of the fire. Where can detect uncontrolled fires at an early step before they turn into the disaster by video-based systems.

## 2 Literature Review

Many authors study smoke and fire detection as:

Jong-Wook Bae et al. developed a system based on a statistical color model to detect fire automatically without the experiential fixed threshold values. HIS (hue, intensity, saturation) color transformation and the mask of binary backgrounds were included


S. Shukla et al. (eds.), *Data Science and Security*, Lecture Notes in Networks and Systems 290,
https://doi.org/10.1007/978-981-16-4486-3_20

in the proposed system. 600 frames together with six different models(fire videotape clips), the analysis shows that the average detection rate was 85% [5]. Punam Patel et al. proposed a method to integrate a number of techniques (color detection, movement detection, and dispersion of the region) in the process of fire video detection video. The proposed method first identified the required color areas of video frames and then determine the moving area follows with pixel area computation. The original video was converted into a number of sequential frames, which consist of fire and non-fire images. The proposed method was based on three steps: fire pixel detection using RGB and YCbCr color model, motion pixel detection, and dispersion of the region in addition to fire shape colored pixels analyzing [6].

Chen Juan et al. discussed two characteristics of fire, any color and flash information to provide basic standards to improve the ability and increase the reliability of flame detection video in complex scenes. The color detection standard was proposed by reconstructing the two- dimensional color space and the appropriate saturation. The sample database is manually collected by a pixel flame. The flame volume and the accumulated gray value are then extracted to construct a time series where the oscillation frequency calculation of flickering flame was arranged into the video sequence of four combustion experiments by Fourier Transform of these time series which detected fire feature successfully [7].

Suphachai Praising et al. have proposed a system to detect fire in the building using HSV and YCbCr color space to isolate orange, yellow, and high lightness from the background. Checked fire growth achieved based on frame differences and the total accuracy of the experiments was more than 90% [8].

Shiping Yea et al proposed a method to detect smoke and flame together for color video sequences obtained from a stationary camera in outdoor. Motion feature One of the types smoke and flame and commonly has been applied in the beginning for extraction from a current frame of candidate areas. The adaptive background subtraction has been used at a stage of moving detection, and optical flow-based movement estimation has been applied to identify a chaotic motion. With the spatial and temporal wavelet analysis, they achieved moving blobs classification. and the total accuracy of the experiments was then 87% for smoke and 92% for fire [9].

Hidenori Maruta et al. proposed smoke detection method involves three steps: analyzing texture features, discrimination of ferret's region using support vector machine, and time accumulation. SVM was used to classify smoke or nonsmoker from the extracted image and progress in their method was obtained [10].

Konstantinos Avgerinakis et al. discussed a method for the smoke detection in videos. They used the smoke localized based on appearance and movement characteristics. The experiments show that it is accurate and achieved temporal and spatial smoke localization. As a result, edge energy in an image is reduced gradually when smoke passes over it. This information, used with append to the information of smoke color to detect smoke. The result of accuracy was more than 84% [11].

Chen Yu Lee et al. developed an approach to detect smoke utilizing temporal and spatial analyses depends on the call block processing technique. To an extracted candidate region, movement feature was adopted such as energy and color features in their spatial, temporal, combine temporal-spatial domains, SVM classifier to determine (edge blurring, gradual energy change), wavelet analysis and gradual chromatic configuration change to reduce false alarm rate and with more than 83% accuracy detect [12].

Hongda Tian et al. proposed a feature to detect smoke using a single image. The image formation model expresses the picture as a linear combination of the components of smoke and non-smoke which are derived based on atmospheric scattering models. The smoke and non-smoke components are separated with the aid of convex optimization. The result shows that the proposed feature is successful for smoke detection with an accuracy of about 84.47% for light and heavy smoke [13].

# 3 Suggested Method

The system is made up of video generated with a fixed camera via video processing to detect smoke. The video data was translated into a sequence of frames and analysed for each frame. Smoke or flame features are hard to determine but, the intensity of fire increases, the matching of properties becomes easy. Image analysis runs the estimation of smoke features. Smoke detection depends on several properties, some characteristics depend on the colours of the smoke and some depend on the greyness of the smoke areas. When the smoke is grey:

$$\mathrm{r} \cong \mathrm{g} \cong \mathrm{b} \tag{1}$$

Either the smoke was colour should be:

$$\mathrm{r} \neq \mathrm{g} \neq \mathrm{b} \tag{2}$$

Varying the motion, all the objects in the image are described. In the next step, each object is tested in the image to determine the smoke behavior on the basis of the following steps.

## 3.1 Cut the Video into Frames

The input video was transformed into a frame array firstly where objects were determined from moving areas in an image. In addition, unwanted light pixels are extracted or minimized and the treasures that meet the threshold and conditions are detected. In this study, several characteristics based on the background image and the image after motion tracking in the video were used. In order to ignore similar video clips, the condition must be:

$$if \sum (f_o - f_n) > 10^{-4} \times sz \quad then\ start\ detect \tag{3}$$

$f_0$ is a background image (first image in the video at $n = 0$), $f_n$ being the sequential images, where $n = 1, 2, 3, \ldots t_n,\ t_{n+1}$ is the total number of the frame and $sz$ represents the size of the frame.

### 3.2 Important Video Frame Determination

In order to extract objects in each frame, motion detection is a required step. In this method, the first frame is stored as a background and the variation between the feature vectors is noticed, based on the specific stage. To raise power efficiency and decrease the time, the subtraction method was used between sequential frames.

The difference between the first image and that is the background and consecutive frames by:

$$df_n = |f_0 - f_n| \tag{4}$$

The difference $df_n$ maybe converted into a binary image with the relation:

$$\begin{aligned} &if\ df_n > 0.1\ then\ Ib_n = 1 \\ &else\ Ib_n = 0 \end{aligned} \tag{5}$$

For such image small areas (less than 500 pixels)was important and can be deleted. The morphological close operation was adpoted to fill in the blanks follows by erosion in the binary image using the same structuring element for both,

$$\mathrm{se} = \begin{bmatrix} 0\ 1\ 0 \\ 1\ 1\ 1 \\ 0\ 1\ 1 \end{bmatrix} \tag{6}$$

### 3.3 Smoke Detection Depending on the Features

After converting the image to the binary, which is generated by the difference in motion between the sequential image. The locations or coordinates are specified in the binary image on the colour image using:

$$if\ Ib_n(x, y) = 1\ \ then\ RC_n = \ c_n(x, y, i) \tag{7}$$

Where $i = 1, 2\ and\ 3$ are red, green, and blue components, $RC_n$ is the important colour vectors that represent the coloured areas of the moving object. The proposed features to distinguish smoke depend on the primary colour space RGB and HSV colour space, $\mathrm{RC_n}$ convert to HSV to obtain $\mathrm{TC_n}$ vector. Let consider $r_{Rn}$, $g_{Rn}$ and $b_{Rn}$ are red, green, and blue channels in the $RC_n$ and $\mathrm{h_{Rn}}$, $\mathrm{s_{Rn}}$ and $\mathrm{v_{Rn}}$ are hue, saturation, and value channels in the $\mathrm{TC_n}$ vector and the background image (first frame $\mathrm{n} = 0$). The following features can be used to detect smoke:

The average of the absolute difference between the two red and blue values are

$$\mathrm{F_1} = \frac{\sum |\mathrm{r_{Rn}} - \mathrm{g_{Rn}}|}{\mathrm{L}} \tag{8}$$

Where L is the length of the vector $\mathrm{RC_n}$. The standard deviation between background images and sequenced images.

$$F_2 = |std(v_{Ro}) - std(v_{Rn})| \tag{9}$$

The average of the absolute difference between the two lighting components in $RC_n$ and $RC_o$.

$$F_3 = \frac{\sum |I_O - I_n|}{L} \tag{10}$$

Where $I_o = \min(r_{Ro}, g_{Ro}, rb_{Ro})$ and $I_n = \min(r_{Rn}, g_{Rn}, rb_{Rn})$. The average of the absolute difference between the two value components in $TC_n$ and $TC_o$

$$F_4 = \frac{\sum |v_{RO} - v_{Rn}|}{L} \tag{11}$$

The average of the absolute difference between the two saturation components in $TC_n$ and $TC_o$

$$F_5 = \frac{\sum |s_{RO} - S_{Rn}|}{L} \tag{12}$$

The average of the absolute difference between the two Hue components in $TC_n$ and $TC_o$

$$F_6 = \frac{\sum |h_{RO} - h_{Rn}|}{L} \tag{13}$$

If these features are smaller than certain threshold values it means that the body is smoke according to the condition:

$$if\ F_i < th_j\ then\ the\ object\ is\ smook \tag{14}$$

Where, $j = 1, 2, 3, \ldots 6$ threshold index and *thi* are (40, 10, 15, 20, 0.15 *and* 0.3). The stages of detection in the video clip illustrated in Fig. 1 in (a) the first image, which was considered a background, smoke image in the same video in (b), the difference between background and smoke image in (c), in (d) convert image (c) to binary, delete small unwanted areas for (d) in (e), fill gaps in (f) and Final detection of smoke zone based on proposed properties in (g).

## 4 Accuracy Meters

It is important to measure accuracy in smoke detection in video clips. In this work, the quality scale was adopted which focuses on early detection, any first detected image contains smoke. The Accuracy of the smoke detection in the video clip given by:

$$A_C = \frac{L - T_o + 1}{L - T_d + 1} \tag{15}$$

Where $L$ is Total frame, $T_o$ is a number of the frame that first detect smoke automatically, $T_d$ is a number of the frame that first true smoke detection.

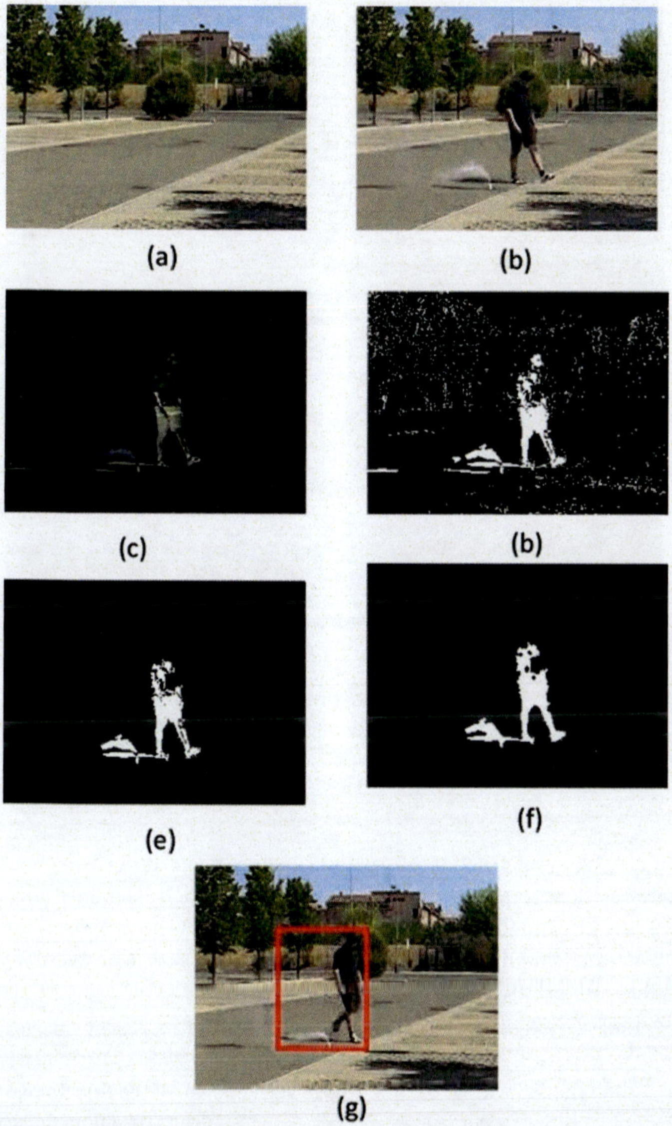


**Fig. 1** The stages of smoke detection, according to the proposed method

## 5 Result and Dissection

In this study, we aimed to detect smoke in video clips based on the gray and chromatic properties in the main color space and space. Six video clips shown in Fig. 2 were used [14, 15] data. The smoke detection algorithm was done with the aid of Matlab (R 2019a) program. Table 1 summarized the accuracy of detection based on the first smoke detected image and the detection rate accuracy found to be around 99.2%. The

**Fig. 2** Frames extracted from the smoke detection videos; the first images of smoke in the videos

proposed automatic smoke detection method using image processing techniques has been compared with several other algorithms [9, 16], which have rate accuracy (71.87 and 98.40%) this illustrated in Table 2. This indicates that the proposed algorithm was successful in automatic and early detection of smoke with detection accuracy better than other algorithms, for all the video data used.

**Table 1** Values of smoke detection accuracy in the selected videos

| Video name | Total frame | No. of the first smoke at true clear detect | No. of the first Smoke detection automatically | $A_C$(%) |
|---|---|---|---|---|
| Video1 | 2346 | 402 | 439 | 98.09 |
| Video2 | 630 | 49 | 53 | 99.30 |
| Video3 | 1200 | 33 | 34 | 99.90 |
| Video4 | 1400 | 3 | 4 | 99.90 |
| Video5 | 900 | 8 | 12 | 99.50 |
| Video6 | 244 | 33 | 35 | 99.05 |

**Table 2** The accuracy rate for automatic smoke detection algorithms

| [16] | [9] | Proposed |
|---|---|---|
| 71.87% | 98.40% | 99.02% |

## 6 Conclusions

Digital image processing techniques are adopting in automatic smoke detection for multiple videos depending on features analysis using RGB and HSV colour spaces. Result analysis illustrate that the proposed detection method succeeds in detecting smoking with high accuracy of 99.2%, compared to the rest of the other algorithms.

**Acknowledgements.** The authors would like to thank the University of Mustansiriyah https://uomustansiriyah.edu.iq/e-newsite.php, for their support in this work.

## References

1. Karam GS et al (2018) Blurred image restoration with unknown point spread function. Al-Mustansiriyah J Sci 29(1):189–194
2. Mirza NA et al (2019) Low lightness enhancement using nonlinear filter based on power function. J Theor Appl Inf Technol 96(1):61–70
3. Daway HG et al (2019) Reconstruction the illumination pattern of the optical microscope to improve image fidelity obtained with the CR-39 detector. In: AIP conference proceedings. AIP Publishing LLC.
4. Gati'Dway H, Mutar AF (2018) Study fire detection based on color spaces. Al-Mustansiriyah J Sci 29(4):93–99
5. Cho B-H et al (2008) Image processing-based fire detection system using statistic color model. In: 2008 international conference on advanced language processing and web information technology. IEEE
6. Patel P, Tiwari S (2012) Flame detection using image processing techniques. Int J Comput Appl 58(18):881–890
7. Chen J, Bao Q (2012) Digital image processing based fire flame color and oscillation frequency analysis. Procedia Eng 45:595–601
8. Seebamrungsat J et al (2014) Fire detection in the buildings using image processing. In: 2014 3rd ICT international student project conference (ICT-ISPC). IEEE
9. Ye S et al (2017) An effective algorithm to detect both smoke and flame using color and wavelet analysis. Pattern Recogn Image Anal 27(1):131–138
10. Maruta H et al (2010) A novel smoke detection method using support vector machine. In: TENCON 2010. 2010 IEEE region 10 conference. IEEE.
11. Avgerinakis K et al (2012) Smoke detection using temporal HOGHOF descriptors and energy colour statistics from video. In: International workshop on multi-sensor systems and networks for fire detection and management
12. Lee YC et al (2012) Smoke detection using spatial and temporal analyses. Int J Innov Comput Inf Control 8(7A):4749–4770

13. Tian H, Li W, Ogunbona P, Wang L (2015) Single image smoke detection. In: Cremers D, Reid I, Saito H, Yang M-H (eds) Computer Vision – ACCV 2014, vol 9004. Lecture Notes in Computer Science. Springer, Cham, pp 87–101. https://doi.org/10.1007/978-3-319-16808-1_7
14. http://signal.ee.bilkent.edu.tr/VisiFire/Demo/SmokeClips/
15. https://aimagelab.ing.unimore.it/visor/video_videosInCategory.asp?iStartFrom=0&idcategory=8
16. Mutar A et al (2018) Smoke detection based on image processing by using grey and transparency features. J Theor Appl Inf Technol 96(21):6995–7005

# A Comparative Study of the Performance of Gait Recognition Using Gait Energy Image and Shannon's Entropy Image with CNN

K. T. Thomas(✉) and K. P. Pushpalatha

School of Computer Sciences, Mahatma Gandhi University, Kottayam, India

**Abstract.** Biometrics is an area where the studies of human behavioral or physiological attributes are performed. The modalities such as voice, keystroke dynamics, handwriting, human gait etc. are categorized as behavioral biometrics, while the attributes of finger print, iris, face, ear etc. are identified as physiological biometrics. Gait is defined as the walking on the foot, which is a result of limb movements. It takes account of the body appearance of a human as well as the dynamics of human walking. Gait is a very effective tool for human recognition in the situations where the passive identification is required. Gait contributes much in such situations where it is really hard to capture clear and useful snap-shot of a face or it is impossible to get specimens of fingerprints or it is difficult to get a subject's voice recorded. The paper compares the performance of two major gait features Gait Energy Image (GEI) and Shannon's Entropy Image using CASIA B Dataset.

**Keywords:** Biometrics · Behavioral biometrics · Stance phase · Swing phase · GEI · Shannon's entropy images · CNN

## 1 Introduction

A biometric is a measure to quantify the identity of a human grounded on a physiological (retina, facial, ear, eyes, iris, geometry of hand, fingerprint, voice etc.) or behavioral (the gesture, speech, gait, signature, keystroke etc.) characteristic. It can be considered as a very efficient human identifier since it is distinctive to and lied originally in each and every human being, so it cannot be stolen, lost or forgotten like other usual identification mechanisms such as passwords, personal identification number, cards etc.

Gait is a passive behavioral biometric methodology which considers the walking style of a person. Human gait represents a persons' way of walking [1]. The term 'walking' can be described as a way of locomotion by lifting a foot and setting it downward one foot at a time, not at all causing both the feet off the surface at any point of time. In formal definition, walking implements a continuous sequence of leg motion to transfer the body frontward at the same time maintaining the stance constancy. Every human being differs in the way they are walking. Hence human identification can be performed using the gait property. Gait recognition falls in the category of behavioral passive biometric method which can be monitored even without the subjects' cooperation.


S. Shukla et al. (eds.), *Data Science and Security*, Lecture Notes in Networks and Systems 290,
https://doi.org/10.1007/978-981-16-4486-3_21

There are various widely considered image based biometric methods such as face, iris, ear, fingerprint. Gait can be considered relatively new, compared to the orthodox approaches such as fingerprint recognition and face recognition. Equated to other biometric methods, gait has some very peculiar characteristics. Its unobtrusiveness is the very best striking feature of gait when used for the purpose of biometrics, i.e., the fact that, many of the other categories of biometrics may need consent of the subject, while a gait could be easily captured from a distance without actually needing the prior consent of the person being observed [2]. Many general biometrics like signature, fingerprints, face, iris, voice etc. cannot be obtained without having physical contact or at a very near proximity from the device being used for the recording.

As terrorist threats are increasing worldwide, the passive human identification mechanisms has attained a major scope of research domain [3]. It can be observed that a large number of cameras for video surveillance are installed in locations such as roads, aerodromes, malls, office facilities and even private buildings, which enables the gait recognition technology to be a very successful tool in the arena of crime prevention and forensic identification. Recognition of a human, grounded on gait has proved its peculiar ability in applications like close observation, especially for suspected spy or a criminal, as it has got numerous striking features [4]. The capture of images representing an individual's gait can be done easily in public zones, with simple camera, and does not even need the support or awareness of the person under observation. Face, ears, eyes etc. are comparatively smaller in size and hence the recognition rate in surveillance depends on the clarity of the images captured. Here gait can contribute much better than others.

Implementation of a gait-based human identification scheme for a very large population in a real application is very much challenging because of the reason that there are a lot of factors which are derived from a source outside the subject being observed, that can adversely have an impact on the imaging of natural gait, such as ambient lighting, structure of scenario, different clothing variations, footwear types, carrying conditions, walking pace, occlusions and camera angles. If we want to get a decent recognition rate, we need to understand that it is utmost important to acquire very good segmentation of the gait silhouettes. Accuracy in segmentation can be adversely affected by lot of factors like similarity in color of the background and the dresses worn by the person under consideration, variations in background that is caused by the change in illumination or appearance of a fresh object in the place etc.

The sections of the paper are arranged as given. Section 2 provides a literature review. Section 3 details the study of the gait, gait cycle as well as gait phase followed by the general procedures used for gait analysis using computer vision. The section also specifies about two major gait features such as GEI and Shannon's entropy image. Section 4 explains the experiments and results of gait recognition using CNN algorithm. Section 5 provides a conclusion for the paper.

## 2 Literature Survey

In paper [2], the author has introduced the basic terminologies in gait recognition including gait recognition, its phases, cycles and the parameters considered for gait recognition.

Khalid Bashir et al. in [6] put forward a fresh idea on gait representation known as the Gait Entropy Image (GEnI) where the main consideration was given features obtained based the randomness of the values of pixels available in the silhouettes over an entire gait cycle. The features helps to identify the particulars related to the motion which really suits the covariate condition variations that may have an impact on appearance. Results proved that proposed method outperformed the existing methods especially where the appearance should be given more prominence.

[7] proposed the concept of the accumulated frame difference energy image (AFDEI), for considering the temporal features along with the dynamic particulars together with the static particulars included in a gait sequence. The gait feature considered here was the fusion of the movement invariant obtained from the GEI and AFDEI. For the purpose of recognition, nearest neighbor classification using Euclidean distance was utilized. The evaluation done on the performance gave a clear cut idea that, the proposed algorithm was efficient when compared with performance of experiments using GEI combined with 2D-PCA and SFDEI combined with HMM perfumed using CASIA–B database for gait.

Chirawat Wattanapanich and Hong Wei [8] worked on the relevance of the usage of lower knee part gait representations of a person in order to recognize the gait. The authors have reviewed the performance of three gait representations, like Gait Gaussian Image, Gait Energy Image as well as Gait Entropy Image. A fresh gait characterization called Gait Gaussian Entropy Image (GGEnI), was put forward by them which combined the benefits of the Entropy images and Gaussian images. Experiments on CASIA B data set proved that the presentation of parts lower to the knee gait does perform as much effectively as the entire body gait presentation and the new methods has better correct classification rate (CCR) than Gaussian image.

In [9], the inadequacy of Silhouette-based gait was identified and a new approach using a gait energy response function is proposed. GERF transforms a gait energy of a silhouette-based gait characteristic to a value which is better acceptable in the case of treating these co-variate conditions. The spatial dependence and optimized vector representation of GERF was considered for improving the efficiency of the system.

Gabor filtering along with spatial metric learning, were included as post-processing techniques, to improve accuracy.

The paper explains about the gait and gait phases and a comparison of two major features of gait images using CNN algorithm.

## 3 Gait and Gait Phases

A gait cycle could simply be mentioned as the time or else the succession of events, during the advancement of body in a direction where a foot contacts the surface to another time where the same foot once again comes into contact with the ground. It comprises of pushing forward of the center of gravity in the route in which the body is moving forward [2].

A solitary gait cycle is popularly referred to as stride. Every stride which is also called as a gait cycle comprises of two distinct phases as shown in Fig. 1.

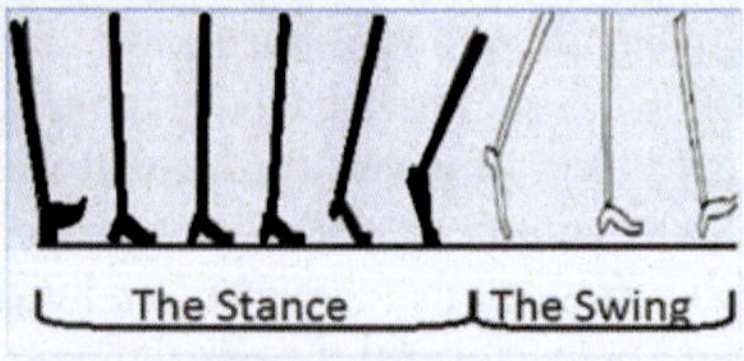


**Fig. 1** Two phases of human gait [2]

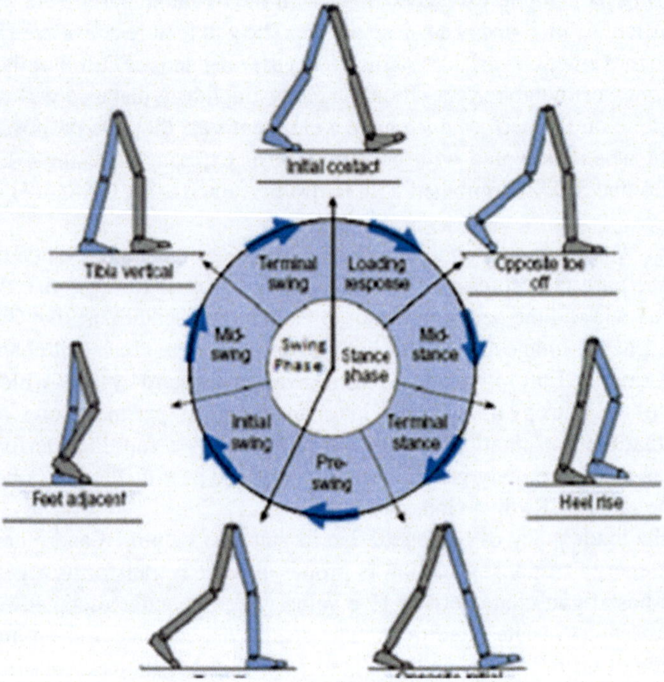


**Fig. 2** Major events in a solitary gait cycle [2]

1) The Stance Phase is the phase when the foot continues to be in touch with the standing ground.
2) The Swing Phase is the phase when foot does not at all touch the surface.

So we can clearly define a gait cycle as the time period flanked by two consecutive incidences of swing or stance repeated during a locomotion by walk. Below mentioned terms could be used to clearly identify important events that happens in a gait cycle (Fig. 2).

We can identify 7 major events in a gait cycle. They are

1) The Initial contact event
2) The Opposite toe off event
3) The Heel rise event

4) The Opposite initial contact event
5) The Toe off event
6) The Feet adjacent event
7) The Tibia vertical event

The seven events mentioned here, splits one gait cycle into seven different stages. The stance phase contains four events; when the foot is touching the surface. The remaining events are contained in the swing phase, where the foot changes its position frontward in the air.

**Common Methodologies in Gait Analysis**

Gait analysis generally follows one of the following two approaches namely the Model-based approach and the Model-Free Approach [5]. Model free approach is based on the appearance of the subject or observation made on the subject while model based approach depends on some measured quantities like spatial-temporal aspects or study of kinetics like force, power, torque etc. The Figs. 3 and 4 shows different methodologies of gait analysis.

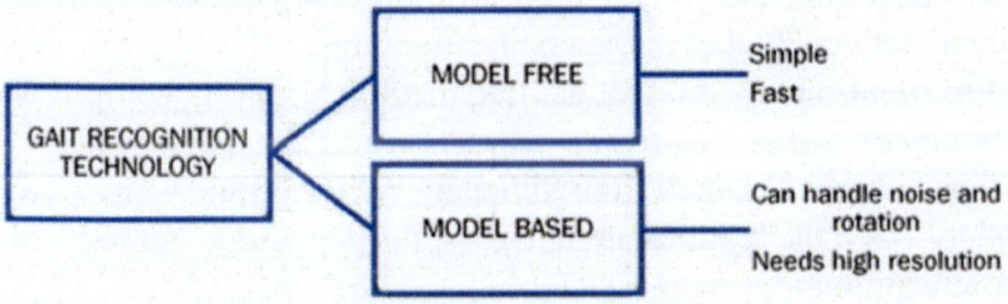


**Fig. 3** The two major methods for gait analysis

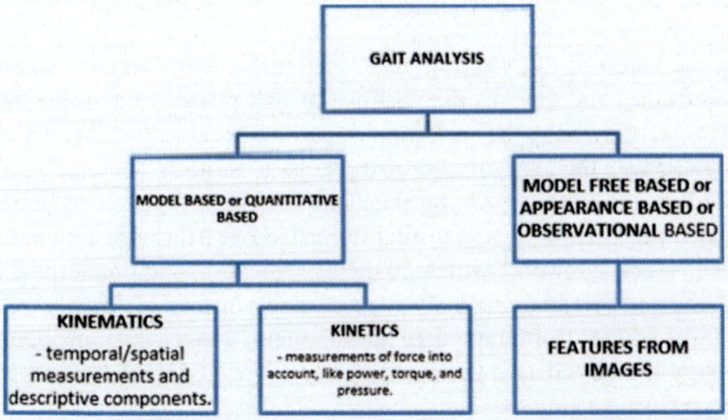


**Fig. 4** Common methodologies in gait analysis

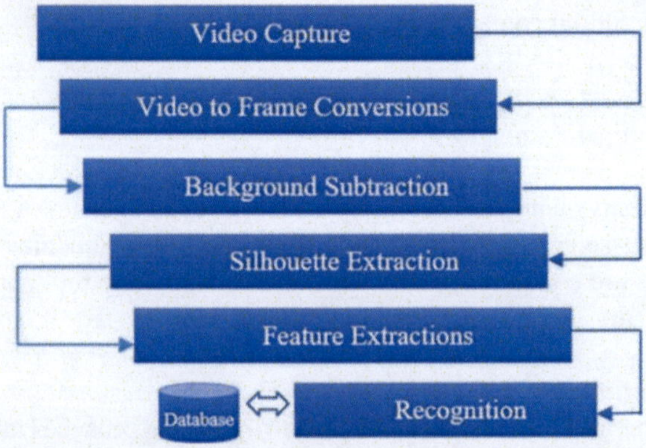


**Fig. 5** Steps in model free gait analysis for human recognition

Gait recognition technologies implemented using model free methods make use of gait features tapped right from the set of images taken from human walking sequences. Full silhouettes of partial silhouettes are used by almost all the model-free techniques in analyzing gait information. Gait recognition methods known as Model-based gait identification/recognition techniques perform the classification based on the skeletal data of a human body and other human body structures.

The Gait recognition based on model free method primarily comprises of detecting human, extracting the silhouette, feature extraction and person identification or classification [13]. The features obtained from Silhouette images could be used for the purpose of classification. Once the features are obtained, deep learning methods can be used for effective classification.

This paper concentrates on model free gait recognition method - Gait Energy Image and Shannon's Entropy Image. The Fig. 5 depicts the steps involved in model free based gait recognition. Basic vision-based gait analysis for biometric recognition follows the following steps [10–12]. We have implemented CNN in this paper, in the following sections.

Initially the video of the walking person captured. The captured video is divided into separate frames and the specific portions of gait phases in a single gait cycle are extracted. This is followed by the background subtraction. During this process, the region of interest (ROI) i.e., the exact image of the human alone is obtained and everything else from the frame is removed. Going ahead the silhouette is obtained from the ROI, in binary digital format. Training and testing is carried out using learning algorithms like convolutional Neural Networks which do the classification and prediction [11].

In many of the model free gait analysis procedures, instead of directly taking features from each of the frames from the video, the silhouette images are considered.

Gait Energy Image (GEI) is the most widely advocated model free representation. GEI depicts gait using a grey scale image obtained by averaging the silhouettes extracted from a video of a complete gait-cycle [6]. We can understand the gait energy image (GEI) is a grey scale image which can represent an entire cycle of the gait sequence implemented by adopting the weighted average method. The sequences of an entire gait

**Fig. 6** Forming gait energy images from silhouettes

cycle are managed to align the silhouette in binary format. The gait energy image can be obtained by the following formula: After the pre-processing, binary gait silhouette images $B_t(x, y)$, at time t in a sequence, gait energy image (GEI) can be calculated as given below [11]:

$$GEI(x, y) = \frac{1}{n}\sum_{1}^{n} B(x, y) \tag{1}$$

Here n represents the frame count in the entire cycle(s) of a silhouette sequence, x and y represents values in the 2D image coordinate. All the silhouette sequence of a complete gait cycle is added to form a Gait Energy image as illustrated in Fig. 6. Gait Energy Image (GEI) is thus one of the model-free gait recognition techniques, which incorporates the use of average image of silhouettes as gait feature [11]. Figure 6 shows the creation of GEI from a set of silhouettes. The last picture in both the rows are a gait energy images obtained from the other images in the row.

In order to implement gait recognition using model free approach, the gait energy image creation can be considered to be a very effective basic algorithm.

Another gait feature, which can be used for model free approach, is the Shannon's entropy Image. Shannon's entropy image could be evaluated from the silhouette images [8]. The Shannon's entropy image encodes the unpredictable randomness of the values of the pixels in silhouettes obtained from an entire gait cycle. From a single image Shannon entropy calculates the uncertainty allied with a random variable. Taking into consideration, the intensity value of each pixels, of the silhouettes, as a discrete random variable, we can calculate the entropy of this variable over an entire gait cycle using the equation given below.

$$H(x, y) = -\sum_{k=1}^{K} p_k(x, y) log_2 p_k(x, y) \tag{2}$$

here x, y represents the pixel coordinates and pk(x, y) represents the probability that the pixel takes on the $k^{th}$ value.

The Observational gait analysis (OGA) can be used as some qualitative measures to perform gait analysis in applications like biometrics or by clinicians in the health care domain. In clinical cases, the deviations of the gait are identified in patients from visual observations and the doctors can identify the problem.

## 4 Experiments and Results

We have considered two types of image features to compare the efficiency of the gait recognition. The characteristics used here are Gait Energy Image (GEI) and Shannon's entropy image.

A deep learning approach was used to perform the person identification. The data set used for the experiments was CASIA data set.

**The Dataset**

The experiments of gait recognition were performed on the CASIA dataset to evaluate the performance of Gait Energy Images and Shannon's Entropy Image. The CASIA-B Gait Database [6] was an indoor gait dataset that contains data of 124 individuals. In the dataset, three walking conditions are considered. They are normal walking condition (nm), clothing condition [6] (cl) and carrying condition (bg).

For each subject there are 11 different views which begins with ("0", "18", "36", "54", "72", "90", "108", "126", "144", "162", "180") walking sequences containing 6 normal sequences where the person under consideration walks without carrying any bag nor he/she wears bulky coat.

There are 11 different views the sequence for each subject. We have used CASIA B dataset. Each subject's image is represented using his/her subject id, with its walking status which can be normal or wearing a bulky coat or carrying a bag. The images in 11 viewing angles starting from frontal view to its extreme opposite back view of the human subject; from left to right images are considered. Three specific variations, the viewing angle, clothing condition and carrying condition changes, are separately considered. From the video file, human silhouettes were segmented out.

**Table 1** Comparison between GEI and Shannon's Entropy Image

| Image feature | No. of training images | No. of validation images | No. of test images | No. of epochs | Accuracy rate | |
|---|---|---|---|---|---|---|
| | | | | | Training accuracy | Testing accuracy |
| GEI | 10254 | 1609 | 1729 | 5 | 99 | 97.8 |
| Shannon's entropy image | 10254 | 1609 | 1729 | 5 | 95.2 | 96.1 |

**The Experiment**

The experiment was done using Python in the Spyder environment. We have considered the Gait Energy Image of CASIA B Dataset. Of the 13,592 images, 75% images are taken as data set incorporated for training and rest 25% as test images. Optimizer RMSProp was used. The training of the network was carried ahead using mini batch size of 16 with 5 epochs.

The Python Keras library with TensorFlow was utilized for executing the experiment. Table 1 shows the major results obtained in the experiment:

Experiments were carried out in order to appraise the performance of the model in varying number of epochs (Figs 7, 8 and 9).

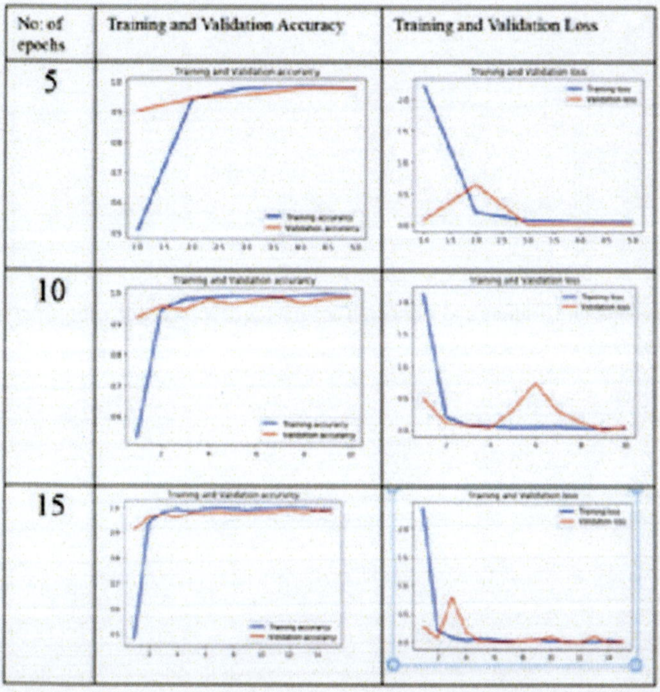


**Fig. 7** Graph depicting training and validation accuracies of recognition using gait energy image

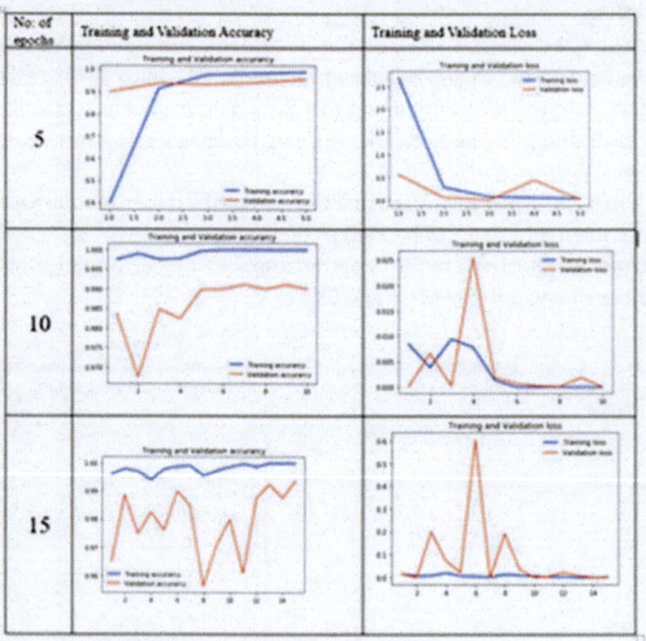


**Fig. 8** Graph depicting training and validation accuracies of recognition using Shannon's entropy image

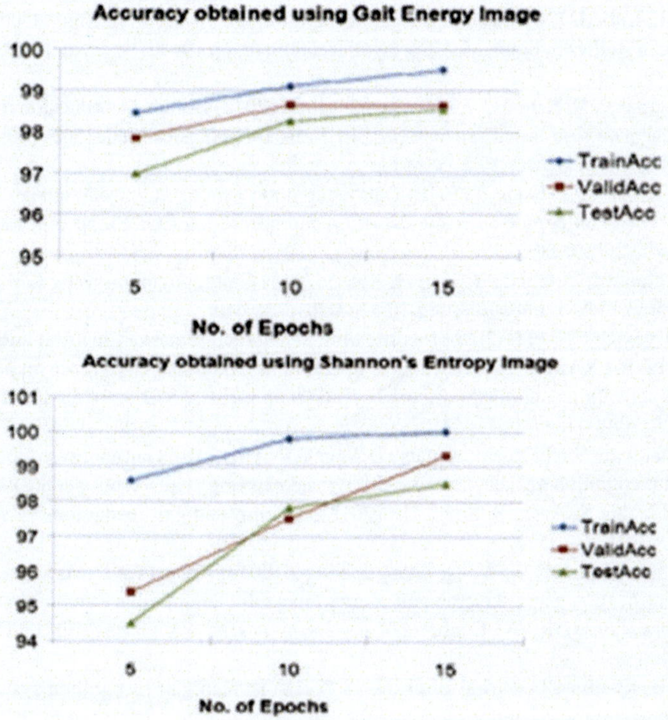


**Fig. 9** Graph showing the obtained accuracies for GEI and Shannon's Entropy Image in varying epochs

## 5 Conclusion

In this paper, a detailed study of human gait, gait cycle, its various phases of gait recognition were performed. We have presented the features that can be extracted for identifying human using human gait as a passive biometric instrument. We presented model free approach which generally uses Gait Energy Images as a source for feature extraction. The challenges in following model free gait biometric analysis are also identified. The performance of two features of gait - Gait Energy Image and Shannon's entropy image is compared using deep learning. It was concluded that that the accuracy rate was slightly higher when GEI was used compared to when Shannon's Entropy Image was used.

## References

1. Andriacchi TP, Ogle JA, Galante JO (1977) Walking speed as a basis for normal and abnormal gait measurements. J Biomech 10(4):261–268
2. Kharb A, Saini V, Jain YK, Dhiman S (2011) A review of gait cycle and its parameters. IJCEM Int J Comput Eng Manage 13:78–83
3. Levine D, Richards J, Whittle M (2012) Whittle's gait analysis. Health sciences. Elsevier

4. Boyd JE, Little JJ (2005) Biometric gait recognition. In: Tistarelli M, Bigun J, Grosso E (eds) Advanced studies in biometrics. Springer, Heidelberg, pp 19–42. https://doi.org/10.1007/11493648_2
5. Khamsemanan N, Nattee C, Jianwattanapaisarn N (2018) Human identification from freestyle walks using posture-based gait feature. IEEE Trans Inf Forensics Secur 13(1):119–128. https://doi.org/10.1109/TIFS.2017.2738611
6. Bashir K, Xiang T, Gong S (2010) Gait recognition using gait entropy image. In: 2009 3rd international conference on crime detection and prevention (ICDP), pp 1–6. https://doi.org/10.1049/ic.2009.0230
7. Luo J, Zhang J, Zi C, Niu Y, Tian H, Xiu C (2015) Gait recognition using GEI and AFDEI. Int J Opt 2015:1–5. https://doi.org/10.1155/2015/763908
8. Wattanapanich C, Wei H (2016) Investigation of gait representations in lower knee gait recognition. In: ICPRAM 2016. Proceedings of the 5th international conference on pattern recognition applications and methods, vol 1, pp 678–683. ISBN 978-989-758-173-1. https://doi.org/10.5220/0005817006780683
9. Li X, Makihara Y, Chi Xu, Muramatsu D, Yagi Y, Ren M (2018) Gait energy response functions for gait recognition against various clothing and carrying status. Appl Sci 8(8):1380
10. Singh JP, Jain S, Arora S, Singh UP (2018) Vision-based gait recognition: a survey. IEEE Access 6:70497–70527
11. Yao L, Kusakunniran W, Wu Q, Zhang J, Tang Z (2018) Robust CNN-based gait verification and identification using skeleton gait energy image. In: Digital image computing: techniques and applications (DICTA), Canberra, Australia, pp 1–7. https://doi.org/10.1109/DICTA.2018.8615802
12. Jawed B, Khalifa OO, Newaj Bhuiyan SS (2018) Human gait recognition system. In: 2018 7th international conference on computer and communication engineering (ICCCE), Kuala Lumpur, pp 89–92. https://doi.org/10.1109/ICCCE.2018.8539245
13. Shirke S, Pawar SS, Shah K (2014) Literature review: model free human gait recognition. In: 4th international conference on communication systems and network technologies, Bhopal, 2014, pp 891–895. https://doi.org/10.1109/CSNT.2014.252

# OntoJudy: A Ontology Approach for Content-Based Judicial Recommendation Using Particle Swarm Optimisation and Structural Topic Modelling

N. Roopak[1(✉)] and Gerard Deepak[2]

[1] Department of Computer Science Engineering, SRM Institute of Science and Technology, Ramapuram, India

[2] Department of Computer Science Engineering, National Institute of Technology, Tiruchirappalli, India

**Abstract.** Under the history of the Judicial Reform of India, major data of judicial cases are commonly used to address the issue of judicial study. Similarity review in legal trials is the foundation of wisdom judicature. The analysis of the Indian Judicial cases in a established format is a central concern by giving necessity for eradicating incompetent information extracting appropriate rules and conditions from the vivid document. Hence, this paper proposes a method to recommend judicial cases to the user, based on the content of the cases. The proposed OntoJudy model uses Static Judicial Domain Ontology with Structural Topic Modelling. The semantic similarity is computed using Particle Swarm Optimization with Jaccard Similarity and SemantoSim. Hybridizing all these helps in yielding better accuracy. To assimilate users preferences for the content recommendations, the CAIL2018 dataset is used which is then classified using Random Forest Classification with the help of extracted query word from the user information. The proposed model has achieved an Accuracy of 95.89% and tends to do better than the other baseline models by attenuating the resilience of the traditional content recommender systems.

**Keywords:** Jaccard similarity · Particle swarm optimization · Random Forest Classification · Static domain ontology · SemantoSim

## 1 Introduction

With the advent of computer science, it has become a very common way of solving certain challenging problems in practice by computer simulation. Meanwhile, with the growth of artificial intelligence, with the help of big data analysis, judicial decision is moving closer to the justice of the law. It should be remembered that the analyzes of parallels in legal trials are the foundation of wisdom. The court, the accuser and the defendant contain a formative legal argument, the fact, and the case's outcome. Jury courts must take all these complex situations into account when relating to comparable


S. Shukla et al. (eds.), *Data Science and Security*, Lecture Notes in Networks and Systems 290,
https://doi.org/10.1007/978-981-16-4486-3_22

cases in order to have legitimacy in culture. When the number of judicial litigation surges, cases without exclusion are impossible to consider. This is why we are seeking to have a modern means of advocating judicial proceedings. While planning or inspecting a legitimate brief, prosecution lawyers invest a lot of energy looking for the most relevant power to support or invalidate a specific mark of law. This includes filtering through an assortment of millions of essential and auxiliary wellsprings of law, just as past briefs and memoranda.

***Motivation:*** When writing or updating a legal brief, litigation attorneys expend a considerable sum of time looking for the most pertinent authority to bolster or contradict a specific point of law. The errand is especially difficult given the requirement for high review; an inadequate legitimate exploration interaction can conceivably miss an exceptionally important wellspring of law that would unfavorably affect the prosecution system. Here plays a major role by the content based recommender systems, where it helps judicials in recommending the contents by user's input information and using CAIL2018 dataset. The proposed hybrid content-based Recommender System model provides the user with the contents of the judicial cases to the user from the static judicial domain ontology created from the Indian context.

***Contribution:*** The proposed model is structured with various concepts like Structural Topic Modelling, Static Judicial Domain Ontology for Indian context, Particle Swarm Optimisation which helped in yielding better accuracy than the baseline models, Jaccard Similarity and SemantoSim for semantic similarity, and Random Forest Classifier for classification of data from the dataset aided with the extracted query word. Static Judicial Domain Ontology helps in getting more accurate recommendations since Structural Topic Modelling and construction of a tree based on semantic re-arrangement with the static ontology paves a way for better recommendations. Experiments were conducted on the data crawled from CAIL2018 and hybridization of various algorithms and concepts into one helped in establishing a higher percentage of precision, recall, accuracy, F-Measure, and very low False Discovery Rate (FDR).

***Organization:*** The remaining sections of the paper are classified as follows. Section 2 provides a description of Related Work. Section 3 profiles the Proposed System Architecture. Section 4 addresses the Implementation of the OntoJudy model. Section 5 depicts the Results and Performance Analysis. Finally, the paper is concluded in Sect. 6.

## 2 Related Work

Radboud Winkels et al. [1] discusses the outcome of current research focused at a statutory recommender process where members of a law-making platform gain advice of other related resources of legislation, for the provided case. Their paper presents an applicable lawsuit, includes a statutory article of focus for a recipient, and adapts the classification of relevant other papers depending on the same case law. Merine Thomas et al. [2] presents Quick Check, a framework that separates the lawful contentions from a client's brief and suggests exceptionally applicable case law cases. Utilizing a blend of full-text search, reference network examination, clickstream investigation, and a pecking order of positioning models prepared on a bunch of over 10K explanations, the framework can viably suggest cases that are comparative in both legitimate issues and realities. Enys Mones et al. [3] utilize the link prediction model that shows the unpredictable organization of references develops in a manner that revamps our capacity to foresee new references which empowers to approve existing references and suggest expected references for future cases inside the court. Simon Philip et al. [4] came up with an algorithm to include or propound suggestions depending on the user's query. Their algorithm uses both a TF-IDF weighting stratagem and cosine similarity tests. Based on the outcome or performance of the method, library users identify the most important research papers to their needs. Charilaos Zisopoulos et al. [5] presents a Content-Based Recommendation System that alludes to frameworks utilized on the Web to prescribe a thing to a user dependent on a portrayal of the item and a profile of the client's advantages. Paul Sheridan et al. [6] presents an ontology-based recommender framework which incorporates the information embodied in a broad ontology to generate quality recommendations. The principle curiosity of this work is a philosophy based strategy for computing similarity among things and incorporation with Item-KNN (K-closest neighbors) calculation. Charbel Obeid et al. [7] addresses a methodology for building up an ontology-based recommender systems improved with AI strategies to arrange understudies in advanced education. The main idea of their paper is to define the student needs, desires, expectations and skills and recommend the required major and university for each one.

## 3 Proposed System Architecture

The proposed system architecture consists of Structural Topic Modelling with Static Judicial Domain Ontology which is then constructed as a tree based on semantic re-arrangement; Semantic Similarity is computed by employing Particle Swarm Optimization under Jaccard Similarity and SemantoSim. Figure 1 depicts the proposed system architecture and the dataset used is CAIL2018_Small. The dataset is initially pre-processed with various techniques as per the requirements for this proposed model. Data cleansing is done so that only the data which is required for the model is taken. Data Transformation is done where removal of capital words from the available dataset, punctuation removal, etc. which helps in deforming the structure of the sentence. Tokenization of the user extracted terms is done with the help of one hot encoder. Word stemming is also done which helps in reducing the number of words that have the same stem words in it. By word stemming the number of words is limited. Lemmatization is done to the separated word with the utilization of a vocabulary and systematic examination of words, which is utilized to eliminate declension endings and to restore the base or dictionary type of words, known as the lemma. Recognition of named entities attempts to recognise and categorize named entities in text into predefined groups, such as names of people, organisations, locations, dates, numbers, numerical amounts, percentages, etc. After the pre-processing of data, the extraction of query words is done from the processed user data. This extracted query word is then used in the classification of the dataset by Random Forest algorithm.

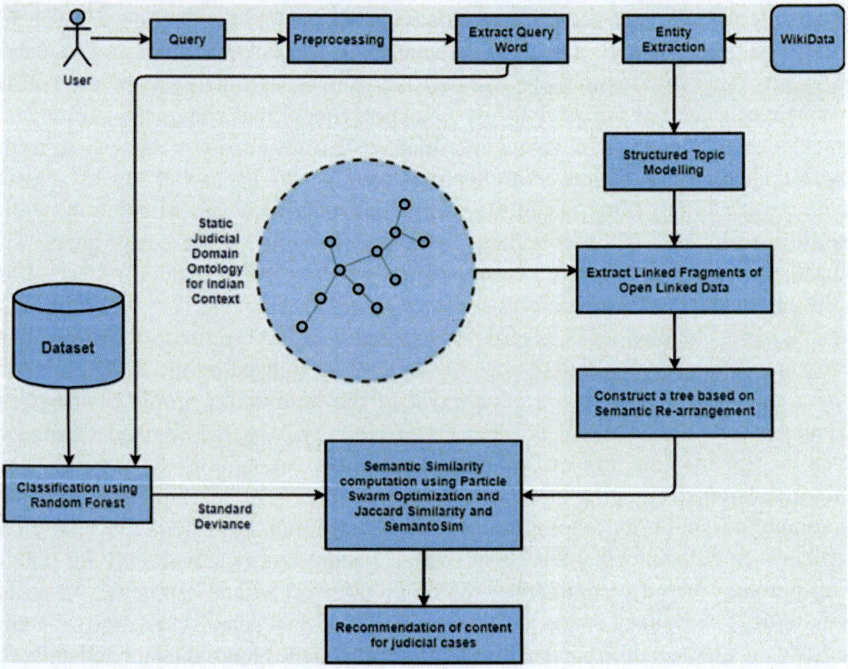


**Fig. 1** Proposed system architecture

Next, entity extraction is done with the help of Wikidata to aggregate knowledge that helps to enhance the quality of the content recommendation. The basic aim of Wikidata is to come up with automated text analysis and provide with the relevant synonyms for the input query. Further the Structural topic modelling (STM) is realised in order to facilitate addition of new and relevant topics into the framework. The STM is a generative model of word tallies by distinguishing a data generation process and afterward utilizing the information to locate the most potential values inside the model for the parameters. The purpose of the STM is to detect topics and approximate their relationship with the help of metadata for the researchers. The model's solutions will be used to do hypothesis tests on these relationships. The STM offers an efficient and versatile text analysis environment that combines both content metadata and topic modeling. This helps to consider which variables are related to various text characteristics within the context of the topic of modeling [8].

Static Judicial Domain Ontology for Indian Context is created. Based on the standard topics of law in standard e-books or prescribed textbooks from the Indian context, the ontologies are modelled. Then the ontology is modelled using WebProtege based on human intervention. The indexes of standard textbooks in the Indian context have been taken for the subjects that are written down in Table 1.

Primary topics were selected and the seed ontology was plotted using WebProtégé. WebProtege is a series of OWL ontologies, personalized user interface design configurations, and teamwork environments. Projects can either be worked without any preparation

**Table 1** List of subjects based on Indian context

| Sl. no. | LLB subjects | Sl. no. | LLB subjects |
|---|---|---|---|
| 1 | Law of TORT | 17 | Law on Education |
| 2 | Constitutional Law | 18 | Law of Trademarks, Design, and Practice |
| 3 | Criminal law | 19 | Civil Law |
| 4 | Law in Changing Society | 20 | Tax Law |
| 5 | Civil Procedure Code and Limitation Act | 21 | Criminal Law |
| 6 | Family Law | 22 | International Law |
| 7 | Contract Law | 23 | Corporate Law |
| 8 | Corporate Law | 24 | Real Estate Law |
| 9 | Law of Evidence | 25 | Labour Law |
| 10 | Law of Pleadings in Civil Matters | 26 | Patent Law |
| 11 | Property Law | 27 | Media Law |
| 12 | Administrative Law | 28 | Competition Law |
| 13 | Labour and Industrial Law | 29 | Intellectual Property Law |
| 14 | Law of Taxation | 30 | Mergers and Acquisitions Law |
| 15 | Copyright Law, Prospects and Protection | 31 | Constitutional Law |
| 16 | Environmental Law | 32 | Consumer Protection Laws and M. V. Act |

or they can be introduced with existing ontologies. A fundamental part of WebProtege is that it underpins multi-user ongoing bilateral ontology creation. Further Dynamic Ontology modelling and generation happens, which is achieved using OntoCollab so that the number of concepts and the individuals are increased. A unique ontology can be made by removing data from information sources which can be applied to the base ontologies without relying upon user contribution, with the respective links to orthodox knowledge [9]. The static domain ontology is manually modelled by taking the terms from the Judicial and Law related e-books which are listed in Table 1, where there are 240 terms then the dynamically generated ontology will have nearly 1000% increase in the terms which is around 2540 approximately. The statically modelled ontology and the Dynamically generated ontology is further axiomated where axioms are introduced to dynamically arrange and the rules are inducted which is then used for recommendation. This static judicial domain ontology for Indian context is then extracted in the form of linked fragments. These linked fragments which are extracted from the Open Linked Data are then constructed as a tree, based on the semantic re-arrangement.

Next step is constructing a tree, based on Semantic Re-arrangement which helps in syntactically analyzing the sentences, identifying their key constituents, and creating hierarchical tree diagrams. Then the semantic similarity is associated with particle swarm optimization, Jaccard Similarity and SemantoSim. Particle Swarm Optimization is a computational technique which advances the issue by attempting iteratively to improve an answer for a given quality measure. It addresses an issue by giving a populace of applicant arrangements called particles and by utilizing fundamental mathematical formulas to move these particles around in the space as far as their position and speed. The motions of each particle are guided by the most prominent local location, but they often move into the best-known search-space positions, which are modified as other particles discover better positions. This should drive the swarm in the direction of the right solutions [10].

The next step of the model is to find the semantic similarity between the input and the available source and give the recommendation for the user's input query. Jaccard Similarity has been used to find similar courses available in the real-world and gives the user several recommendations of courses. This Jaccard Similarity uses the vectorization method to vectorize a token and finds similarities available for that token in the courses available. Depending on their SemantoSim values, the ontologies which are explicit were grouped and suggested between the axioms of concepts and individuals. SemantoSim is a measure for processing the semantic relatedness between the extracted query word and the initial concepts corresponding to the ontologies. Classification of the pre-available dataset is done by Random Forest method with the extracted query words from the user given query. Random Forest and Random Decision Forests comprise a group method for learning the classification, regression and other tasks. Random Forest Classification done by building a large number of decision-making trees during training the dataset and output the class like classification or regression of the individual trees [11]. The classified data from the available dataset with the help of extracted query word from the user input query is then semantically computed using Particle Swarm Optimisation and also using Jaccard Similarity and SemantoSim. The content for the judicial cases is then recommended to the user.

## 4 Implementation

The proposed model is implemented using Java in Windows 10 Operating system. The implementation process is realized in an i5 Intel Core processor with 32 GB of RAM and 8 GB of Nvidia graphics card. Java Compiler is preferred for the implementation of this proposed model. The dataset used in this model is taken from CAIL2018, which contains 204, 231 documents in total. Ontology is created using agents which are written using JADE for modelling. OntoCollab was used for creation of Static Domain Judicial Ontology. The proposed model algorithm is depicted as Algorithm 1.

**Algorithm 1: Proposed algorithm for OntoJudy model.**

**Input:** Queries collected from the user and the CAIL2018 dataset.
**Output:** Recommendation of content of judicial cases based on user inputs.

***Begin***

**Step 1:** Preprocessing of Data obtained from the user.

**Step 2:** Extracting the query words and passing it for classification of Dataset using Random Forest Classifier.

**Step 3:** Entity Extraction from the extracted Query Words using Wikidata.

**Step 4:** Structural Topic Modelling is done for the extracted entity of the query words.

**Step 5:** Constructing a Static Judicial Domain Ontology for Indian Context using WebProtege, which is then extracted in the form of linked fragments from the Open Linked Data.

**Step 6:** Constructing a tree based on the Semantic Re-arrangement of the linked fragments obtained from the Open Linked Data.

**Step 7:** CAIL2018 Dataset is classified using Random Forest Classifier aided with the Extracted Query Word from the user input information.

**Step 8:** Semantically aware by Particle Swarm Optimisation with Jaccard Similarity and SemantoSim is calculated between the Classified dataset and the linked ontological terms.

Algorithm :

**1.** for each particle $i = 1, \ldots, W$ do
Loading a uniformly distributed random vector along with the particle's position: $x_i \sim D(b_{lo}, b_{up})$.
Initialize the initial position of the particle as the particles best known position : $a_i \leftarrow x_i$
if $f(a_i) < f(b)$ then,
Updation of the swarm's best known position: $b \leftarrow a_i$
Loading the velocity of the particle: $v_i \sim UD(-|c_{up}-c_{lo}|, |c_{up}-c_{lo}|)$

**2.** while not equal to the criterion:
for every particle $j = 1, \ldots, W$
for every dimension $d = 1, \ldots, m$
Picking a indiscriminate values: $z_p, z_g \sim Y(0,1)$
Updating velocity of the particle: $v_{i,d} \leftarrow \omega\, v_{i,d} + \varphi_p\, z_p\, (p_{i,d}-x_{i,d}) + \varphi_g\, z_g\, (g_d-x_{i,d})$
Updating the position of the particle: $x_i \leftarrow x_i + lr\, v_i$
if $f(x_i) < f(a_i)$ then,
Updating best known position of the particle: $a_i \leftarrow x_i$
if $f(a_i) < f(b)$ then,
Updating best known position of the swarm: $b \leftarrow a_i$

**Step 9:** Recommending contents to the user based on users interests.

***End***

## 5 Results and Performance Evaluation

Any Recommender System aims to provide users with the best practicable and adequate recommendations. To accomplish this objective, the efficiency of this content based recommendation system is estimated. The metrics used for performance evaluation are Precision, Recall, Accuracy, F-Measure and FDR. Equation (1) is used to calculate the precision percentage of the model. Equation (2) calculates the recall percentage of the OntoJudy model. Equation (3) computes the percentage of accuracy. Equations (4) and (5) estimates the F-Measure and FDR of the proposed approach. Accuracy is used to find the accuracy of the model predicted. F-Measure is defined as the weighted harmonic mean of the precision and recall of the test and is used to evaluate a test's accuracy. FDR is the proportion of hypotheses that is true. Lesser the FDR, better the model. The proposed approach has been baselined with CBRF [12] which uses a TF-IDF and LDA topic model for diversification of the tag recommendation. Since the proposed model is an incremental model which uses Synonymization with Word Embeddings and Ontology with Knowledge Generation with honey bee optimization.

$$Precision\% = \frac{Retrieved \cap Relevant}{Retrieved}. \quad (1)$$

$$Recall\% = \frac{Retrieved \cap Relevant}{Relevant}. \quad (2)$$

$$Accuracy\% = \frac{Proportion\ Corrects\ of\ each\ query\ passed\ ground\ truth\ test}{Total\ No.\ of\ Queries}. \quad (3)$$

$$F - Measure\% = \frac{2(Precision * Recall)}{(Precision + Recall)}. \quad (4)$$

$$False\ Discovery\ Rated = 1 - Precision. \quad (5)$$

The comparison of the performance metrics in terms of Precision, Recall, Accuracy, F-Measure, and FDR of the proposed OntoJudy model with the baseline models is depicted in Fig. 2. From Fig. 2, it is indicative that the proposed OntoJudy model is more coherent than the baseline models. The proposed OntoJudy model is composed of Structural Topic Modelling, Static Judicial Domain Ontology for Indian context, Particle Swarm Optimisation which helped in yielding better accuracy than the baseline models, Jaccard Similarity and SemantoSim for semantic similarity, and Random Forest Classifier for classification of data from the dataset aided with the extracted query word. The baseline algorithms for content based recommendations are CBRF, Eliminating Static Ontology and Wikidata from OntoJudy, Eliminating STM from OntoJudy, Absence of PSO driven Semantic Similarity using Jaccard and SemantoSim from OntoJudy, and Eliminating Classification from OntoJudy. The content-based recommendation using CBRF yields a precision of 88.14 and a recall of 92.28% and an accuracy percentage of 90.35 with an F-Measure of 90.21% and FDR of 0.12. The content-based recommendation by eliminating Static Ontology and Wikidata from OntoJudy furnishes a precision of 90.47, a recall of 94.17%, an accuracy percentage of 92.36, and an F-Measure of 92.28% with an FDR of 0.1. The content-based recommendation by eliminating STM

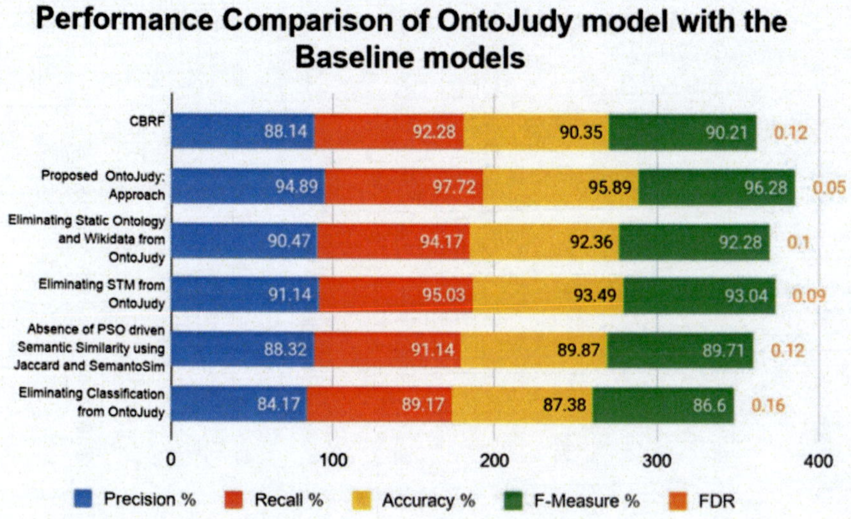


**Fig. 2** Performance comparison of OntoJudy model with baseline approaches

from OntoJudy yields a precision of 91.14, a recall of 95.03, an accuracy of 93.49, and F-Measure of 93.04% with an FDR of 0.09. The content-based recommendation in the absence of PSO driven Semantic Similarity using Jaccard and SemantoSim from OntoJudy produces a precision of 88.32, a recall of 91.14, an accuracy of 89.87, and F-Measure of 89.71% with an FDR of 0.12. The content-based recommendation by eliminating Classification from OntoJudy yields a precision of 84.17 and a recall of 89.17% and an accuracy percentage of 87.38 with an F-Measure of 86.60% and FDR of 0.16. The proposed OntoJudy model yields a precision of 94.89, a recall of 97.72, with an accuracy of 95.89%, and an F-Measure of 96.28% with an FDR of 0.05. The goal of the research done in this paper is to assess the virtue of content based recommendations for judicial cases using this hybrid methodology.

Figure 3 depicts the Accuracy Vs Number of Recommendations in steps of 10 up to 50 recommendations. From Fig. 3, it is evident that OntoJudy has a higher Accuracy distribution when compared to the other baseline systems. The proposed OntoJudy model has an accuracy of 97.89% for 10 numbers of recommendations but the other baseline models yield less accuracy for 10 numbers of recommendations. From Fig. 2, the FDR is the least for the proposed OntoJudy model which shows that the other baseline models available are not efficient when compared to this fusion model. From Fig. 3, it is inferred that the proposed model has proved to be efficient in terms of Accuracy for an individual number of recommendations in numbers from 10 to 50. The OntoJudy is efficient in comparison with the baseline models for content-based recommendations for the reason that it uses Static and Dynamic Entity Addition using Ontologies and Wikidata with Structural Topic Modelling, and Classification using Random Forest with Semantic Similarity using Jaccard and SemantoSim under Particle Swarm Optimisation.

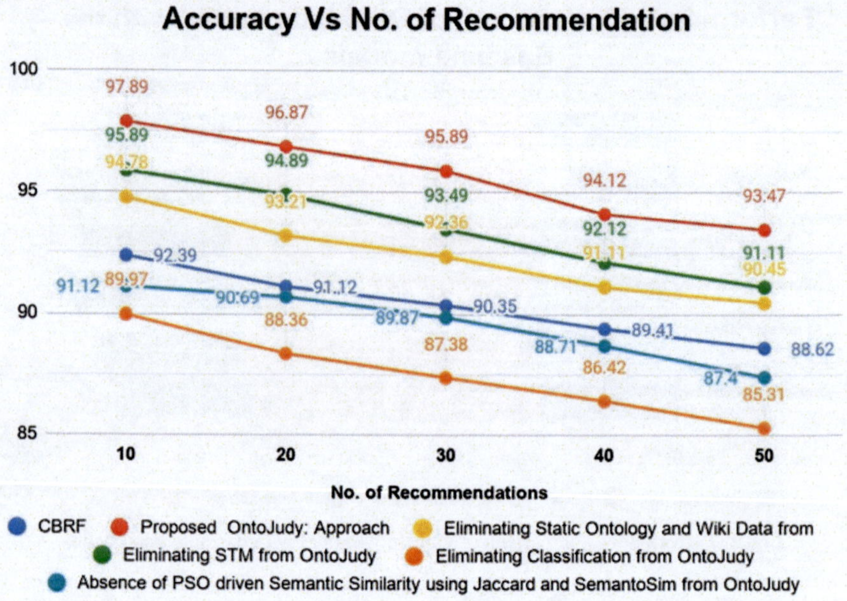


**Fig. 3** Accuracy vs no. of recommendations

## 6 Conclusions

The proposed model employs the Hybridized Deep Learning strategy for content-based Judicial recommendation. The recommendations are achieved based on the user's input queries and CAIL2018 dataset which are met perfectly to provide higher accuracy and F-Measure in comparison to other baseline models. The proposed model uses a hybrid model that consists of Static and Dynamic Entity Addition using Ontologies and Wikidata with Structural Topic Modelling, and Classification using Random Forest with Semantic Similarity using Jaccard and SemantoSim under Particle Swarm Optimisation when ensembled yields better precision, recall, accuracy, and F-Measure percentages with a low FDR. Experiments have demonstrated that the usage of the Ontologies, Structural Topic Modelling, Classification using Random Forest, Semantic Similarity using Jaccard and SemantoSim under Particle Swarm Optimisation helps in bringing better recommendations. OntoJudy has achieved an overall accuracy of 97.89% with a low FDR 0.05 which makes OntoJudy the best in class system for judicial case recommendation.

## References

1. Winkels R, Boer A, Vredebregt B, van Someren A (December 2014) Towards a legal recommender system. Leibniz Center for Law, University of Amsterdam. https://doi.org/10.3233/978-1-61499-468-8-169
2. Thomas M et al (2020) Quick check: a legal research recommendation system. In: NLLP @ KDD 2020, San Diego, US, 24 August 2020
3. Mones E, Sapieżyński P, Thordal S, Olsen HP, Lehmann S (2021) Emergence of network effects and predictability in the judicial system. Sci Rep 11(1):2740

4. Shola PB, Philip S, John AO (2014) Application of content-based approach in research paper recommendation system for a digital library. Int J Adv Comput Sci and Appl (IJACSA) 5(10):37–40
5. Zisopoulos C, Karagiannidis S, Antaris S, Zisopoulos C (10 June 2014) Content-based recommendation systems
6. Sheridan P, Onsjö M, Becerra C, Jimenez S, Dueñas G (2019) An ontology-based recommender system with an application to the star trek television franchise. Fut Internet 11(9):182. https://doi.org/10.3390/fi11090182
7. Obeid C, Lahoud I, El Khoury H, Champin P-A (2018) Ontology-based recommender system in higher education. In: 2018 IW3C2. International world wide web conference committee. ACM. Published under creative commons CC BY 4.0 license. ISBN 978-1-4503-5640-4/18/04. https://doi.org/10.1145/3178876.3191533
8. Roberts M, Stewart B, Tingley D (2019) stm: R package for structural topic models. J Stat Softw 91(2):1–40. https://doi.org/10.18637/jss.v000.i00
9. Anitha Kumari K, SudhaSadasivam G, Aruna T, Christie Sajitha S (2013) Dynamic ontology construction for e-trading. In: Meghanathan N, Nagamalai D, Chaki N (eds) Advances in intelligent systems and computing. Advances in computing and information technology, vol 178. Springer, Heidelberg
10. Golbon-Haghighi M-H, Saeidi-Manesh H, Zhang G, Zhang Y (2018) Pattern synthesis for the cylindrical polarimetric phased array radar (CPPAR). Prog Electromagnet Res 66:87–98. https://doi.org/10.2528/PIERM18011016
11. Ho, TK (1995) Random decision forests. In: Proceedings of the 3rd international conference on document analysis and recognition, Montreal, QC, 14–16 August 1995, pp. 278–282. Archived from the original (PDF) on 17 April 2016. Retrieved 5 June 2016
12. Guo Z, He T, Qin Z, Xie Z, Liu J (2019) A content-based recommendation framework for judicial cases. In: Cheng X, Jing W, Song X, Zeguang Lu (eds) ICPCSEE 2019. Data science: 5th international conference of pioneering computer scientists, engineers and educators, Guilin, China, 20–23 September 2019, Proceedings, Part I. Springer, Singapore, pp 76–88. https://doi.org/10.1007/978-981-15-0118-0_7

# Classifying Emails into Spam or Ham Using ML Algorithms

Gopika Mohanan(✉), Deepika Menon Padmanabhan, and G. S. Anisha

Amrita School of Arts and Sciences, Amrita Vishwa Vidyapeetham, Kochi, India
gopikamohan61@gmail.com, menondeepika1@gmail.com, anisha@asas.kh.amrita.edu

**Abstract.** Spam is commonly used to conduct fraudulent activities. Users should be cautious enough not to open spam emails and not to respond to them. Otherwise, this could lead to cyber criminals gaining control to access our devices. Emails are one of the most secure ways of communication. But due to its rapid growth the chance of it getting affected by spam emails increases. To prevent this we need to come up with an efficient system for email spam detection. In this paper, we try to implement machine learning algorithms using Scikit-learn in Colaboratory and we aim to find the best algorithm. From Kaggle, a publicly accessible email dataset is taken. In order to compute the output of these algorithms, the dataset is being introduced to classification models like Decision Tree, SVM, Naive Bayes, and Random Forest. This paper can play a significant role by eliminating junk emails, Viruses etc.

**Keywords:** Spam email classification · Colab · Naive Bayes · Decision Tree · Support Vector Machine · Random Forest

## 1 Introduction

Spam e-mails are meaningless messages sent to tons of users in detail. Spam emails prove to be the online medium's most critical challenge. In order to solve this, filtering methods such as using unusual sender addresses and arbitrarily adding characters to the beginning or end of the subject line of the letter, spammers started using a few precarious techniques [1].

Spam emails are emails sent by spammers over the internet, known as junk emails or unsolicited messages. In the case of spam emails, users face many problems such as storage space constraints, computing capacity, and it is becoming an obstacle to finding the extra email, wasting time for users, and is also a hazard to user safety [2].

Email filtering is, therefore, necessary to make email more safe, reliable, and suitable. Spam filtering is a way of identifying unsolicited messages and stopping them from accessing the inbox of the recipient. There are different systems nowadays to produce an anti-spam approach to avoid unsolicited bulk email [3].

The research aims to implement the SVM, Random Forest, Naive Bayes, Decision tree algorithms for e-mail spam filtering on a publicly available email dataset and to find the best classifier among them based on their performance. We implement these


S. Shukla et al. (eds.), *Data Science and Security*, Lecture Notes in Networks and Systems 290,
https://doi.org/10.1007/978-981-16-4486-3_23

algorithms in Colaboratory using Scikit-learn in Python. The Dataset used in this paper has two variables in which the first column represents the content of emails and the second represents whether the email is ham or spam.

Section 2 of this paper is about related works, Sect. 3 is about the methodology used in classifying emails into spam or ham, Sect. 4 deals with the experiment and result analysis of the three algorithms on the email dataset followed by Sect. 5 that presents the conclusion by obtaining the best classifier for email spam detection and it also mentions the future work.

## 2 Related Works

### 2.1 Impact of Feature Selection Technique on Email Classification

**Aakanksha Sharaff, Naresh Kumar Nagwani, and Kunal Swami [4].** The paper contrasts and addresses the efficacy of two methods of feature selection, i.e. To create a classifier for spam email filtering, For machine learning methods such as the Bayes algorithm, tree-based algorithm, and SVM, Chi-square and Info-gain are used. Using Cross-Validation tenfold, the experiment is carried out and performance metrics are used to compare the effects, such as accuracy, precision, recall. According to the paper, SVM is the best performer of all classification techniques and provides overall best results in all cases without using any feature selection techniques with consistent 98% accuracy. The next best classifier is the 97.5% accuracy consistency of the Naïve Bayes classifier in all situations. Info-Gain function selection technique works better with the classifier of Naive Bayes. With role range, J48 demonstrates minor enhancement. Info-Gain performs better among feature selection techniques compared to Chi- square's selection technique.

### 2.2 A Hybrid Algorithm for Malicious Spam Detection in Email through Machine Learning

**Prabha Pandey, Chetan Agrawal, Tehreem Nishat Ansari [1]** This study explores a few of the most prevalent techniques of machine learning in the spam e-mail classification context. Logistic Regression, SVM, and Naive Bayes techniques seemed very satisfying efficiency in terms of accuracy among the 5 techniques. Through enhancing the efficiency of the Naive Bayes classification model, they also suggested that further research should be carried out through a hybrid approach.

### 2.3 Study on the Effect of Preprocessing Methods for Spam Email Detection

**Fariska Ruskanda [5].** The preprocessing measure's influence on classifiers is studied in this research paper. The experiments are carried out using the SVM and Naive Bayes. The Ling-spam corpus dataset is the data collection used. In order to test their impact on the classification results, they used combinations of 5 preprocessing methods on 2 spam classifiers, i.e. Naive Bayes and SVM. This research suggests that the appropriate use (or no use) be made of the appropriate depending on the classifier used, preprocessing techniques on each classifier will result in better precision. The combination of the elimination of stop words and stemming gave better results than other combinations for the Naive Bayes classifier. However, the preprocessing stage often does not provide an improvement in classification results for the SVM classifier. The Naive Bayes classifier, which is a probabilistic classifier, is susceptible to the presence of stop words and word forms.

### 2.4 Review Web Spam Detection Using Data Mining

Swathi Raj, K. Raghu veer [6]. This paper offers a system for the delivery of unsolicited mail classes using Bernoulli's and non-stop possibility distribution. On 3 datasets, the Spam classifiers are tested and it is revealed that the classifier offers better than continuous chance distribution with Bernoulli's chance distribution. Between Naive Bayes and Decision Trees, it is found that there may be no advanced classifier version. Together with the opportunity distribution used, dataset, and the problem involved, the overall output of classifier fashions varies and depends on many elements. The classifier must learn information by hand in Naïve Bayes, at the same time as the classifier chooses the pleasant characteristic in Decision Tree by looking at the desk or version text.

### 2.5 Machine Learning-Based Spam Email Detection

Priti Sharmal, Uma Bhardwaj [7]. A hybrid bagging approach of J48 and Naive Bayes is used in this paper for detecting junk mails. Based on different performance measures, three tests are carried out and the results obtained are compared. The two measures are conducted independently using Naive Bayes and J48 algorithms and the third is the combination of both Naive Bayes and J48. It is found that the efficiency of individual J48 is more efficient. For the extension of this analysis, the idea of boosting strategy is proposed.

## 3 Methodology

In this paper, we categorize emails based on multiple machine learning algorithms into spam and ham. For this purpose, As training data, 80% of the data collection is used as a training set and the rest 20% as test data. At first, the dataset is collected. Here we use a publicly available email dataset from Kaggle. Then, the data is preprocessed. Data preprocessing is the most vital step in classification models. It removes ambiguities, errors, redundancy present in the raw data. After preprocessing, feature selection is applied. A bag-of-words model is used for extracting features from the text. This is accompanied by a classification that uses different machine learning classification algorithms to process spam emails. The contents of the emails are converted to a vector using CountVectorizer of Scikit-learn library. It represents the count of each word that appears in the text in a vector format and thus helps in the classification process by highlighting the best features for email spam detection.

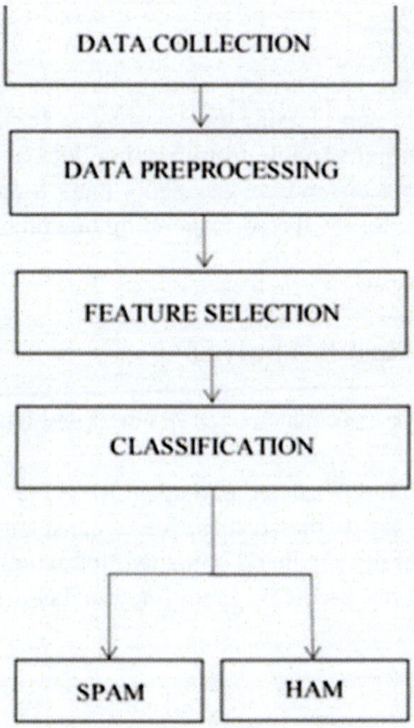


**Fig. 1** Proposed flow diagram

### 3.1 Naive Bayes

As stated by Bayes' Theorem, based on conditional probability, the Naive Bayes classifier is developed. It is simple and easy, and the model's output results are dependent on the dataset used [8].

### 3.2 Support Vector Machine (SVMs)

SVM are relatively modern techniques that have quickly gained popularity due to the detailed results in a wide range of machine learning problems, they have accomplished [9]. The SVM is an algorithm that discovers the hyperplane in a way that distinguishes the 2 classes extremely well. Linear and nonlinear classification problems can be carried out by SVM [8].

### 3.3 Random Forest

In each iteration of the bagging algorithm, Random Forests randomly creates decision trees and often produces accurate predictors. The Random forest is a meta-learner that comprises many individual trees. The random forest algorithm uses a voting technique that chooses with the most votes the individual classification [10].

### 3.4 Decision Tree

Decision tree algorithms are most widely used, unlike other supervised learning algorithms, to solve regression and classification problems. It starts to develop a training model that predicts the output variable from training data sets. In the case of Decision Trees, from the root of the decision tree, a category mark for a study is expected. While continuing to compare, we follow the corresponding branch of the value and leap to the next node.

## 4 Experiment and Result Analysis

The terms of a Confusion matrix that is used in our experiment are as follows:

True Negative: The model has predicted non-spam for actual non-spam mails.
True Positive: The model has predicted spam for actual spam mails.
False Negative: The model has predicted non-spam for actual spam mails.
False Positive: The model has predicted spam for actual non-spam mails.

### 4.1 Dataset

In this paper, a publicly available email dataset from Kaggle is used. It consists of 5728 spam and ham emails. The sources of this dataset are from Enron Corpus, Spam-Assassin, it comprises messages contributed by Internet users and mails taken from a project that collects spam mails for identifying spammers. The dataset has two variables i.e., Text and Spam. The text represents the email text, and the spam represents a binary number (0/1) that denotes whether the email is spam or ham.

Naive Bayes: From the test data containing a total of 1139 mails, with 870 ham and 269 spam mails. Naive Bayes classifier correctly classifies 862 ham emails as ham and 268 spam mails as spam.8 ham mails out of 870 ham emails are wrongly classified as

```
              precision    recall  f1-score   support

           0       1.00      0.99      0.99       870
           1       0.97      1.00      0.98       269

    accuracy                           0.99      1139
   macro avg       0.98      0.99      0.99      1139
weighted avg       0.99      0.99      0.99      1139

Confusion Matrix:
 [[862   8]
 [  1 268]]

Accuracy:  0.9920983318700615
```

**Fig. 2** Classification report of Naive Bayes

| | precision | recall | f1-score | support |
|---|---|---|---|---|
| 0 | 0.96 | 0.99 | 0.98 | 870 |
| 1 | 0.98 | 0.86 | 0.91 | 269 |
| accuracy | | | 0.96 | 1139 |
| macro avg | 0.97 | 0.93 | 0.95 | 1139 |
| weighted avg | 0.96 | 0.96 | 0.96 | 1139 |

```
Confusion Matrix:
 [[865   5]
 [ 38 231]]

Accuracy:  0.9622475856014048
```

**Fig. 3** Classification report of SVM

| | precision | recall | f1-score | support |
|---|---|---|---|---|
| 0 | 0.96 | 0.98 | 0.97 | 870 |
| 1 | 0.93 | 0.88 | 0.91 | 269 |
| accuracy | | | 0.96 | 1139 |
| macro avg | 0.95 | 0.93 | 0.94 | 1139 |
| weighted avg | 0.96 | 0.96 | 0.96 | 1139 |

```
Confusion Matrix:
 [[853  17]
 [ 32 237]]

Accuracy:  0.9569798068481123
```

**Fig. 4** Classification report of Decision Tree

spam and 1 spam mail out of 269 spam mails is wrongly classified as ham or legitimate mail. An accuracy of 99.20% is produced by this classifier.

SVM: From the test data containing a total of 1139 mails, with 870 ham and 269 spam mails. SVM classifier correctly classifies 865ham emails as ham and 231 spam mails as spam.5 ham mails out of 870 ham emails are wrongly classified as spam and 38 spam mails out of 269 spam emails are wrongly classified as ham or legitimate mail. An accuracy of 96.22% is produced by this classifier.

Decision Tree: From the test data containing a total of 1139 mails, with 870 ham and 269 spam mails. Decision Tree classifier correctly classifies 853 ham emails as ham and 237 spam mails as spam.17 ham mails out of 870 ham emails are wrongly classified as spam and 32 spam mails out of 269 spam emails are wrongly classified as ham or legitimate mail. An accuracy of 95.69% is produced by this classifier.

```
              precision    recall  f1-score   support

           0       1.00      1.00      1.00       560
           1       1.00      1.00      1.00       579

    accuracy                           1.00      1139
   macro avg       1.00      1.00      1.00      1139
weighted avg       1.00      1.00      1.00      1139

Confusion Matrix:
 [[560   0]
 [  2 577]]

Accuracy:  0.9982440737489026
```

**Fig. 5** Classification report of Random Forest (RF)

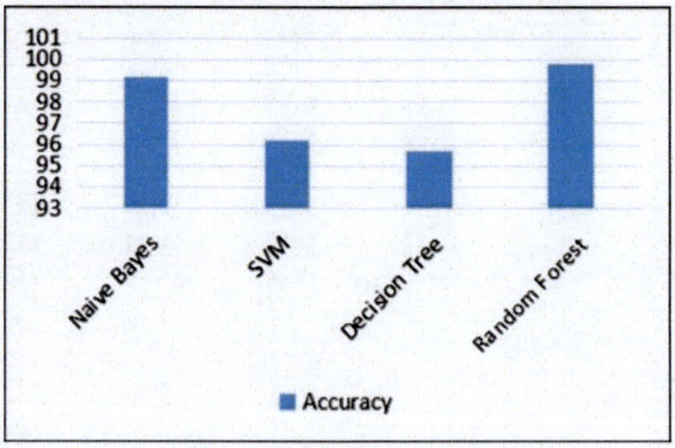


**Fig 6** Comparison of accuracy

Random Forest: From the test data containing a total of 1139 mails, with 560 ham and 579 spam mails. Random Forest classifier correctly classifies 560 ham emails as ham and 577 spam mails as spam.2 spam mails out of 577 spam emails are wrongly classified as ham or legitimate mail. An accuracy of 99.82% is produced by this classifier.

The experimental result obtained on Naive Bayes, Random Forest, Decision Tree, and SVM algorithm on email dataset is discussed here. The accuracy obtained by Naive Bayes, SVM, Decision Tree, Random Forest is 99.2, 96.22, 95.69 and 99.82% respectively. The Random Forest classifier performs well relative to a Decision Tree with 20 random trees. With regard to precision and other performance measures, such as recall, accuracy, f-measure, etc., Random Forest is found to be the optimum classifier and is accompanied by Naive Bayes. A very large number of input variables are managed by Random Forest, thus producing good results for large datasets. It is also successful in estimating missing data and retains high accuracy despite the lack of a significant proportion of data.

## 5 Conclusion

From our research based on email spam detection using machine learning algorithms, we come up with Random Forest Algorithm to filter spam emails based on certain performance measures. Here we use and compare these algorithms to know which will give the best accuracy. We find that the highest accuracy is in the Random Forest classifier. Ensemble learning provides more reliable results, thus we can conclude that Random Forest is the best classifier for email spam detection. Based on this algorithm we can do accurate spam filtering. It can be further extended by using various classifiers and by testing it on different datasets.

## References

1. Pandey P, Agrawal C, Ansari TN (2018) A hybrid algorithm for malicious spam detection in email through machine learning. Int J Appl Eng Res 13(24):16971–16979. ISSN 0973-4562
2. Sahami M, Dumais S, Heckerman D, Horvitz E (July 1998) A Bayesian approach to filtering junk e-mail. In: Learning for text categorization. Papers from the 1998 workshop, vol 62, pp 98–105
3. Bhuiyan H, Ashiquzzaman A, Juthi TI, Ara SBJ (2018) A survey of existing e-mail spam filtering methods considering machine learning techniques. Glob J Comput Sci Technol C Softw Data Eng 18(2):21–29. Version 1.0
4. Sharaff A, Nagwani NK, Swami K (2015) Impact of feature selection technique on email classification. Int J Knowl Eng IACSIT 1(1):59–63
5. Ruskanda F (2019) Study on the effect of preprocessing methods for spam email detection. Indonesian J Comput (Indo-JC) 4(1):109
6. Raj S, Raghuveer K (2020) Review web spam detection using data mining. Int Res J Eng Technol (IRJET) 7:4040–4044
7. Sharma P, Bhardwaj U (2018) Machine learning based spam e-mail detection. Int J Intell Eng Syst 11(3):1–10
8. Goswami V, Malviya V, Sharma P (2020) Detecting spam emails/SMS using Naive Bayes, support vector machine and Random Forest. In: Raj JS, Abul Bashar SR, Ramson J (eds) Innovative data communication technologies and application: ICIDCA 2019. Springer, Cham, pp 608–615. https://doi.org/10.1007/978-3-030-38040-3_69
9. Deepika M, Rani S (2017) Performance of machine learning techniques for email spam filtering. In: CETCSE-2K17. National conference on convergence of emerging technologies in computer science engineering
10. Mishra R, Thakur R (2013) Analysis of Random Forest and Naïve Bayes for spam mail using feature selection catagorization. Int J Comput Appl 80(3):42–47

# Rice Yield Forecasting in West Bengal Using Hybrid Model

Aishika Banik(✉), G. Raju, and Samiksha Shukla

Department of Data Science, Christ University, Bangalore, India
aishika.banik@science.christuniversity.in

**Abstract.** Agriculture in India is the primary source of revenue, yet farmers still face challenges. The primary goal of agricultural development is to produce a high crop yield. The Datasets collected for the study of real-world time series include a blend of linear and nonlinear patterns. A mixture of linear and non - linear models, rather than a single linear or non - linear model, gives a more precise forecasting models for time series data. The ARIMA and ANN prediction models are combined in this paper to create a Hybrid model. This model is used to predict rice yield for all 18 West Bengal districts during the Kharif season, based on 20 years of information(2000–2019) collected from various sources such as India Meteorological Department, Area, and production Statistics, DAV from NASA, etc. The hybrid model aims to enhance efficiency indicators such as MSE, MAE, and MAPE, demonstrating excellent performance for rice yield prediction in all the districts of West Bengal. In the future, it can be applied to other crops that can support farmers in their farming.

**Keywords:** Rice yield · Forecasting · Hybrid model · ARIMA · ANN

## 1 Introduction

Between the northern latitudes of 21°25′24" and 27°13′15" and the eastern longitudes of 85°48′20" and 89°53′04" lies West Bengal, a critical state in Eastern India. Nearly 30% of the state's income is from agriculture. The state provides more than 15% of the total rice production of the country.

As a nation, India steadily ranks first globally, followed by the United States and China. Despite the increase of the country's broad-based economic development, the efficient supply of agriculture to India's GDP is decreasing rapidly. With the advent of technology, farmers are also facing many problems. Achieving a high crop yield is the key priority in the agricultural field. Recognition and oversight of factors that influence crop yield will help farmers make better decisions. Large numbers of crop harvest estimation models take either statistical or crop parameterized models into account. In the last decade, regression and Artificial Intelligence (AI) approaches were used to estimate crop yield under various cropping conditions.

Time series prediction or forecasting in which historical findings are collected and evaluated to construct a model that describes the dynamic relationship plays a critical role in better understanding cultural, financial, epidemiological, and organizational


S. Shukla et al. (eds.), *Data Science and Security*, Lecture Notes in Networks and Systems 290,
https://doi.org/10.1007/978-981-16-4486-3_24

conduct. To predict linear time series data, conventional stochastic time series models, such as Auto-Regressive Integrated Moving Average (ARIMA), are highly efficient. The Artificial Neural Network (ANN) model performed better in modeling data from non-linear time series. As time series data sets include both linear and non-linear patterns, it would be critical to make decisions focused on an individual model. Hybridization can minimize the bias and the variance of the prediction error of the linear and nonlinear models.

In recent years, Hybrid models have gained much popularity for prediction purposes, not only in agriculture but also in other domains. This paper presents a Hybrid Neural Network-Based model to forecast the rice yield of 18 districts in West Bengal during the Kharif season. Based on that, Hybrid architecture is proposed for the rice yield dataset of West Bengal collected from various sources.

Section 2 presents the analysis of the significant prototypes used to predict rice crop yield. Section 3 describes the data collection method and the techniques used as part of the Hybrid approach. Section 4 addresses the experimental design and findings derived from the trial, accompanied by a conclusion and future work in Sect. 5.

## 2 Related Works

Various algorithms and approaches are designed for the prediction of rice crop yield in India. There is also an extensive collection of ARIMA, ANN, and Hybrid-based prediction papers, and it is gaining popularity these days on a considerable scale.

In [1], Aditi Chandra et al. developed prediction models of Kharif rice yield for the Purulia and Bankura districts in West Bengal using the Artificial Neural Network and Random Forest. These were established for 2006 to 2015 by consolidating the monthly NDVI with the weather and non-weather variables.

Kasampalis et al. [2] highlighted the high temporal frequency and broad spatial coverage of low-cost satellite data, which affects the preferred option.

The Artificial Neural Network (ANN), one of the most prevailing modeling and prediction methods, is used by Giritharan Ravichandran et al. [3], which guides farmers to understand the state of the land and to help them know the crops that could profit them.

Laxmi Goparaju et al. [4] used long-term (1970–2000) monthly real datasets. They analyzed the seasonal precipitation trends, possible evapotranspiration, and aridity index to scale India's district-level crop production.

The crop's water demand is influenced by pre-rainfall, possible evapotranspiration, and aridity across various seasons or areas indicated by Ashaolu and Iroye [5].

In paper [6], Surjeet Kumar et al. used models such as Artificial Neural Network (ANN), Statistical Equations, Genetic Algorithm (GA), and Fuzzy Logic to build a predictive model for rice production in India. They created a hybrid model using previous time-series data to get the optimum outcome.

SB Satpute et al. [7] used the CERES-Rice (DSSAT 4.5) crop simulation model to forecast the South Dinajpur District of West Bengal monsoon rice yield 2013 by simulating the development, growth, and yield of two prevalent rice cultivars. They studied the 29 years (1983–2011) daily weather data about the maximum and minimum temperature, rainfall, and bright sunshine hours to identify the variability in precipitation and temperature.

Adaptive Neuro-Fuzzy Inference Method (ANFIS), Interval Type n Fuzzy Auto-Regressive Integrated Moving Average (ITnARIMA), and Modified Regularized Least Squares Fuzzy Support Vector Regression (MRLSFSVR) were used by Arindam Chaudhuri [8] to estimate the Productivity Index Percent (PI percent) of the time series data for rice production and compared with the typical Multiple Regression Statistical Tool.

Dr. S. A. Jyothi Rani and N. Chandan Babu [9] used Convolution Neural Networks (CNN), Recurrent Neural Network, Multilayer Perceptron (MLP) and, Auto-Regressive Integrated Moving Averages (ARIMA) to forecast rice development methods.

Kiran Kumar Paidipati and Arjun Banik [10] carried out forecasts based on historical rice cultivation data from 1950–51 to 2017–18 with the assistance of Auto-Regressive Integrated Moving Average (ARIMA) and Long Short-Term Memory Neural Network (LSTM-NN).

## 3 Methodology

### 3.1 Data Collection

The data is collected for the years 2000–2019 from various sources to forecast rice yield. The Total Area, Production, and Yield of Rice is collected from the "Area and Production Statistics" issued by the Ministry of Agriculture and Farmers' Welfare for the years 2000–2019 for the districts of West Bengal.

The Rainfall data is gathered from "Customized Rainfall Information System (CRIS)" given by the India Meteorological Department, Ministry of Earth Sciences for the years 2014–2018 by putting the state and the district required. For the years 2011–2013, it is collected from the Statistical Abstract 2015 given by West Bengal Agricultural Statistics, and also from the years 2000–2002 and 2004–2010 it was collected from India Meteorological Department directly. Maximum temperature, minimum temperature, temperature, precipitation, and relative humidity data are acquired by "Power Data Access Viewer (DAV)" from NASA upon placing each district's longitude and latitude for the same time.

And the rest of the variables data like HYV used, Chemical fertilizers, and Insecticides are collected from "Statistical Abstract 2015" given by West Bengal Agricultural Statistics for 2010–2015. and 2000–2005 from "Statistical Abstract 2004". All the above data is gathered for all the 18 districts from the years 2000–2019. The final dataset has 18 district-wise datasets containing 21 rows and 14 attributes all over.

### 3.2 ARIMA Model

ARIMA is a classical time series model used in stationary time series data to track linear tendencies. ARIMA (p, d, q) is the acronym for ARIMA model. The constants p and q are the order of the respective AR model and the MA model, and d is the degree of differentiation. Mathematically, ARIMA model can be indicated as:

$$Y_t = \mu + \sum_{i=1}^{p} \varphi_i Y_{t-1} + \varepsilon_t - + \sum_{j=1}^{q} \theta_j \varepsilon_{t-j} \tag{1}$$

In Eq. (1), $Y_t$ denotes the actual value of the considered variable at time t, which is the random error at time t. The coefficients of the ARIMA model are the $\varphi_i$ and $\theta_j$. The ARIMA model is predicated on the premise that the error series assumes a zero mean with constant variance, satisfying the i.i.d. condition. Three basic three iterative steps are required to build an ARIMA model for time series data: (a) model recognition (achieving stationarity), (b) model parameterisation ('best' choices of p and q), and (c) model diagnostics evaluation (finding the 'best' fit ARIMA model using Akaike's Information Criterion).

## 3.3 ANN Model

An artificial neural network's most significant advantage is its potential to model dynamic non-linear relationships without making presumptions about the relationship's essence. ANN contains an input layer, a hidden layer, a nonlinear activation function, and a linear transfer function output layer. A non-linear functional mapping from past observations $\left(X_{t-1}, X_{t-2}, \cdots\cdots, X_{t-p}\right)$ to the future value $X_t$, i.e., is performed by the ANN model.

$$X_t = f\left(X_{t-1}, X_{t-2}, \cdots\cdots, X_{t-p}, w\right) + \varepsilon_t \tag{2}$$

In Eq. (2), w indicates a vector of all parameters, and f is a function defined by the configuration of the network and the weights of connections. The ANN is therefore analogous to an autoregressive nonlinear model. For a time series, the critical task of ANN modeling is to select an adequate amount of hidden nodes q that is data-dependent, and in determining this parameter, there is no systematic law. Additionally, in choosing the required number of hidden nodes, another crucial work for ANN modeling of a time series is in selection the number of lagged observations, p, and the input vector dimension. In an ANN model, this is probably the most critical parameter to be calculated since it plays a significant role in deciding the time series' (nonlinear) autocorrelation structure.

## 3.4 Hybrid Model

The two-step approach is the hybrid ARIMA-ANN model. Decomposition techniques separate the time series into a linear and non-linear component, respectively. It is $Y_t = L_t, + N_t$ where $L_t$ denotes the linear element, and the non-linear part is marked by $N_t$. From the data, these two elements have to be estimated. First, ARIMA is allowed to model the linear component, and after that only, the non-linear relationship will contain the residuals from the linear model. Let the error (residual) from the linear model at time t be denoted by $e_t$, then, $e_t = Y_t - \hat{L}_t$, where, $\widehat{L_t}$, is the predicted relationship's forecast value for time t. ANNs can be used to model residuals to find nonlinear associations. For the error terms, an ANN model for n input nodes is used, $e_t = f\left(e_{t-1}, e_{t-2}, \cdots, e_{t-n}\right) + \varepsilon_t$, where f represents a non-linear neural network function and $e_t$ is random error. The residual is not inherently random if model f is not a suitable one. The combined forecast would therefore be, $\widehat{Y_t} = L_t + \widehat{N}_t$

Figure 1 shows the architecture of Hybrid Model-I. ARIMA model building is the first part, which takes the input of the outcome variable. The parameters p, d, and q are determined from the ACF plot, Augmented Dickey-Fuller Test, and PACF plot,

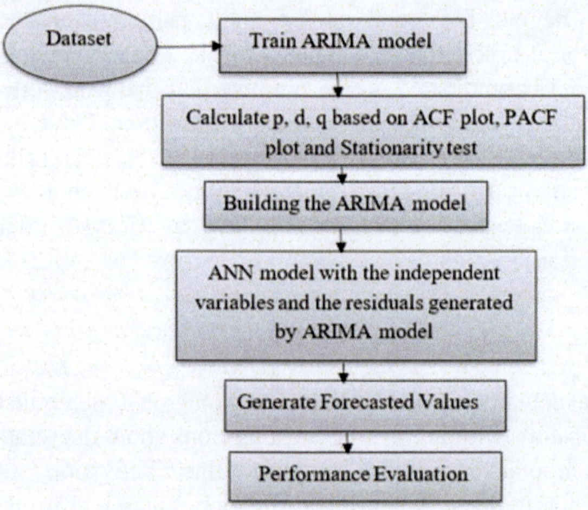


**Fig. 1** Hybrid model-I architecture

respectively. Then using those parameters, the ARIMA model is built on the training set. The residual values will then be obtained by calculating the difference between predicted ARIMA values and original time series values in the training set.

Similarly, for the testing set, the forecasted values are generated along with the residuals. The other independent features are normalized and then passed in the ANN model along with the ARIMA model's training residuals. The outcome variable is the same as the ARIMA model which is the yield. The final forecasts are obtained from the ANN model on the testing test.

The architecture of Hybrid Model-II is shown in Fig. 2. The ARIMA model is same as mentioned above. After ARIMA model building, the forecasted values are generated, and the residuals are computed for the testing set. The other features in the dataset are normalized and then passed in the ANN model as independent variables to estimate the training residuals obtained from the ARIMA model acting as the dependent variable. The error forecasts are obtained from the ANN model on the testing set. Both the ARIMA forecasts and ANN forecasts from both models are combined, and the final predictions are computed.

### 3.5 Performance Metrics

Equation (3), (4), (5) represents performance metrics such as Mean Absolute Percentage Error (MAPE), Mean Absolute Error (MAE), and Mean Square Error (MSE) which are computed to evaluate the performances of different forecasting models for the above dataset of 18 districts.

$$\text{MAPE} = \frac{1}{n}\sum_{i=1}^{n}\left|\frac{Y_i - \widehat{Y}_i}{Y_i}\right| \quad (3)$$

$$\text{MAE} = \frac{1}{n}\sum_{i=1}^{n}\left|Y_i - \widehat{Y}_i\right| \quad (4)$$

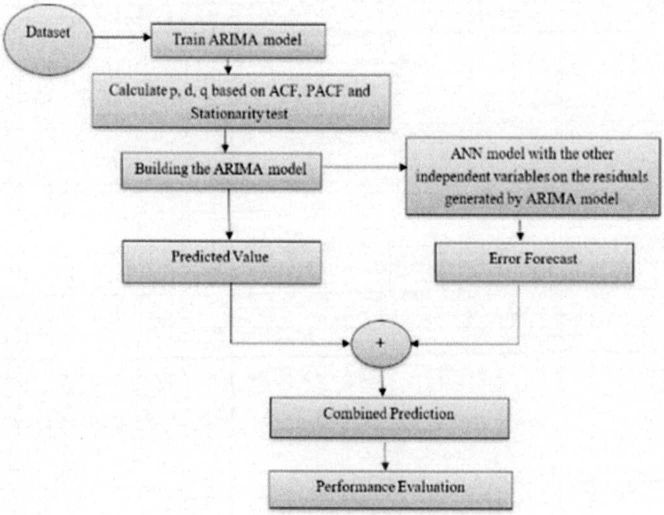


**Fig. 2** Hybrid model-II architecture

$$\text{MSE} = \frac{1}{n}\sum_{i=1}^{n}\left(Y_i - \widehat{Y}_i\right)^2 \quad (5)$$

where, $\widehat{Y}_i$ and $Y_i$ represent the predicted value and actual value of outcome variable respectively, and n represents the number of data points. These metrics are used to understand the forecasting capabilities of the models efficiently. Minimum the values of these metrics, the better is the model prediction.

## 4 Experiments

The hybrid models with two architectures are implemented through Python 3.0+ due to their flexible nature. The Python libraries required for the implementation are NumPy, Pandas, Scikit-learn, Statsmodel, and Keras. Before constructing the models, the data was pre-processed by handling the missing values and feature scaling by the MinMaxScaler technique. There are missing values of four variables in all the districts' datasets in specific years, and those are imputed by mean and median.

The models are evaluated on a rice yield prediction dataset collected from multiple sources. Three experiments are performed on the 18 districts. The main difference in the experiment is the type of model used for training purposes. For experiment 1, the ARIMA model is considered based on the parameters p, d, and q determined from the ACF plot, ADF test, and PACF plot, respectively. The ACF and PACF plots of 5 districts of West Bengal are shown below in Fig. 3.

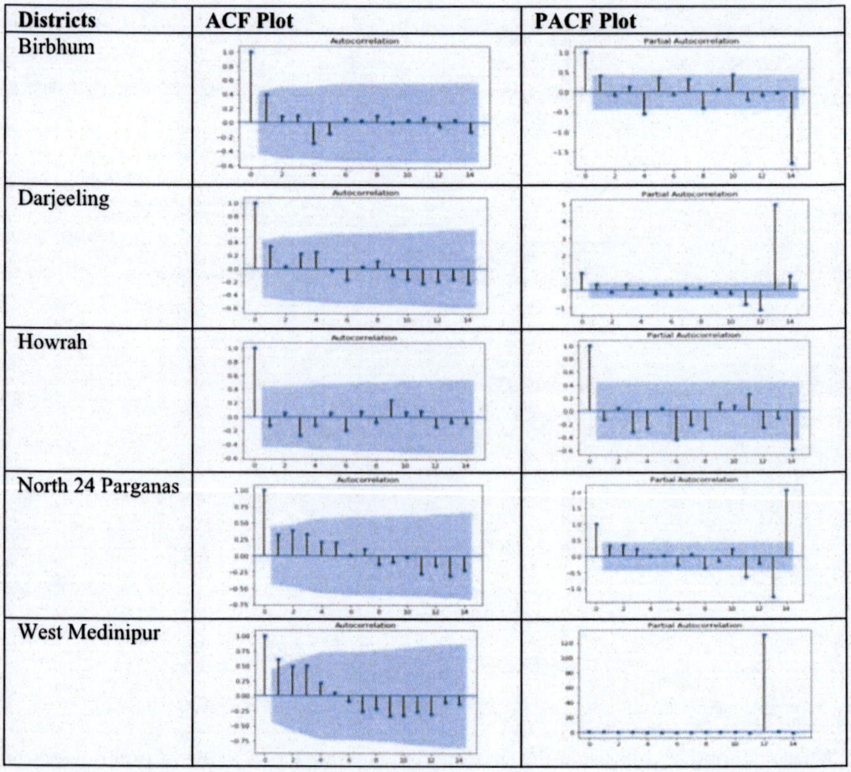


**Fig. 3** ACF and PACF plots for Birbhum, Darjeeling, Howrah, North 24 Parganas, and West Medinipur districts of West Bengal

For experiment 2, the ANN model is built with the 'YIELD' column, which is the response variable and the other variables are the explanatory variables. For experiments 3 and 4, the Hybrid models are used where first experiment 1 is performed followed by experiment 2, where the dependent and independent variables are considered. For training, 70% of the entire data is used, and the remaining 30% is used for testing in all the experiments. A maximum of 95 epochs is considered for experiments 2 and 3. There are one hidden layer and one output layer in all the neural network architecture. The hidden layer has the activation function ReLu, and the final output layer has a linear activation function because it is a regression problem. The model uses the 'adam' optimizer as the minimization algorithm and Mean Squared Error as the loss function. Table 1 shows the performance metrics obtained for the four experiments considering the 18 districts of West Bengal.

**Table 1** Performance metrics of the models

| Districts | Performance metrics | ARIMA | ANN | Hybrid model-I | Hybrid model-II |
|---|---|---|---|---|---|
| Bankura | MSE | 0.1063 | 0.4447 | 0.7921 | 0.0499 |
| | MAE | 0.2234 | 0.6108 | 0.6485 | 0.1912 |
| | MAPE | 7.1494 | 24.3345 | 22.6516 | 7.0331 |
| Bardhaman | MSE | 0.1161 | 0.5009 | 0.4702 | 0.0596 |
| | MAE | 0.1954 | 0.5701 | 0.5772 | 0.1886 |
| | MAPE | 6.2769 | 21.1702 | 17.2429 | 5.7469 |
| Birbhum | MSE | 0.1259 | 0.3316 | 0.2277 | 0.0618 |
| | MAE | 0.3042 | 0.5621 | 0.4185 | 0.1674 |
| | MAPE | 9.8632 | 26.7144 | 17.5397 | 8.3359 |
| Coochbehar | MSE | 0.0767 | 0.1217 | 0.0229 | 0.0598 |
| | MAE | 0.2565 | 0.2628 | 0.1268 | 0.1893 |
| | MAPE | 10.894 | 16.7943 | 11.6358 | 8.5743 |
| Dakshin Dinajpur | MSE | 0.2465 | 0.8319 | 0.5699 | 0.1272 |
| | MAE | 0.3609 | 0.7533 | 0.6134 | 0.2797 |
| | MAPE | 26.2651 | 36.3296 | 28.928 | 21.7417 |
| Darjeeling | MSE | 0.6013 | 0.1366 | 0.1733 | 0.1058 |
| | MAE | 0.2846 | 0.2917 | 0.2897 | 0.2726 |
| | MAPE | 20.0852 | 18.5217 | 15.6682 | 13.8743 |
| East Medinipur | MSE | 0.1808 | 0.8152 | 0.6392 | 0.1694 |
| | MAE | 0.526 | 0.6565 | 0.6752 | 0.3352 |
| | MAPE | 17.5678 | 28.5741 | 26.8346 | 20.6585 |
| Hooghly | MSE | 0.1499 | 0.1828 | 0.4254 | 0.078 |
| | MAE | 0.3007 | 0.3809 | 0.5512 | 0.2405 |
| | MAPE | 9.3028 | 18.4273 | 20.4581 | 8.6704 |
| Howrah | MSE | 0.4684 | 0.3159 | 1.363 | 0.3087 |
| | MAE | 0.51 | 0.5516 | 1.0288 | 0.4645 |
| | MAPE | 43.9822 | 47.2486 | 49.2926 | 35.4873 |
| Jalpaiguri | MSE | 0.7118 | 0.4329 | 0.5201 | 0.3931 |
| | MAE | 0.483 | 0.53 | 0.4808 | 0.4774 |

(*continued*)

**Table 1** (*continued*)

| Districts | Performance metrics | ARIMA | ANN | Hybrid model-I | Hybrid model-II |
|---|---|---|---|---|---|
| | MAPE | 23.4017 | 28.9675 | 20.4581 | 19.8988 |
| Maldah | MSE | 0.362 | 0.474 | 0.2927 | 0.2836 |
| | MAE | 0.5682 | 0.6516 | 0.4771 | 0.4562 |
| | MAPE | 17.4613 | 27.4609 | 21.3287 | 13.2015 |
| Murshidabad | MSE | 0.705 | 0.3696 | 0.3706 | 0.2301 |
| | MAE | 0.8192 | 0.5074 | 0.4693 | 0.4264 |
| | MAPE | 17.3135 | 19.4688 | 20.9274 | 17.2843 |
| Nadia | MSE | 0.0608 | 0.1704 | 0.1377 | 0.0558 |
| | MAE | 0.2344 | 0.2615 | 0.3397 | 0.1843 |
| | MAPE | 7.3591 | 15.3082 | 20.7911 | 6.7041 |
| North 24 Parganas | MSE | 0.0651 | 0.2117 | 0.0827 | 0.0519 |
| | MAE | 0.1751 | 0.3409 | 0.2219 | 0.1666 |
| | MAPE | 8.171 | 18.1439 | 8.4358 | 7.775 |
| Purulia | MSE | 0.6358 | 1.214 | 0.5706 | 0.374 |
| | MAE | 0.7115 | 0.8088 | 0.67 | 0.4776 |
| | MAPE | 26.1139 | 36.7164 | 47.1983 | 16.7543 |
| South 24 Parganas | MSE | 0.1558 | 0.5632 | 0.4355 | 0.2047 |
| | MAE | 0.3296 | 0.6674 | 0.6167 | 0.2434 |
| | MAPE | 14.8823 | 27.7594 | 17.1755 | 14.0367 |
| Uttar Dinajpur | MSE | 0.2518 | 0.4541 | 0.2375 | 0.1677 |
| | MAE | 0.429 | 0.4774 | 0.381 | 0.3015 |
| | MAPE | 27.9599 | 25.5225 | 26.0957 | 22.4061 |
| West Medinipur | MSE | 0.0563 | 0.2594 | 0.3573 | 0.0417 |
| | MAE | 0.2126 | 0.4276 | 0.4438 | 0.1658 |
| | MAPE | 6.9478 | 21.9277 | 18.1217 | 6.7181 |

## 5 Conclusion and Future Work

In the research work, it is observed that all the performance metrics MSE, MAE, and MAPE, are less for the Hybrid Model-II compared to the other models. It shows that the Hybrid Model-II is an efficient prediction model for the rice yield of West Bengal. The primary limitation of this study is the data availability. For the Artificial Neural Network (ANN) model, the algorithm requires sufficient data for training. The data used in the experiments was only limited to 20 years (i.e., 2000–2019).

Hybrid models with two different approaches are built and experimented with rice prediction dataset collected from various sources and merged. This work is limited to the prediction related to rice crops in all the districts of West Bengal. This study can be extended to other geographies and yields in India.

## References


1. Chandra A, Mitra P, Dubey SK, Ray SS (2019) Machine learning approach for kharif rice yield prediction integrating multi-temporal vegetation indices and weather and non-weather variables. In: The International Archives of the Photogrammetry, Remote Sensing and Spatial Information Sciences, vol. XLII-3/W6, February 2019
2. Kasampalis DA, Alexandridis TK, Deva C, Challinor A, Moshou D, Zalidis G (2018) Contribution of remote sensing on crop models: a review. J Imaging 4(52)https://doi.org/10.3390/jimaging4040052
3. Ravichandran G, Koteeshwari RS (2016) Agricultural Crop Predictor and Advisor using ANN for Smartphones. IEEE, February 2016
4. Goparaju L, Ahmad F (2019) analysis of seasonal precipitation, potential evapotranspiration, aridity, future precipitation anomaly and major crops at district level of india. KN – J Cartograph Geograph Inf 69:143–154. https://doi.org/10.1007/s42489-019-00020-4
5. Ashaolu ED, Iroye KA (2018) Rainfall and potential evapotranspiration patterns and their effects on climatic water balance in the Western Lithoral Hydrological Zone of Nigeria. Ruhuna J Sci 9(2):92–116. https://doi.org/10.4038/rjs.v9i2.45
6. Kumar S, Sanyal MK (2019) A soft computing model to predict the rice production in India. Int J Eng Adv Technol (IJEAT) 8(6) ISSN: 2249–8958
7. Satpute SB, Rai A, Bandyopadhyay S, Mahata D, Halder D, Gupta DS, Bandyopadhyay S (2018) Forecasting of rice yield of South Dinajpur district of West Bengal using CERES-rice (DSSAT 4.5) model. Int J Chem Stud 6(3):2542–2546
8. Chaudhuri A (2013) Forecasting rice production in West Bengal State in India: statistical vs. computational intelligence techniques. Int J Agric Environ Inf Syst 4(4):68–91
9. Rani SAJ, Babu NC (2020) Forecasting production of rice in India–using Arima and deep learning methods. Int J Math Trends Technol (IJMTT) 66(4)
10. Paidipati KK, Banik A (2019) Forecasting of rice cultivation in India–a comparative analysis with ARIMA and LSTM-NN models. EAI Endorsed Trans Scalable Inf Syst

# An Inventory Model for Growing Items with Deterioration and Trade Credit

Ashish Sharma[1](✉) and Amit Kumar Saraswat[2]

[1] School of Sciences, CHRIST (Deemed to be University), Delhi-NCR, Ghaziabad 201003, India
ashish.sharma@christuniversity.in

[2] Department of Mathematics, Institute of Applied Sciences and Humanities, GLA University, NH-2, Mathura 281406, India
amit.saraswat@gla.ac.in

**Abstract.** Growing items industry plays a vital role in the economy of most of the countries. Growing item industries consists of live stocks like sheep, fishes, pigs, chickens etc. In this paper, we developed a mathematical model for growing items by considering various operational constraints. The aim of the present model is to optimize the net profit by optimizing decision variables like time after growing period and shortages. Also, the delay in payment policy has been used to maximize the profit. A numerical example is provided in support of the solution procedure. Sensitivity analysis provides some important insights.

**Keywords:** Growing items · Inventory management · Deteriorating items · Mortality · Delay in payment · Price dependent demand

## 1 Introduction

Procurement management of the growing items is difficult as the inventory increases by the time without actual purchasing new items. Poultry items, live stocks etc. are considered as growing items. In this paper we focused on poultry items. As per the report of Department of Animal Husbandry, Dairying & Fisheries, Ministry of Agriculture & Farmers Welfare, Government of India for National Action Plan for Egg & Poultry-2022 [5, 6], the consumption of poultry products will increase approximately 6 kg by year 2030. In India poultry sector is valued more than 80,000 crore which is further divided into two sectors: (i) organized commercial sector whose market share is about 80% and (ii) unorganized sector (backyard farming) whose market share is about 20%. The needs of unorganized and organized sectors are different and an extensive research is required in this area.

Rezaei [8] provided a model for growing items by considering the growth function given by Richards [9]. He developed a model to optimize the total cost by optimizing the total time used in consumption period and shortage period and the shortages occur. This model was further developed by Zhang et al. [15] by incorporating carbon constrained for environmental sustainability. Nobil et al. [7] extended this model by assuming linear rate of growth and partial shortages.


S. Shukla et al. (eds.), *Data Science and Security*, Lecture Notes in Networks and Systems 290,
https://doi.org/10.1007/978-981-16-4486-3_25

Sebatjane and Adeutnji [11] developed a model for growing items with incremental quality discount. This paper was further extended by Hidayat et al. [4] by adding constraints in budgets.

Tread credit is one of the best policies for both buyer and seller. It helps in increasing sells volume in limited financial opportunity. In this policy buyer has an interest free payment option up to certain time limit, which make buyer more interested in purchasing product. Goyal [3] formulated an economic order quantity (EOQ) model to discuss the impact of tread credit policy. After that Agrawal and Jaggi [1] extended it for deteriorating items. Many authors Shaikh et al. [12], Agrawal et al. [2], Saren et al. [10], Taleizadeh et al. [14] etc. worked in the same line. Sharma and Kaushik [13] also considered delay in payment for the deteriorating items.

Thus, in the light of the aspects discussed above, in present article, an EOQ model has been presented for the growing items, with constant demand, deterioration and credit policy for buyer. In this paper, we assumed that a fixed quantity of growing items have been procured at the beginning of each cycle and start feeding them. After the completion of growing period all the items are sold after slaughtering them in order to satisfy the demand. The costs considered in this paper are feeding cost, holding cost, ordering cost and purchasing cost. Our objective is to find the optimal profit by finding the optimal values of the decision variables.

The rest of the paper organized as follows; in Sect. 2 model formation, notations and the assumptions. Analysis of the model are presented in Sect. 3. Section 4 contains particular cases and solution procedure is explained in Sect. 5. Sensitivity analysis presented in Sect. 6. Final concluding remarks are provided in Sect. 7.

## 2 Modal Formation, Notations and Assumptions

In this paper, we assumed that company purchased few days' old items and grown up in the controlled environment up to the optimum weight but some of the item lost due to mortality. After getting optimum weight, items are available to slaughter and sell. Shortages are allowed and in order to increase the sell volume, tread-credit policy has been considered.

Following notations have been considered in the formation of mathematical model. Notation is defined in bracket ( ).

Initial ordered quantity per cycle (Y), Growth rate of weight per item per unit time in grams (k), Initial weight of newborn item in gm ($w_1$), Final weight of newborn item in gm ($w_2$), Weight of entire inventory at any time t (wt), Length of Growing Period ($t_1$), Length of consumption period ($t_2$), Length of shortage period (t3), Shortage quantity per period in gm (S), Selling price per unit weight (p), Purchase cost per unit weight (c), Shortage cost per unit weight per unit time (f), Ordering cost (Oc), Growing cost per unit weight per unit time (r), Annual rate of interest earn (Ie), Holding cost per unit weight per unit time (h), Annual rate of interest charged (Ic), permissible delay in time to settle the account (td), Setup time (TS).

$$T = t_2 + t_3 \text{ (Decision variable)}$$

Assumptions used throughout the model are

(1) Demand is constant. (2) Deterioration in inventory for a cycle is fixed.

(3) Growth rate is a linear approximation of Richard growth function. (4) $Ts + t_1 < t_2 + t_3$ i.e. the total time up to the starting of consumption period must be less than the total time from the starting of consumption period up to the end of cycle. (5) $cIc > pIe$ i.e. annual interest rate must satisfy this inequality. (6) Growth rate does not change from one cycle to another cycle.

Our objective is to optimize the net profit by optimizing the number of ordered items (Y), length of consumption ($t_2$) and shortage period ($t_3$).

## 3 Analysis

As per our notation. The initial weight ($W_1$) and Weight of inventory at the end of growth period ($W_2$) of the inventory is given by

$$W_1 = Y \times w_1 \tag{1}$$

$$W_2 = Y \times w_2 \tag{2}$$

Thus, after satisfying shortages 'S' of the previous cycle the residual inventory is given by

$$W_t = w_2 - s \tag{3}$$

The growing period $t_1$, the selling period $t_2$ and the shortage period $t_3$ can be defined as:

$$t_1 = \frac{w_2 - w_1}{k} \tag{4}$$

$$t_2 = \frac{w_2 Y - S}{a + \theta} \tag{5}$$

$$t_3 = \frac{S}{a} \tag{6}$$

Thus, by using Eq. (2), (3) and Eq. (5) we get Y in terms of T as,

$$Y = \frac{T(a + \theta)}{w_2} - \frac{S\theta}{aw_2} \tag{7}$$

Total Earning (TE) is given by

$$T_E = \int_0^{t_2} pd\, dt + pS \tag{8}$$

Ordering cost ($O_C$) per period is given by

$$O_c = Ycw_1 \tag{9}$$

Feeding cost (Fc) which includes the cost of vaccination, drinking water, medicated food etc. is given by

$$F_c = r\int_0^{t_1} \{(t-0)Yk\}dt \tag{10}$$

The holding cost ($H_C$) associated to the available items in the period $t_2$ is given by

$$H_c = h\int_0^{t_2} \{(t-0)(d+\theta)\}dt \tag{11}$$

Shortage Cost ($S_C$) is given by

$$S_c = f\int_0^{t_3} \{(t_3-t)d\}dt \tag{12}$$

Now interest earn ($I_E$) and total interest charge ($I_C$) for inventory model is given by

$$I_E = pI_e\left[\int_0^{t_2} \{(t_2-t)d\}dt + S\right] \tag{13}$$

$$I_C = cI_c\int_{t_p}^{t_2} \{(t-0)(d+\theta)\}dt \tag{14}$$

Thus, we can define net profit per unit per unit time as;

$$NP = (I_E - I_C + TE - (O_C + F_C + H_C + S_C))/(t_1 + T) \tag{15}$$

Now, we establish a theorem to demonstrate the concaveness of NP,

**Theorem:** Net profit is concave with respect to S.

**Proof.** On differentiating NP with respect to S we get

$$\frac{\partial NP}{\partial S} = \frac{\begin{pmatrix} 2a^2k((h+cI_c)T + I_e(p-pT))w_2 \\ -2(h+cI_c)kSw_2\theta + a\begin{pmatrix} -2fkSw_2 - 2cI_ckSw_2 + 2I_ekpSw_2 \\ +2ckw_1\theta + rw_1^2\theta \\ +2cI_ckTw_2\theta - 2rw_1w_2\theta \\ +rw_2\theta - 2hkw_2(S-T\theta) \end{pmatrix} \end{pmatrix}}{2a^2w_2(kT - w_1 - w_2)} \tag{16}$$

and

$$\frac{\partial^2 \mathrm{NP}}{\partial \mathrm{S}^2} = -\frac{k(a(f + h + cI_c - I_e p) + (h + cI_c)\theta)}{\mathrm{a}^2(\mathrm{kT} - \mathrm{w}_1 - \mathrm{w}_2)} \tag{17}$$

As $\frac{\partial^2 \mathrm{NP}}{\partial \mathrm{S}^2} < 0$ hence the NP is concave with respect to S.

On differentiating NP one time with respect to T we get

$$\frac{\partial NP}{\partial T} = \frac{1}{2a^2 \mathrm{w}2(kT - \mathrm{w}1 + \mathrm{w}_2)^2}\left[\begin{array}{l} -a^3\left\{\begin{array}{l} -r(\mathrm{w}_1 - \mathrm{w}_2)^3 + ck\left(\begin{array}{l} -2\mathrm{w}_1^2 - 2(-1 + \mathrm{I_c}T)\mathrm{w}_1\mathrm{w}_2 \\ +\mathrm{I_c}\mathrm{w}_2\left(k\left(T^2 + \mathrm{tp}^2\right) + 2T\mathrm{w}_2\right) \end{array}\right) \\ +k\mathrm{w}_2\left((h + I_e)kT^2 - 2(\mathrm{w}_1 - \mathrm{w}_2)(h - p - 2I_e T)\right) \end{array}\right\} \\ +(h + c\mathrm{I_c})k^2S^2\mathrm{w}_2\theta \\ +a^2\left\{\begin{array}{l} 2Ak^2\mathrm{w}_2 - (h + cI_c - pI_e)2kS\mathrm{w}_2(\mathrm{w}_1 - \mathrm{w}_2) - 2I_e k^2 pS\mathrm{w}_2 \\ +2ck\mathrm{w}_1(\mathrm{w}_1 - \mathrm{w}_2)\theta + r\left(\mathrm{w}_1^3 - \mathrm{w}_2^2\right)\theta - cI_c k^2 tp^2\mathrm{w}_2\theta \\ -3r\mathrm{w}_1\mathrm{w}_2(\mathrm{w}_1 - \mathrm{w}_2)\theta + (h + cI_c)kT\mathrm{w}_2\theta(2\mathrm{w}_1 - kT - \mathrm{w}_2) \end{array}\right\} \\ +akS\left\{\begin{array}{l} (f + cI_c - I_e p)kSw_2 - 2ck\mathrm{w}_1\theta - r\theta\left(\mathrm{w}_1^2 + \mathrm{w}_2^2\right) \\ -2cI_c(\mathrm{w}_1 - \mathrm{w}_2)w_2\theta + 2r\mathrm{w}_1 w_2\theta \\ +hw_2(kS - 2(\mathrm{w}_1 - \mathrm{w}_2)\theta) \end{array}\right\} \end{array}\right] \tag{18}$$

On differentiating NP second time with respect to T. We find that it is difficult to check the sign of second derivative of NP with respect to T analytically so we will check the concavity with respect to T numerically.

## 4 Particular Case

For some particular values of the parameter we can get the following published models.

1. If we consider $\theta = \mathrm{I_c} = \mathrm{I_e} = 0$ then we will get the model presented by Nobil et al. [7].
2. If we consider $\mathrm{I_c} = \mathrm{I_e} = 0$ then it will reduce to a model without permissible delay in payment.

## 5 Solution Procedure

In this section a mathematical solution procedure of the model with the support of an example has been provided. The solution procedure is as follows:

Step1: Insert the values of the parameters of Ic, Ie, $\mathrm{t_d}$, a, $\theta$, k, $\mathrm{w}_1$, $\mathrm{w}_2$, Ts, c, h, f, r, p.

Step2: Solve $\frac{\partial NP}{\partial S} = 0$ *and* $\frac{\partial NP}{\partial T} = 0$ from expressions (16) and (18) to find the optimal value of (S, T), let it be $(S^*, T^*)$.

Step 3: For $(S^*, T^*)$. calculate the values of $\mathrm{t}_2$, $\mathrm{t}_3$, Y from expressions (5), (6) and (7) respectively and NP from 15.

Step 4: Check the condition of assumption 4, if it is satisfied, process end and if it is not satisfied change the values of parameters and go to step 1.

**Example 1.** Consider the values of a = 10,000, θ = 1000 deterioration rate per consumption cycle, k = 15,330 g per item per unit time, $w_1 = 84$ (g), $w_2 = 1260$ (g), $T_s =$ 0.01 year, c = \$0.3 per unit weight, h = \$0.4 per unit weight per unit time, f = \$2 per unit weight per unit time, $O_c =$ \$1000 per unit weight per unit time, r = \$0.8 per unit weight per unit time, p = \$2 per unit weight, $I_e =$ \$0.04 per year, $I_c =$ \$0.06 per year.

**Result:** From the solution procedure, the optimal values of $(S^*, T^*)$. are $S^* = 229.19$ and $T^* = 1.17335$. By using expression (6) and (7) and the relation $T = T_1 + T_2$, the optimal values of $T_3 = 0.222919$, $T_2 = 950431$ and $X = 10.0666$. The optimal value of the Net profit is $NP = 15855.20$. $\frac{\partial^2 NP}{\partial T^2} = -3038.25$.

Now we will carry out the sensitivity analysis of the numerical example.

## 6 Sensitivity Analysis

Table 1 shows the effect of change in NP in terms of percent loss occurred due to change in the decision variables and parameters.

From Table 1 we can observe that, the increase in the value of T results in a decrease in the value of net profit, but it is more sensitive to the negative change as compare to the positive changes for instance −75% change in T profit function shows loss of 25.07% while for +75% change in T net profit shows loss of 4.35%. The net profit shows % loss with the positive and negative percent change in the value of S. Also increase in p increases the net profit, the small change in p results in the significant change in net profit. The net profit increases/decreases with decrease/increase in θ, $w_1$ and k. Selling price is the most sensitive parameter therefore the accurate value of p is of great importance.

**Table 1** Percent loss in NP with respect to change in parameters and decision variable

| % change | Variable | | | | | |
|---|---|---|---|---|---|---|
| | S | T | p | θ | $w_1$ | k |
| −100 | 2.9833 | 214.9525 | 121.1233 | 1.184 | −0.568 | −1.833 |
| −75 | 1.6781 | 25.0698 | 90.8424 | −0.888 | −0.428 | −1.375 |
| −50 | 0.7458 | 6.2164 | 60.5616 | −0.592 | −0.287 | −0.916 |
| −25 | 0.1865 | 1.0782 | 30.2808 | −0.296 | −0.144 | −0.458 |
| 0 | 0 | 0 | 0 | 0 | 0 | 0 |
| 25 | 0.1865 | 0.667 | −30.281 | 0.296 | 0.1457 | 0.4582 |
| 50 | 0.7458 | 02.243 | −60.562 | 0.592 | 0.2929 | 0.9164 |
| 75 | 1.6781 | 4.3524 | −90.842 | 0.888 | 0.4415 | 1.3746 |
| 100 | 2.9833 | 6.8016 | 121.123 | 1.184 | 0.5917 | 1.8328 |

## 7 Conclusion

In this paper a new mathematical model was presented for growing items by considering several effective constraints. The model aimed to optimize net profit by optimizing the decision variables, ordered quantity and shortages. Concavity of the profit function with respect to cycle length discussed numerically and with respected shortages discussed analytically. Analytical results show that the profit function is concave with respect to shortage period while concavity with respect to cycle length after growing period is discussed numerically. Sensitivity analysis shows that selling price is the most sensitive parameter. In future model can be extended for the probabilistic demand.

## References

1. Aggarwal SP, Jaggi CK (1995) Ordering policies of deteriorating items under permissible delay in payments. J Oper Res Soc 46(5):658–662
2. Agrawal S, Gupta R, Banerjee S (2020) EOQ model under discounted partial advance—partial trade credit policy with price-dependent demand, optimization and inventory management, pp 219–237
3. Goyal SK (1985) Economic order quantity under conditions of permissible delay in payment. J Oper Res Soc 36:335–338
4. Hidayat YA, Riaventin VN, Jayadi O (2020) Economic order quantity model for growing items with incremental quantity discounts, capacitated storage facility, and limited budget. Jurnal Teknik Industri 22(1):1
5. http://dadf.gov.in/sites/default/filess/BAHS%20%28Basic%20Animal%20Husbandry%20Statistics-2019%29_0.pdf. Accessed 5 Apr 2020
6. http://www.dahd.nic.in/sites/default/filess/Seeking%20Comments%20on%20National%20Action%20Plan-%20Poultry-%202022%20by%2012-12-2017.pdf. Accessed 5 Apr 2020
7. Nobil AH, Sedigh AHA, Cárdenas-Barrón LE (2019) A generalized economic order quantity inventory model with shortage: case study of a poultry farmer. Arab J Sci Eng 44:2653–2663
8. Rezaei J (2014) Economic order quantity for growing items. Int J Prod Econ 155:109–113
9. Richards FJ (1959) A flexible growth function for empirical use. J exp Bot 10(2):290–301
10. Saren S, Sarkar B, Bachar RK (2020) Application of various price-discount policy for deteriorated products and delay-in-payments in an advanced inventory model. Inventions 5(3):50
11. Sebatjane M, Adetunji O (2019) Economic order quantity model for growing items with incremental quantity discounts. J Ind Eng Int 15:545–556
12. Shaikh A, Panda GC, Sahu S, Das AK (2019) Economic order quantity model for deteriorating item with preservation technology in time dependent demand with partial backlogging and trade credit. IJLSM 32:1–24
13. Sharma A, Kaushik J (2020) Inventory model for deteriorating items with ramp type demand under permissible delay in payment. Int J Procurement Manage. https://doi.org/10.1504/IJPM.2020.10033289
14. Taleizadeh AA, Sarkar B, Hasani M (2020) Delayed payment policy in multi-product single-machine economic production quantity model with repair failure and partial backordering. J Ind Manage Optim 16(3):1273
15. Zhang Y, Li LY, Tian XQ, Feng C (2016) Inventory management research for growing items with carbon-constrained. In: 2016 chinese control conference (CCC), vol 35. IEEE, pp 9588–9593

# A Deep Learning Based Approach for Classification of News as Real or Fake


Juhi Kumari[1](✉), Raman Choudhary[2], Swadha Kumari[1], and Gopal Krishna[1]

[1] Department of Computer Science and Engineering, Netaji Subhas Institute of Technology, Amhara, Bihta, Patna, India
juhikumari6928@gmail.com , gopalk.phd20.cs@nitp.ac.in

[2] Department of Computer Science and Engineering, Nalanda College of Engineering, Chandi-Jalalpur Road, Bihar Sharif, Bihar, India
ramanc.phd18.cs@nitp.ac.in



**Abstract.** In recent past, the growth of both the printed and digital media has greatly facilitated the business and the society. On account of the reach of social media, even the smallest news or event could be spread like wildfire. Often due to this, the news gets amplified and distorted drastically resulting in generation of fake news. This fake news not only misleads the masses but also causes severe impacts in real world. The rapid increase in the area of fake news and its abrasion to judiciary, democracy and trust in the public made the development of a system for detection of fake news vital. Here, in this paper, we have dealt with the proposal of a model in order to detect fake news, by the use of deep learning algorithms to predict whether the given data is real or fake. The experiments were executed using various deep learning algorithms like Convolutional Neural Network (CNN), Long short-term memory (LSTM) and Bidirectional LSTMs (Bi-LSTM). Thereby we compared the results obtained. The model proposed here, has a great accuracy of about 99%.



**Keywords:** Fake news detection · Natural language processing · Deep learning


## 1 Introduction

Fake News is a terminology used to depict altered news or agendas consisting of misleading information channeled through the conventional media such as television and print and also unconventional media medium like social media. The mainspring of spreading altered news is to delude the commoners, hamper stature of a syndicate, or to benefit from emotionalism. Fake news is abundantly being shared and can be prominently seen on internet communities such as Twitter and Facebook. These platforms act as a general medium for the commoners to express and channelize their views and opinions in an unprocessed and unedited way.



S. Shukla et al. (eds.), *Data Science and Security*, Lecture Notes in Networks and Systems 290,
https://doi.org/10.1007/978-981-16-4486-3_26

One of the sectors majorly influenced by this new model is the information industry. This is when the concept of fake news arises. The big technological conglomerates (Google, Facebook, Twitter) have sensed this issue and have already started to work on systems that can detect the fake news on their platforms. Even though, this method evolves rapidly, this is a very complicated situation that needs further analysis. The main motive of this work is to acquire several models based on deep neural networks aided for detection and classification of the fake news so that the people can have some indicator for the degree of trust of the information they ingest and thus avert, to such extent possible, bias and misconceptions.

Technologies such as Natural Language Processing (NLP) Artificial Intelligence (AI) tools are a great resource for researchers for the establishment of automated system having the self-sufficient capability to sense and classify fake news. Though, detection of altered news is an exigent task to achieve as it demands model for summarization of news, then further comparison with the authenticated news to be able to classify it as fake. Also, the notion of comparison of proposed news with the actual news is an intimidating task as its highly referential.

The primary objective of the model proposed here is to detect the fake news which further in addition to the algorithms of the media platforms would increase trust in society. Herein, thorough experimentation was carried out using several deep neural networks models and then results of performance of models was compared. Through experimental procedures, a model was proposed which detects fake news and have proved to obtain a very high accuracy. Different models are studied which affected the efficiency of the model and the details for future work have been summarized.

The paper is presented in various segments. Here, Sect. 2 contains of the work that have already been achieved for the fake news detection. Section 3 describes the approach and the model proposed. The different algorithms used and the different techniques used for preprocessing and improving accuracy is pointed out by Sect. 4. It also beholds the comparison of result and accuracy obtained by all the different approaches and a detailed analysis of the model that has the best accuracy. Lastly, Sect. 5 concludes the paper further discussing the future scope of the model.

## 2 Related Work

The problem of detecting fake news is much more arduous than that of the deceptive reviews. The wide circulation of fake news causes negative impact on masses and they can even influence public events.

Generous amount of research has been done for the feature analysis for fake news. Jin et al. in their work performed rigorous analysis of images of news articles using multimedia datasets basically for fake news detection. Herein, they also explored through innumerable statistical and visual images for the prediction of realism of the articles. Besides, they also proposed a method so as to detect altered news using the credibility propagation network fabricated by using conflicting viewpoints which were extracted from the tweets [1].

The early identification of fake news was done by Yang et al. by proposing a productive model via the paths of news propagation utilizing the multivariate time series. They put forth a new model of deep learning, consisting of four vital components, namely, propagation path representation based on CNN, that based on RNN, construction and transformation of propagation path and lastly, the classification of propagation path. These collaborated with each other for the detection of fake news at its very early stage [2].

Recently, surface-level linguistic patterns have proved to be prevalent for the identification of fake news. It incorporated of classifiers to detect if the tweets are factual or altered [3,4].

Ruchansky et al. in their work came up with aa model comprising of three vital parts; one module was responsible for learning, on the basis of the behaviour of users,the source characteristic; another module was based on Recurrent Neural Network (RNN) and captured the transitory pattern based on the activity of users over different articles; these two modules were brought together with the third one for an effective classification of an article into fake or not [5].

Another fake news detection model was proposed by Wang et al. using Event Adversarial Neural Network (EANN), basically comprising of three major components: one was the fake news detector, other was feature extractor for multi-modals and lastly, the event differentiator. Herein, the different textual and visual features were drawn out from the social media posts by the multi-modal feature extractor [6]. Hardalov et al., in their work, used linguistic features, different semantic features and credibility. These incorporated normalized quantity of distict words and n-grams per article [7].

The different credibility features consists of features like pronoun usage, capitalization, sentiment polarity and punctuation features which were obtained via lexicons. Embedding vectors methods were used to analyse text semantics [8,9].

The objective of the work here, is to build a primarily new solution for addressing the detection of fake news based on the textual content of the news.

## 3 Proposed Methodology

A real and fake news dataset was obtained from Kaggle which contained two csv files, fake and true. The dataset contains several features like id, title, author, text. In this paper, however, only the titles and contents were used.

The flow of the model proposed is shown below in Fig. 1, wherein, firstly, both the dataset had to be merged keeping different labels for fake and real. Then, the dataset was pre-processed with the use of various algorithms. Tokenization, glove embedding were done for better and improved accuracy. Then, different models based on deep learning, like CNN, LSTM and Bi-LSTM, were used. Thereafter, a comparative study was done to figure out the model and approach that proves to be the best with its own combinations of hyper parameters for the dataset.

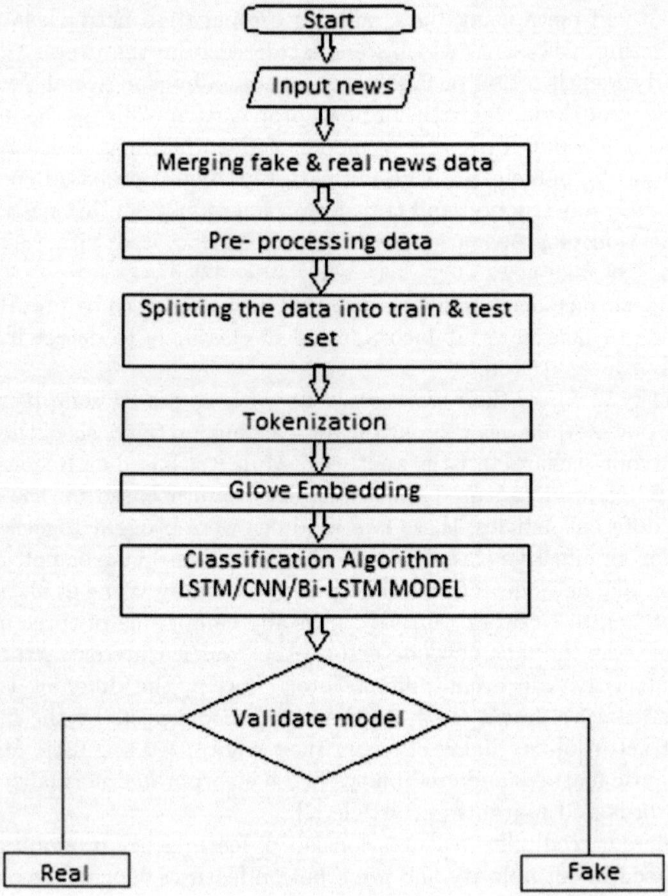


**Fig. 1** Proposed model

# 4 Experiment Analysis

## 4.1 Data Pre-processing

Pre-processing of data is a data mining technique that involves transformation of raw data into easily understandable format. Though, data based on actuality is frequently inconsistent, insufficient and/or missing specific traits/functionalities as well as errors in such circumstances are probable. There are various steps involved in pre-processing of data, as:

**Step 1:** Importing the required libraries - Different libraries such as numpy, pandas, seaborn, matplotlib, nltk, beautifulsoup, re, keras, os were imported.
**Step 2:** Dataset Importation - Textual datasets are present in Comma-Separated Value(.csv) format. In CSV files, tabular data is stored in plain text style. Each line of a Csv file represents a data record. A local Csv file is referred to as the data

frame. Our dataset comprises of 2 csv files, fake and true, and then downloaded and merged into one. Before merging datasets into one, there was a need to label them. In data1 dataframe fake.csv file was kept. In data2 dataframe true.csv file was kept. 0 was used for True and 1 for False. To do that, a column was added into each Dataset called label and assign the value as 0 for real and 1 for fake.
**Step 3:** Data Cleaning - Before use-casing the data by the use of deep neural networks model, data must be rectified using specific algorithms such as html parsing, square brackets removal, URL's removal, removal of stop words, tokenization, conversion to lower case, and removal of punctuation. This will help us in refining the data in a way, that only the relevant information of the data stays present and everything else is eliminated thus helping in reduction in the size of data [10].
**Step 4:** Splitting of the data into test and training sets - The data was segmented into separate training dataset and testing dataset. The training dataset comprises of a known output and the model learns on the basis of the provided data in order to generalize and classify the data later on. The text dataset then is used for testing the model' prediction on this subset. For initialization of the internal random number generator, the random state parameter is used, which will then be used for deciding and splitting the dataset into train and test subsets. It just makes sure that the same split is obtained every time on running the code [11].
**Step 5:** Tokenize the data - The data was tokenized to transform words into numbers.
**Step 6:** Word embedding - For obtaining vector representation for words 300-dimensional GloVe embeddings were used. The calculation of index mapping words was done to acknowledged embeddings. The embedding index vocabulary were leveraged and index of words were used to determine embedding matrix. The embedded matrix was then loaded into an embedding layer [12,13].

### 4.2 Model Analysis

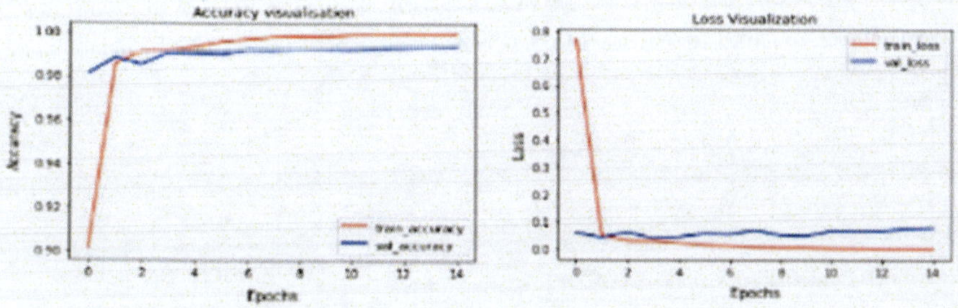


(a) Epoch vs. Accuracy (b) Epoch vs. Loss

**Fig. 2** CNN

**CNN Model.** CNN Model Batch size was kept equal to 128, the number of epochs is 15 and the embed size is 300. Models usually gain from the depreciation of learning rate by the factor of 2–10 once per learning stagnates. The CNN model was built and was compiled using Binary Cross-Entropy loss function and Adam Optimizer. Then a call back was added to improve the time of the train from our Model. 99.95% accuracy is obtained in 15 epochs. Figure 2 explains the graph of Epoch vs. Accuracy and Epoch vs. Loss of the CNN model.

**LSTM Model.** LSTM model is made up of numerous LSTM layers. A single LSTM layer above acts as a sequential output rather than a single value output to the LSTM layer. The LSTM Model was implemented. The model was executed by the use of optimizer function- Adam and the loss function here was Binary Cross-Entropy. 99.96% accuracy was obtained in 15 epochs. Figure 3 explains the graph of Epoch vs. Accuracy and Epoch vs. Loss of the LSTM model.

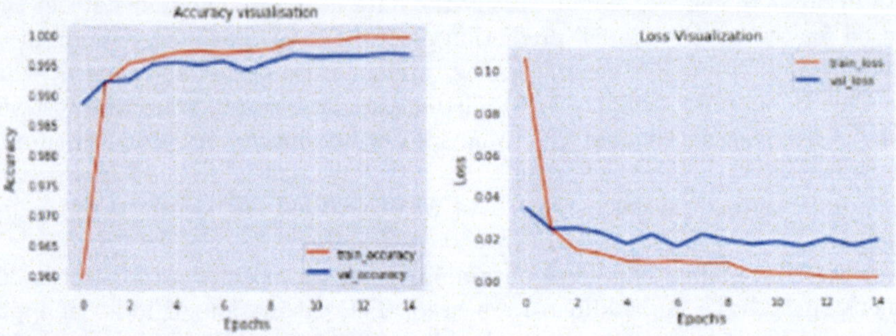


(a) Epoch vs. Accuracy (b) Epoch vs. Loss

**Fig. 3** LSTM

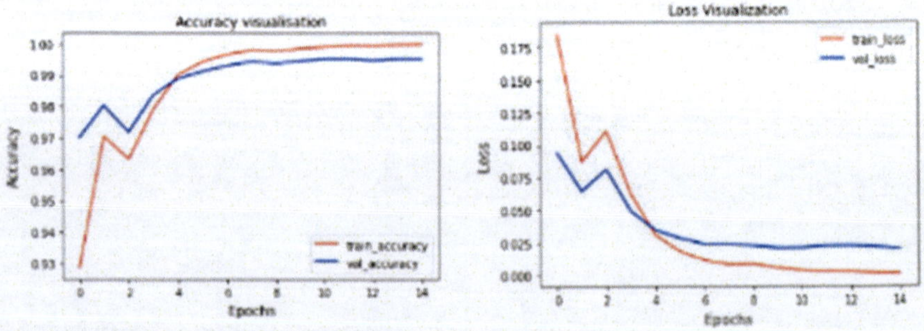


(a) Epoch vs. Accuracy (b) Epoch vs. Loss

**Fig. 4** Bi-LSTM

**Table 1** Performance of models

| Model | Epochs | Training accuracy | Testing Accuracy | F1-Score | Precision | Recall |
|---|---|---|---|---|---|---|
| CNN | 15 | 99.99 | 99.36 | 0.99 | 0.99 | 0.99 |
| LSTM | 15 | 99.99 | 99.67 | 1 | 1 | 1 |
| Bi-LSTM | 15 | 99.99 | 99.52 | 1 | 1 | 1 |

**Bi-LSTM.** Bi-LSTM Bidirectional LSTMs are supported in Keras library via the use of Bidirectional layer wrapper. This wrapper uses the recurrent layer as an argument. The Bi-LSTM Model was built. Here again, the model execution took place Binary Cross-Entropy as loss function and Adam optimizer. A call back was added to improve the time of the train from our Model. 99.99% accuracy was obtained in 15 epochs. Figure 4 explains the graph of Epoch vs. Accuracy and Epoch vs. Loss of the Bi-LSTM model.

### 4.3 Experimental Result

Now, the results obtained by each one of the three trained models are compared (Table 1).

The table shown above proves that the majority baseline on this dataset gives about 99% accuracy on the training and test sets respectively. The LSTM surmounted all models, showing an accuracy of 99.67% on the test set. However, CNN and BILSTM determined noticeably good results too: CNN reached 99.36% accuracy on the test set and BILSTM achieved 99.52% accuracy on test set. Overall, the utmost accuracy was reached using LSTM as 99.67%.

## 5 Conclusion and Future Aspects

The paper proposes model for detection of fake news using long short-term memory algorithm. Various other algorithms like convolution neural network and bidirectional long short-term memory were also experimented on the dataset. Epoch vs. accuracy and epoch vs. loss graphs were obtained for each algorithms. During the compared analysis, long short-term memory algorithm proved to be able to provide the best accuracy. Owing to this, it is supposable to train highly effective neural networks using textual features. Moreover, revolutionary and intricate results could be achieved using the proposed strategies. Gaining the experience in the process of development of the model, permits to affirm that this task would be highly beneficial for the society and the masses with the motive to eradicate the fake news on the media. The collection of more data in reference to a particular period of time is important when concerning about the future improvisation of the model. For the masses to actually use these models, certain methods, like integrating the model with media platforms, mobile applications and browser extensions, are very much required. The model proposed in this paper, has proven itself to have a greater accuracy and therefore, if put to use would be much fruitful for the society.

## References


1. Jin Z, Cao J, Zhang Y, Zhou J, Tian Q (2016) Novel visual and statistical image features for microblogs news verification. IEEE Trans. Multimedia 19(3):598–608
2. Liu Y, Wu YF (2018) Early detection of fake news on social media through propagation path classification with recurrent and convolutional networks. In: Proceedings of the AAAI Conference on Artificial Intelligence, vol 32, no 1
3. Zhou X, Reza Z, Kai S, Huan L (2019) Fake news: fundamental theories, detection strategies and challenges. In: Proceedings of the Twelfth ACM International Conference on Web Search and Data Mining, pp 836–837
4. Kaliyar RK, Goswami A, Narang P, Sinha S (2020) FNDNeta deep convolutional neural network for fake news detection. Cogn Syst Res 61:32–44
5. Natali R, Seo S, Liu Y (2017) Csi: a hybrid deep model for fake news detection. In: Proceedings of the 2017 ACM on Conference on Information and Knowledge Management, pp 797–806
6. Wang Y, et al (2018) Eann: event adversarial neural networks for multi-modal fake news detection. In: Proceedings of the 24th ACM Sigkdd International Conference on Knowledge Discovery Data Mining, pp 849–857
7. Momchil H, Koychev I, Nakov P (2016) In search of credible news. In: International Conference on Artificial Intelligence: Methodology, Systems, and Applications, pp 172–180. Springer, Cham
8. Ma J, Gao W, Wei Z, Lu Y, Wong KF (2015) Detect rumors using time series of social context information on microblogging websites. In: Proceedings of the 24th ACM International on Conference on Information and Knowledge Management, pp 1751–1754
9. Singhal S, Kabra A, Sharma M, Shah RR, Chakraborty T, Kumaraguru P (2020) Spotfake+: a multimodal framework for fake news detection via transfer learning (student abstract). In: Proceedings of the AAAI Conference on Artificial Intelligence, vol 34, no 10, pp 13915–13916
10. Ciampaglia GL, Shiralkar P, Rocha LM, Bollen J, Menczer F, Flammini A (2015) Computational fact checking from knowledge networks. PloS one 10(6):e0128193
11. Conroy NK, Rubin VL, Chen Y (2015) Automatic deception detection: methods for finding fake news. Proc Assoc Inf Sci Technol 52(1):1–4
12. Joulin A, Edouard G, Piotr B, Tomas M (2016) Bag of tricks for efficient text classification. arXiv preprint arXiv:1607.01759
13. Ferreira W, Andreas V (2016) Emergent: a novel data-set for stance classification. In: Proceedings of the 2016 Conference of the North American Chapter of the Association for Computational Linguistics: Human Language Technologies, pp 1163–1168

# User Authentication with Graphical Passwords using Hybrid Images and Hash Function

Sachin Davis Mundassery(✉) and Sreeja Cherillath Sukumaran

Department of Computer Science, CHRIST (Deemed to be) University, Bangalore, India
sachin.mundassery@mca.christuniversity.in,
sreeja.cs@christuniversity.in

**Abstract.** As per human psychology, people remember visual objects more than texts. Although many user authentication mechanisms are based on text passwords, biometric characteristics, tokens, etc., image passwords have proven to be a substitute due to its ease of use and reliability. The technological advancements and evolutions in authentication mechanisms brought greater convenience but increased the probability of exposing passwords through various attacks like shoulder-surfing, dictionary, key-logger, and social engineering attacks. The proposed methodology addresses these vulnerabilities and ensures to keep up the usability of graphical passwords. The system displays hybrid images that users need to recognize and type the randomly generated alphanumeric or special character values associated with each of them. A mechanism to generate One Time Password (OTP) is included for additional security. As a result, it is difficult for an attacker to capture and misuse the password.

**Keywords:** Graphical passwords · User authentication · Hybrid images · Hash · Security · Shoulder-surfing attack · Key-logger attack · Brute force attack

## 1 Introduction

In security, authentication is the process of recognizing a user's identity. User authentication with alphanumeric passwords was user friendly and easy to remember. But the problem with these forms of authentication mechanisms occurs when people use weak passwords or use the same password in other authentication systems or write it down somewhere. If the password is leaked, the attacker can log in to every other system with the same credentials. Users tend to take simple and easy to remember passwords, as strong passwords are hard to remember [1].

The major threats that text passwords encounter are shoulder surfing, brute force, key-logger, screen-capture and dictionary attacks. User authentication using image passwords was introduced to encounter these flaws of text-based passwords. The graphical passwords became famous due to their usability and reliability. It uses images for authentication, inspired by the fact that human beings remember visual objects more quickly than strings of characters [2]. Greg Blonder [3] proposed this novel technique of authentication which is more user friendly and an increased level of security compared to textual passwords.


S. Shukla et al. (eds.), *Data Science and Security*, Lecture Notes in Networks and Systems 290,
https://doi.org/10.1007/978-981-16-4486-3_27

Generally, graphical passwords belong to either recognition or recall based approach. For recognition, different images are presented to the user. The user has to recognize the right images that he or she selected during the registration process. Whereas in recall-based technique, a user is asked to reproduce something in the correct sequence of what they chose or created during the registration process [2].

Although numerous image-based password authentication schemes were developed to overcome the shortcomings of textual passwords, most of them were vulnerable to the attacks faced by text passwords. Authentication schemes that claim to be safe have usability limitations or small password space. Most of the existing authentication mechanisms using graphical passwords aim at preventing one major attack, for example, shoulder-surfing attack. The attacker could easily gain access to the system by some other means.

The paper is organized as Sect. 2 describes about the literature review. Section 3 is about the proposed system and the algorithm. The working example is explained in Sect. 4. Section 5 presents the security analysis. Section 6 concludes the paper with future work. The references are mentioned in references section.

## 2 Literature Review

Researches conducted on graphical passwords' security engrossed on either specific schemes or concrete attacks. Even though various schemes attempt to overcome existing image-based user authentication threats, many still face issues like shoulder-surfing, brute force, key-logger, screen-capture attacks, etc. Encountering all these issues stays as a challenge to a researcher.

A shoulder-surfing resistant graphical password scheme using hybrid images was proposed by Basak et al. [4]. The proposed method is an innovative idea where the password depends on the image's perception. But it has a time-consuming registration phase because users need to type a story for the selection made.

Wiedenbeck et al. [8] proposed Pass-points that allow users to use any image and click on the desired position to create their password. However, it is difficult for users to ensure accurate clicks within tolerance, which is time-consuming. Since the attacker can see the points that the user clicked directly, it is also vulnerable to shoulder surfing.

In [10], users select several options for pass-objects variant, each assigned a unique code. During the authentication phase, the user must provide a unique pass-objects variant code. Also, code must be provided that indicates the relative position of pass-objects that will travel around the pair of eyes during the system's many rounds. While the proposed scheme is resistant to shoulder navigation, there is a fair amount of code to consider with these pass-objects variants.

As an extension of DAS, [11] proposed a method BDAS that uses background images to reduce the predictable characteristics like passwords tending to the centre of the grid, symmetry and helping users to have stronger passwords. But the dependability of the user's drawings on such background images was not mentioned.

In [12], authors proposed a Hash Visualization technique Déjà Vu, a graphical password scheme that needs users to create image portfolios and stores the seeds of portfolio images in the database. The weakness is that it needs to store them as clear text and the process is time-consuming as the user undergoes a training phase where he or she needs to identify his or her portfolio from a set of decoy images.

Ahmad Almuhem [13] proposed a scheme where a user selects an image of their choice and then chooses several point-of-interest (POI) regions on the image and type in their associated words or phrase. The system faces issues like the user have to remember all those words or phrases. Also, the procedure is time-consuming as typing of all those words takes time.

A hybrid system combining recognition and recall-based schemes was proposed by Wazir et al. [14], in which users need to type their username and textual password in the registration phase. Then the user needs to select a minimum of 3 objects from the displayed set, and he or she needs to draw them, which will be saved in the database. In the authentication phase, the user has to draw the pre-selected objects after giving his username and password. If the drawn sketch completely coordinates with templates of objects stored in the database, it will authenticate the person. The drawbacks of the proposed system are that it is time-consuming and challenging for the user to draw a particular object with a stylus or mouse and get verified. This proposed item is also prone to shoulder surfing attack, as the objects drawn by the user are visible.

The shoulder surf resistance scheme was developed by Roth et al. [15] using a PIN that increases noise for intruders. PINs are displayed randomly in black or white for each round. The user's answer depends on the color of each digit of the password. After a series of binary options, the system determines the user's PIN to be entered and crosses the user's settings in the process. This confuses the attacker when looking at the screen without a video capture device. However, if an attacker managed to capture the entire loop, they could obtain the password.

T. Takada proposed a scheme called FakePointer [16] in which users are given a unique secret consisting of a PIN and a response indicator. Each time the user logs in, the user is presented with an image of the 10-digit number pad. A randomly selected shape is displayed under each number. The user must rotate the numbers in a circle with the arrow keys to the left or right. The process is repeated until the first digit of the PIN matches the keypad response indicator's first form. For successful authentication, the process is repeated until all PIN digits have been entered and verified.

Misbah et al. [17], proposed a hybrid authentication scheme using image and text passwords that is resistant to shoulder-surfing attack. The proposed scheme makes use of a $6 \times 6$ grid of images that represents the English alphabet. After entering his username, the user builds the desired text password by selecting images whose first alphabet would be the letters of the user's text password.

Taking the advantage of user authentication using graphical passwords, the proposed method addresses the above stated vulnerabilities and ensures usability and security.

## 3 Proposed System

Based on the study of existing authentication systems, hybrid images found to ensure better secure. Considering this the proposed approach uses hybrid images. These images change their interpretation depending on their viewing distance. It combines the low spatial frequencies of one picture with another picture's high spatial frequencies, producing an image that can be perceived in two different ways. The proposed system also uses the bcrypt library of Node Package Manager (NPM) to get the key's hash value during the registration. For additional security, the system incorporates two-factor authentication where an OTP will be sent to the user's registered mobile number for each new login. The system ensures an approach towards a more secure, robust, reliable, and user-friendly authentication mechanism. The hybrid images and the alphanumeric or special character values associated with them change for each login attempt, thus preventing various attacks like key-logger, shoulder-surfing, and dictionary attacks.

## 3.1 Algorithm

The following flowchart demonstrates the flow of the proposed scheme (Figs. 1 and 2).

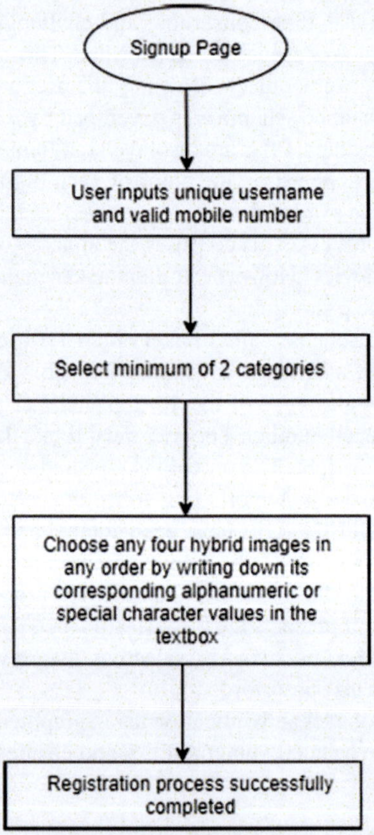


**Fig. 1** Registration process

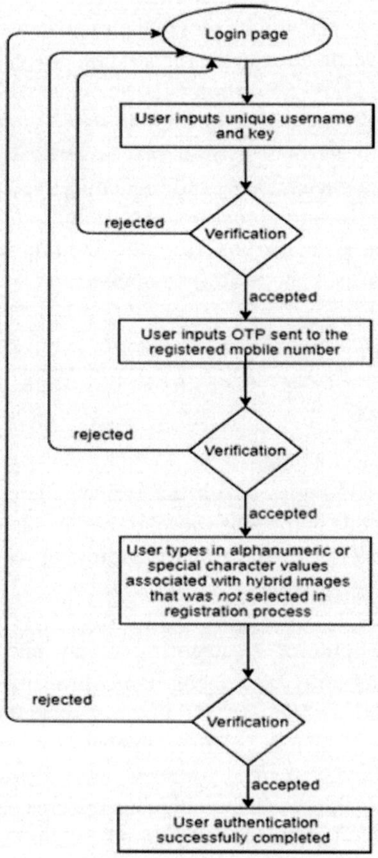


**Fig. 2** Login process

### 3.1.1 Registration Process

Step 1: User need to provide a unique username and a registered mobile number. Also, select the option whether he or she requires an extra OTP verification for additional security.

Step 2: Select a minimum of 2 categories (example: animals, birds, etc.) from the 5 categories displayed. Each category has 5 hybrid images.

Step 3: Now users should select 4 hybrid images from the displayed set of hybrid images by typing its corresponding alphanumeric or special character values in the text box which will be their unique key for the login process.

Step 4: With this the user has successfully completed his or her registration process.

### 3.1.2 Login Process

Step 1: Users need to give their unique username and key. If failed, he or she will be redirected back to the login page.

Step 2: If successful, an OTP will be sent to their registered mobile number for those users who opted for OTP verification in the registration process. Here, users need to give that OTP in the textbox. If OTP verification fails, he or she will be redirected back to the login page. For other users, they will be directed to step 3 without going through step 2.

Step 3: Now from the set of 9 hybrid images vertically aligned and associated alphanumeric or special character users need to identify those hybrid images that he or she did not select in the registration process and type in its corresponding alphanumeric or special character values in the textbox, from the first images to last sequentially.

Step 4: With this the user is successfully authenticated.

# 4 Working Example

## 4.1 Registration Process

The registration process has four steps.

Step 1: When new users register with the system, they need to provide a unique username, a registered mobile number as shown in Fig. 3. Also, users can select whether he or she requires an extra OTP verification for additional security after clicking on the signup button.

Step 2: The system displays different categories of hybrid images from which the users should select a minimum of 2 categories of their choice as depicted in Fig. 4. Now the system creates an array of objects: five hybrid images belonging to each of the selected categories and random alphanumeric or special character values for each hybrid image. The order in which the random alphanumeric or special character values and hybrid images are arranged changes for each new registration.

Step 3: Now, five hybrid images from each of the selected categories will be displayed along with random alphanumeric or special character values associated with them as in

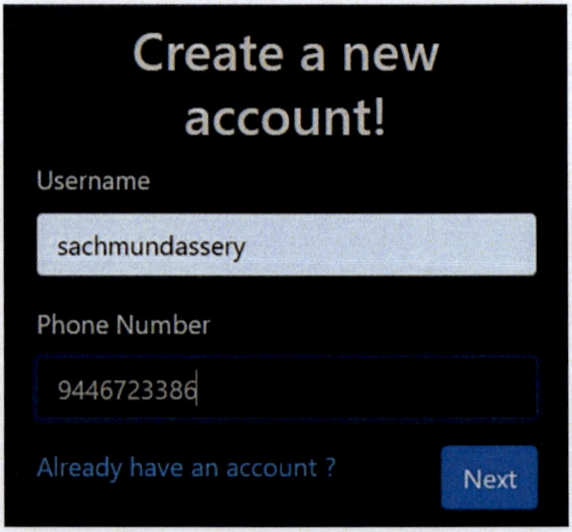


Fig. 3 Username and phone number

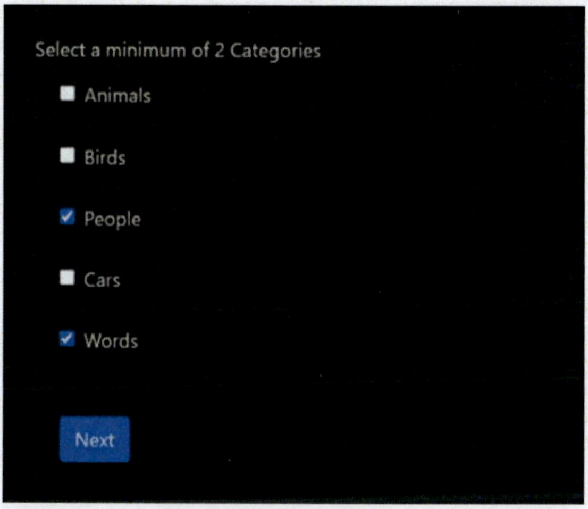


**Fig. 4** Category selection

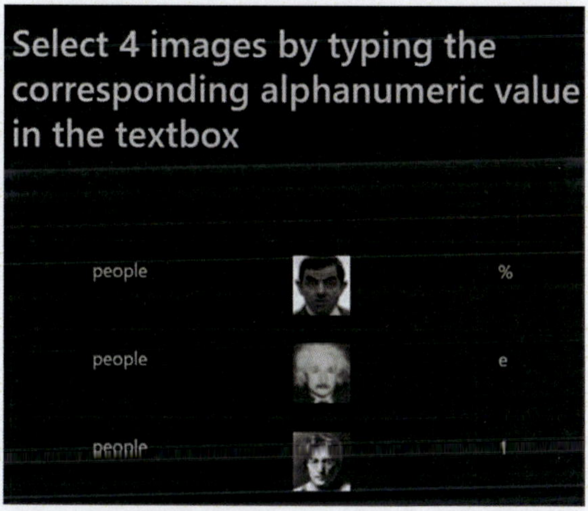


**Fig. 5** Images from selected categories

Fig. 5. Users need to choose any four hybrid images from the displayed set of hybrid images in any order by writing down its corresponding alphanumeric or special character values in the textbox, Fig. 6, which will be their unique key that they should provide to log in.

Step 4: With this the user has successfully completed his or her registration process. The username, mobile number, categories and the corresponding images selected are stored in the database. The generated unique key is hashed using bcrypt library of Node Package Manager (NPM) and stored in the database.

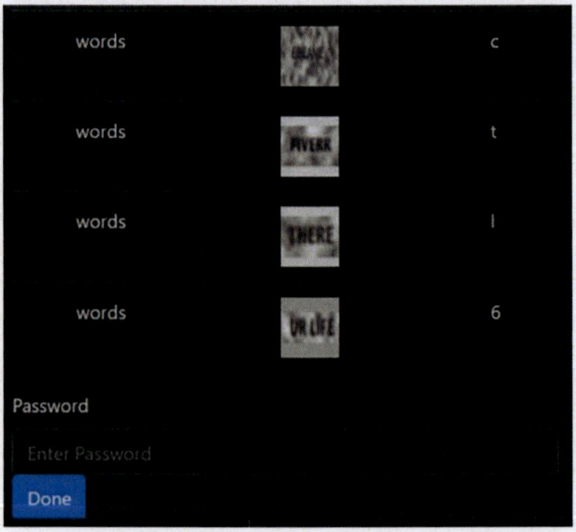


**Fig. 6** Textbox to type the alphanumeric values associated with the selected hybrid images

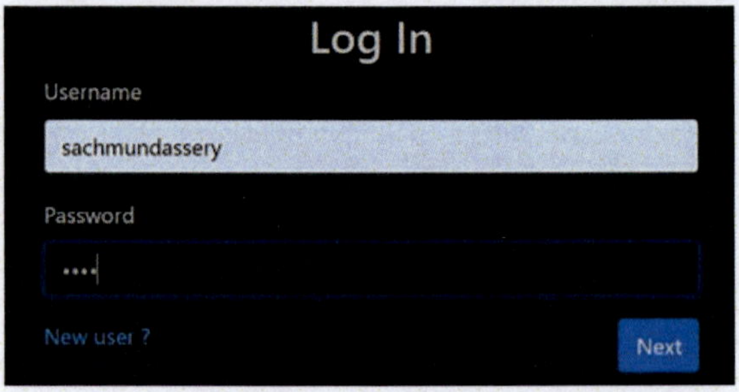


**Fig. 7** Username and key

## 4.2 Login Process

The login process has five steps.

Step 1: On clicking the sign in button users enter into step 1 of the login process. Users need to provide their unique username and key that they generated in step three of the registration process as shown in Fig. 7. The system checks for a user with that username and compares the hash value of the key that the user typed in with the one saved inside the database under that particular user. If the two matches, the system will identify the user trying to log in and move on to step 2 of the login process. Otherwise the user will be redirected back to the login page.

Step 2: After successful completion of the step 1, the system has identified the user who is trying to login. Now the system sends an OTP to the registered mobile number, for those users who selected for additional OTP security. Here, users need to provide that OTP in the textbox as in Fig. 8. The system verifies the OTP. If a user fails in this

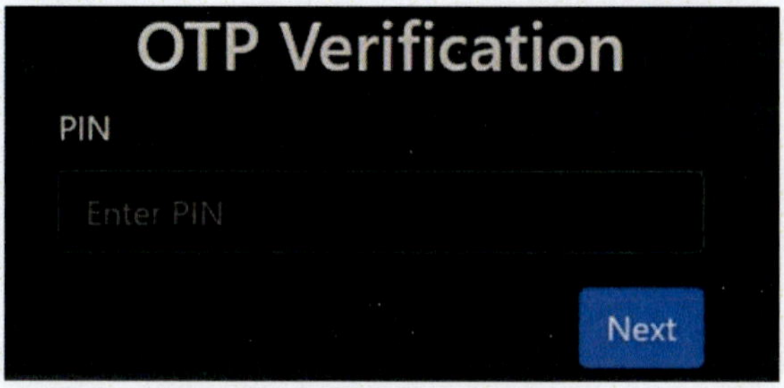


**Fig. 8** OTP verification

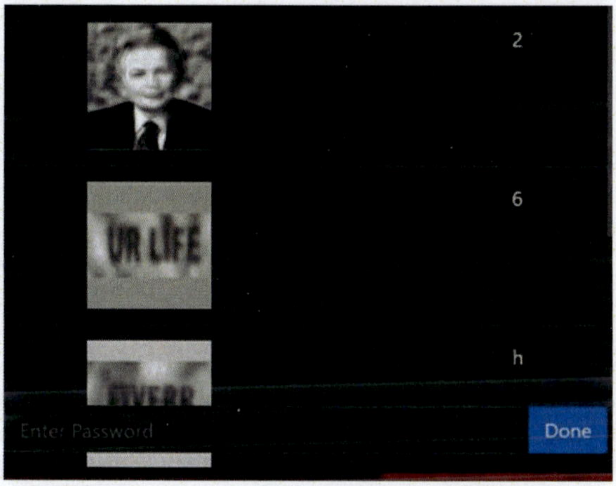


**Fig. 9** Recognizing the image and form the password

step, he or she will be redirected back to the login page, else to step 3. For users who did not select the option for OTP will be directly taken to step 3.

Step 3: Now an array of objects is created. This array of objects contains nine objects, which are hybrid images taken from the database and random alphanumeric or special character values, everything arranged randomly. The hybrid images set contains all the four hybrid images that the user had selected in the registration process and five random hybrid images.

Now the system runs a loop through this array of objects and appends the alphanumeric or special character values associated with the hybrid images into an empty temporary string. The appended alphanumeric or special character values belong to those hybrid images that the user did not select during the registration process.

Step 4: At the user interface, that set of 9 hybrid images vertically aligned with associated alphanumeric or special character values and a textbox will be displayed. The users need to identify those hybrid images that they did not select in the registration process and type in their corresponding alphanumeric or special character values, from the first image to the last sequentially, inside the text box, Fig. 9.

Step 5: The response from the user is compared with that temporarily stored string. If it matches, the user is successfully authenticated. Else the user is redirected to the login page.

## 5 Security Analysis

The proposed methodology addresses the existing vulnerabilities of using image passwords for user authentication. The usage of hybrid images helps to reduce shoulder-surfing attack to a very good extent. The image displayed will be identified differently by the user and the imposter. Since the users type the alphanumeric or special character values associated with those hybrid images that they did not select during registration, which change for each new login, it will be a complex task for the attacker to capture the password. This helps in minimizing shoulder-surfing attacks.

In brute force attack, the attackers guess all possible combinations of passwords hoping that eventually a combination would be correct. Since the password that the user types in step 3 of the login process changes for each new login, it is difficult to carry on a brute force attack. The images shown, and their corresponding alphanumeric or special character values change every time when the user tries to log in. So, to have a brute force attack is a difficult task. The password is also hashed and then compared with the hashed password that is there with the database, which enhances the security.

A dictionary attack is an attack where the attacker tries many passwords, such as words in a dictionary or previously used passwords. The usage of random alphanumeric or special character values helps to prevent this attack. In step 3 of the login process algorithm, users type the alphanumeric or special character values associated with the hybrid images that the user did not select in the registration process. Also, these associated values change for every login process making them unique. This makes it difficult for the attacker to guess passwords. The system also addresses key logger attacks which records information that a user identifies on a website or app and sends back to a third party. As the user recorded references change with each new authentication method, it is difficult for an attacker to trace them.

## 6 Conclusion and Future Work

Today, achieving information security even with the most reliable, user friendly and secure systems are rising as a major challenge. Although many authentication systems are available, each has its own pros and cons. The proposed scheme based on recognition with the usage of hybrid images, is useful in overcoming major attacks of using image passwords like shoulder surfing attack, dictionary attack, key – logger attack to a very good extent and ensures usability. Randomizing the hybrid images and its associated alphanumeric or special character values helped in enhancing the security of the proposed technique without compromising its usability and reliability.

The future work will address the security and storage of the images used for authentication. Perceptual hashing could be applied to multimedia content identification, retrieval, authentication, etc. It needs to store these images as image hash. Also randomize the combination of images that form hybrid images.

## References


1. Sood S, Sarje A, Singh K (Dec 2009) Cryptanalysis of password authentication schemes: current status and key issues. In: Methods and Models in Computer Science. ICM2CS 2009. Proceeding of International Conference on, pp 1–7
2. Paivio A, Rogers T, Smythe P (1968) Why are pictures easier to recall than words? Psychono Sci 11(4):137–138
3. Blonder GE (1996) Graphical password. US Patent 5,559,961, 24 Sept 1996
4. Bilgi B, Tugrul B (2018) A shoulder-surfing resistant graphical authentication method. In: International conference on artificial intelligence and data processing (IDAP). IEEE, pp 1–4
5. Yeung ALC, Wai BLW, Fung CH, Mughal F, Iranmanesh V (2015) Graphical password: shoulder-surfing resistant using falsification. In: 9th Malaysian software engineering conference (MySEC), pp 145–148
6. Jermyn I, Mayer A, Monrose F, Reiter M, Rubin A (Aug 1999) The design and analysis of graphical passwords. In: 8th USENIX security symposium
7. Thorpe J, van Oorschot PC (Aug 2004) Towards secure design choices for implementing graphical passwords. In: USENIX security symposium. IEEE, pp 50–60
8. Wiedenbeck S, Waters J, Birget J, Brodskiy A, Memon N (2005) PassPoints: design and longitudinal evaluation of a graphical password system. Int J Hum Comput Stud 63(1–2):102–127
9. Sacha Brostoff M (2000) Angela sasse: are passfaces more usable than passwords? a field trial investigation. Springer, London
10. Man S, Hong D, Mathews M (2003) A shoulder surfing resistant graphical password scheme. In: Proceedings of international conference on security and management. Las Vergas, NV, pp 105–111
11. Paul D, Yan J (2007) Do background images improve draw a secret graphical password? In: Proceedings of the 14th ACM conference on computer and communications security (CCS), pp 36–47
12. Dhamija R, Perrig A (2000) Déjà vu: a user study using images for authentication. In: Proceedings of 9th USENIX security symposium, vol 9, p 4
13. Almulhem A (2011) A graphical password authentication system. In: World congress on internet security (WorldCIS)
14. Khan WZ, Xiang Y, Aalsalem MY, Arshad Q (2011) A hybrid graphical password based system. In: ICA3PP 2011
15. Roth V, Richter K, Freidinger R (2004) A pin-entry method resilient against shoulder surfing. In: Proceedings of the 11th ACM conference on computer and communications security, ser. CCS 2004, pp 236–245. ACM, New York
16. Takada T (2008) FakePointer: an authentication scheme for improving security against peeping attacks using video cameras. In: Mobile ubiquitous computing, systems, services and technologies, 2008. UBICOMM 2008. The second international conference on. IEEE, pp 395–400
17. Siddiqui MU, Umar MS, Siddiqui M (2018) A novel shoulder-surfing resistant graphical authentication scheme. In: 4th international conference on computing communication and automation (ICCCA)

# UAS Cyber Security Hazards Analysis and Approach to Qualitative Assessment

Yuliya Averyanova[1](✉), Olha Sushchenko[1], Ivan Ostroumov[1], Nataliia Kuzmenko[1], Maksym Zaliskyi[1], Oleksandr Solomentsev[1], Borys Kuznetsov[2], Tatyana Nikitina[3], Olena Havrylenko[4], Anatoliy Popov[4], Valerii Volosyuk[4], Oleksandr Shmatko[4], Nikolay Ruzhentsev[4], Simeon Zhyla[4], Vladimir Pavlikov[4], Kostiantyn Dergachov[4], and Eduard Tserne[4]

[1] National Aviation University, Huzara Avenue 1, 03058 Kyiv, Ukraine
ayua@nau.edu.ua

[2] State Institution "Institute of Technical Problems of Magnetism of the National Academy of Sciences of Ukraine", Industrialna Street 19, Kharkiv 61106, Ukraine

[3] Kharkiv National Automobile and Highway University, Ya. Mudroho Street 25, Kharkiv 61002, Ukraine

[4] National Aerospace University H.E. Zhukovsky "Kharkiv Aviation Institute", Chkalov Street 17, Kharkiv 61070, Ukraine

**Abstract.** In this paper, some cybersecurity vulnerabilities of modern UASs were considered. Vulnerabilities of communication, navigation, control, and surveillance equipment of modern UASs were analyzed. The cyber threats analysis and assessment algorithm were presented. The proposed algorithm is focusing on possible intruders' portrait that can be important when qualitative risk assessment. The approach to assess the UASs cyber hazards connected with considered vulnerabilities and threats was made. It was stressed that the particular assessment highly depends on the properly defined mission, potential intruders, and even social and political situation. A short overview of the recommendations to mitigate risk connected with cyber hazards was done.

**Keywords:** UAS · Cybersecurity · Cyber hazards assessment

## 1 Introduction

Modern remotely piloted aviation is characterized by multitasking, the difference in types of aircraft as well as their possibilities to perform different tasks, technical peculiarities, and modes of operation.

A lot of unmanned aerial vehicles (UAV) are remotely piloted vehicles. They are operated by the pilot-operator within the visual line of sight. Many UAVs that are used nowadays and are under development are designed as autonomous systems. The operation of such type of UAV is realized when they are beyond the visual line of sight of the pilot-operator. In both cases, it is reasonable to use the term unmanned aerial systems (UAS) as it comprises all the necessary facilities for UAV operation, including UAV


S. Shukla et al. (eds.), *Data Science and Security*, Lecture Notes in Networks and Systems 290,
https://doi.org/10.1007/978-981-16-4486-3_28

itself, GPS, Ground control system, equipment for particular tasks, special software, tools for maintenance, as well as prepared and skilled remote pilots.

As UAS is the component of the aviation system [1], the safety program should be established in order to achieve an acceptable level of safety performance in civil aviation. In turn, it requires the hazards identification and risk assessment for Safety Risk Management, according to [2]. Considering the diverse and increasing application of UAS, including civilian and military, and the shift to the use of autonomous systems for many applications, it is crucial to understand existing and potential threats connected with UAVs operation to ensure flight safety. Moreover, as indicated in [3], cybersecurity was not a design priority when developing current UAS' autopilot systems. This fact makes the UASs vulnerable to malicious cyber-attacks and requires the analysis of current and potential vulnerabilities and hazards to develop resilient engineering decisions for future UASs.

In this paper, the approach to UAS hazards identification and qualitative assessment is made focusing on cyber threats to UAS. The possible recommendations to threats mitigation are proposed.

## 2 Cyber Security Aspects

Modern UAS can be considered as a cyber-physical system as it consists of a UAV that is an example of a physical component and different elements for data storage, transfer, processing, and representation for physical component control.

These elements can be separate devices or complex systems from different traders and developers. The security requirements of the complex cybersecurity system are complicated by the demands to take into account the peculiarities of each of the elements or subsystems. Some of the requirements were identified and given in [4].

As the operation of the modern UASs requires diverse information from different systems, the information becomes the key element and resource that should be protected against various threats. The main threats to information according to terminology that is used in Information Security are revealed in spoiling confidentiality, integrity, and availability [5]. In [5], the first aspect, confidentiality, is defined as ensuring that only authorized users can access information; integrity is defined as ensuring completeness, accuracy, and an absence of unauthorized modifications in all its components of the system; and the third aspect, the availability guarantees that system and all system's components are available and operational when required by any authorized users.

Commonly, cyber threats are realized by finding the vulnerabilities in the systems, their operations, or software. The threats aim to spoil integrity, confidentiality, and availability of information when data is in different states - at rest, in use, and in motion. In paper [6], the simulation of strategy that exploits some vulnerability in the system to use one of the three primary security aspects is considered.

When UASs operation the example of data at rest can be the information that is stored in an onboard computer or control system. The data in use can be the data that are used by the computer, for example, to correct flight trajectory, or read from storage, or taken from network to perform particular tasks. Data in motion can be represented by the information that is transferred via communication channels or transferred to the short-term memory of the onboard information system. The subsystems of UASs, including UAV and subsystems of mission control, communication, navigation, and surveillance, are very sensitive to cyber threats and have the vulnerabilities provided by the technological and functional peculiarities. In [7], it is indicated that still, it is not enough attention to cybersecurity concerns when UASs design and development.

When developing solutions for security enhancement, it is reasonable to understand and analyze threats and vulnerabilities connected with UAVs operation firstly. Then to assess the risks and make the decision on risks acceptance.

## 3 Cyber Threats Analysis and Assessment Algorithm

The risk assessment is a multifarious task that involves analyzing many different factors. Taking into account this fact let us represent the stages of cyber risk assessment in the process of UASs risk management with the algorithm shown in Fig. 1.

Actually, Fig. 1 represents the common steps required for the risk management process. It is like a strategy that is aimed to understand, assess and manage the risks. The risk assessment is part of the risk management process and usually comprises the steps from Identification to Risk assessment. As it is shown in the diagram, it was proposed to add additional steps for precise consideration – "Identification 1" and "Analysis of Intruder" as they can significantly influence the final decision on ratings of threats.

The first block of the diagram shown in Fig. 1 is the first stage of the analysis of the cyber threat, and it is the identification of the mission planned to be performed with

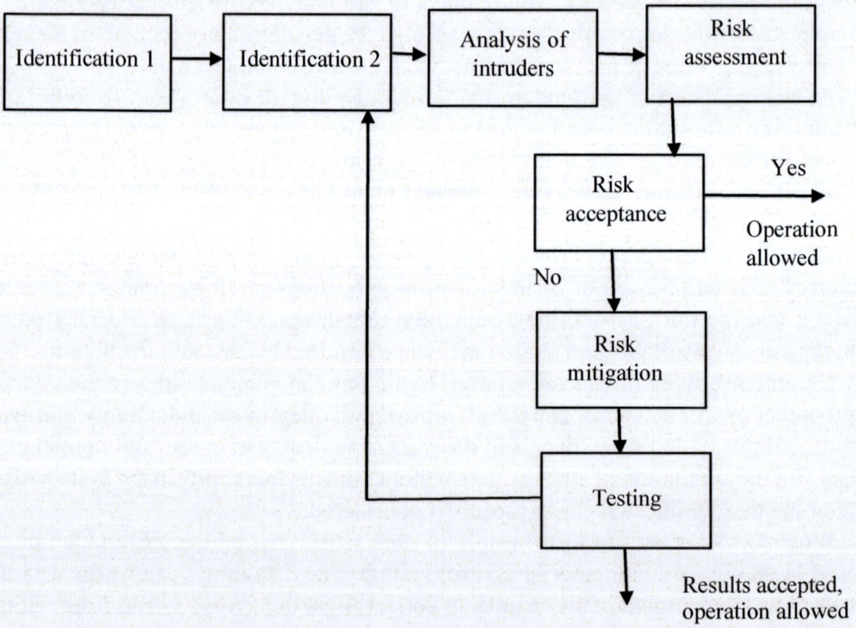


**Fig. 1** Block-diagram of the cyber threats analysis and assessment algorithm

UAS, performances of aircraft, and peculiarities of subsystems of navigation, control, communication, and surveillance. This is an important stage as the identification of the mentioned parameters helps to identify the vulnerabilities of the subsystem in the context of planned mission and technical abilities of the cyber-physical system.

The next step of Cyber Threats Analysis is "Identification 2". It includes the identification of vulnerabilities, threats, and potential hazards.

For risk assessment, it is also essential to understand the abilities and reasons of potential intruders. Understanding is important as intruders' motives can vary significantly from the ordinary familiarization with information about UAS mission to the intention to make terroristic action using with intercepted UAS system. The probability of realization of the malice depends on the level of preparation and technical equipment available for the intruder. Therefore, the next step is the analysis of the potential intruder portrait.

The next step is the Risk Assessment. It is indicated [2] that it is recommended to use the risk assessment matrix based on two parameters: the severity of consequences and the probability (frequency) of threats occurrence. Also, it is noted in [2] that it is important to distinguish the difference between threats and risk. The threat can be defined as the potential to cause harm, and the risk can be defined as the likelihood of that harm being realized during a specified amount of risk exposure.

Despite the paper's primary aim that is the analysis of the cyber threats to UASs, we demonstrated the further steps of the risk management process to show the place of risk mitigation actions. This is an important step if the identified previously risk cannot be accepted and additional measures to reduce the severity of consequences or probability of threats occurrence are required. Some suggestions on risk mitigation will be done in this paper in Sect. 5.

After Risk assessment, it is recommended to compose the Risk acceptance matrix to define whether it is possible to continue the task fulfillment or some additional actions are required to mitigate the undesirable probability of threat realization. After the implementation of measures on risk mitigation, the testing is done. In case of positive results of the tested mission, the decision to perform the task can be taken. Otherwise, the process of risk assessment should be revised.

## 4 Cybersecurity Threats and Vulnerabilities Analysis and Assessment

As it was mentioned in the introductory part of the paper, the UASs are the cyber-physical system and can be considered as a target of cyber threats. The short statistics of accidents connected with UAV are presented in [8]. The taxonomy of attack is also considered in [9]. Some statistical information focusing on cyber threats to UASs is presented in [10].

On the basis of statistical data and expert opinion, let us compose the set of the vulnerabilities of the system of communication, control, navigation, and surveillance. It is obvious that the list of vulnerabilities can be subjective and limited by the considered cases. We compose the information to connect vulnerabilities with the subsystem that can be attacked and the threats that can be realized due to exploiting the subsystem's weakness. The aims of the threats are also presented in the table (Table 1).

**Table 1** Identification 2 step. Threats identification

| Vulnerability | Subsystem | Aim of the threat | Threat |
|---|---|---|---|
| Listening | Communication lines | Examination with information | Breach in Confidentiality |
| Low, poor signal, jamming | Communication lines | Information destruction | Breach in availability |
| Spoofing | GPS, Radar signals GPS | Information destruction | Breach in integrity |
| Interception | Communication lines | Examination with information | Breach in Confidentiality |
| Information system intrusion | GPS, Radar signals Control systems, Computers | Modification and destruction | Breach in integrity and availability |

**Table 2** Cyber hazards severity of consequences

| Vulnerability | The severity of a consequences |
|---|---|
| Listening | Insignificant |
| Low, poor signal, jamming | Minor |
| Spoofing | Major |
| Interception | Insignificant |
| Information system intrusion | Catastrophic |

**Table 3** Cyber hazards probability of occurrence

| Vulnerability | Probability of occurrence |
|---|---|
| Listening | Periodically |
| Low, poor signal, jamming | Periodically |
| Spoofing | Rare |
| Interception | Rare |
| Information system intrusion | Very rare |

According to the methodology given in [2] and the proposed algorithm (Fig. 1), we evaluate the hazards according to the severity of consequences and the probability of threats occurrence. Table 2 presents the cyber vulnerabilities assessment on the severity of consequences, and Table 3 shows the cyber vulnerabilities assessment on the probability of threats occurrence. The criteria for assessment are taken from [2].

Listening is assessed as a relatively probable situation because it can be realized by a wide range of intruders. It can be realized by the passive observer who has a private interest, by the amateur, who wants to be recognized, or even cybercriminal who intends to make influence or extract profit. Obviously, the equipment used by different intruders also varies from the relatively cheap radio receivers, common websites, and mobile applications to the professional apparatus. The consequences are assessed as low because active actions from intruders are rather a rare situation in this case.

Jamming is assessed as the event with relatively high probability taking into account the nature of the signal propagation, interferences, and noises. Intentional interference is rather a rare event, but mutual electromagnetic radiation together with the different nature factors more often and probable situation. The consequences are assessed as relatively low (minor) because when jamming is implemented for all communication channels simultaneously, it is a rather low probability.

The probability of spoofing is assessed as medium (rare). During the latest period in Ukraine, the specialists who worked with UASs and were interviewed as experts indicated the increased attempts of spoofing when UAV operation. The object of spoofing can be GPS (Global Positioning System) signal. In this case, spoofing is interference that represents a signal similar to the navigation satellite signal. The navigation receiver perceives the false information of spoofing as a useful signal. The result of spoofing is the generation of false coordinate solutions. Thus, the object moves along the trajectory set by the intruder. Due to the spoofing, the time can be synchronized with large errors. The spoofing can be realized with relatively expensive equipment that constricts this technique's use by many intruders. The consequences are assessed as Major. This is an average assessment because the spoofing can be realized by the amateur who use the gaps in the security. These actions can be revealed and mitigated with additional surveillance and navigation systems. Hazards of spoofing realized by the cyberterrorists of professional intruders can have more serious consequences. But at the same time, this is not a very often situation. Anyway, for a more precise assessment, the identification of the primary goal of UAS operation (Identification 1) as well as the social and even political situation can affect the final assessment significantly.

Information interception, as well as listening, leads to the familiarization with information. The probability is assessed as medium (Rare) because the interception is required special equipment. This makes the interception realization a relatively complex task. The consequences are assessed as low because intruders' active actions are rather rare at this type of attack.

Intrusion into information systems can be realized by insiders or cybercriminals and terrorists. The insiders can be the business partner or officer who is dissatisfied by the salary or position. Insiders can have access to the available resources, key programs, and elements of the system to reach the aim. This makes the probability of realization relatively high. But the experience shows that proper selection of candidates for the critical position, in turn, decreases the probability of occurrence of such type of attack. Therefore, the average probability is relatively low (very rare), but the consequences can be rather serious (Catastrophic). On average, it gives a medium level of risks due to the intrusion into Information systems.

## 5 Suggestions on Cyber Hazards Mitigation

According to [2], if the results of the assessment are of low probability and/or have insignificant consequences on the system operation and mission fulfillment, the operation of the system is allowed under the condition of accepted risk. In other cases, measures to mitigate risk should be taken. Based on the analysis and assessment performed in the previous section, we concentrate our attention on measures that are appointed to mitigate hazards of jamming, spoofing, and information system intrusion.

The measures to mitigate interference and gamming can be differentiated depending on the nature of the interference. The interference can be self-produced due to space limitation when different apparatus arrangements, including antenna and other electronic systems and devices. So, the problem of electromagnetic compatibility can arise. In paper [11, 12] it is presented some algorithms to mitigate the self-interference to UASs. In [12], it is noted that interference can be mitigated with proper RF filters in the antenna and receiver. Paper [13] describes GNSS (Global Navigation Satellite System) gamming countermeasures that allow the control radio to be used as a backup navigation source.

The method and algorithm to mitigate the risk of spoofing with adaptive antenna array were presented in [14]. The model of the adaptive antenna array that includes 3D radiation pattern, elements that control the signal amplitude and signal phase, linear signal adder was demonstrated and studied in [14]. The GPS spoofing attack detection and mitigation technique were also presented in [15].

The measures to decrease the intrusion into information systems can include the malware scanners installed on computers that are intended to detect and block malware.

## 6 Conclusions

In this paper, the security vulnerabilities that allow the realization of cyber threats to modern UASs were considered. The assessment of the UASs cyber hazards connected with considered vulnerabilities and threats was made. It was noted that one of the key points is understanding the aims, reasons, and potentials of the possible intruders.

Because they can influence significantly onto the particular assessment of risk. For example, spoofing can be realized by cyberterrorists or by professional intruders. In this case, the consequences will rather serious. But this is not a very often situation. At the same time, for the particular planned mission of UAS this can appear rather probable, and thus, identification can be as "periodical" or even "often". So, additional measures to mitigate risks should be taken.

Anyway, for a more precise assessment, the identification of the primary goal of UAS operation (Identification 1) as well as the social and political situation should be taken into account, as they influence the evaluation of the hazards according to the chosen criteria.

The mentioned factors were taken into account when developing the algorithm of Cyber Threats Analysis and Assessment. The algorithm represents the strategy that is aimed to understand, assess and manage the risks. The algorithm considers the mentioned factors additionally to the accepted and recommended in [2]. For future research, it would be interesting to compare the statistics of risks shift depending on factors of the stage "Identification 1" in different regions. This can be important for the improvement of evaluation of the potential cyber hazards for UAS and the development of relevant recommendations for the risk management process.

## References


1. Manual on Remotely Piloted System (RPAS) (2015) International Civil Aviation Organization, Doc 119
2. Safety Management Manual (2017) ICAO Doc 9859
3. Kim A, Wampler B, Goppert J, Hwang I Cyber attack vulnerabilities analysis for unmanned aerial vehicles. Am Inst Aeronaut Astronaut 1–30. https://static1.squarespace.com/
4. Cyber Security Research Alliance (Apr 2013) Designed-in cyber security for cyber-physical systems - workshop report by the cyber security research alliance. Cyber Security Research Alliance. http://www.cybersecurityresearch.org/documents/CSRA_Workshop_Report.pdf. Accessed 31 Jan 2015
5. Watkins SG (2013) An Introduction to Information Security and ISO27001:2013. IT Governance Publishing, Ely
6. Schneier B (1999) Attack trees – modeling security threats. Dr Dobbs J 24:1–9
7. Bharat BM, Manoj B, Doina B (2016) Securing unmanned autonomous systems from cyber threats. J Def Model Simul Appl Methodol Technol 16(2):1–17
8. Averyanova Y, Blahaja L (2019) A study on unmanned aerial system vulnerabilities for durability enhancement. In: Proceedings of the 5th International Conference on Actual Problems of Unmanned Aerial Vehicles Development (APUAVD-2019) Oct 22–24, Kyiv, Ukraine
9. Best K, Schmid J, Tierney S, Awan J, Beyene N, Hollida M, Khan R, Lee K (2020) How to Analyze the Cyber Threat from Drones. RAND Corporation, Santa Monica
10. Krishna CG, Murphy RR (2017) A review on cybersecurity vulnerabilities for unmanned aerial vehicles. In: IEEE International Workshop on Safety, Security, and Rescue Robotics (SSRR), pp 1–26
11. DME/TACAN interference mitigation for GNSS: algorithms and flight test results, Grace Xingxin Gao (2013). http://gracegao.ae.illinois.edu/publications/journal/2013_GPS%20solutions_DME.pdf
12. Wilde W, Cuypers G, Sleewaegen J-M, Deurloo R, Bougard B (2016) GNSS Interference in Unmanned Aerial Systems. https://www.septentrio.com/sites/default/files/gnss_interference_in_unmanned_aerial_systems_final.pdf
13. Bergh B, Pollin S (2018) Keeping UAVs under control during GPS jamming. IEEE Syst. PP(99):1–12
14. Averyanova Y, Kutsenko O, Konin V (2020) Interference Suppression at Cooperative Use of GPS, GLONASS, GALILEO, BEIDOU. In: Proceedings of Ukrainian Microwave Week, pp 44–48
15. Javaid A, Jahan F, Sun W (2017) Analysis of global positioning system-based attacks and a novel global positioning system spoofing detection/mitigation algorithm for unmanned aerial vehicle simulation. SAGE J 93(5), pp 427–441

# Some New Results on Non-zero Component Graphs of Vector Spaces Over Finite Fields

Vrinda Mary Mathew(✉) and Sudev Naduvath

Deaprtment of Mathematics, CHRIST (Deemed to be University), Bangalore, India
vrinda.mathew@res.christuniversity.in

**Abstract.** The non-zero component graph of a vector space with finite dimension over a finite field $F$ is the graph $G = (V, E)$, where vertices of $G$ are the non-zero vectors in $V$, two of which are adjacent if they have at least one basis vector with non-zero coefficient common in their basic representation. In this paper, we discuss certain properties of the non-zero component graphs of vector spaces with finite dimension over finite fields and their graph invariants.

**Keywords:** Non-zero component graph · Graph invariants · Domatic number algorithm

## 1 Introduction

We refer to [1–3] for terminology and concepts in graph theory, and [4–6] for more topics in algebraic graph theory. We consider [7–9] for linear algebra topics.

Let $V$ be a vector space over a field F with the basis $\{\alpha_1, \alpha_2, \ldots, \alpha_n\}$ and $\theta$, the null vector. Any vector $a \in V$ can be represented uniquely as a linear combination of the form $a = a_1\alpha_1 + a_2\alpha_2 + a_3\alpha_3 + \ldots + a_n\alpha_n$. With respect to $\{\alpha_1, \alpha_2, \ldots, \alpha_n\}$, we call this representation of $a$ as its basic representation. With regard to $\{\alpha_1, \alpha_2, \ldots, \alpha_n\}$, the notion of *non-zero component graph* of a vector space with finite dimension, denoted by $G(V_\alpha) = (V, E)$, is defined as follows in [10]: $V = V - \{\theta\}$ and for $a, b \in V$, $a \sim b$ or $(a, b) \in E$ if a and b have at least one $\alpha_i$ with non-zero coefficient common in their basic representation.

If context is clear, we can represent the non-zero component graph of a vector space $V$ with finite dimension over a finite field $F$ by $G$ instead of $G(V_\alpha)$. Also, throughout this discussion, by the term non-zero component graph, we mean the non-zero component graph of a given vector space with finite dimension over a finite field.

## 2 Distances in Non-zero Component Graphs

Remember that the eccentricity $ecc(v)$ of a vertex $v$ in a connected graph $G$ is the maximum of the distances between $v$ and the rest of the vertices of $G$. In the following result, the eccentricity of a vertex of a non-zero component graph $G$ is discussed:


S. Shukla et al. (eds.), *Data Science and Security*, Lecture Notes in Networks and Systems 290,
https://doi.org/10.1007/978-981-16-4486-3_29

**Proposition 2.1.** *The eccentricity of any vertex in G is either* 1 *or* 2.

*Proof.* The universal vertices of the non-zero component graph are the vertices corresponding to the vectors of the form $c_1\alpha_1 + c_2\alpha_2 + \cdots + c_n\alpha_n, \forall c_i \neq 0$. The eccentricity of these vertices would be 1 since they are adjacent to all of the other vertices in the graph. Let $S$ be the set of vertices corresponding to the vectors of the form $c_1\alpha_1 + c_2\alpha_2 + \cdots + c_n\alpha_n$ with some $c_i = 0$. There is a vertex $u$ in $S$ that is not adjacent to $v$ for every vertex $v \in S$. Since these vertices are adjacent to all universal vertices of the graph, the distance between them is 2. Therefore, the eccentricity of $v$ is 2. This completes the proof.

The minimum of the maximum distances between any two vertices in the graph is called the *radius* of the graph. The following proposition discusses the radius of non-zero component graphs.

**Proposition 2.2.** *The radius of the non-zero component graph G is 1.*

*Proof.* The proof is immediate from the fact that the non-zero component graph $G$ has at least one universal vertex.

The *radial graph* of a graph $G$ is denoted by $R(G)$, and two vertices in $R(G)$ are adjacent if their distance in $G$ is equal to the radius of $G$ (see [11]). A *self radial graph* is a graph whose radial graph is isomorphic to itself.

**Proposition 2.3.** *The radial graph of the non-zero component graph is the graph itself.*

*Proof.* The radius of the non-zero component graph $G$ is 1. Also, $R(G) = G$ if and only if $G \in F_1$, a set of all connected graphs of radius 1. Since $rad(G) = 1$, $G \in F_1$ and hence $R(G) = G$.

The vertex whose eccentricity equals the graph's radius is the graph's *central point.*

**Proposition 2.4.** *Every universal vertex of a non-zero component graph G is its centre point.*

*Proof.* Note that the universal vertices of the graph $G$ are those vertices corresponding to the vectors of the form $c_1\alpha_1 + c_2\alpha_2 + \ldots + c_n\alpha_n$; $\forall\, c_i \neq 0$. Also, the eccentricity of these vertices is 1. Therefore, the eccentricity of these vertices is same as the radius of the graph $G$ and hence these vertices are centre points of the graph $G$.

The *graph center* of a graph $G$, denoted by $GC$, is the collection of all central points of $G$. The following theorem discusses the size of center of the non-zero component graph $G$.

**Proposition 2.5.** *If* $dimV = n$ *and* $|F| = q$*, the cardinality of graph center is* $|GC| = (q-1)^n$.

*Proof.* By Proposition 2.4, the elements in the graph center are the vertices corresponding to the vectors of the form $c_1\alpha_1 + c_2\alpha_2 + \cdots + c_n\alpha_n, \forall c_i \neq 0$. The number of vertices of this type depends on the number of non-zero elements in the field. Therefore, each scalar $c_i$ in the above expression can be any one of the $q - 1$ non-zero elements in the field and hence by the fundamental theorem of counting, the number of central points is equal to $(q-1)^n$. This completes the proof.

## 3 Connectivity in Non-zero Component Graphs

Note that the non-zero component graph of a vector space $V$ is always a connected graph which has no bridges and articulation points. As a result, in this section, we look at the connectivity and related properties of a non-zero component graph $G$.

**Proposition 3.1.** *The complement of the non-zero component graph of a vector space $V$ with dimension $n$ over a field $F$ with $q$ elements will always be a disconnected graph with $(q-1)^n + 1$components and $\omega(\bar{G}) = dimV$.*

*Proof.* In the non-zero component graph, the vertex of the form $c_1\alpha_1+c_2\alpha_2+\cdots+c_n\alpha_n$, where $c_i \neq 0$ for all $1 \leq i \leq n$, will be adjacent to all the other vertices in the graph. Hence in the complement these vertices will remain as isolated vertices making the graph disconnected. Since there are $(q-1)^n$ such vertices in $G$, all these vertices will remain as isolated vertices in the complement of $G$. Hence, we can say that the complement of non-zero component graph will be a disconnected graph with $(q-1)^n + 1$ components. Moreover, the basis vectors which were independent in the original graph forms a clique in the complement and thereby the clique number $\omega(\bar{G}) = n$.

The following theorem gives a lower bound on a non-zero component graph's vertex connectivity.

**Theorem 3.2.** *If a non-zero component graph $G$ is $k$-connected, then $k \geq (q-1)^n$.*

*Proof.* We prove that the removal of $(q-1)^n - 1$ vertices from $G$ does not make $G$ disconnected. By Proposition 2.5, there are $(q-1)^n$ vertices which are adjacent to all the remaining vertices in the graph. The removal of all these vertices may result in a disconnected graph. The presence of any of these vertices in the reduced graph makes it connected. Thus, we can say that the removal of $(q-1)^n - 1$ vertices never makes the resultant graph disconnected due to the presence of a universal vertex.

**Theorem 3.3.** *Let $V$ be a vector space with dimension $n$ over a finite field $F$. The non-zero component graph $G$ is bipartite if and only if $|F| = 2$ and $dimV \leq 2$.*

*Proof.* First, assume that $G$ is bipartite. Then, we have to show that $|F| = 2$ and $\dim V \leq 2$. Assume the contrary. In this context, the following two possibilities must be considered:

*Case 1.* Let $\dim V = n$, where $n > 2$. Let $\{\alpha_1, \alpha_2, \ldots \alpha_n\}$ be the basis of $V$. Then, any set of vertices corresponding to the vectors of the form $c_l\alpha_i, c_l\alpha_i + c_m\alpha_j, c_l\alpha_i + c_m\alpha_k$ where $1 \leq l \leq q$, $1 \leq m \leq q$ will form a cycle of length 3, a contradiction to the hypothesis that $G$ is bipartite.

*Case 2.* Let $|F| \neq 2$. Then the vertices corresponding to the vectors of the form $c_i\alpha_i, c_j\alpha_i, c_k\alpha_i$, where $c_i, c_j, c_k \in F$, form an odd cycle, which is again a contradiction to the hypothesis.

From the above two cases, we have, $|F| = 2$ and $\dim V \leq 2$ if $G$ is bipartite.

Conversely, assume that $|F| = 2$ and $\dim V \leq 2$. We have to show that $G$ is bipartite. If possible, assume that $G$ be a non-bipartite graph. Therefore, it contains at least one odd cycle. This occurs only when $|F| \geq 2$ or $\dim V \geq 2$, a contradiction to the hypothesis. Thus, $G$ is bipartite, completing the proof.

**Theorem 3.4.** *Let $V$ be a vector space with dimension $n$ over a finite field $F$. Then, its non-zero component graph $G$ is perfect if and only if $\dim V \leq 4$.*

*Proof.* Let $G$ be a perfect graph. We have to show that $\dim V \leq 4$. Assume the contrary. That is, $\dim V > 4$. Then there exists vertices corresponding to the vectors of the form $c_1\alpha_1 + c_2\alpha_2, c_2\alpha_2 + c_3\alpha_3, c_3\alpha_3 + c_4\alpha_4, c_4\alpha_4 + c_5\alpha_5$ in $G$ such that they form an induced odd cycle contradicting the fact that $G$ is perfect.

Conversely, assume that $\dim V \leq 4$. We have to show that $G$ is perfect. If possible, assume that $G$ is not perfect. Then, $G$ will have an induced odd cycle of length at least 5, which is possible only when $\dim V \geq 5$, so that the vertices of the cycle will be of the following form, as shown in Fig. 1.

Thus, $\dim V \leq 4$ so that $G$ is perfect.

**Corollary 3.5.** *The vertex covering number of the non-zero component graph of a vector space with dimension $n$ over a finite field having $q$ elements is $q^n - 1 - n$.*

*Proof.* The order of the non-zero component graph of a vector space with dimension $n$ over the field $F$ with $|F| = q$ is $q^n - 1$ and the independence number of $G$ is $\dim V = n$. Since the sum of the vertex covering number and the independence number of any graph is equal to the order of the graph, we conclude that the vertex covering number of $G$ is equal to $q^n - 1 - n$.

A matching in $G$ is a collection of pairwise independent edges of $G$, and the cardinality of the maximum matching is the matching number of $G$, denoted by $\alpha'(G)$. The following theorems discusses the matching number of non-zero component graph.

**Theorem 3.6.** *The matching number of the non-zero component graph of a vector space with dimension $n$ over the field $F$ with $q$ elements is*

$$\alpha'(G) = \frac{q^n - 1}{2}$$

*Proof.* The non-zero component graph has $q^n - 1$ vertices. Each vertex $v$ in $G$ can be paired up with some vertex, say $u$, in $G - \{v\}$. Since order of a finite field is always a power of prime, $q = p^n$.

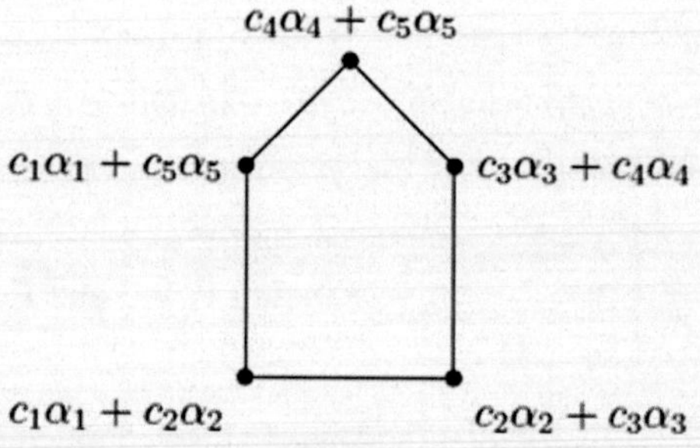


**Fig. 1** .

*Case 1:* When $p = 2$, the order of the graph $G$ will be odd and hence while pairing up the vertices, one vertex will be left out. Thus, the matching number in this case will be $\lfloor \frac{1}{2}[q^n - 1] \rfloor$.

*Case 2:* When $p \neq 2$, the order of $F$ is always an odd number. In this case, the number of vertices of the graph is always an even number and independent edge set can be obtained by pairing up the vertices in the graph. Hence, the matching number is equal to half the number of vertices in $G$ i.e., $\frac{1}{2}[q^n - 1]$.

**Corollary 3.7.** *The edge covering number* $\rho(G)$ *of the non-zero component graph G is* $\left\lceil \frac{q^n-1}{2} \right\rceil$.

*Proof.* Since the sum of the edge covering number and matching number is equal to the number of vertices in $G$ i.e., $q^n - 1$, by Theorem 3.6 it is clear that $\rho(G) = \left\lceil \frac{q^n-1}{2} \right\rceil$.

A perfect matching of a graph $G$ is one that saturates all the vertices of $G$. The following result investigates when a non-zero component graph has a perfect matching.

**Theorem 3.8.** *The non-zero component graph has a perfect matching except when order of the field is even.*

*Proof.* As mentioned in Proposition 2.5, we are able to find an independent set of edges which covers all the vertices in $G$ when the order of the field is odd, resulting in a perfect matching of $G$. The order of the field is always a power of a prime, say $q = p^n$, and it becomes even only when $p = 2$ in which case a perfect matching could not be found out. This completes the theorem.

## 4 Domination in Non-zero Component Graphs

If for any vertex v in V, either $v \in D$ or $v \sim u$ for any $u \in D$, a subset D of the vertex set V of a graph G is said to be a *dominating set* of G. A *minimal dominating set* of G is one whose proper subsets aren't also dominating sets of G. The minimum cardinality of a minimal dominating set of G is the *domination number* of a graph G. Note that the domination number of G is 1 as it has a universal vertex (see [12]). The domination number of the complement of a non-zero component graph is discussed in the following theorem.

**Proposition 4.1.** *The domination number of the complement of the non-zero component graph G is*

$$\gamma(\bar{G}) = \begin{cases} q-1, & \textit{when } \dim V = 1 \\ 2, & \textit{when } \dim V = 2 \text{ and } F = \mathbb{Z}_2 \\ (q-1)^n + n, & \textit{otherwise} \end{cases}$$

*Proof.* When $\dim V = 1$, $G$ is complete and hence $\bar{G}$ is a completely disconnected graph, and hence $\gamma(\bar{G})$ equals the number of vertices in the graph G. When $\dim V = 2$ and $F = Z_2$, $G$ is isomorphic to $P_3$ and hence $\gamma(\bar{G}) = 2$. In all the other cases, the universal vertices in $G$ becomes the isolated vertices in $\bar{G}$ and hence they dominate themselves. Also the basis vectors dominates the other vertices in $\bar{G}$. Thus, the domination number is $(q-1)^n + n$.

In the following discussion, we investigate some other domination parameters of the non-zero component graphs of vector spaces of finite dimension over finite fields.

Recall that the maximum cardinality of a minimal dominating set of the graph $G$ is called the *upper domination number* of G. The upper domination number of non-zero component graphs is estimated in the following theorem.

**Proposition 4.2.** *Let V be vector space with dimension n over a finite field F and G be the non-zero component graph associated with it. Then* $\Gamma(G) = \alpha(G)$.

*Proof.* For the non-zero component graph $G$, the set of vertices corresponding to the basis vectors itself forms a minimal dominating set with maximum possible cardinality $\dim V$. Also, we note that these vertices also forms a independent set with the maximum cardinality in the non-zero component graph $G$. Hence, we have $\Gamma(G) = \alpha(G)$, completing the proof.

A graph's *independent dominating set* is a set that is both a dominating and an independent set. The size of the smallest dominating set that is an independent set is denoted by the *independent dominance number*, denoted by $i(G)$. In the following theorem, the independence dominance number of a non-zero component graph is discussed.

**Proposition 4.3.** *The independent domination number i(G) of the non-zero component graph is* 1.

*Proof.* Let $G$ be the non-zero component graph of a finite dimensional vector space V over the finite field $F$. The vertex corresponding to the vector of the form $c_1\alpha_1 + c_2\alpha_2 + \ldots + c_n\alpha_n$, where $c_i \neq 0$ for $1 \leq i \leq n$, will be adjacent to all the other vertices in the non-zero component graph. This vertex is a minimal dominating set and is also an independent set since it dominates the graph. Hence, we have $i(G) = 1$.

The maximum cardinality of the minimal dominating set that is an independent set is the *independent upper domination number*, denoted by $i_\Gamma(G)$.

**Theorem 4.4.** *The independent upper domination number of the non-zero component graph G of vector space V with dimension n is n.*

*Proof.* In $G$, the vertices corresponding to the basis vectors form a minimal dominating set which is an independent set also. Since every vertex of the non-zero component graph $G$ corresponds to an $r$-element subset of $V$, where $1 \leq r \leq n$, Any disjoint r-element subset of the basis of $V$ corresponds to every minimal independent dominating set of $G$. Then the number of elements in a minimal dominating set is $\lceil \frac{n}{r} \rceil$. Thus a minimal dominating set of $G$ attains maximum cardinality when $r = 1$. Therefore $i_\Gamma(G) = n$.

A *total dominating set* of vertices in a graph $G$ is one in which every vertex of $G$ is adjacent to at least one vertex in the set. The *total domination number* of the graph $G$ is the minimum cardinality of a total dominating set of $G$. In the following result, the total domination number of a non-zero component graph is calculated.

**Proposition 4.5.** *The total domination number of the non-zero component graph G is 2.*

*Proof.* Let $u$ be a universal vertex in $G$. But, the set $\{u\}$ alone will not form a total dominating set since $u$ does not dominate itself. Hence, we need to add one more vertex $v$ to get a total dominating set in $G$ with the least cardinality. This completes the proof.

If the dominating set $S$ of $G$ is also a dominating set of the complement $\bar{G}$ of $G$, so it is a *global dominating set* of $G$. The *global domination number* of $G$ is the minimum cardinality of a global dominating set. The global domination number of a non-zero component graph of vector space is discussed in the following theorem.

**Theorem 4.6.** *The global domination number of the non-zero component graph G equals the domination number of* $\bar{G}$.

*Proof.* The universal vertex in $G$ dominates the graph $G$. But, the universal vertices in $G$ are isolated vertices in $\bar{G}$ and the other vertices form a connected graph in $\bar{G}$. Thus the dominating set of $\bar{G}$ contains the universal vertices of $G$ and the basis vectors. Hence we can say that this set forms a global dominating set of $G$. Since this set is same as the dominating set of $\bar{G}$, global domination number of $G$ is equal to the domination number of $\bar{G}$.

The maximum number $k$ such that $V$ can be partitioned into $k$ disjoint dominating sets is the domatic number of a graph $G$, denoted by $DN(G)$. We suggest the following algorithm for determining the domatic number of a non-zero component graph:

**Algorithm 1:** Domatic Number Algorithm

```
Let V be a vector space over field F.
Input: dim V = n, |F| = q
Output: Domatic Number = DN
Data: The graph G
A = 0
B = 0
C = 0
if n is odd then
    DN = (q − 1)^n
    for i ← 0 to ⌈n/2⌉ do
        A = (n choose i) * (q − 1)^(n−i)
        B = (n choose n−i) * (q − 1)^i
        C = (A − B)/2
        DN = DN + B + C
else
    DN = (q − 1)^n + ((n choose n/2) * (q − 1)^(n/2))/2
    for i ← 0 to ⌊n/2⌋ do
        A = (n choose i) * (q − 1)^(n−i)
        B = (n choose n−i) * (q − 1)^i
        C = (A − B)/2
        DN = DN + B + C
```

**Theorem 4.7.** *The above algorithm gives the domatic partition of a non-zero component graph G.*

*Proof.* Since $V$ is a vector space with finite dimension, say, $n$, the basis of the vector space contains $n$ linearly independent vectors. Let $\{\alpha_1, \alpha_2, \ldots, \alpha_n\}$ be the basis of $V$. Also given that $|F| = q$. Since $\dim V = n$, any element in $V$ will be of the form $c_1\alpha_1 + c_2\alpha_2 + \ldots + c_n\alpha_n, \forall c_i \in F$. The vertex set can be divided into $n$ distinct sets. $A_i, i = 0, 1, 2, \ldots, n-1$ as follows:

$$\begin{aligned} A_0 &= \left\{\textstyle\sum_{i=1}^{n} c_i\alpha_i\text{: } \forall c_i \neq 0\right\} \\ A_1 &= \left\{\textstyle\sum_{i=1}^{n} c_i\alpha_i\text{: one } c_i = 0\right\} \\ A_2 &= \left\{\textstyle\sum_{i=1}^{n} c_i\alpha_i\text{: two } c_i = 0\right\} \\ \vdots \;\; &\;\; \vdots \quad \ddots \\ A_{n-1} &= \left\{\textstyle\sum_{i=1}^{n} c_i\alpha_i\text{: (n − 1) } c_i = 0\right\} \end{aligned}$$

The cardinality of the above-mentioned sets are determined as follows:

$$\begin{aligned} |A_0| &= (q-1)^n \\ |A_1| &= \binom{n}{1}(q-1)^{n-1} \\ |A_2| &= \binom{n}{2}(q-1)^{n-2} \\ \vdots \;\; &\;\; \vdots \quad \ddots \\ |A_{n-1}| &= \binom{n}{n-1}(q-1) \end{aligned}$$

Since the vertices in $A_0$ are linear combinations of all the basis vectors, they are adjacent to all the vertices in the graph. There are $(q-1)^n$ such vertices and each of this vertex can be taken as a dominating set.

Next, when we consider the vertices in $A_{n-1}$ and $A_1$, any vertex in $A_{n-1}$ will be of the form $c_i\alpha_i$ and that in $A_1$ will be of the form $\sum_{i=1}^{n} c_i\alpha_i$ one $c_i = 0$. In this case the vertices corresponding to the vectors $c_j\alpha_j$ and $\sum_{i=1}^{n} c_i\alpha_i$, $c_j = 0$ may be used to form a dominating set. Therefore $\binom{n}{n-1}(q-1)$ dominating sets can be formed. This completes all the vertices in $A_{n-1}$. But there are $\binom{n}{1}(q-1)^{n-1} - \binom{n}{n-1}(q-1)$ vertices remaining in $A_{n-1}$. These vertices will be adjacent to each other and hence they can be paired up to form a dominating set. In total to the vertices in $A_1$ and $A_{n-1}$ forms $\frac{1}{2}\left[\binom{n}{n-1}(q-1) + \binom{n}{1}(q-1)^{n-1} - \binom{n}{n-1}(q-1)\right]$ disjoint dominating sets. In a similar way, the vertices in $A_2$ and $A_{n-2}$ forms $\frac{1}{2}\left[\binom{n}{2}(q-1)^{n-2} - \binom{n}{n-2}(q-1)^2\right] + \binom{n}{2}(q-1)^{n-2}$ disjoint dominating sets.

In general the vertices in $A_i$ and $A_{n-i}$, where $i > 0$ forms $\binom{n}{i}(q-1)^{n-i} + \frac{1}{2}\left[\binom{n}{i}(q-1)^{n-i} - \binom{n}{n-1}(q-1)^{i}\right]$ disjoint dominating sets.

*Case 1:* When $n$ is odd, initially the vertex set is initially partitioned into odd number of $A_i$'s and except $A_0$, the vertices in $A_i$ is paired up with the vertices in $A_{n-i}$. Since $n$ is odd, all the sets get paired up and correspondingly we can find the number of disjoint dominating sets.

*Case 2:* When $n$ is even, initially the vertex set is initially partitioned into odd number of $A_i$'s and except $A_0$, the vertices in $A_i$ is paired up with the vertices in $A_{n-i}$. Since $n$ is even, except the set $A_{n/2}$ all the sets get paired up and correspondingly we can find the number of disjoint dominating sets. The vertices in $A_{n/2}$ can be paired up to form disjoint dominating sets. Thereby we can get the number of disjoint dominating sets.

**Theorem 4.8.** *The domatic number of the non-zero component graph is equal to* $\frac{1}{2}[(q-1)^n + q^n - 1]$

*Proof.* As per the above algorithm, we are pairing up the vertices other than the universal vertices, hence we can directly get the domatic number as $(q-1)^n + \frac{1}{2}\left[q^n - 1 - (q-1)^n\right]$. This competes the proof.

A graph represents a network, with vertices and edges representing access points and communication links. A transmitting group of access points will send messages to all points in the network, and this transmitting group is a dominating set in the graph. This promotes the corresponding graph to find a domatic partition.

## 5 Conclusion

In this paper, we looked at some of the properties of non-zero component graphs, which are derived from vector spaces of finite dimensions over finite fields. In the first section, fundamental structural parameters such as radius, diameter and center of the non-zero component graphs and following to it, certain graph invariants such as certain types of domination numbers, matching and covering numbers, domatic numbers and connectivity are also investigated. The study offers a lot for future research as many other graph parameters of this type of graphs are yet to be explored in detail. Another interesting problem in this area is to study the effect of operations of vector spaces on the corresponding non-zero component graphs and their operations and structural characteristics. All these facts point to the wide scope for further research in this field.

## References

1. Bondy JA, Murty USR (2008) Graph theory. Springer, New York
2. Harary F (1967) Graph theory and theoretical physics. Academic Press, Cambridge
3. West DB (2001) Introduction to graph theory. Prentice hall, Upper Saddle River
4. Beineke LW, Wilson RJ, Cameron PJ (2004) Topics in algebraic graph theory, vol 102. Cambridge University Press, Cambridge

5. Biggs N (1993) Algebraic graph theory. Cambridge University Press, Cambridge
6. Godsil C, Royle GF (2001) Algebraic graph theory, vol 207. Springer Science & Business Media, Berlin
7. Beezer R A (2015) A first course in linear algebra. Independent
8. Hoffman K, Kunze R (2015) Linear algebra. Pearson India, New Delhi
9. Strang G (1993) Introduction to linear algebra, vol 3. Wellesley-Cambridge Press, Wellesley
10. Das A (2016) Nonzero component graph of a finite dimensional vector space. Comm Algebra 44(9):3918–3926
11. Kathiresan K, Marimuthu G (2010) A study on radial graphs. Ars Combin 96:353–360
12. Das A (2017) On non-zero component graph of vector spaces over finite fields. J Algebra Appl 16(01), 1750007:1–10

# Multiple Approaches in Retail Analytics to Augment Revenues

Haimanti Banik(✉) and Lakshmi Shankar Iyer

Department of School of Business and Management, CHRIST (Deemed to be University), Bangalore, India
haimanti.banik@mba.christuniversity.in, lakshmi.iyer@christuniversity.in

**Abstract.** Knowledge is power. The retail sector has been revolutionized around the clock by the plentiful product knowledge available to customers. Today, customers can use the knowledge available online at any time to study, compare and purchase products from anywhere. Retail companies can stay ahead of shopper trends by using retail information analytics to discover and analyze online and in-store shopper patterns. A product recommender will suggest products from a wide selection that would otherwise be very difficult to locate for the customer. The algorithm would recommend various products, increase the sales of items that would otherwise be difficult to sell. Market basket analysis is a common use scenario for the search for frequent patterns, which involves analyzing the transactional data of a retail store to decide which items are bought together. To do so data from online resource has been taken, which is analyzed and several conclusions were made.

**Keywords:** Product recommender · Skip-gram · word2vec · Association mining · ECLAT

## 1 Introduction

An optimal change is demonstrated by the retail industry. Day by day, changing consumer preferences and technological advances are gradually changing the retail landscape. Consumers expect rich, hyper-connected, personalized and engaging shopping experiences.

As the retail market becomes increasingly competitive, it has never been more important to be able to maximize the operation of business processes while meeting consumer expectations. In order to thrive effectively, it is also important to control and channel data to work towards consumer delight and produce healthy profits. Data analytics are now being applied at every stage of the retail process in the case of large retail players both internationally and in India—monitoring emerging popular goods, forecasting sales and future demand via predictive simulation, optimizing product placements and offers through customer heat mapping and many more [1]. In addition, what forms the center of data analytics is that determines consumers likely to be interested in specific product


S. Shukla et al. (eds.), *Data Science and Security*, Lecture Notes in Networks and Systems 290,
https://doi.org/10.1007/978-981-16-4486-3_30

categories based on their previous purchasing habits, works out the best way to reach them through targeted marketing campaigns and eventually works out what to sell them next. This project focuses on the approaches in retail analytics in shaping appropriate marketing strategies.

The issue of unpredictability in consumer behavior in retail stores prompted our research, as well as the need for online retailers to be able to distinguish current or future consumers based on their online behavior. In reality, only a small percentage of people who enter a shop want to buy something. As a result, by concentrating on a community or groups of key customers, the online retailer's ability to predict purchasing sessions improves the likelihood of gaining a competitive edge [5].

## 2 Literature Review

### 2.1 Retail Analytics: Driving Success in Retail Industry with Business Analytics

As the retail market becomes increasingly competitive, it has never been more important to be able to maximize the operation of business processes while meeting consumer expectations. Data or rather big data analytics are now being applied at every stage of the retail process in the case of large retail players both internationally and in India—monitoring emerging hot goods, predicting sales and future demand via predictive simulation, optimizing product placements and offers through consumer heat mapping and many more. It's what decides the buyers are likely to be involved in particular product categories based on their past buying patterns, determines the optimum way to reach them through targeted marketing strategies, and ultimately determines what to offer the customers next.

### 2.2 ECLAT Based Market Basket Analysis for Electronic Showroom

Market basket analysis is a technique of data mining to identify associations between datasets. Since a large amount of data is constantly collected and stored in databases, many businesses are becoming concerned about mining association rules from them. Analysis of market baskets explores consumer purchasing habits by establishing connections between different items put in their shopping baskets by consumers. Examining the buying habits of clients and helping to improve revenue is beneficial. Therefore, this method is intended to build market basket analysis system that will produce association rules between item sets using the ECLAT (Equivalence Class Transformation) algorithm.

### 2.3 From Word Embeddings to Item Recommendation

Platforms for social networks may use the knowledge provided by their users to better serve them. Recommendation services are one of the services that these sites offer. Using their past habits, recommendation systems will predict the future preferences of users. A collection of widely known NLP techniques, like Word2Vec, is implemented to the domain of recommendation systems in this work. Unlike former works that applied Word2Vec for recommendation, this work uses non-textual features, check-ins, and recommends venues for the target users to visit/check-in. A Foursquare check-in data collection is used for the experiments. The findings show that it is promising to use continuous vector space representations of things modelled by Word2Vec techniques to make recommendations.

## 3 Methodology

The industry standard procedure used for data mining is the CRISP-DM Technique. Business Understanding, Data Understanding, Data Preparation, Modelling, Evaluation and Deployment are the various phases involved in the CRISP-DM approach. The data is obtained from online resource and data mining method is used to establish an understanding of this. The data is cleaned before transformation in order to boost data consistency, and attributes can be extracted from the same. Modelling is done for the next step. The next step involves constructing the model based on set criteria that are to be evaluated by the output and defining the model that is constructed matches almost perfectly to the set target. For the last stage, evaluation, the model is tested in real time from the products predicted by the model.

### 3.1 Business Understanding

Each retail company's mission is to attract new customers, retain current customers and sell more to each customer. A retail company has to sell consumers the goods they want at the correct prices to ensure this. In addition, the correct customer experience needs to be guaranteed. In order to achieve the above targets, many retail organizations use data analytics today. Two major factors driving the retail industry are consumer sentiment and brand power. For this reason, for companies in the retail sector, data analytics is extremely relevant. Customer data worth is something that one cannot afford to neglect. Retail companies can increase their value and succeed in this dynamic environment through knowing the feelings of clients. The retail industry has always appealed to researchers because of its scale, multi-faceted and competitive nature, the opportunity for researchers to apply their own domain expertise, and comprehensive coverage by business analysts [2].

### 3.2 Problem Statement

The world of retail is changing rapidly. However, while the vast range of products is something that drives customers to a particular retail store, a lot of these stores fail to sell through a high percentage of their merchandise. This is often due to poor product knowledge experience. Customers spend hours going through hundreds, sometimes thousands of items of merchandise never finding an item they like. Shoppers need to be provided suggestions based on their likes and needs in order to create a better shopping environment that boosts sales and increases the time spent in a store.

Many businesses generate massive amounts of data in the course of their daily operations. For example, large quantities of consumer purchase data are collected daily at grocery store checkout counters. Retailers are interested in analyzing data in order to learn more about their customers' shopping habits. In the form of association rules or collections of frequently occurring objects, association analysis is useful for uncovering interesting associations concealed in large datasets. Retailers may use these kinds of rules to help them find new ways to cross-sell their goods to their buyers. Thus, the objectives are:

1. To recommend similar products to the customers from the vast variety of items.
2. To determine product affinity of the items present in the dataset to associate the items and cross-sell them.

### 3.3 Data Understanding

The data consists of eight columns which consist of the different transactions from the customers. The different variables are as follows:

**Invoice No:** The invoice No is the customers' transaction id containing the details of items purchased by the customer.
**Stock Code:** A stock code is a description of a product or service that is given to a customer by your company.
**Description:** The aim of a product description is to give consumers enough detail about the product's features and benefits to make them want to purchase it.
**Quantity:** The number of the items purchased by the customers.
**Invoice Date:** The date for which the invoice of an item is issued. It is a document that holds a record of a transaction between a buyer and a seller.
**Unit Price:** Unit Price is a measurement used for indicating the price of particular good.
**Customer ID:** It's a unique number assigned to each customer irrespective of frequency they use the service for. For purposes of order processing, monitoring, and customer account management, a customer ID is assigned to each of our accounts.
**Country:** The region from which each customer belongs to. The dataset is dominated by customers from United Kingdom.

#### Exploratory Data Analysis

On Thursdays, which has the maximum sales, the store can have special offers for their customers. On those days, the retailers can also upsell the slow-moving products to their customers by bundling them with the products that are bought more frequently (Fig. 1).

Customers mostly visit the store in the afternoon and more sales is generated in the initial half of the day than in the latter half. Some flash sales can be given in the evenings to improve the sales in that time (Fig. 2).

The most popular product is Paper Craft, Little Birdie. Since the product is bought the greatest number of times, the unit price of the product can be increased so that it might help in generating higher revenue to the company (Fig. 3).

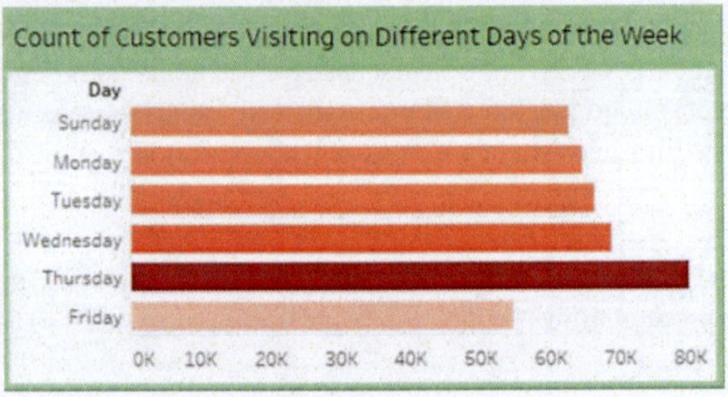


**Fig. 1** Count of customers on different days of week

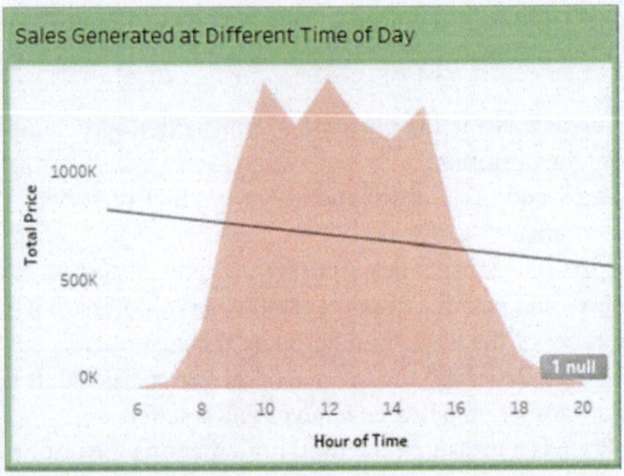


**Fig. 2** Sales Generated at different time of day

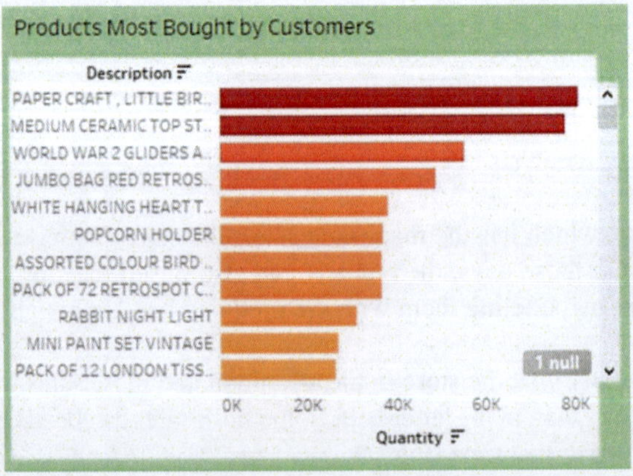


**Fig. 3** Products most bought by customers

### 3.4 Data Preparation

The data preparation process includes all activities from the initial raw data to creating the final data set. The rows having negative quantity, indicating returned or damaged products were removed. There were null values in columns like Invoice No and Customer ID, which were removed. Column representing total price was introduced.

## 4 Model Building

### 4.1 Modelling

**Product Recommendation**
Mikolov et al. created a collection of models called Word2Vec. It entails two distinct techniques, skip-gram and CBOW, which produce word embedding, or distributed word representations. While the CBOW technique predicts the current word by looking at the words around it, the skip-gram technique predicts the words around the current word by looking at the current word itself [4]. Skip-gram technique was used for modelling and generate recommendations.

The measures in the proposed recommendation process are as follows: First, Word2Vec techniques are used to model the data taken as input. The output model is then used to implement the recommendation process. A word2vec model is a basic neural network model with just one hidden layer. The weights learned by the model's secret layer after it has been conditioned are wanted. These weights can then be used as term embeddings. The model has a vocabulary of 3,151 distinct words, each with a vector size of 100. The vectors of all the words in the vocabulary would then be obtained and stored in one location for easy accessibility.

**ECLAT Association Mining**
The association rule mining method ECLAT (Equivalence Class Transformation) generates frequent items only once. The items that often appear in the database are frequent item sets. For finding frequent item sets, there are a variety of algorithms. Apriori is a simple algorithm for the discovery of frequent item sets. But it takes more time to locate the frequent item sets, which is a time-consuming operation, as it needs to search the database repeatedly. The algorithm ECLAT is designed to eliminate the shortcomings of the algorithm Apriori.

The vertical database algorithm is used by ECLAT. This necessitates only scanning the database once. First, the horizontally structured data is converted to a vertical structure by transforming the data set once. By intersecting the TID-sets of each pair of a frequent single item on this data set, mining can be performed. When two TID-lists are intersected, ECLAT uses a technique known as "fast intersection," in which the resulting TID-list is only considered if its cardinality exceeds minimum support. To put it another way, each intersection is removed as soon as it fails to fulfill the minimum support requirements [3].

## 4.2 Evaluation

**Product Recommendation**

A product is given as input and the similar products existing in the store is generated as output (Table 1).

While the vast range of products drives customers to a particular retail store, a lot of these stores fail to sell through a high percentage of their merchandise. This is often due to poor product knowledge experience. Customers spend hours going through hundreds, sometimes thousands of items of merchandise never finding an item they like. Thus, word2vec algorithm will help the retailers know similar items related to a particular product in just few seconds.

**ECLAT Association Mining**

Some of the most frequent item sets based on their scores are generated (Table 2).

The Association Mining Rules do not extract an individual's personal opinion; instead, they look for connections between a collection of elements in each transaction. Investing time and resources in deliberate product placements like this not only reduces a buyer's purchase time, but also advises the consumer to what important items he may be willing to purchase, thus assisting cross-selling in the shop. Rules of association help to reveal all such connections between things from huge databases.

It is thus seen that both the models address to the problem statements discussed above.

**Table 1** Product recommendations

| | |
|---|---|
| REGENCY CAKESTAND 3 TIER | REGENCY TEAPOT ROSES |
| | PINK OVAL SHAPE TRINKET BOX |
| | REGENCY CAKE SLICE |
| | ROSES REGENCY TEACUP AND SAUCER |
| | SILVER OVAL SHAPE TRINKET BOX |
| HEART T-LIGHT HOLDER | STAR T-LIGHT HOLDER |
| | SET/9 CHRISTMAS T-LIGHTS SCENTED |
| | CHERRY BLOSSOM LUGGAGE TAG |
| | SMALL ZINC/GLASS CANDLEHOLDER |
| | CHRISTMAS TREE T-LIGHT HOLDER |
| DOLLY GIRL LUNCH BOX | SPACEBOY LUNCH BOX |
| | CIRCUS PARADE LUNCH BOX |
| | LUNCH BOX I LOVE LONDON |
| | CHILDREN'S APRON DOLLY GIRL |
| | CHILDRENS DOLLY GIRL MUG |

**Table 2** Frequent items

| | |
|---|---|
| SET/6 RED SPOTTY PAPER PLATES | SET/6 RED SPOTTY PAPER CUPS |
| PLASTERS IN TIN WOODLAND ANIMALS | PLASTERS IN TIN SPACEBOY |
| SET/20 RED RETROSPOT PAPER NAPKINS | SET/6 RED SPOTTY PAPER CUPS |
| SET/6 RED SPOTTY PAPER PLATES | SET/20 RED RETROSPOT PAPER NAPKINS |
| ALARM CLOCK BAKELIKE RED | ALARM CLOCK BAKELIKE GREEN |
| SPACEBOY LUNCH BOX | DOLLY GIRL LUNCH BOX |

### 4.3 Deployment

These models can be deployed in real time on new datasets for increasing the revenue of the store. While deploying the results obtained from the algorithms of recommender system and ECLAT association technique will be matched with the ones that will obtained in real time. At last, the efficiency of the two models can be evaluated based on these results.

## 5 Conclusion

We may conclude that recommender systems are a powerful new technology for extracting additional value for a company based on their customer databases. Recommendation systems support consumers by helping them to discover products that they like. In e-commerce, recommendation systems are increasingly becoming a key instrument. It is very necessary to get frequent item sets. We see that the best way to get the products needs to be chosen. We get confused about what should be bought if we go to some store. Because of the large number of database data stores. Therefore, many algorithms are used by shopkeepers to find the best way to provide consumers or customers with goods. We use the Eclat algorithm. The Eclat algorithm helps to find common sets of items. With less time, it finds common item sets and occupies less memory as it follows depth first search.

A retail company depends on its clients. Retailers must continually come up with systems that can draw buyers in order to keep customers interested in their products, which can result in stronger profits for them. The market is governed by data in this digital age. A retailer will recommend the best choice to clients to increase the profits of the store with precise analysis of the data. With the support of the AI-driven recommendation engine, better and reliable suggestions can be made to the consumer, enabling the retailer to engage the customer both online and in-store.

## References

1. Chandramana SB (2017) Retail analytics: driving success in retail industry with business analytics. Research J Soc Sci Manag 7:159–166

2. Dekimpe MG (2020) Retailing and retailing research in the age of big data analytics. Int J Res Mark 37(1):3–14
3. Moe HM (2019) ECLAT based market basket analysis for electronic showroom. Int J Adv Res Dev 4:25–28
4. Ozsoy MG (2016) From word embeddings to item recommendation
5. Suchacka G, Chodak G (2017) Using association rules to assess purchase probability in online stores. Inf Syst e-Bus Manag 15:751–780

# Factors Influencing Online Shopping Behaviour: An Empirical Study of Bangalore

Hemlata Joshi, Anubin Binoy, Fathimath Safna(✉), and Maria David

Department of Statistics, CHRIST (Deemed to be University), Bangalore, India

**Abstract.** Online shopping is growing rapidly in India, predominantly driven by tremendous and substantial divulgatory activities among millennial consumers. Online shopping is becoming more popular and attracts significant attention because it has excellent potential for both consumers and vendors. The convenience of online shopping makes it more successful and makes it an emerging trend among consumers. When all the companies are striving against one another, certain factors influence the behavior of customers. This paper analyses the relationship between the critical, independent variables, including consumer behavior, cultural, social, personal, psychological, and marketing mix factors. The results revealed that the influence of Brand as a factor had positively influenced the customer's decisions in shopping online and evaluates the customer's level of satisfaction with Online shopping. Results provided in this research could be employed as reference information for Ecommerce app builders and marketers regarding such issues in the city.

**Keywords:** Marketing · E-Commerce · Online shopping · Consumers · Logit

## 1 Introduction

Technology today is termed as tools and machines used to solve real-world problems. Technology is a set of skills, techniques, methods, and processes used according to the requirement. Nowadays mobiles, tablets, computers and most importantly the internet has become a part of our lives without which we cannot live. Technology has its advantages and disadvantages. It has helped develop the global economy and has brought in many innovations. Over the last few years, e-commerce has become an indispensable part of international retailing, and it has also undergone several changes and will keep changing according to the trends. This research's motivation is based on the factors that influence consumers' online shopping behavior and better understand e-commerce platforms and the consumers. Online shopping has been rapidly increasing in India. There are over 75 million online buyers in India, and the number of online shoppers is over 2.05 billion people, or 26.28% of the world's population of 7.8 billion. But this study contains the online shopping behavior of consumers from Bangalore, Karnataka. Online shoppers have been growing at a tremendous rate, and consumers' annual shopping has been increasing. But there are advantages and disadvantages for the same, and several factors are being considered by the consumers while doing online shopping. This study


S. Shukla et al. (eds.), *Data Science and Security*, Lecture Notes in Networks and Systems 290,
https://doi.org/10.1007/978-981-16-4486-3_31

mainly focuses on understanding these factors such as Brand, quality, price, review, etc. [2, 3, 5]. The disadvantage is that consumers may stay online and shop but do not purchase any product, resulting in wastage of time. Online shopping is not possible without a gadget and internet services [1], and only those who can enjoy this who has the privilege of both of these. This has increased because detailed product information and improved services attract more and more people to change their consumer behavior from traditional mode to online shopping. Companies have realized that the factors affecting consumer behavior are unavoidable, and they also have to keep changing their marketing strategy. The consumer can purchase products online 24X7, and it also provides the minimum possible price with various cash backs and discounts. Sometimes faulty products are delivered, and it consumes time for exchanging the product too. There are delivery charges, and the products arrive in exceptionally preserved packages, so one cannot touch and feel them [4]. This study endows a deeper understanding of the effect of different factors on consumer buying behavior. This study is organized around customer behavior, specifically customers of online shopping, to analyze and evaluate the factors influencing online shopping behavior, including understanding the usage of different apps for Online shopping, Studying the customer's level of satisfaction about online shopping, and analyzing the dependability of age groups in online shopping during the Covid19 pandemic.

## 2 Methodology

Data employed in this study were obtained from a structured questionnaire design. The questionnaire was distributed among Bengalureans. The questionnaire was divided into two sections. The first part includes socioeconomic and demographic variables. The second part presents the factor that leads to customer shopping online. The first five questions were the demographic information of the respondents. Then each of the remaining questions presents the factor that leads to customer shopping online. All statements were formulated on a 5-point Likert-type scale between "strongly disagree (1)" to "strongly agree (5)" and "Yes" or "No." Since the response variable 'brand' is binary, the study employed Logistic Regression in predicting the influence of Brand in online shopping by developing the relationship between the factors/predictors like gender, monthly salary, level of education, and age group. Let $\pi$ denote the probability of influence of Brand on online shopping when p predictor variables are given, and the relationship between the probability $\pi$ and p predictors is represented in the form of a logistic model i.e.

$$\pi = Pr(Y = 1|X_1 = x_1, \ldots X_p = x_p) = \frac{e^{\beta_0+\beta_1 x_{1+\ldots+}\beta_p x_p}}{1 + e^{\beta_0+\beta_1 x_{1+\ldots+}\beta_p x_p}} \tag{1}$$

The function given in Eq. (1) is the logistic regression function. Among the non-linear regression coefficients $\beta_0$, $\beta_1 \ldots \beta_p$, it is linearised by the logit transformation, i.e., if the probability of Brand influencing online shopping is $\pi$, then the resulting ratio $\frac{\pi}{1-\pi}$ is the log odds of Brand influencing online shopping.

Here,

$$1 - \pi = Pr(Y = 0|X_1 = x_1, ..., X_p = x_p) \\ = \frac{1}{1 + e^{\beta_0 + \beta_1 x_1 + ... + \beta_p x_p}}. \quad (2)$$

Then,

$$\frac{\pi}{1 - \pi} = e^{\beta_0 + \beta_1 x_1 + ... + \beta_p x_p} \quad (3)$$

Take natural log on both sides,

$$\text{logit}(\pi) = \log(\frac{\pi}{1 - \pi}) \\ = \beta_0 + \beta_1 x_1 + \ldots + \beta_p x_p \quad (4)$$

In Eq. (4), logit($\pi$) is a linear function in terms of the regression parameters called the logit function and the range of $\pi$ in Eq. (1) is between 0 to 1, and the range of values of $log\left(\frac{\pi}{1-\pi}\right)$ is between $-\infty$ and $\infty$, which makes the logits quite suitable for linear regression model and the error term 'e' satisfies all of the basic assumptions of ordinary least squares.

The percentage of respondents on each row who agree with the statement is shown to the right of the zero lines in the survey responses of the factors affecting online shopping shown in Fig. 1. The counts (or percentages) who disagree are shown to the left of the zero lines. The number of respondents who don't agree or disagree is divided in half and represented by a neutral colour. Here, we can see that Brand, accessibility, and price

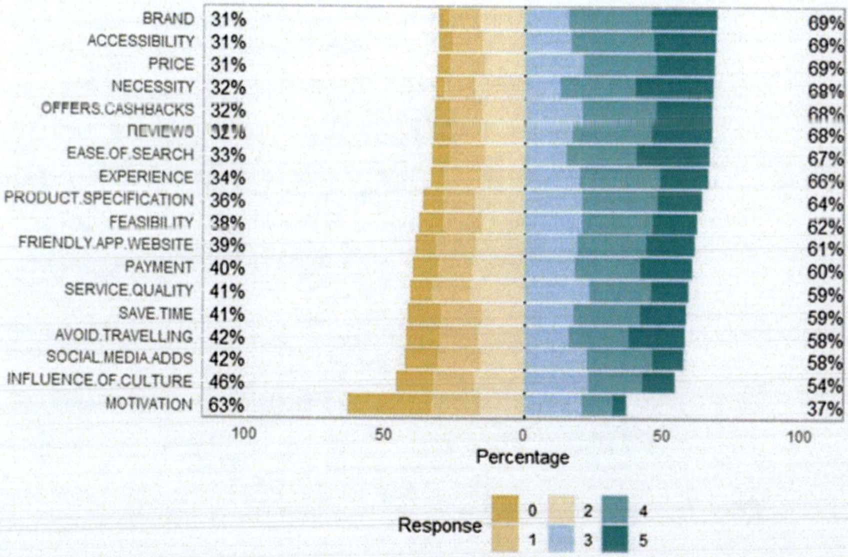


**Fig. 1** Bar plot of the Likert scales to survey responses of the factors influencing online shopping

are the most influencing factors, with 69% of the respondents agreeing. The factors Necessity, Offers/Cashbacks, and reviews have 68% of respondents. The factor that had the highest percentage (63%) of disapproval was "motivation," which is followed by the "influence of culture"(46%). Therefore, the top five factors influencing online shopping are Brand, accessibility, price, necessity, and offers/cashback.

From the Polychoric correlations of the factors influencing online shopping as given in Fig. 2, the dark blue shaded values indicate a high correlation (>0.72), and a lighter shade of blue shows a moderate correlation between the factors. The lightest shade of blue indicates a weak correlation (<0.45) between the factors. From our study's responses, we found a high correlation (>0.82) between the factors- "product specification" and "service quality," "friendly app website," "Accessibility," "Offer cashback and price," "Reviews," and "experience."

The Chi-square test of independence was done with the x variable (did you rely more on shopping during covid 19) and the variable y(Age) to find a relationship among the 2 variables.

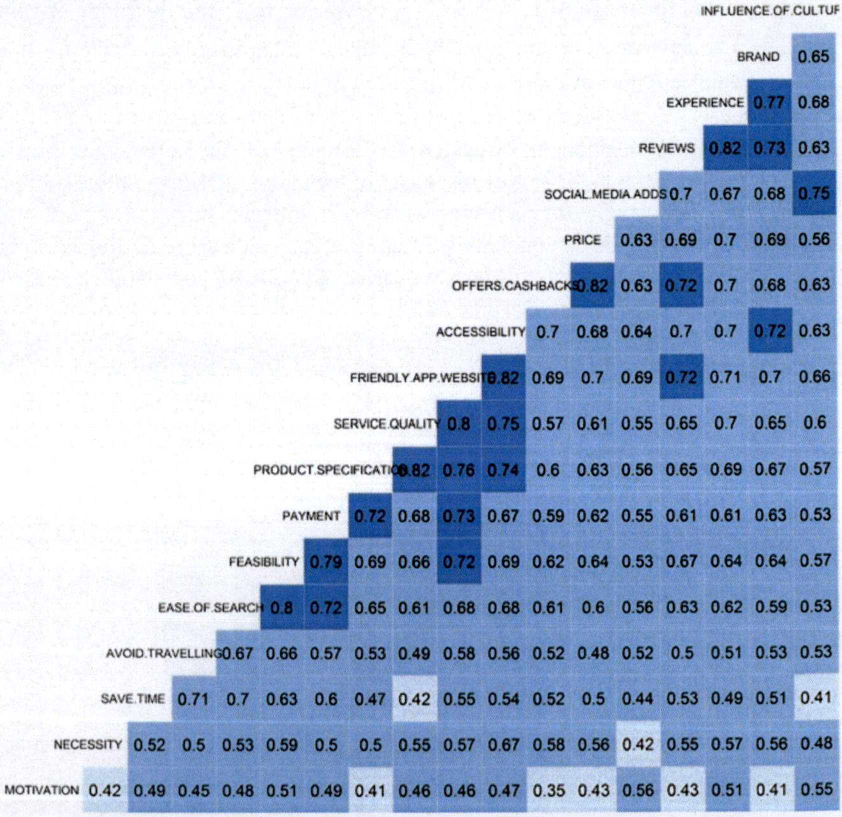


**Fig. 2** Polychoric correlations of the factors influencing online shopping

**Table 1** Chi-squared test

| | DF | Chi-squared value | P-value | LOS |
|---|---|---|---|---|
| Pearson's chi-squared test | 5 | 15.15 | 0.009739 | 0.05 |

Chi-square test of independence is defined as,

$$X^2 = \sum_{i=1}^{r} \cdot \sum_{j=1}^{c} \frac{(O_{ij} - E_{ij})^2}{E_{ij}}$$

In the survey conducted, We had asked our respondents, "Did you rely more on shopping during the Covid-19 pandemic". This question was answered with either Yes or No. The respondents' age was also recorded in the variable "Age." The variable Age had six groups 18–24, 25–31, 32–38, 39–45, 46–52 and above 52. The hypothesis was constructed to test if respondents relying more on shopping online during covid-19 had a significant difference between the age groups. Null Hypothesis: Respondents relying on online shopping during the Covid-19 pandemic were independent of the respondents' age groups.

Alternate Hypothesis: Respondents relying on online shopping during the Covid-19 pandemic were dependent on the respondents' age groups.

From contingency Table 1, the p-value (0.009739) obtained is less than 0.05, indicating a relationship between the variables x and y. Therefore, we reject $H_0$ (Null hypothesis) and fail to reject $H_1$ (Alternate hypothesis). From the chi-square test, Respondents relying on online shopping during the Covid-19 pandemic is dependent on the age groups of respondents.

According to the findings of the survey on the missing factors in online shopping, the word cloud in Fig. 3 answers to "What is missing in online shopping?" The words "nothing," "products," "quality," "delivery" are the words that are highly common from the responses. The word with the highest frequency is "nothing," which signifies that nothing is missing in online shopping and are satisfied with the current shopping experience. The word with the second-highest frequency is found to be "quality," which indicates that respondents are concerned about the quality of the products, which is missing in online shopping. "Products" And "product" also appear frequently in the responses, which indicates that they are less satisfied with the products available in online shopping.

The bar graph in Fig. 4 shows us the top ten frequent words that appeared in the responses to the survey question, "What are the changes to be made in online shopping?" The bar graph clearly shows us that "products," "nothing," "quality," "delivery," "online," "time" are the words with the highest frequencies. The changes expected by the respondents are mainly dealt with the products, quality of the products, and the delivery of the products. At the same time, the word with the second-highest frequency is "nothing" which implies that most of the respondents do not expect any changes in online shopping and are satisfied with the present online shopping experience.

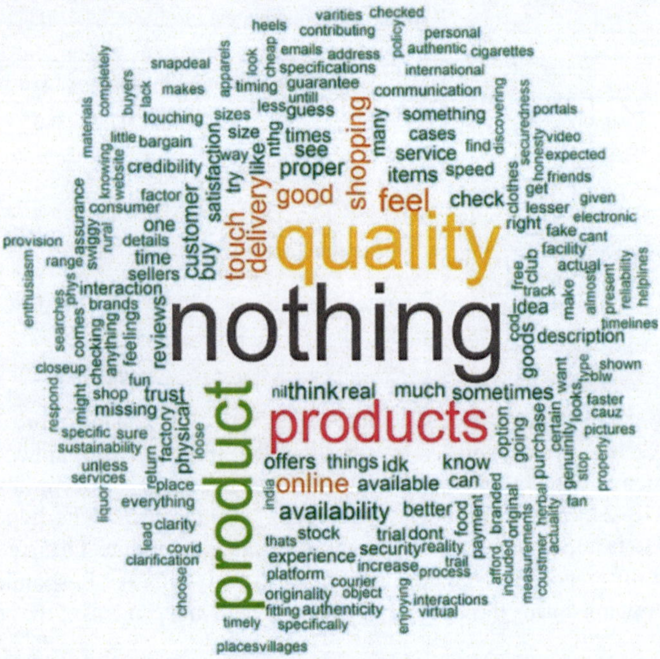


**Fig. 3** World cloud of survey responses to the missing factors in online shopping

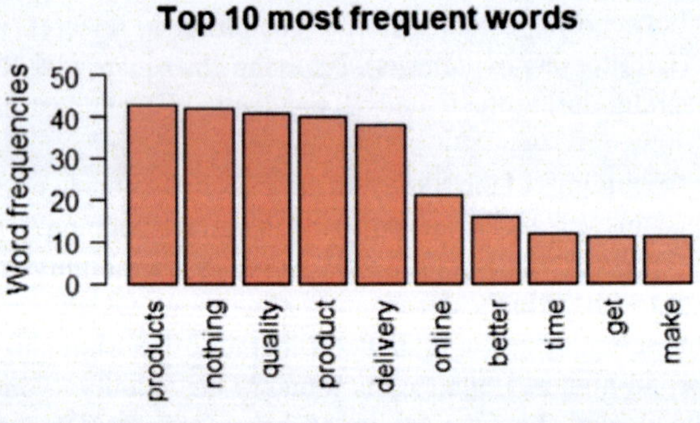


**Fig. 4** Bar chart showing frequent words for the changes expected by respondents in online shopping

This study employed Logistic Regression in predicting the factors influencing consumers' online purchasing behaviours. The model was created based on the response variable "Brand," a factor that influences online shopping, and eight independent variables education, time, necessity, experience, accessibility, the influence of culture, gender, and social media.

Quality control checks by ensuring all of the factor levels are represented by people shopping online on the response variable Brand and otherwise. For example, as listed a few in Table 2, The model is employed to predict whether consumers purchase only

**Table 2** Cross tabs for quality control check

| | Education | | | Necessity | | Accessibility | |
|---|---|---|---|---|---|---|---|
| Brand | Other | Postgraduate | Undergraduate | No | Yes | No | Yes |
| No | 13 | 14 | 75 | 30 | 72 | 87 | 15 |
| Yes | 28 | 23 | 178 | 77 | 152 | 16 | 213 |

branded products online concerning influencing the decision. We can fit a generalized linear model as the variables are of a categorical type. The model provides us with the respective z-values, the null deviance, and residual deviance by which we found the *Mc Fadden's Pseudo* $R^2$.

McFadden's R-square is defined as:

$$R^2_{MF} = 1 - \frac{LL_{full}}{LL_0}$$

where $LL_{full}$ is the full log-likelihood model, and $LL_0$ is the log-likelihood function of the model with the intercept only. The value of McFadden's R-square is found to be approximately equal to 79%. And this depicts that the model is a good fit. We can plot the predicted probabilities for each sample having shopped online based on Brand and color by whether or not they shopped online.

Logistic regression makes profound predictions and finds the best model to produce high-quality, reliable outcomes of 79%. In Fig. 5, we found that more than 200 people have shopped online based on the factor brand. A c-statistic is found using the model and is 0.9869 or 99%, which indicates that the model is almost perfect. Further, we will work on the Hosmer–Lemeshow statistic. To compute HL test, the data is divided into

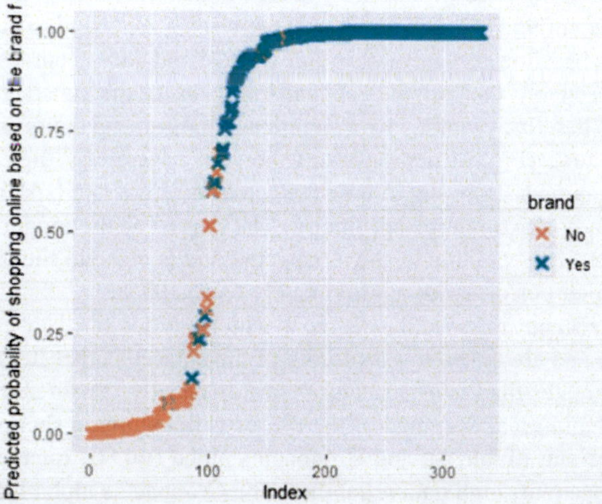


**Fig. 5** Scatter plot showing the predicted probability of shopping online based on the factor "brand"

**Table 3** Hosmer and Lemeshow goodness of fit

| Hosmer and Lemeshow goodness of fit (GOF) test | | |
|---|---|---|
| data: logistic$y, fitted(logistic) | | |
| **X-squared** = 2.9718 | **df** = 8 | **p-value** = 0.9361 |

several subgroups, and the observed and expected frequencies in each of the group are calculated, and then the Chi-squared statistic calculated gives the Hosmer–Lemeshow statistic as

$$\chi_{HL}^2 = \sum_{g=1}^{G} \frac{(O_g - E_g)^2}{E_g\left(1 - \frac{E_g}{n_g}\right)}$$

Og is the observed events, Eg is the expected events, and ng signifies the number of observations for the g' th group, and G is the number of groups.

The data is grouped into ten groups, and the Hosmer and Lemeshow (HL) goodness of fit test is obtained as $\chi$ 2 HL = 2.9718, df = 8, p−value = 0.9361, from Table 3 it can be seen that the at 95% confidence level, the p-value is greater than 0.05. and it shows that the model is a good fit.

Models that predict categorical labels are called classification models. The training dataset will be used to construct our model, and the test dataset will be used to evaluate it. This method of evaluating model efficiency is known as holdout-validation. The data is split into two partitions, 70% for training and 30% for tests. A baseline accuracy exists for the majority class of the target variable of 69%. The training set and Confusion matrix predictions on training data are found by dividing the number of rows of training data to get an accuracy of 95.66%. Predictions on the test set are found to get an accuracy of 94%. The data's baseline accuracy was 69%, with 96 and 94% accuracy on training and test results, respectively. On both the train and test datasets, the logistic regression model outperforms the baseline accuracy by a significant margin, and the results are excellent.

The ROC Curve (Fig. 6) demonstrates the area covered by the predictive model graphically. It can be seen in Fig. 6 how the actual positive rate (specificity) is plotted against the false positive rate (1-specificity).The curve for the model associated with Brand as the predictor variable is above baseline and is close to the upper left corner, which the classifier performance approximately equals 97%.

Internal consistency refers to the degree to which entities in a dataset evaluate a comparable variable and are linked to the affinity of the dataset's other items. A Cronbach's alpha () value greater than 0.70 and closer to 1 is usually considered a reliable score (acceptable). The internal accuracy of the use of various online shopping applications is taken into account (Table 4), and Cronbach's alpha (=0.80) for the dataset used in this paper demonstrates high data reliability. The Cronbach's alpha for all of the listed subscales was greater than 0.70. Table 5 displays the Cronbach's alpha values for the subscales.

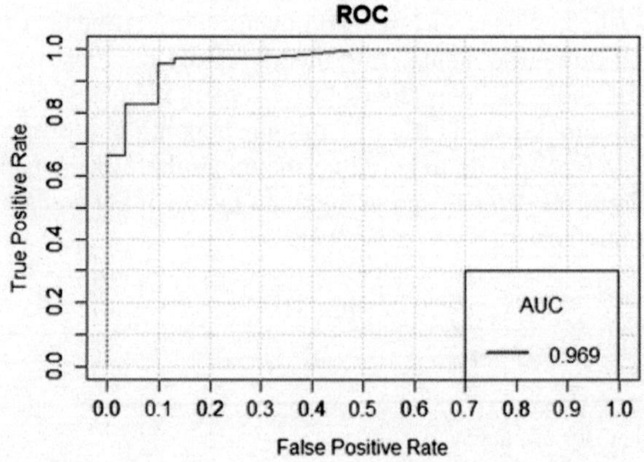


**Fig. 6** Receiver operating characteristic curve by the predicted model

**Table 4** Cronbach's reliability test results

| Cronbach's reliability test results | | |
|---|---|---|
| Cronbach's alpha | Cronbach's alpha based on standardized items | Number of items |
| 0.8 | 0.8 | 6 |

**Table 5** Cronbach's alpha test for subscales

| Cronbach's alpha test for the subscales | |
|---|---|
| Factor (usage of different apps | Cronbach alpha |
| Amazon | 0.77 |
| Myntra | 0.76 |
| Flipkart | 0.8 |
| Zomato | 0.76 |
| Big Basket | 0.76 |
| Medlife | 0.74 |

## 3 Conclusion

Nowadays, online shopping has become a regular part of an individual's life; optimizing e-commerce stores is essential to have a better experience anticipated by potential customers. This analysis will assist the e-commerce industry in developing a favourable environment that will contribute to increased revenue; negative experiences will likely lead to irreversible consumer loss. The primary aim of this study was to identify factors that influence consumers' willingness to buy goods from online retailers. Based on theoretical study, we surveyed Bengalureans and analyzed the users' perceptions, which were influenced by the factors when buying products online. Amid the COVID-19 spread in 2020, Online consumption habits are growing rapidly in cities like Bangalore. The satisfaction level of using online products by customers increases each day because of the innovative technologies being used by online shopping platforms. In this study, it is observed that many people do trust online shopping platforms, which is a plus point for such online platforms. This is one of the main factors influencing online shopping: whether to shop online or not.

## References

1. Bucko J, Lukas K, Martina F (2018) Online shopping: factors that affect consumer purchasing behaviour. Cognet Bus Manag 5(1):1–15
2. Chayapa K, Wang CL (2011) Online shopper behaviour: influences of online shopping decision. Asian J Bus Res 1(2):66–74
3. Khanh NTV, Gim G (2014) Factors affecting the online shopping behaviour: an empirical investigation in Vietnam. Int J Eng Res Appl 4(2):388–392
4. Sivakumar A, Gunashekharan A (2017) An empirical study on the factors affecting online shopping behaviour of millennial consumers. J Internet Commer 16(3):1–15
5. Pandey A, Parmar J (2019) Factors affecting consumer's online shopping buying behavior. In: Proceedings of 10th international conference on digital strategies for organizational success, pp 541–548

# Smart Electricity and Consumer Management Using Hyperledger Fabric Technology

Faiza Tahreen[1](✉), Sushil Kumar[2], Gopal Krishna[1], and Filza Zarin[1]

[1] Department of Computer Science, Netaji Subhas Institute of Technology, Patna, Bihar, India
[2] Department of Computer Science, Lok Nayak Jaiprakash Institute of Technology, Chapra, Bihar, India

**Abstract.** Blockchain technology comes up with the great potential to foster various section like Banking, medical, judicial, and education, with its distinctive combination of features, such as, decentralization, immutability, and transparency, etc. [7]. This paper propose an innovative, cost-effective solution, to optimize electricity connection and consumer management system, with minimum physical consumer interactions, and the least negative impact on his/her lifestyle, comfort, and convenience. Besides, it would also provide data privacy as well as transparency with high-security services and ease the workload of the employees in the electricity department by integrating their workspace with hyperledger fabric [1]. Here, developed a consummated end-to-end typical business blockchain application that have the client application where the outside world communicates using the chaincode which is deployed and executed inside a fabric network. This system will save much of the employee's time, which was earlier taken by the traditional electricity management system.

**Keywords:** Chaincode · CouchDB · Client-application · Shell-script · fabricCA · Docker-compose · Orderer-Peer

## 1 Introduction

Now a days, we're seeing increase in the development of blockchain-based applications supported by emergence of various blockchain platforms, as it promises benefits in trustability, transparency, association, collaboration, organization, recognition, and reliability [3]. There are many blockchain platform available in the market which help in the growth of the industry, and hyperledger fabric is one of them. Hyperledger Fabric with modular architecture is a project of Hyperledger [6], proposed for building blockchain based solutions or applications. The architecture modularity allows network developers to plug in their favour modules like consensus, and membership services, distinguishing it from other blockchain solutions [2]. As we all know, in today's digital age, electricity has materialized as the most significant and censorious input for sustaining the process of economic as well as social development. For the public(consumer), it speaks for the face of the utility. To sustain the growth of the economy the efficient functioning of this department is essential. The public grievances or complaining, mostly relating to


S. Shukla et al. (eds.), *Data Science and Security*, Lecture Notes in Networks and Systems 290,
https://doi.org/10.1007/978-981-16-4486-3_32

wrong or inflated billing, defective meter and non-replacement of the defective meter, timely maintenance and disconnection, new connections and extension of loads besides unscheduled outages etc. This paper proposed a system which represent how the electricity department get benefited from hyperledger technology and its characteristics. The modular architecture of hyperledger contains various modules formed of layer architecture, discuss follows uses for better system designing [5]. *The consensus layer* responsible for generating agreements and conforming to the correctness to achieve reliability in network. *The smart-contract layer(chaincode)* is responsible for processing request or demand as well as determining the validate condition, and also allows querying and updating the ledger details. *The communication layer* helps in establishing communication between every node and organization which took part in the network configuration to store all details. *Identity services* allows the creation of trust during the setup of a blockchain instance. *Policy services* manages multiple policies specified in the system, such as endorsement policy, consensus policy, or group management policy which provide better interaction. *The CouchDB* non-relational database to store the queries, and details.

The paper bringing all the above modules together, proposed a systematic network which is a fabric network. This network is launch using shell script which will accelerate up a blockchain network consist of peers, orderers, certificate authorities, and more. Also, access the ledger, install and instantiate chainodes that will be used by application programs. The Paper first having the model approach of the system, like about the stakeholders, the information we want to store, and the possible services that consumer can get. Then Paper move forward with implementation part, and dive into the directory of the model which contains various supporting tools, APIs programs, files, and folders. After this showcase of analyzing what components the fabric network container contains along with understanding the chaincode. The rest of the paper contains bringing up the first-network, and deploying the client application along with the observation of the implementation on the terminal with desire output. At last Paper is Concluded with the scope of better electricity management system using hyperledger fabric technology.

## 2 Proof- of- Concept

Smart electricity & consumer management is a database of electrical department records stored in the ledger of a fabric network [11]. In the world state database inside the ledger the data are stored. The data interaction is through chaincode. In Smart electricity and consumer management the functions which can communicate with the stored data in the ledger is written in chaincode. These function are for initiation, query, and update the database. The world state is queried or updated only through the chaincode functions, and any update is logged in the blockchain inside the ledger is as a tamper-resistant record [10]. Above mentioned happening inside the fabric network. SDK'S (Software Development Kit) are utilize by the Client application to communicate with the deployed chaincodes, and the fabric network. The client application program is written to match the functions defined inside the chaincode. Here, for the chaincode interaction and user enrollment the smart electricity and consumer management have four sets of JavaScript codes that run in Node.

Let's deep-dive into the smart electricity and consumer management Development approach.

## 3 Proposed Model

1. **The Stakeholders**

    1. **Consumers:** They are the ones who will avail of all the services like getting new connections, paying the bill, retrieving bill details, changing ownership, changing personal information or details, etc., shown in Fig. 2.
    2. **Sub-Division Office:** They will provide all the information regarding the services to the customer so that the consumer avails all the services, collect all the documents and related details, and finally they will forward these documents to the Division Office, shown in Fig. 2.
    3. **Division Office:** They will work on the services that the consumer wants to avail, continue the process and finally complete the task for the consumer, shown in Fig. 2.

2. **Information we want to store**

    1. **Consumer Details:-** It consists of all the personal and connection details of the consumer like: name, address, Mobno, aadhar no, meter no, etc., shown in Fig. 1.
    2. **Bill Details:-** It consists of all the bill details of the consumer like:- name on the bill, billno, amount of bill, pre meter &curr meter reading details etc., shown in Fig. 1.

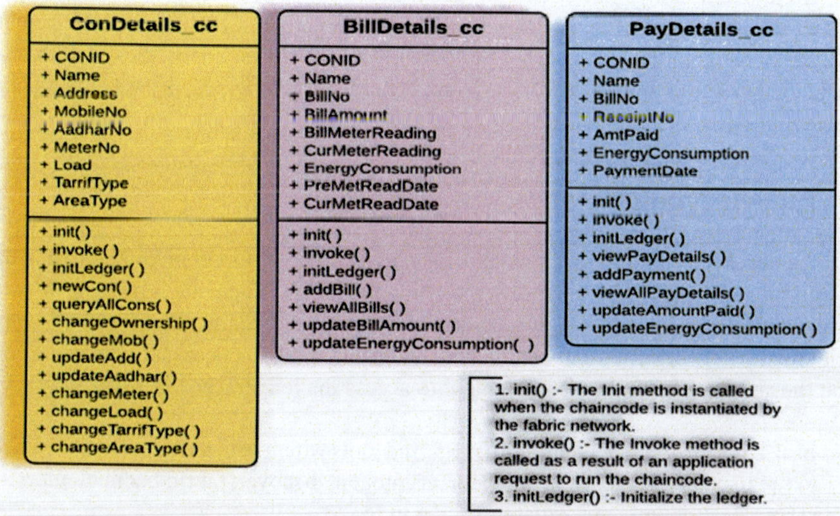


**Fig. 1** Chaincode structure contains information we want to store

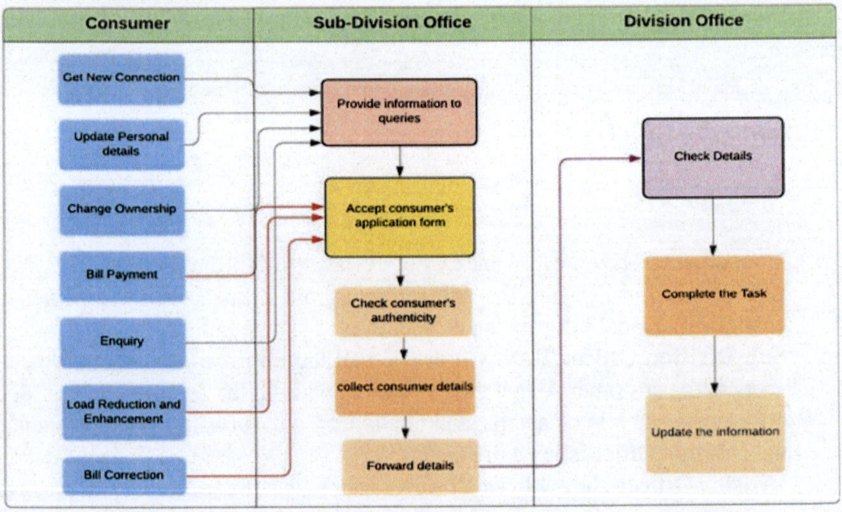


**Fig. 2** Architecture and workflow of the model

3. **Payment Details:-** It consists of all the payment details of the consumer like:- Name of the consumer, Address of the consumer, Bill number, The receipt number of the bill, Amount paid by the consumer, The energy utilized by the consumer, The date on which bill amount is paid by the consumer, shown in Fig. 1.

3. **Possible Services**

Getting new connection, Changing the load, Paying a bill with transparency, Updating or changing the mobile number, Updating the Email-Id, Change or update personal address, Changing the ownership, Changing the meter, Correction in the bill with transparency, View the payment details, View the issued bill details, Viewing the energy consumption.

## 4 Implementation

For the development of the system there need to have some important directory and their supporting tools which participate in modeling the model. There are four basic directory in the model these are bin, first-network, electricity, and chaincode. These all help in bringing up network environment, install and instantiating the necessary tools and codes that the application programs implement to access the ledger, shown in Fig. 3.

1. **bin:** The bin directory contains the required tools which help the hyperledger network to run like configtxgen, configtxlator, cryptogen, discover, fabric-caclient, fabric-ca-server, peer, idemixgen, orderer, shown in Fig. 3.
2. **first-network:** The first-network directory contains all the folders and files which are used to build the network while deploying, byfn.sh folder in this is a shell script

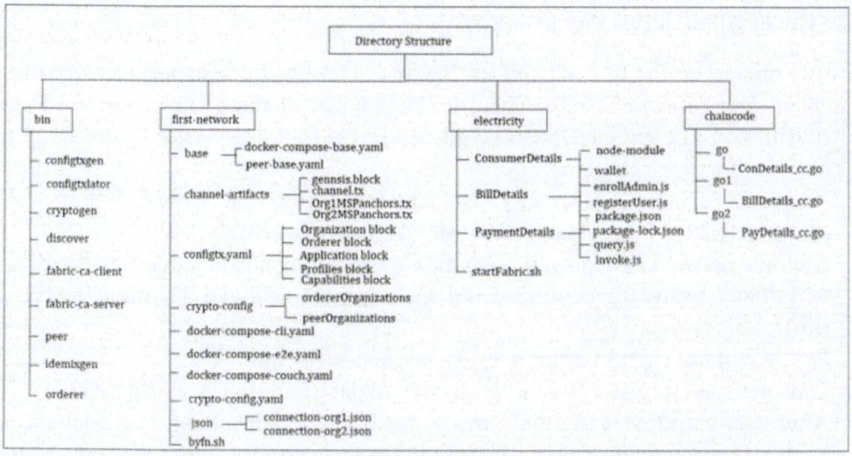


**Fig. 3** Directory structure of the model with supporting folders

file used to run the network. first-network directory is responsible for generating the network artifacts, running up the network and stopping it in hyperledger fabric, shown in Fig. 3.

3. **electricity:** The electricity directory contains application programs which are written in node.js. In this folder, there are three sub-folders each of which contain API programs for their respective chaincode stored in the chaincode directory and one shell script startFabric.sh required for deploying.The three sub-folders are BillDetails, ConsumerDetails, and PaymentDetails each contains node-modules folder, wallet folder, enrollAdmin.js file, registerUser.js file, package.json file, package-lock.json file, query.js file and invoke.js file, shown in Fig. 3.
4. **chaincode:** The chaincode directory contains three sub-folders go, go1 and go2 each of which contains a chaincode written in go language.

## 4.1 First Network of Smart Electricity and Consumer Management

The fabric network of smart electricity and consumer management is the first network having:- North (org1) and South (org2) two organization each consequent with two peer nodes (Division (peer0) and SubDivision (peer1)). One ordered organization and one ordered node using SOLO as the ordering method. SOLO ordering method is used by one ordered organization and one orderer node. CouchDB database is run by each peer node, and each organization is with CA (Certificate Authority), running fabric-CA software with proper configuration. There is a client communicating with first network as a CLI (command-line interface). All these modules deployed as containers, and running on a host.

### 4.2 Client Application Deployment

Firstly bringing up the first network of Smart electricity & consumer management by command *first-network/byfn.sh.* Then, for deploying all three chaincode ConDetails cc.go, BillDetails cc.go, PayDetails cc.go, follow the following steps:-

1. Navigate to the electricity folder and run the startFabric.sh shell script**./startFabric.sh,./startFabric.sh go1,./startFabric.sh go2.**
2. After the network is deployed go to the ConsumerDetails, PaymentDetails, BillDetails folder using the command: **cd ConsumerDetails**, **cd PaymentDetails**, **cd BillDetails** respectively.
3. First, use **node enrollAdmin** in every respective folder.
4. Then use **node registerUser** in respective folder to register & enroll user1.
5. After user enrollment use **node query** for chaincode invoking. Use contract API *evaluateTransaction()* for query with the arguments. For eg: *const res = await contract.evaluateTransaction ('veiwAllBills');*
6. Use **node invoke**. Use contract API *submitTransaction()* to invoke with the arguments. For eg: *await contract.submitTransaction ('addBill', 'CON6', 'Kakashi', 'sp-161', '1611', '1423', '1497', '74', '18-07-2020', '17-08-2020');*

## 5 Result on the Terminal

Result Contain all the Billdetails of the consumers (shown in Fig. 4) which is temperproof, transparent, and unalterable, as saved in private distributed ledger, showcase in tabular form below.

| Cons -umer | Bill-Amt | BillNo | CurrMet -Read | CurrMet-ReadDate | EnergyCo - sumption | Name | PreMet -Read | PreMet-ReadDate |
|---|---|---|---|---|---|---|---|---|
| CON0 | 1654 | da-345 | 1853 | 13-08-2020 | 85U | Shirish | 1768 | 12-07-2020 |
| CON1 | 2887 | ew-234 | 2513 | 09-08-2020 | 172U | Goku | 2341 | 10-07-2020 |
| CON2 | 889 | wa-287 | 3278 | 11-08-2020 | 65U | Naruto | 3213 | 08-072020 |
| CON3 | 1256 | kj-419 | 2986 | 10-08-2020 | 112U | Sasuke | 2874 | 13-07-2020 |
| CON6 | 1611 | sp-161 | 1423 | 17-08-2020 | 74U | Kakashi | 1497 | 18-07-2020 |

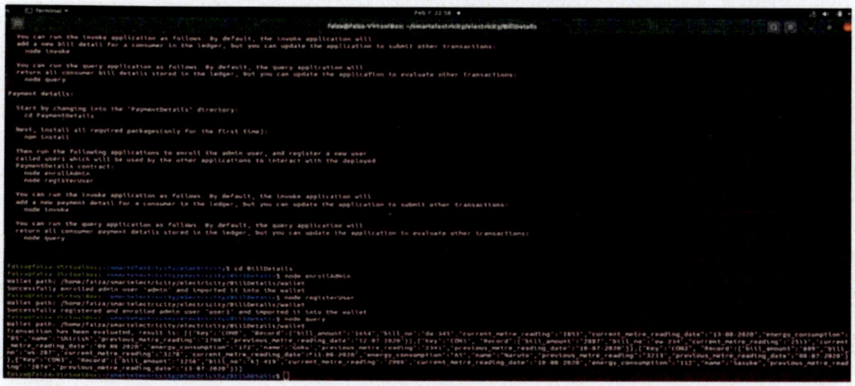

**Fig. 4** Output

# 6 Conclusion

In the electricity management system, there are many challenges and issues which are faced by the consumers in getting connections, updating information, paying bills, etc. One of the elucidation to resolve these issues is the implementation of blockchain architecture. The hyperledger fabric technology [9] equips electricity connection and consumer management system with a peer-to-peer network where non-trusting nodes communicate with each other without trusted emissary in a valid manner, saving all the data and details related to the consumers in a ledger which is cryptographically secure and cannot be tampered with, therefore it becomes easy for the employees of the electricity department to manage everything and consumers don't have to face the problems which occur due to the manual errors done by the department's staff like extra bill charge, paying fine instead of paying bills, changing the ownership, delays in getting a new connection, and in updating the personal details, etc. The hyperledger fabric's modular architecture [8] makes electricity connection and consumer management system robust, flexible and reliable as well as economizing the department's staff time which is wasted by today's electricity management system architecture which follows the long procedure in order to meet with the consumer demands.

# References


1. Cachin C, et al (2016) Architecture of the hyperledger blockchain fabric .In: Workshop on Distributed Cryptocurrencies and Consensus Ledgers, vol 310, p 4, Chicago, IL
2. Nasir Q, Qasse IA, Talib MA, Nassif AB (2018) Performance analysis of hyperledger fabric platforms. In: Security and Communication Networks, vol 2018, Hindawi
3. Thakkar P, Nathan S, Viswanathan B, (2018) Performance benchmarking and optimizing hyperledger fabric blockchain platform. In: 2018 IEEE 26th International Symposium on Modeling, Analysis, and Simulation of Computer and Telecommunication Systems (MASCOTS), pp 264–276. IEEE press
4. Androulaki E, et al (2018) Hyperledger fabric: a distributed operating system for permissioned blockchains. In: Proceedings of the Thirteenth EuroSys Conference, pp 1–15

5. Javaid H, Hu C, Brebner G (2019) Optimizing validationphase of hyperledger fabric. In: 2019 IEEE 27th International Symposium on Modeling, Analysis, and Simulation of Computer and Telecommunication Systems (MASCOTS), pp 269–275, IEEE press
6. Yamashita K, Nomura Y, Zhou E, Pi B, Jun S (2019) Potential risks of hyperledger fabric smart contracts. In: 2019 IEEE International Workshop on Blockchain Oriented Software Engineering (IWBOSE), pp 1–10, IEEE press
7. Pilkington, M (2016) Blockchain technology: principles and applications: Research handbook on Digital Transformations. Edward Elgar Publishing
8. Sukhwani H, Wang N, Trivedi KS, Rindos A (2018) Performance modeling of hyperledger fabric (permissioned blockchain network). In: 2018 IEEE 17th International Symposium on Network Computing and Applications (NCA), pp 1–8. IEEE press
9. Foschini L, Gavagna A, Martuscelli G, Montanari R (202) Hyperledger fabric blockchain: chaincode performance analysis. In: ICC 2020–2020 IEEE International Conference on Communications (ICC), pp 1–6. IEEE press
10. Hackernoon: https://hackernoon.com/hyperledger-fabric-installationguide-74065855eca9.
11. HyperledgerFabrics-docs: https://hyperledger-fabric.readthedocs.io/en/release1.4/understandfabcarnetwork.html

# Early Prediction of Plant Disease Using AI Enabled IOT

S. Vijayalakshmi[1], G. Balakrishnan[2], and S. Nithya Lakshmi[2](✉)

[1] Christ University, Bangalore, India
[2] Fatima Michael College of Engineering and Technology, Madurai, India

**Abstract.** India is an industrialized country, and about 70% of the residents rely on agriculture. Leaves are damaged by chemicals, and climates issues. An unknown illness is found on plants leads to the lowering of quality of produced. Internet of Things is a practice of reinventing the wheel agriculture by enabling farmers to tackle the problems in the industry with practical farming techniques. IoT helps to inform knowledge about factors like weather, and moisture condition. We proposed IoT, ML, and image processing based method to identify the infection. IOT enabled camera to capture the image then required region of interest is extracted. After ROI extraction, image is enhanced to remove the unwanted details form the image and to improve image quality. We compute image features. At the end we do the classification which is a twostep process training and testing and done by SVM. Our proposed method gives 92% accuracy.

**Keywords:** Internet of Thing · Plant disease · Artificial neural network · Image features · Region of interest · Support vector machine

## 1 Introduction

Agriculture is very important for sustaining human life and existence. India is a famous for its food and agriculture production. The majority of the population relies on farming for survival. Farmers have various methods of cultivation. To increase productivity, automation has been introduced into this industry. Besides automating, a few issues need to be solved by policy makers. To test the requirement of the restriction like the disease detection and weed plants growth inhibiting, this project has established the automatic detection of weed plants and these findings will aid in detecting weed plants. When there is disease to the plants, we can normally say that leaves have prominent symptoms. The patches on the leaves of it come from the disease. When there is too much disease in the plant then the whole leaf is covered with spots. Widespread detection early on might help prevent the disease from spreading to the entire crop. The crop will not grow well because of the presence of weed. Thus, the removal of plants would promote plant health. There are many factors behind this project, including these. If a plant gets this disease from one plant, then the disease will reach many different plants, and will be spread quicker at a larger rate. As a result, the disease detection plays a very important role in the field of agriculture. By using the developed methodology, though we could not entirely prevent


S. Shukla et al. (eds.), *Data Science and Security*, Lecture Notes in Networks and Systems 290,
https://doi.org/10.1007/978-981-16-4486-3_33

the infection from spreading to the plant, it could at least decrease the occurrence in which the disease has infected the plants. Agricultural parameters using IoT technology and device availability for these knowledge interconnections and encounters. IoT enables selected objects that must be known or that can be applied remotely Creation is turned into a time frame for all large categories of electronic systems, including, for example, smart grids, smart houses, mobile and smart city communities, as IoT is carefully extended with sensors and activators. To solve the above discussed problem we develop a AI and IOT based method that can help in the prediction of disease at very early stage so that it help the farmers to control the disease spread to the entire crop. In this, image is captured by IOT enabled camera and AI helps in classification according to calculated features value.

## 2 Literature Survey

Plant pests can cause significant impacts in terms of crop yields [1]. It is estimated that the total damage to the economy is around $20 billion throughout the world every year [2, 3]. Geographical location is among the most challenging factors for scientists. On the other hand, they are inefficient and time-consuming [4]. Therefore, a rapid and accurate identification of plant diseases needs to be advanced for the benefit of business and ecology to agriculture [5]. Lots of research papers have been reviewed during the literature review to help explain the various applications of computers in allied fields beside the work carried out by this project. In [6] authors have introduced a tool for monitoring plant growth at a low cost using color sensors. A device is built that can calculate productive shade of plant foliage to look at plant health status. And also discussed about a color sensor for growing plant growth that is affordable and made from a cheap material. Deep learning applied to plant leaf identification for speed and accuracy. It involves segmentation, retrieval and recognition which involves the use of deep learning algorithms throughout the entire process. The crucial first step is to get good plants [7]. The researchers reported a new recognition system for plant disease leaf images based on a hybrid cluster analysis. authors also describe a CBIR system to retrieve texture characteristics and means value to analyze color features, and a Classification method was used for classification.

In order to identify grapevine disease, authors introduced new method by applying image processing and machine learning techniques including Support Vector Machine (SVM), Random Forest, and AdaBoost classifiers. The authors have categorized the leaves for the presence of different infectious diseases. This study concludes that the overall accuracy of SVM is higher than Random Forest and AdaBoost classifiers [8]. A device is producing images of different diseases so that therapeutic choices are evaluated. Image processing comprises of some basic steps in the acquisition, processing of images, pre-processing, fine-tuning of images, segmentation of images, extraction of features, statistical analysis and detection and prediction. The K-Means Clustering technique was used to detect and classify plant disease in an information-theoretic environment [9].

## 3 Proposed Method

In addition to all the common features of smart farming a novel method for early prediction system for identifying the plant disease has been proposed by integrating the machine learning algorithm with IoT which will support the farmers to predict whether the plants are having the possibility of getting affected by any disease or not in this process image processing is combined with ML and IoT. The system is equipped with powerful camera for image capturing as well as a control system which can enable the farmers to focus on any plant in the field for capturing the image. Initially the algorithm is using the SVM classifier for identifying the plant status based on various features such as the shape, colour, growth and texture with respect to temperature, humidity, and light. Here a novel method which will be able to relate and compare these features against the age of that portion of the plant is being proposed. Currently this algorithm has been tested only with the leaves, but this can be extended for any portion like fruits, flowers and so on to analyse the healthiness of the plant. The proposed flowchart is shown in Fig. 1:

- Leaf Image Acquisition
- Preprocessing
- Leaf Region Extraction
- Features Extraction
- Classification

### 3.1 Leaf Image Acquisition

IOT system is enabled with camera and sensors that is used to capture the leaf image this captured image is input for our method. After capturing the image, it will be stored into database for further processing.

### 3.2 Preprocessing

Preprocessing of image means to improve the quality of image by removing unwanted area, and noise. It is beneficial to both of segmentation and features extraction processes. Reliability and accuracy. Here we used fuzzy histogram equalization that comprises two phases. First, a fuzzy set based histogram will handle the in approximation of grey levels in an improved manner. The second step involves dividing the histogram of the pixels

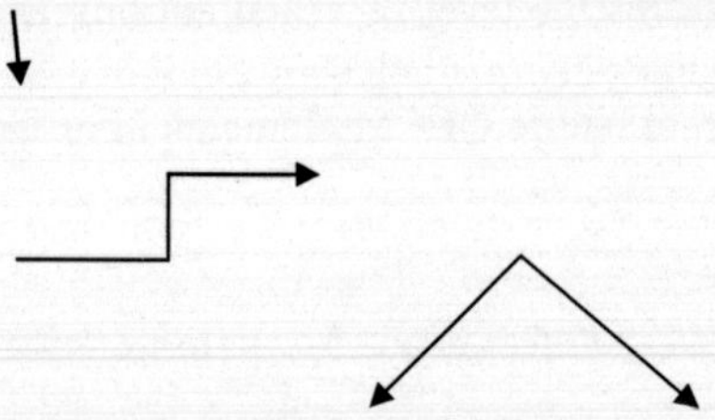

**Fig. 1** Plant infection detection system

of the same type into the two sub histograms based on the median values of the initial picture and then equalizing them separately to protect image brightness.

### 3.3 Leaf Region Retrieval

From the enhanced image, we extract only the required portion so that no need to process the entire image and it also save the time. When image is captured through camera then only 40% information is related to infection rest 60% information is unnecessary which is related to background and not so important. So to improve the efficiency of the result we cropped the image.

### 3.4 Features Extraction

This is a process of dimensional reduction that captures an image's interesting parts. Different types of information are available in an image but only few of them can be used to define the image. Various features like texture, shape, and color can be computed and used to distinguish or to classify the image as normal and diseased image. The texture feature represents the color distribution, roughness, and hardness in an image. And out of the features it was found out that texture feature provides the good result as compared to others and used to identify the infection in plant leaf. The GLCM is taken into account in order to acquire the color and texture characteristics by setting the range of gray levels to 8 and the offset to [0,1]. GLCM extract 2$^{nd}$ order statistical texture feature. In this no. of row and column are equal to no. of grey levels. We compute some texture features are:

- *Image Contrast*: - intensity local variation into pixel and its neighbouring pixel can be computed as:

$$contrast = \sum_{i=0,j=0}^{n-1} p_{ij}(i-j)^2 \quad (1)$$

- *Variance:* It is measurement of heterogeneity and computed as

$$var = \sum_{i=0,j=0}^{n-1} (1-\mu_i)^2 p_{ij} \quad (2)$$

- *Entropy:* Entropy measures the level of visual entropy in an image and rises to its largest value when the intensity of all elements in P matrix is equal. When the image does not have a uniform texture, GLCM elements have large values, implying very high entropy.

$$E_{ntropy} = \sum_{i,j=0}^{n-1} -(p_{ij})p_{ij} \quad (3)$$

**Table 1** Computed value of image feature

| Feature | Feature value | |
|---|---|---|
| | Minimum | Maximum |
| Contrast | 0.1236 | 1.7130 |
| Variance | 3.437 | 2.311 |
| Entropy | 0.6378 | 2.8198 |
| Energy | 0.3523 | 0.8797 |
| Correlation | 0.4394 | 0.9482 |

- *Energy*: It is a measurement of global uniformity in image and is calculated as:

$$E_{nergy} = \sum_{i,j=0}^{n-1} (p_{ij})^2 \tag{4}$$

- *Correlation:* It is the measurement of linear dependency between image grey level and can be found by (Table 1):

$$correlation = \sum_{i=0,j=0}^{n-} p_{ij} \frac{(i-\mu)(j-\mu)}{\sigma^2} \tag{5}$$

## 3.5 Classification

After feature extraction, machine learning approach is applied for pattern identification. For disease classification two stages are involved training and testing. In our method we used 80% of samples for training and 20% for testing (Fig. 2).

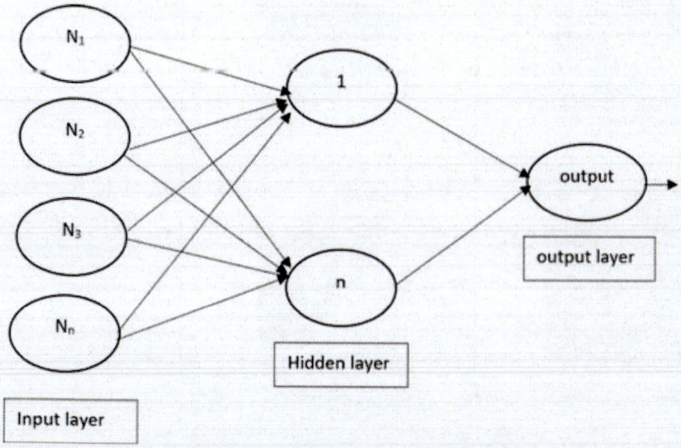


**Fig. 2** Artificial neural network

Here we use artificial neural network for training the samples. Back propagation network trained multilayered network that have backward and forward pass. In forward, first calculate output to compared with preferred output result. Error that is used alter the network weight to reduce the dimension error is performed in forward pass. This iteration process is performed until the low error.

When the training is completed, the classification technique use the rest of the sample to test.in our method we used support vector machine to classify the image as diseased or not affected. In SVM, quadratic kernel function and the box constraints level is 4.0 is used.

Total no of image = 120.

Training sample = 96 (80% of total sample)

**Table 2** Disease detection

| Captured Image | Leaf ROI | Result |
|---|---|---|
|  |  | No Disease |
|  |  | Disease infected leaf |
|  |  | No disease |
|  |  | Disease infected leaf |

Testing sample = 24 (20% total sample)

$$proposed\ method\ accuracy = \frac{TP + TN}{TP + FP + TN + FN} \tag{6}$$

We applied our method on some different leaf to check whether it is infected or normal. The result of our proposed method is shown in Table 2, in this some leaves are infected, and some leaves are normal. Our method can work on all types of plants leaf to check whether it is infected or normal.

## 4 Conclusion

Farmers can take up the machinery leverage to evaluate the crop, to identify the diseases at early stage which help them to decide on possible treatment. The accurate identification of infection or disease is very important for stopping the spread of the infection to the entire crops. Here we used IOT, machine learning and image processing techniques to identify the crop infection at very early stage which is done by capturing leaf by using IOT techniques, machine learning and classification method to classify the leaf whether it is normal and infected. We calculate the different features, but texture feature provides us the correct result, so we take it for further processing. Currently this algorithm has been tested only with the leaves, but this can be extended for any portion like fruits, flowers and so on to analyze the healthiness of the plant. Our proposed method accuracy is 92% and in future we tried to improve the accuracy by doing some improvement in the method.

## References

1. Ampatzidis, Y, De Bellis L, Luvisi A (2017) iPathology: robotic applications and management of plants and plant diseases. Sustainability 9(6): 1010
2. Ghosal S, Blystone D, Singh AK, Ganapathysubramanian B, Singh A, Sarkar S (2018) An explainable deep machine vision framework for plant stress phenotyping. In: Proceedings of the National Academy of Sciences, vol 115, no 18, pp 4613–4618
3. Barbedo JGA (2018) Factors influencing the use of deep learning for plant disease recognition. Biosyst Eng 172: 84–91
4. Geetharamani G, Arun Pandian J (2019) Identification of plant leaf diseases using a nine-layer deep convolutional neural network. Comput Electr Eng 76: 323–338
5. Singh V, Misra AK (2017) Detection of plant leaf diseases using image segmentation and soft computing techniques. Inf Process Agric 4(1): 41–49
6. Seelye M, Gupta GS, Bailey D (2011) Low cost colour sensors for monitoring plant growth in a laboratory. In: Indian Maritime Technology Conference (IMTC) Mark V5
7. Amanda R, Kelsee B, Peter MC, Babuali A, James L, Hughes DP (2017) Deep learning for image-based cassava disease detection. Front Plant Sci 8: 1852
8. Jaisakthi SM, Mirunalini P, Thenmozhi D (2019) Grape leaf disease identification using machine learning techniques. In: Proceedings of the 2019 IEEE International Conference on Computational Intelligence in Data Science (ICCIDS), Chennai, India, 21–23 February 2019, pp 1–6
9. Sachin Khirade SD, Patil AB (2015) Plant disease detection using image processing. In: 2015 International Conference on Computing Communication Control and Automation. https://doi.org/10.1109/iccubea.2015.153

# Feature Selection Based on Hall of Fame Strategy of Genetic Algorithm for Flow-Based IDS

Rahul Adhao(✉) and Vinod Pachghare

Department of Computer Engineering and IT, College of Engineering Pune (COEP), Shivajinagar, Pune 411005, India
{rba.comp,vkp.comp}@coep.ac.in

**Abstract.** The Feature engineering refers to the use of domain knowledge of the available data to selects the features that simplify the machine learning algorithm. It is essential for the implementation of machine learning and is both difficult and costly. Automated feature learning can obviate the need for manual feature engineering. The feature engineering is the next trend after big data, and feature selection and extraction of features are the subpart of it. Feature selection is pa procedure of choosing group of relevant features for use in model constructions. Genetic Algorithm is used for feature selections in the proposed model, that to be in particular Hall of Fame Method of Genetic Algorithms is used for feature ranking, and out of these ranked features, top set features (reduced feature) will be utilized for network traffic classification using the decision tree. The experimentation of the proposed model is carried out using the CICIDS2017 Dataset. The experimentations result prove that the proposed feature selection algorithm is better at accuracy, considering fewer features than the original.

**Keywords:** Feature engineering · Genetics algorithm · Intrusion detection system · Machine learning · CICIDS2017

## 1 Introduction

Feature engineering utilizes domain knowledge of the provided information to select the features that successively assist Machine Learning (ML) algorithms to supply economical results. Feature Selection (FS) is a procedure for choosing a set of relevant model building features. The target of FS is to optimize relevance and scale back redundancy. It is a procedure of finding a feature set comprising only relevant features [1]. For instance, in medical diagnosis, the objective is to deduce the connection between the symptoms and their analogous diagnosis. If accidentally we include the patient identification number (PID) as one of the input features, an over-turned machine learning process may conclude that the illness is dependent on the PID number [2].


S. Shukla et al. (eds.), *Data Science and Security*, Lecture Notes in Networks and Systems 290,
https://doi.org/10.1007/978-981-16-4486-3_34

### 1.1 Hall of Fame Strategy in Genetic Algorithm

The Genetic Algorithm (GA) is a stochastic method for feature optimization based on natural genetics and biological evolution concepts. Seeking the best response in the world is an evolutionary heuristic technique (in the proposed approach, it used to select the best subset of features). The [3–5] provides in detail working of genetic algorithms. The GA used hall of fame (a type memory or storage element) as a container which keeps the best person who ever lived during the population's during the genetic evolution. It is sorted at all times so that the first element of the hall of fame is the entity with the highest fitness. It will keep track of the best individual to appear in the evolution [6]. Each population contributes its top 'n' (where n < = N, N total number of features) to a hall of fame. Feature selection aims to select an optimal set of features that produces the highest prediction accuracy, so the fitness function adopted in the proposed approach is the classification accuracy. In the proposed approach decision tree is used as classifiers. During each generation, an elite subset of features that gives the highest accuracy is kept aside in the hall-of-fame. A list named 'topBestFeature' is maintained to count each feature's occurrence as the best feature in each elite subset across all generations across the GA. At the end of the Genetic Algorithm, the topBestFeature list provides a list of sorted feature performing best for the given datasets. In this proposed approach, we have used these topBestFeature lists for feature selections.

The proposed approach has reduced features from 78 features present in the CICIDS2017 dataset to five features only. These reduced features are providing better accuracy and also require less time. The paper's organization is as follows: Sect. 2 discusses the current state of the same domain's art research. Section 3 proposed model is presented, and experimentations and result analysis are provided in the subsequent sections. The last section provides the conclusions of the work.

## 2 Literature Review

Intrusion detection (ID) is a process of identifying unnecessary traffic on a networked device. An intrusion Detection System (IDS) is a part of installed software or physical appliance that scrutinizes network traffic to detect extreme events, activities, illegal and malicious traffic, traffic that infringes on security policy, and traffic that infringes acceptable usable policies. The goal of ID lies in classifying the normal flow from the anomaly. Mahendra Prasad et al. [7] have presented a feature selection technique using Bayes theorem combined with rough set theory. The estimated probability was used here for feature ranking. In this approach, the feature count was reduced from 80 to 40 for the CICIDS2017 Dataset. The accuracy, precision, and F-Measure of the system were 97.95, 96.37, and 96.37%, respectively. The author mentioned that the system's main drawback was an optimal subset of features and a range of estimated probability of relevant and irrelevant features needing manual interventions.

Arif Yulianto [8] used Ensemble Feature Selection (EFS), Principal component analysis (PCA), and Synthetic Minority Oversampling Technique (SMOTE) to improve the performance of AdaBoost-based IDS on the CICIDS2017 Dataset. This author claims to improve the imbalance of training data (SMOTE) and improper selection of the older studies' classification methods. Here PCA, EFS, and SMOTE were used for feature

**Table 1** AdaBoost performance on CICIDS2017 with PCA, EFS, and SMOTE

| | EFS with AdaBoost | EFS with SMOTE and AdaBoost | PCA and AdaBoost | PCA, SMOTE and AdaBoost |
|---|---|---|---|---|
| Accuracy (%) | 81.47 | 81.33 | 81.47 | 81.47 |
| # of features | 25 | 25 | 16 | 16 |
| Precision (%) | 85.15 | 81.33 | 81.49 | 81.69 |
| Recall (%) | 94.92 | 100 | 99.93 | 95.76 |
| F-measure | 89.77 | 90.01 | 89.78 | 88.17 |

selections, and AdaBoost is used as a classifier. The performance of the approach is summarized in Table 1:

The results of the proposed system are meager when compared with other similar work.

Jamal et al. [9] evaluated various classification algorithms' performance on the NSL-KDD dataset, KDD'99 Dataset, and noise added Dataset. Multiple classification algorithms from the family of classification algorithms were tested and compared. The authors selected the top six algorithms NN (SOM), JRip, NBTree, J48, and RF-based on the performance evaluation matrices. Here the original 41 feature size of NSL-KDD is reduced to 16 feature size. The author also concluded that the NSL-KDD Dataset depicts a realistic environment for evaluating classification algorithms compared to the KDD'99 Dataset.

Uzair Bashir et al. [10] proposed a data mining algorithm and machine learning algorithm to implement the Intrusion Detection System. The authors used J48 Decision Tree and Naive Bayes for implementing IDS. The author measured the Algorithm's performance with NSL-KDD based on Detection Rate, False Positive Rate, Kappa Statistics, and F-Measure. The J48 Decision Tree performance is better than Naive Bayes in terms of Detection Rate, False Positive Rate, Kappa Statistics, and F-Measure.

The author of this paper used Genetic Algorithms (GA) with Principal Component Analysis (PCA) for feature selections [11]. Here PCA is used only for feature transformation purposes. After this, normalized features are fed to GA for feature selection. The Decision Tree (DST) is used as a classifier for this experimentations. This hybrid model of PCA-GA-DST reduced the CICIDS2017 Dataset's features to 40 features with an accuracy of 99.53%. The author further reduced the feature count to 5, 10, and 15 for all files of the CICIDS2017 Dataset in the presented approach.

## 3 Proposed Model

In the proposed model CICIDS2017 dataset is used for the evaluations. Here all seven files of the selected Dataset are used for the experimentations [12, 13]. The following steps depict proposed models, and Fig. 1 shows a diagrammatic representation of the proposed model.

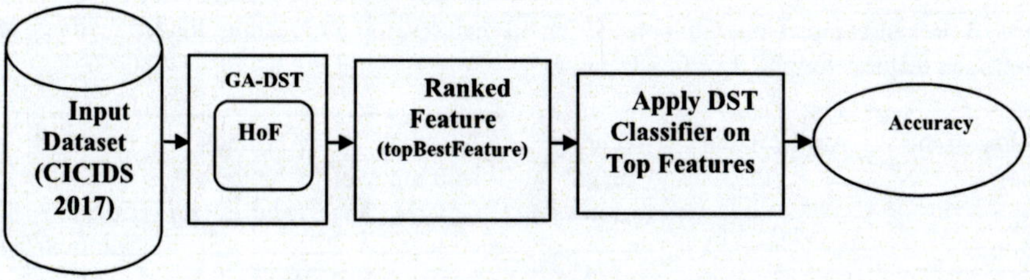


**Fig. 1** Flow chart for the proposed approach

**Table 2** System setup and algorithmic parameter configuration of the experimentation

| Parameter | Corresponding value |
|---|---|
| System used | Windows 7 with 64-bit Operating System |
| Hardware specification of the system used | 32 GB RAM with Intel Core (TM) i7 3.60 GHz processor |
| Nor of generation of genetic algorithm | 10 |
| Population size of genetic algorithm | 100 |
| Mutation probability of genetic algorithm | 0.2 |
| Cross over probability of genetic algorithm | 0.5 |

1. Select a file from the CICIDS2017 Dataset and input it to the Genetic Algorithm.
2. Genetic Algorithm creates populations and evaluates the fitness of subsets of features from the populations.
3. During fitness evolution, the genetic algorithm stores the top best performing subsets of features in the hall of fame. (Genetic Algorithm arrange all these best performing subsets as per their accuracies (higher the accuracies topper the subsets)
4. Create a list of topBestFeature to maintain a count of each feature stored in the hall of fame.
5. Arrange these features as per their count in the hall of fame.
6. Calculate accuracies of these top 5/10/15 features from topBestFeature using decision tree in weka on the corresponding file.

## 4 Experimental Setup and Result Analysis

The proposed model have used CICIDS2017 Dataset for the experimentations. The system setup and algorithmic parameter configuration of the experimentation is as per Table 2.

All the algorithmic parameter of the GA are find out through the series of experimentations. An open-source machine learning WEKA tool is used to evaluate feature group combinations using a decision tree classifier with a 60–40% spilt. Table 3 shows

**Table 3** Result comparison with considering all features, top five feature, top ten features, and top fifteen features for CICIDS2017 Dataset

| Name of the files | Attack found | Accuracy with all features (%) | Accuracy with top five features (%) | Accuracy with top ten features (%) | Accuracy with top fifteen features (%) |
|---|---|---|---|---|---|
| Tuesday file | "Benign, FTP-Patator, SSH-Patator" | 98.40 | 99.76 | 99.77 | 99.97 |
| Wednesday file | "Benign, DoS GoldenEye, DoS Hulk, DoS Slowhttptest, DoS slowloris, Heartbleed" | 93.68 | 97.52 | 98.64 | 99.80 |
| Thursday web attacks file | "Benign, Web Attack – Brute Force, Web Attack – Sql Injection, Web Attack – XSS" | 99.31 | 99.04 | 99.04 | 99.50 |
| Thursday infiltration file | "Benign, Infiltration" | 99.97 | 99.99 | 99.99 | 99.99 |
| Friday bot file | "Benign, Bot" | 98.92 | 99.57 | 99.59 | 99.65 |
| Friday port scan file | "Benign, PortScan" | 93.41 | 99.93 | 99.93 | 99.98 |
| Friday DDos file | "Benign, DDoS" | 82.93 | 86.30 | 86.70 | 86.64 |

accuracy comparison by considering all features with top five feature, top ten features, and top fifteen features. We can see with the top fifteen features accuracies improve for all seven files. Figure 2 shows precision, recall, and F-score with all features (AF), top five features, top ten features, and top fifteen features. The results clearly show that the hall of fame produced the top best feature list gives good accuracy than all features considered.

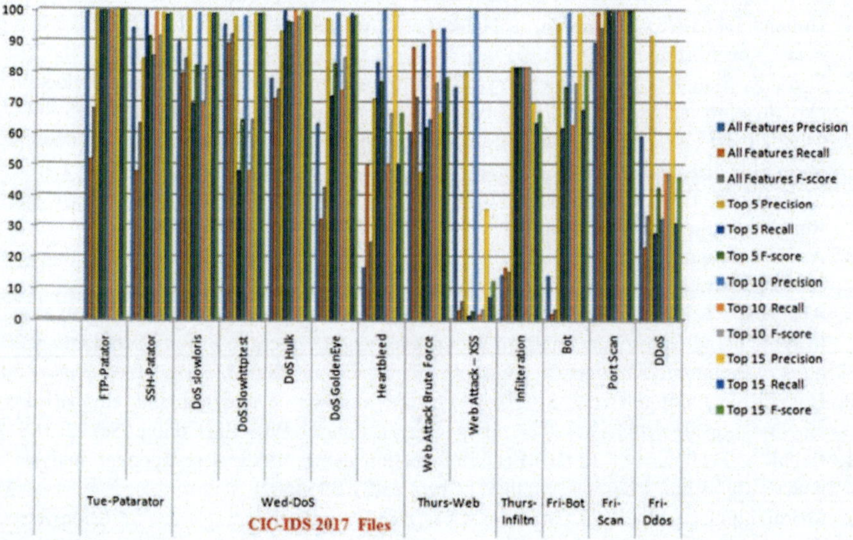


**Fig. 2** Precision, recall, and F-score with all features (AF), top five features, top ten features, and top fifteen features

## 5 Conclusions

In this study, the authors have proposed a model on Hall-of-fame based genetic algorithm model for feature selection and feature ranking. We have done experiments on the Dataset provided by the Canadian Institute of Cybersecurity (CICIDS2017). The experiments show that the proposed feature selection algorithm is better at accuracy, precision, recall, and f-score rate than considering all original features. In the future, we would like to experiment with the other latest available datasets to compare this approach with the state of the art research. Building more rigorous mathematical formulas to design more reliable and faster-converging models will be the work forward.

**Acknowledgements.** The authors wish to acknowledge Information Security Education and Awareness Project, Department of Electronics and Information Technology, Ministry of Communications and Information Technology, Government of India which has made it possible to undertake this research.

## References

1. Brownlee J (2021) An introduction to feature selection. https://machinelearningmastery.com/an-introduction-to-feature-selection/. Accessed 5 Feb 2021
2. Sidey-Gibbons J, Sidey-Gibbons C (2019) Machine learning in medicine: a practical introduction. BMC Med Res Method 19(1): 64
3. Rosin CD, Belew RK (1997) New methods for competitive coevolution Evol Comput 5(1): 1–29

4. Holland JH (1975) Adaptation in Natural and Artificial Systems. University of Michigan Press, Ann Abrbor
5. Das AK, Sengupta S, Bhattacharyya S (2018) A group incremental feature selection for classification using rough set theory based genetic algorithm. Appl Soft Comput 65: 400–411
6. Mitchell GG (2007) Evolutionary computation applied to combinatorial optimisation problems. PhD diss. Dublin City University
7. Mahendra P, Tripathi S, Dahal K (2020) An efficient feature selection based Bayesian and Rough set approach for intrusion detection. Appl Soft Comput 87: 105980
8. Arif Y, Sukarno P, Suwastika NA (2019) Improving adaboost-based intrusion detection system (IDS) performance on CICIDS 2017 dataset. J Phys Conf Ser 1192(1): 1–9
9. Abuzneid AA, Faezipour M, Abdulhammed R, Abu Mallouh A, Musafer H (2019) Machine learning based feature reduction for network intrusion detection, Faculty Scholar Day 2019. https://scholarworks.bridgeport.edu/xmlui/handle/123456789/4134. Accessed 30 Nov 2020
10. Hussain J, Lalmuanawma S (2016) Feature analysis, evaluation and comparisons of classification algorithms based on noisy intrusion dataset. Procedia Comput Sci 92: 188–198
11. Rahul A, Pachghare V (2020) Feature selection using principal component analysis and genetic algorithm. J Discrete Math Sci Crypt 23(2): 595–602
12. Sharafaldin I, Lashkari AH, Ghorbani AA (2018) Toward generating a new intrusion detection dataset and intrusion traffic characterization. In: International Conference on Information Systems Security and Privacy (ICISSP), Portugal, pp 108–116
13. Kshirsagar D, Kumar S (2020) An ensemble feature reduction method for web-attack detection. J Discrete Math Sci Crypt 23(1):283–291

# Consecutive Radio Labelling of Graphs

Anna Treesa Raj and Joseph Varghese Kureethara(✉)

Christ University, Bangalore, India
annatreesa.raj@res.christuniversity.in, frjoseph@christuniversity.in

**Abstract.** Radio labelling or radio colouring is an assignment of positive integers to the vertices of a graph such that the difference between labels of any two vertices must be at least one more than the difference between the diameter of the graph and the distance between the vertices themselves. A graph $G$ admits consecutive radio labelling when the radio number of the graph equals the order of the graph. In this paper, we study certain graphs admitting consecutive radio labelling and identify certain properties of such graphs. Moreover, we characterize the graphs with diameter two admitting consecutive radio labelling and examine certain properties of the labelling under some graph operations.

## 1 Introduction

Graph labelling is one among the major attractions of research in graph theory. Its relevance in the real-life situations and its challenges as a combinatorial problem intensify the interest for the researchers. Radio labelling is motivated from the channel assignment problem that deploys an efficient way to assign frequencies to transmitters that minimize the interference and optimize the number of frequencies involved [4].

Channel assignment problem can be efficiently modelled using graphs where the vertices of the graphs correspond to the transmitters and two vertices are adjacent if there is a possibility of frequency interference in the corresponding transmitters. Radio labelling or radio colouring is an assignment of positive integers (some authors include zero also) to the vertices of a graph such that difference between the labels of any two vertices must be at least one more than the difference between the diameter of the graph and the distance between the vertices themselves. This can be expressed mathematically as follows.

Let $G = (V, E)$ be a graph. Let $c : V \to \mathbf{N}$ be the labelling of the vertices such that for any $u, v \in V$

$$|c(u) - c(v)| \geq diam_G - d(u, v) + 1.$$

Here, $diam_G$ and $d(u, v)$ are the diameter of the graph and the distance between $u$ and $v$, respectively. The challenge in radio labelling is to find the radio number of a given structure. The span of the radio labelling of a graph is the largest label used to radio label a graph. The radio number of a graph $rn(G)$ is defined as the minimum span where the minimum is taken over all the radio labellings of


S. Shukla et al. (eds.), *Data Science and Security*, Lecture Notes in Networks and Systems 290,
https://doi.org/10.1007/978-981-16-4486-3_35

$G$ [2]. Radio number has been found out for several classes of graphs such as paths and cycles [8], square paths and square cycles [6,7], gear graphs [3], trees [5] etc.

We examine those graphs whose radio number is the order of the graph and some of their properties. We define consecutive radio labelling as follows.

**Definition 1.** *Consecutive radio labelling is the radio labelling whose radio number is the order of the graph.*

That is, a graph $G$ is said to admit consecutive radio labelling when $rn(G) = |V|$.

By the definition of radio labelling, no two vertices of the same graph receive identical labels. Therefore, the radio number of a graph coincides with its order only if the vertices receive consecutive labels from 1 to $n$ which is the order of the graph.

There are many graphs that exhibit this property. Figure 1 shows the radio labelling of the Petersen graph. Since the radio number of the Petersen graph is 10, it admits consecutive radio labelling. In this work, we investigate certain necessary and sufficient conditions for graphs that admit consecutive radio labelling. Moreover, we characterize the graphs with diameter two admitting consecutive radio labelling. We also examine certain properties of the labelling under some graph operations.

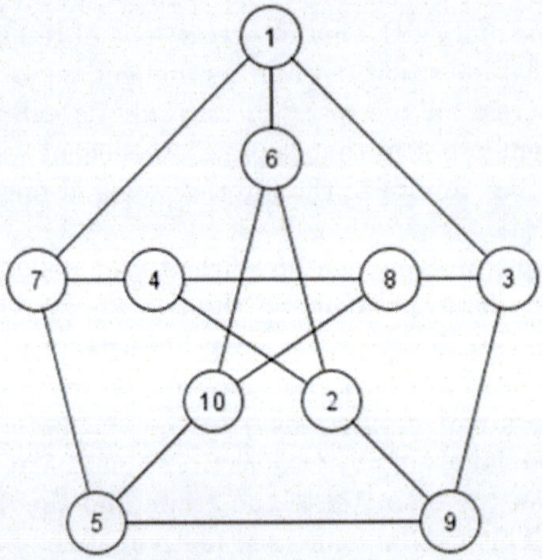


**Fig. 1** Petersen graph admits consecutive radio labelling

## 2 Basic Results

**Proposition 1.** *Complete graphs admit consecutive radio labelling.*

*Proof.* Let the order of the complete graph be $n$. By the definition of radio labelling the colour difference between any two vertices must be at least one more than the difference between diameter of the graph and the distance between the vertices. Since the diameter of complete graph is 1, the colour difference between any two vertices should be at least one. Hence the vertices of a complete graph can be labelled using consecutive integers from 1 to $n$. Therefore complete graphs admit consecutive radio labelling.

**Theorem 1.** *Let $G$ be a graph such that there exist two vertices having exactly one vertex at diametric distance. If $G$ admits consecutive radio labelling then the diametric vertices of the two vertices should be distinct.*

*Proof.* Assume if possible both the vertices say $v$ and $u$ having same vertex $w$ at diametric distance. Suppose $w$ receives label $c$. Then either $v$ or $u$ receives the colour $c+1$. But, then $c+2$ will be forbidden. If we start with either $v$ or $u$ then the label 4 is forbidden. Hence the diametric vertices of the two vertices should be distinct.

**Proposition 2.** *A cycle $C_n$ admits consecutive radio labelling if and only if it is $C_3$ or $C_5$*

*Proof.* The consecutive radio labellings of $C_3$ and $C_5$ are given in Fig. 2.

Let us consider $C_n$ such that $n > 2$ is even positive integer. For an even cycle, every vertex has exactly one vertex at diametric distance. Hence, none of the even cycles admits consecutive radio labelling.

Let us consider $C_n$ where $n$ is odd and $n \geq 7$. Assume, if possible, the $C_n$ admits consecutive radio labelling. Let us denote the vertices with label $k$ as $v_k$. There will be a vertex $v_k$ having label $k$. Clearly, $v_{k-1}$ having label $k-1$ will

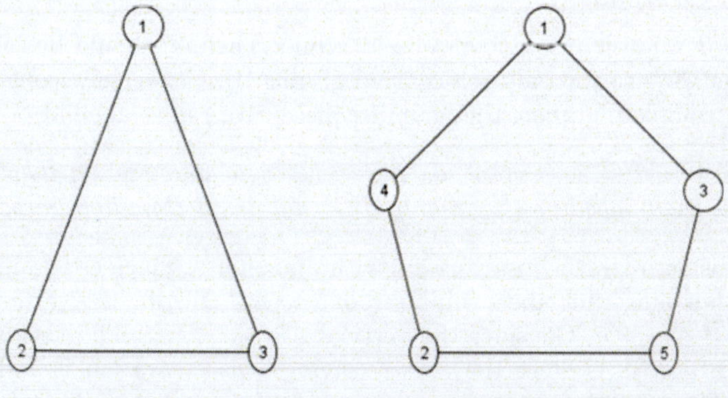


**Fig. 2** Consecutive radio labelling of $C_3$ and $C_5$

be at diametric distance from $v_k$. The vertex $v_{k+1}$ having label $k+1$ is again at diametric distance from $v_k$. But every vertex of an odd cycle has exactly two vertices at diametric distance and the vertices which are at diametric distance for a given vertex are adjacent. Therefore the distance between $v_{k-1}$ and $v_{k+1}$ is 1. By the definition of radio labelling, their label difference should be at least 3 which is the diameter of the graph. This is a contradiction. Hence the proof.

## 3 Some Structural Properties

**Theorem 2.** *If a graph $G$ admits consecutive radio labelling then at least $n-2$ vertices of $G$ have at least two vertices at diametric distance.*

*Proof.* Suppose $G$ be a graph on $n$ vertices satisfying consecutive radio labelling. Let $v_1$ be the vertex having label 1. By the definition of radio labelling the vertex having label 2, say, $v_2$ should be at diametric distance from $v_1$. Again the vertex having label 3, i.e., $v_3$ should be at diametric distance from $v_2$. Continuing like this, the vertex having label $k$, say, $v_k$ should be at diametric distance from $v_{k-1}$ and the vertex $v_{k+1}$ with label $k+1$ is also at diametric distance from $v_k$. Therefore, for any $k \in \{2, 3, ..., n-1\}$ there exist at least two vertices $v_{k-1}$ and $v_{k+1}$ at diametric distance. Hence the theorem.

**Theorem 3.** *If a graph $G$ admits consecutive radio colouring then $G$ is self-centred.*

*Proof.* Consecutive radio labelling refers to assigning consecutive integers to the vertices of a graph under the condition of radio labelling. Assuming assigned label $c$ to a vertex $v$, then another vertex $u$ receives $c+1$ only if $v$ and $u$ are at diametric distance. Therefore, to assign consecutive labels, each vertex should have a vertex at a diametric distance. Therefore, every vertex of $G$ is peripheral and thereby $G$ is self-centred.

**Corollary 1.** *A tree admits consecutive radio labelling if and only if it is $K_1$ or $K_2$.*

*Proof.* Let $T$ admit consecutive radio labelling. Then, $T$ should be self-centred. But $K_1$ and $K_2$ are the only self-centred graphs [1]. Conversely, consider $K_1$ or $K_2$. Being complete graphs, they admit consecutive radio labelling.

**Theorem 4.** *Let $G$ be a graph on $n$ vertices with diameter 2, then $G$ admits consecutive radio labelling if and only if $\overline{G}$ contains a Hamiltonian path.*

*Proof.* Consider a graph on $n$ vertices with diameter 2. Since $G$ admits consecutive radio labelling, the labels will be from 1 to $n$. Let $v_i$ be the vertex receiving label $i$ for $1 \leq i \leq n$. Consider the vertices $v_1$, $v_2$, ..., $v_n$. The vertex $v_i$ is not adjacent to $v_{i+1}$ in $G$ since they are receiving consecutive labels and diameter of $G$ is 2. Therefore, $v_i$ is adjacent to $v_{i+1}$ in $\overline{G}$ for $1 \leq i \leq n$. Therefore, $v_1$, $v_2$, ..., $v_n$ is a Hamiltonian path in $\overline{G}$.

Conversely, suppose that there is a Hamiltonian path in $\overline{G}$. Let $v_1, v_2, \ldots, v_n$ be the Hamiltonian path in $\overline{G}$. Assign the label $i$ to $v_i$ for $1 \leq i \leq n$. Since $v_i$ is adjacent to $v_{i+1}$ in $\overline{G}$, $v_i$ is not adjacent to $v_{i+1}$ in $G$. Therefore, adjacent vertices are not receiving consecutive labels in $G$. Hence, the above labelling is a radio labelling in $G$. Since the span of the labelling is $n$, it is a consecutive radio labelling. Therefore, $G$ admits consecutive radio labelling.

## 4 Consecutive Radio Labelling of the Join and the Cartesian Products

We present two results that show that consecutive radio labelling can be extended to the graphs products. At first, we consider the join of graphs.

**Theorem 5.** *Let $G_1$ and $G_2$ be two graphs admitting consecutive radio labelling. The join of graphs, $G = G_1 + G_2$ admits consecutive radio labelling if and only if $G_1$ and $G_2$ are complete graphs.*

*Proof.* Suppose $G$ admits consecutive radio labelling. $G$ being the join of $G_1$ and $G_2$, there should be a vertex in $G_2$ having the label $o(G_1) + 1$. This is possible only when the diameter of graph $G$ is 1. Therefore, $G_1$ and $G_2$ are complete.

Conversely, suppose that $G_1$ and $G_2$ are complete. The graph $G$ is complete since it is the join of two complete graphs. Hence $G$ admits consecutive radio labelling.

When it comes to the labelling of the Cartesian product of graphs, the outcome is not that immediate. Hence, it is worth exploring the consecutive radio labelling of the Cartesian product of graphs. We present here a particular case.

**Theorem 6.** *If $G = K_n \times K_m$ where $m \neq 2$, then $G$ admits consecutive radio labelling.*

*Proof.* The graph $K_1 \times K_n$ admits consecutive radio labelling for all $n$ since it is the complete graph on $n$ vertices.

Let $m > 1$ and $n > 1$.

Consider the graph $G = K_n \times K_m$. Let $w_{i,j} \in V(K_n \times K_n)$ then $w_{i,j} = (u_i, v_j)$ where $u_i \in V(K_n)$ and $v_j \in V(K_m)$. To prove that $G$ admits consecutive radio labelling, consider the following cases.

**Case 1:** $m = n \neq 2$

Consider the following labelling.

Let $c(w_{i,i} = i)$ for $1 \leq i \leq n$. For $k \in \{1, 2, \ldots, n-2\}$, $c(w_{i,i+k}) = kn - \frac{k(k-1)}{2} + i$ for $1 \leq i \leq n-k$. Thus, $c(w_{i,i-k}) = \frac{n^2+n-2}{2} + kn - \frac{k(k+1)}{2} - (i-k-1)$ for $k+1 \leq i \leq n$. i.e., $c(w_{1,n}) = n^2 - 1$ and $c(w_{n,1}) = n^2$.

To prove that the above labelling is a consecutive radio labelling, we have to show the following.

- All labels are distinct.
- Adjacent vertices are not receiving consecutive labels.

- Span of the labelling is $n^2$.

To prove that the labels are distinct, consider any two distinct vertices $w_{a,b}$ and $w_{x,y}$. If $a = b$ and $x = y$, then by the labelling $c(w_{a,a}) = a$ and $c(w_{x,x}) = x$. Hence, we get distinct labels since the vertices are distinct. If $a = b$ and $x \neq y$ then $c(w_{a,a}) \neq c(w_{x,y})$ since $c(w_{x,y}) \geq n+1$ and $c(w_{a,a}) \leq n$ by the labelling defined.

Now, consider the case where $a \neq b$ and $x \neq y$. Here, it will satisfy any of the following cases.

(i) $b = a + r$ and $y = x + s$
(ii) $b = a + r$ and $y = x - s$
(iii) $b = a - r$ and $y = x - s$

Without loss of generality let us assume $r > s$. Then, $c(w_{a,b}) = c(w_{a,a+r}) = rn - \frac{r(r-1)}{2} + a$. i.e., $c(w_{x,y}) = c(w_{x,x+s}) = sn - \frac{s(s-1)}{2} + x$. If it is $(i)$ then,

$$
\begin{aligned}
|c(w_{a,b}) - c(w_{x,y})| &= \left| rn - \frac{r(r-1)}{2} + a - sn + \frac{s(s-1)}{2} - x \right| \\
&= \left| (r-s)n + \frac{s(s-1)}{2} - \frac{r(r-1)}{2} + a - x \right| \\
&\neq 0
\end{aligned}
$$

This is because $a \neq x$.

Similarly, we can prove the other two conditions. Hence, the labels are distinct for distinct vertices.

To prove that the adjacent vertices are not receiving consecutive labels, again consider any two vertices $w_{a,b}$ and $w_{x,y}$. These two vertices are adjacent if and only if $a = x$ or $b = y$. Let us consider $a = x$. Similarly, we can prove the other one too. Since $n \neq 2$, vertices adjacent to $w_{i,i}$ will not receive consecutive labels. Hence, let us consider $a \neq b$ and $x \neq y$. This will also satisfy the above three conditions. Without loss of generality let us assume $r > s$.

If it is $(i)$ then,

$$
\begin{aligned}
|c(w_{a,b}) - c(w_{a,y})| &= \left| rn - \frac{r(r-1)}{2} + a - sn + \frac{s(s-1)}{2} - a \right| \\
&= \left| (r-s)n + \frac{s(s-1)}{2} - \frac{r(r-1)}{2} \right| \\
&\geq 3
\end{aligned}
$$

This is because, we obtain the extreme value when $r = n-2$ and $s = n-3$. Similarly, we can prove for the other conditions also. Hence adjacent vertices will not receive consecutive labels.

Now let us prove that the span of the labelling is $n^2$.

From the given labelling it is clear that $c(w_{i,i}) < c(w_{i,i+k}) < c(w_{i,i-k})$. The maximum value for $c(w_{i,i-k})$ is obtained when $i = n-1$ and $k = n-2$. Thus, $c(w_{n-1,n-2}) = n^2 - 2 < n^2$.

Hence, the highest value allotted by $c$ is $n^2$ for $w_{n,1}$. Hence, the span of $c$ is $n^2$. So the given labelling is a consecutive radio labelling. Therefore, $K_n \times K_n$ admits consecutive radio labelling.

**Case 2:** $m > n$

For $K_n \times K_m$, consider the labelling as follows.

$c(w_{i,i}) = i$ for $1 \leq i \leq n$.

For $k \in \{1, 2, \ldots, m - n\}$, $c(w_{i,i+k}) = kn + i$ for $1 \leq i \leq n$.

For $l \in \{1, 2, \ldots, n - 2\}$, $c(w_{i,i+m-n+l}) = (m - n + l)n - \frac{l(l-1)}{2} + i$ for $1 \leq i \leq n - l$.

For $k \in \{1, 2, \ldots, n - 2\}$, $c(w_{i,i-k}) = \frac{2mn-n^2+n-2}{2} + kn - \frac{k(k+1)}{2} - (i - k - 1)$ for $k + 1 \leq i \leq n$. Hence, $c(w_{1,m}) = mn - 1$, $c(w_{n,1}) = mn$.

Proceeding like **Case 1**, we can show that the above labelling is a consecutive radio labelling. Therefore, $K_n \times K_m$ admits consecutive radio labelling.

## 5 Conclusion

We have introduced the concept of consecutive radio labelling and identified certain graphs and their properties that admitted the same. We have characterized the graphs with a diameter of 2 satisfying consecutive radio labelling. We have also identified the consecutive radio labelling for some graph operations.

## References


1. Buckley F (1989) Self-centred graphs. Ann N Y Acad Sci 576(1):71–78
2. Chartrand G, Erwin D, Zhang P (2005) A graph labelling problem suggested by FM channel restrictions. Bull Inst Combin Appl 43:43–57
3. Fernandez C, Flores A, Tomova M, Wyels C (2008) The radio number of gear graphs. arXiv preprint arXiv:0809.2623
4. Hale WK (1980) Frequency assignment: theory and applications. Proc IEEE 68(12):1497–1514
5. Liu DD-F (2008) Radio number for trees. Discret Math 308(7):1153–1164
6. Liu DD-F, Xie M (2004) Radio number for square of cycles. Congr Numer 169:105–125
7. Liu DD-F, Xie M (2009) Radio number for square paths. ARS Combin 90:307–319
8. Liu DD-F, Zhu X (2005) Multilevel distance labellings for paths and cycles. SIAM J Discret Math 19(3):610–621

# Blockchain-Enabled Midday Meal Monitoring System

Ritika Kashyap(✉), Neha Kumari, Sandeep Kumar, and Gopal Krishna

Department of Computer Science and Engineering, Netaji Subhas Institute of Technology, Patna, Bihar, India

**Abstract.** The Midday meal is a program run by the Government of India. It is planned to strengthen the nutritional level of school-age children. The program encourages attendance in the school nationwide. The government of India reviews the mission on the Midday meal program, comprising members from the central government, state government, UNICEF, and the Supreme Court commissioner. But the management of the existing program has some limitations. It is based on the traditional way of keeping records of the students which can be manipulated. Delay in payment to various schools enrolled with this program is also seen over time. This paper proposes a model to strengthen the management of the existing Midday meal program using the smart contract in the Ethereum environment of blockchain. Decentralization, transparency, and immutability are the three prime features that have helped in overcoming the loopholes of the existing Midday meal management program. Our main purpose is to remove the intermediaries and to bring transparency and trust in the existing system using blockchain. The smart contract here shows how to improve authenticity, efficiency and bring transparency among schools and concerned government authorities in the whole Midday meal management program.

**Keywords:** Blockchain · Smart contract · Midday meal · Decentralized application · Ethereum

## 1 Introduction

The dropping rate of children in school is remarkably high in developing or underdeveloped countries. Considering this, the Government of India has launched the Midday meal program aiming at both education and health of the children. It seeks to attract students to the school to increase enrollment, retention, and attendance. It also focuses on improving nutritional levels among school-age children nationwide.

The existing general management structure of the Midday meal program can be understood from Fig. 1. Here we can see that there are various stages involved in the Midday meal program. The initial stage starts with the School level where the Head of the school has to submit its report related to the students to avail the benefit of the Midday meal program. The report then goes to various levels for verification such as Block level, District level, and State level where verification is done by the Block Resource Person,


S. Shukla et al. (eds.), *Data Science and Security*, Lecture Notes in Networks and Systems 290,
https://doi.org/10.1007/978-981-16-4486-3_36

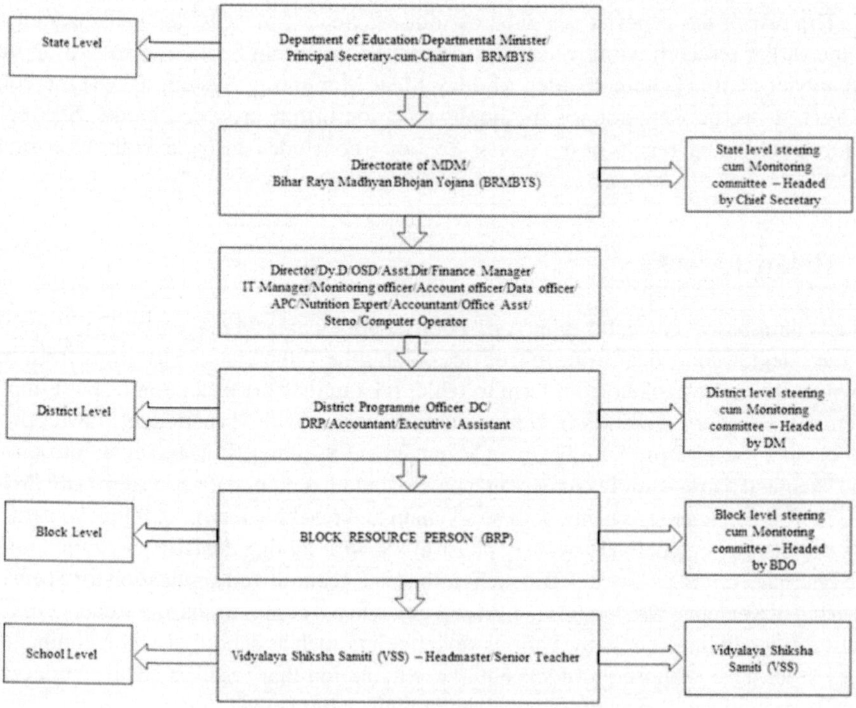


**Fig. 1** Midday meal general management structure

District Resource Person and State Resource Person, respectively. The fund for the Midday meal is released by the Government to school only after successful verification is done at various levels.

Through this program, free lunches are provided on all working days. Children of the primary and upper primary class in government and various concerned government learning hubs are benefited from this program. However, the management of the Midday meal has some loopholes which have not been overcome yet. Due to the lack of technical solutions and manual maintenance of budgets, there arise many problems in the execution. Some of the problems are:-it becomes very difficult to manage the records, to identify any discrepancy in the system, flow of money. As a result, some people in the middle may perform any corrupt activities.

The loopholes in the existing system may be deciphered through the implementation of smart contract, that is one of the broad applications of blockchain technology. Blockchain is an immutable, anonymous, unhackable, and decentralized ledger (log). It is a shared database of accounts of all transactions or digital events that have been accomplished and distributed among striving parties. Blockchain technology incurs only infrastructure cost and no transaction cost. The smart contract used here is a computer protocol. It is aimed to verify, digitally expedite, or apply the compromise or execution of a contract. Without mediators, we can do trustworthy transactions that are tangible and immutable.

The rest of the paper is designed as follows: Sect. 2 presents a concise summary of the earlier research works related to the smart contract. In Sect. 3 we have proposed our model of Blockchain-enabled Midday Meal Monitoring System using Ethereum. In Sect. 4, we have mentioned the implementation of our proposed model. Section 5 comprises terminal results of our model. Section 6 concludes the paper with some future tasks.

## 2 Related Works

Blockchain-based Traceability in Agri-Food Supply Chain Management [1]. It results in the integration of data from IoT devices along the value chain. The crops from IoT devices are being tracked from farm to table. It examines the execution of both implementations, one of which is in Ethereum and another is in Hyperledger Sawtooth. A Blockchain-based Drug Supply Chain Management System [2]. Abbas et al. introduced and executed a drug supply chain management and recommendation system (DSCMR). DSCMR is implemented using both blockchain and machine learning. It involves tracking and tracing drug delivery at every phase along with solving the issue of duplication in the Pharmaceutical Industry. A Blockchain-based Decentralized application for Securely Sharing of Student's Credentials [3]. At present, when it comes to sharing student credentials, it is a tedious process for various stakeholders such as school, teacher, companies. Raaj et al. have proposed a blockchain-based solution that resolves security concerns turning around the sharing of students' credentials. It has helped to strengthen the current educational system. A Blockchain-based Land Registration System Proposal for Turkey [4]. The author has discussed that how we can resolve some serious problems that arise in this process of land registry such as false pricing, and high physical transactions. The steps for developing a Blockchain-based Land Registration System have also been discussed in this paper. A Blockchain for Supply Chain and Manufacturing Industries and Future [5]. In this paper, we come to know that how blockchain could help in the manufacturing and machine tool industry, how it improves the business relationship between partners, and how it is different from the traditional system.

## 3 Proposed Model

Our proposed model consists of four major stakeholders as shown in Fig. 2. The four major stakeholders include School, Block, District, and State. Figure 2 provides an overview of how our Midday Meal System will execute. Using smart contracts in this proposed model, we can provide a transparent, secure, and conflict free method to share all the data related to the Midday meal program. This contract will automatically be triggered if the specified status matches the corresponding database events. In our decentralized application, a role-based modifier is inherited by the role-based access control. The access control consists of four contracts for each stakeholder (school, block, district, and state). Each contract contains a function that allows an address to be added to the role. That function is only permitted by the contract owner. A contract modifier is used to enforce access controls within the blockchain.

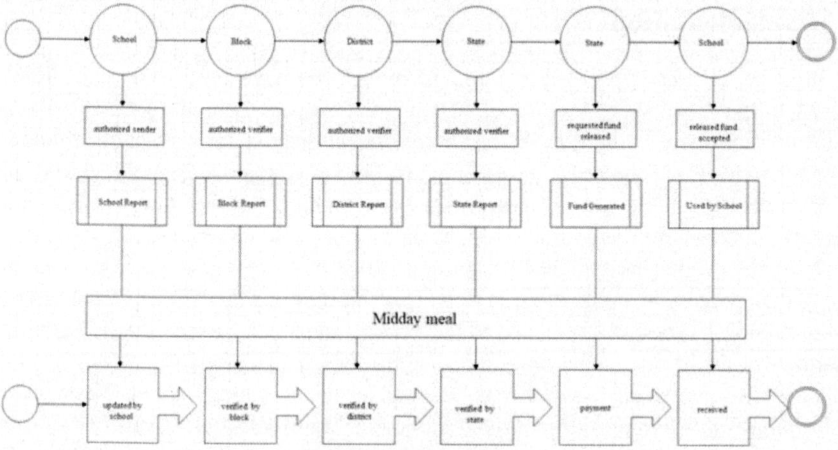


**Fig. 2** Proposed model

The roles of each of the stakeholders in sequence are:

- School: The primary step of the Midday meal starts with the school. The school will provide data of the students enrolled in that school using "School Report" function. Only the authorized member of the school (say Principal) can update the data. The Principal of the school will have to enter some specified credentials such as unique ID, school code, Midday meal ID to update the data. After entering all the credentials, the state of the Midday meal process will be changed to "updated by school". The other stakeholders can get the update by calling a public function that will track the current state of the Midday meal process.
- Block: As soon as the data is updated by the school, the Block Officer can access the updated data by entering the required credentials. The authorized Block Officer will verify the data and update about the successful or unsuccessful verification process using "Block Report" function. The verification will be successful only if the data seems to be correct. The state of the Midday meal process will be updated to "verified by block" after successful verification by the Block Officer.
- District: After successful completion of the work of Block Officer, the role of District Officer comes. The authorized District Officer can now access the data updated by both the block and the school by entering the required credentials. Now, he will verify at his level and update about the verification using the "District Report" function. Here also, the verification will be successful only if the data seems to be authentic to the District Officer. The state of the Midday meal process after a successful verification will be updated to "verified by district".
- State: Similarly, the authorized State Officer will do the verification and update it using the "State Report" function. After successful verification, the government will release the requested fund for the school and without any delay, the fund will be transferred to the school by the State. The State will transfer the fund using the "Fund Generated" function. Now, the school can use that fund for the Midday meal. The debit of the fund by the school can be seen using the "Used by school" function.

# 4 Implementation

To produce an archetype of our proposed model, we have used Ethereum, Solidity [6], MetaMask, and Web3.js. First of all, four stakeholders (school, block, district, and state) are considered where each stakeholder has a distinct set of functionalities and confinements. The School dashboard has two functions. One is to update the Midday meal credentials and another is to access the fund released by the state. The Block and District Officers have the option to see the data along with the sender hash value and to verify those updated credentials. The State dashboard also has two functions. One is to verify the credentials and another is to release the fund for the Midday meal program to the school. Every change related to the Midday meal can be seen by the authorized members of the Midday meal program using the "fetch data" function.

The development environment can be set up on any operating system. We begin here by installing the Ethereum decentralized application framework called truffle. Along with it, we must install the node js, npm packages, Web3, and a browser extension wallet called MetaMask. For the implementation and deployment of the smart contract on Ethereum, we have used the Solidity language. Afterward, we have employed the Web3.js library, to develop a user-friendly interface that permits users to quickly communicate with the smart contract.

## 4.1 Tools, Techniques, and Languages Used

**Ganache** is a simulator for developing Ethereum decentralized applications faster. It includes all popular events. It can be run smoothly to make development without waiting to sync.

**Web3.js** is a JavaScript library, that allows calling smart contract functions. To communicate with the smart contract, it needs a provider object. We only call the methods of Web3 using code. For execution, Web3 connects to the provider which sends the API code to the Ethereum blockchain.

**Ethereum Virtual Machine** Also known as EVM, is the runtime environment used in Ethereum to translate solidity language into a readable language. It is completely isolated. It specializes in halting Denial-of-service attacks. It assures that no programs have a way to any other's state, ensuring communication can be built without any potential intervention.

**MetaMask** is a software cryptocurrency wallet used to interact with the Ethereum blockchain. It allows users to securely manage identities and sign the transaction. The MetaMask extension turns the normal browser into an Ethereum browser.

**Solidity** is a statically typed, Curly-braces programming language for implementing smart contract on the Ethereum network. It has been designed for developing Smart Contracts that run on the Ethereum Virtual Machine (EVM).

**Truffle** is a development environment. We use it in Ethereum for examining the framework. It has built-in feathers for the assemblage of smart contracts, deployment, and network management for extending to numerous public & private chains.

## 5 Experimental Results

The experimental results of the terminal for our Blockchain-enabled Midday Meal Monitoring System are shown in Fig. 3 and Fig. 4. Figure 3 and Fig. 4 shows the deployment cost of functions and the cost of functions, respectively. Here the cost is calculated in Gas. Gas [7] refers to the amount of computational effort required to execute the transactions on the Ethereum network. Gas prices are denoted in Gwei, each Gwei is equal to 0.000000001 ETH (10–9 ETH).

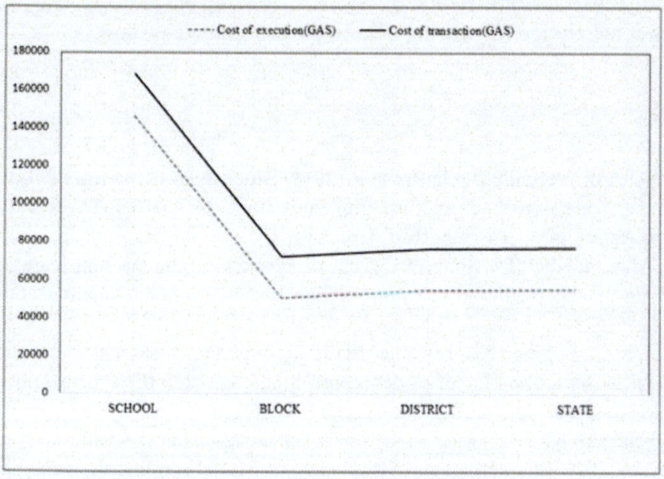


**Fig. 3** Deployment cost of functions

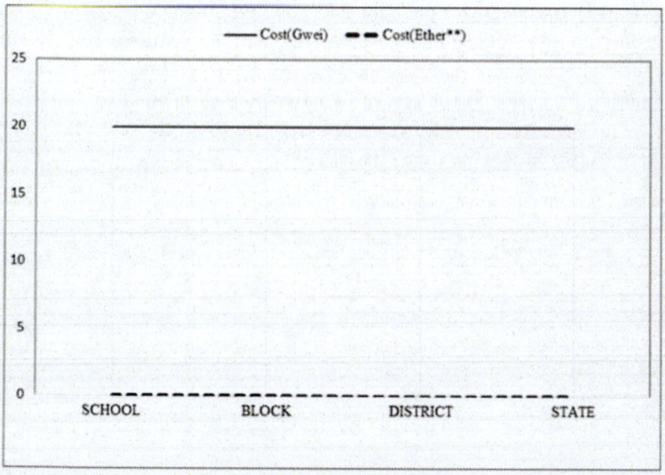


**Fig. 4** Cost of functions

## 6 Conclusion and Future Scope

In this paper, we have proposed a decentralized traceable smart contract based on blockchain technology. This smart contract will deliver real-time data to every concerned government authority of the Midday meal program. Our smart contract can significantly advance the performance and transparency in the existing management of the Midday meal program. Some of the current problems which we have tried to resolve here are the issue of delay in payment, manipulation of records related to payment and students, lack of transparency.

In the future, we can improve the storage size by using the InterPlanetary File System. It may be an ideal choice for storing data in the distributed system.

## References


1. Caro MP, Ali MS, Vecchio M, Giaffreda R (2018) Blockchain-based traceability in agri-food supply chain management: a practical implementation. In: 2018 IoT Vertical and Topical Summit on Agriculture - Tuscany (IOT Tuscany)
2. Abbas K, Afaq M, Khan TA, Song W-C (2020) A blockchain and machine learning based drug supply chain management and recommendation system for smart pharmaceutical industry. Electronics 9:852
3. Mishra RA, Kalla A, Singh NA, Liyanage M (2020) Implementation and analysis of blockchain based DApp for secure sharing of students credentials. In: 2020 IEEE 17th Annual Consumer Communications and Networking Conference (CCNC)
4. Mendi AF, Sakakli KK, Cabuk A (2020) A blockchain based land registration system proposal for Turkey. In: 2020 4th International Symposium on Multidisciplinary Studies and Innovative Technologies (ISMSIT)
5. Jain VN, Mishra D (2018) Blockchain for supply chain and manufacturing industries and future it holds! Int J Eng Res 7:1–8
6. Dannen C (2017) Introducing Ethereum and solidity: foundations of cryptocurrency and blockchain programming for beginners. Apress, Berkeley. https://doi.org/10.1007/978-1-4842-2535-6
7. Bouraga S (2020) An evaluation of gas consumption prediction on Ethereum based on transaction history summarization. In: 2020 2nd Conference on Blockchain Research & Applications for Innovative Networks and Services (BRAINS)

# On Equitable Near Proper Coloring of Mycielski Graph of Graphs

Sabitha Jose(✉) and Sudev Naduvath

Department of Mathematics, CHRIST (Deemed to be University), Bangalore, Karnataka, India
sabitha.jose@res.christuniversity.in

**Abstract.** When the available number of colors are less than that of the equitable chromatic number, there may be some edges whose end vertices receive the same color. These edges are called as bad edges. An equitable near-proper coloring of a graph $G$ is a defective coloring in which the number of vertices in any two color classes differ by at most one and the resulting bad edges is minimized by restricting the number of color classes that can have adjacency among their own elements. In this paper, we investigate the equitable near-proper coloring of Mycielski graph of graphs and determine the equitable defective number of those graphs.

**Keywords:** Equitable coloring · Near-proper coloring · Equitable near-proper coloring · Mycielski graph

## 1 Introduction

For general concepts and definitions in graph theory, we refer to [1, 2]. Unless specified otherwise, all graphs mentioned in this paper are finite, simple, connected and undirected. A graph is said to be *equitably colorable* if the adjacent vertices receive different colors and the number of vertices in any two color classes differ by at most one. The smallest integer $k$ for which $G$ is equitably $k$ -colorable is called the *equitable chromatic number* of $G$ and is denoted by $\chi_e(G)$. The notion of equitable coloring was introduced by Walter Meyer [3] in 1973.

An *improper coloring* or a *defective coloring* of a graph $G$ is a vertex coloring in which adjacent vertices are allowed to have the same color. The edges whose end vertices receive the same color are called *bad edges*.

A *near-proper coloring* of $G$ is a coloring which minimizes the number of bad edges by restricting the number of color classes that can have adjacency among their own elements. The number of bad edges resulting from a near-proper coloring of $G$ is denoted by $b_k(G)$. Some interesting studies in this direction can be viewed in [4]. Near-proper coloring of graphs have significant role in real-life situations. Non-availability of sufficient number of colors (quantity of resources) leads us to different defective coloring problems.


S. Shukla et al. (eds.), *Data Science and Security*, Lecture Notes in Networks and Systems 290,
https://doi.org/10.1007/978-981-16-4486-3_37

An equitable near-proper coloring of graphs are introduced in [5] and defined as follows.

**Definition 1.1** An *equitable near-proper coloring* of a graph G is an improper coloring in which the vertex set can be partitioned into $k$ color classes $V_1, V_2, ..., V_k$ such that $\big||V_i|-|V_j|\big| \leq 1$ for any $1 \leq i \neq j \leq k$ and the number of bad edges is minimised by restricting the number of color classes that can have adjacency among their own elements.

**Definition 1.2** The minimum number of bad edges which result from an equitable near-proper coloring of $G$ is defined as *equitable defective number* and is denoted by $b^k_{\chi_e}(G)$.

Motivated by the studies mentioned above, in this paper, we discuss the equitable near-proper coloring of some Mycielski graph of graphs.

## 2 Equitable Near Proper Coloring of Mycielski Graph of Graphs

Mycielski introduced the transformation of a graph $G$ into a new graph denoted by $\mu(G)$ in 1955. If $G$ has $n$ vertices then $\mu(G)$ consists of $2n+1$ vertices. We construct the Mycielski graph as follows (see [6, 7]).

Let $G$ be a graph with vertex set $V(G) = \{v_1, v_2, \cdots, v_n\}$. The *Mycielski graph* of a graph or the Mycielskian of a graph $G$, which is denoted by $\mu(G)$ is the graph with vertex set $V(\mu(G)) = \{v_1, v_2, \cdots, v_n, u_1, u_2, \cdots, u_n, w\}$ such that $v_iv_j \in E(\mu(G)) \Leftrightarrow v_iv_j \in E(G)$, $v_iu_j \in E(\mu(G)) \Leftrightarrow v_iv_j \in E(G)$ and $u_iw \in E(\mu(G))$ $\forall i = 1, 2, \cdots, n$.

**Theorem 2.1** *The equitable defective number of Mycielskian of paths $\mu(P_n)$ is given by*

$$b^k_{\chi_e}(\mu(P_n)) = \begin{cases} \lfloor \frac{n}{k} \rfloor & \text{if } k = 2 \\ \lfloor \frac{2n+1}{k} \rfloor - \lceil \frac{n}{2} \rceil - 1 & \text{if } k = 3. \end{cases}$$

**Proof.** Let $\mu(P_n)$ be the Mycielski graph of paths on $2n+1$ vertices. Let $\{v_1, v_2, \cdots, v_n\}$ be the vertices of the path $P_n$ and $\{u_1, u_2, \cdots, u_n\}$ be the vertices corresponding to $\{v_1, v_2, \cdots, v_n\}$. Let $w$ be the vertex adjacent with all $u_i$'s where $1 \leq i \leq n$. We see that the equitable chromatic number of Mycielskian of paths is given by 3 when $n \leq 11, n \neq 10$ and 4 when $n \geq 10, n \neq 11$ (see [8]). Hence, in an equitable near-proper coloring we consider two cases as $k = 2$ and $k = 3$.

*Case-1*: When $k = 2$, we have two available colors say $c_1$ and $c_2$. Assign all $v_i$'s $(1 \leq i \leq n)$ with the available colors $c_1$ and $c_2$. Also assign the corresponding $u_i$'s with the same colors in such a way that both $u_i$ and the corresponding vertex $v_i$ receive the same color. (We observe that when $n$ is even, among the $2n$ vertices $n$ number of vertices receive color $c_1$ and also $n$ number of vertices receive color $c_2$. And when $n$ is odd, $n+1$ number of vertices receive color $c_1$ and $n-1$ number of vertices receive color $c_2$). Now assign the vertex $w$ with color $c_2$. Since $w$ is adjacent with all $u_i$'s and $\lfloor \frac{n}{2} \rfloor$ number of $u_i$'s receive color $c_2$, we observe that we obtain $\lfloor \frac{n}{2} \rfloor$ number of bad edges.

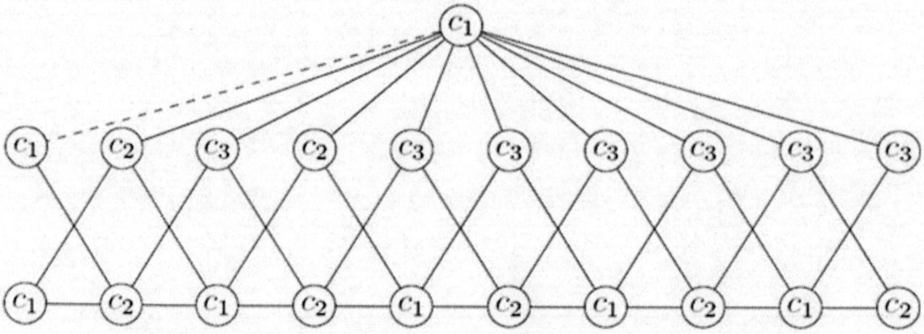


**Fig. 1** Mycielskian of paths with 3-equitable near proper coloring

*Case-2*: When $k = 3$, in an equitable near-proper coloring of Mycielskian of paths there is one color class contains $\lceil \frac{2n+1}{k} \rceil$ vertices and two color classes contain $\lfloor \frac{2n+1}{k} \rfloor$ vertices. Now, as in Case-1 assign all $v_i$'s with colors $c_1$ and $c_2$ alternatively. We choose the minimum cardinality color class that is, the color class with $\lfloor \frac{2n+1}{k} \rfloor$ vertices for color $c_1$. Here we observe that $\lceil \frac{n}{2} \rceil$ number of $v_i$'s receive color $c_1$. Now assign the vertex $w$ with color $c_1$. Now we see that $\lfloor \frac{2n+1}{3} \rfloor - \lceil \frac{n}{2} \rceil$ number of $u_i$'s should receive color $c_1$ in order to satisfy the equitability condition. The remaining $u_i$'s can be assigned with other two colors in an equitable manner. Since $w$ is adjacent with all $u_i$'s and $\lfloor \frac{2n+1}{k} \rfloor - \lceil \frac{n}{2} \rceil$ number of $u_i$'s receive color $c_1$, we conclude that we obtain $\lfloor \frac{2n+1}{3} \rfloor - \lceil \frac{n}{2} \rceil - 1$ bad edges in this case. (See Fig. 1 for illustration.)

Figure 1 depicts a 3-equitable near proper coloring of Mycielskian of paths.

**Theorem 2.2** *The equitable defective number of Mycielskian of cycles $\mu(C_n)$ is given by*

1. If $k = 2$, then $b^k_{\chi_e}(\mu(C_n)) = \begin{cases} \frac{n}{2} & \text{if } n \text{ is even} \\ \lfloor \frac{n}{2} \rfloor + 3 & \text{if } n \text{ is odd} \end{cases}$
2. If $k = 3$, then $b^k_{\chi_e}(\mu(C_n)) = \begin{cases} \lceil \frac{2n+1}{3} \rceil - \lfloor \frac{n}{2} \rfloor - 1 & \text{if } n = 3, 5 \\ \lfloor \frac{2n+1}{3} \rfloor - \lfloor \frac{n}{2} \rfloor - 1 & \text{otherwise.} \end{cases}$

**Proof.** Let $\mu(C_n)$ be the Mycielski graph of cycles with $2n + 1$ vertices. Let $\{v_1, v_2, \cdots, v_n\}$ be the vertices of the cycle $C_n$ and $\{u_1, u_2, \cdots, u_n\}$ be the vertices corresponding to $\{v_1, v_2, \cdots, v_n\}$. Let $w$ be the vertex adjacent with all $u_i$'s where $1 \le i \le n$. We see that the equitable chromatic number of Mycielskian of cycles is given by 3 when $n = 4, 6, 8$ and 4 in all other cases. Thus, we consider two cases as $k = 2$ and $k = 3$.

*Case-1*: When $k = 2$, we consider two subcases as below.

*Subcase-1.1*: When $k = 2$ and $n$ is even, assign the vertices of the cycle $C_n$ with the available colors $c_1$ and $c_2$ alternatively. Now assign the corresponding vertices with the same colors such that both $u_i$ and $v_i$ receive the same color. Now assign the vertex $w$ with either $c_1$ or $c_2$. Since $\frac{n}{2}$ number of $u_i$'s receive color $c_1$ and $\frac{n}{2}$ number of $u_i$'s receive color $c_2$ we obtain $\frac{n}{2}$ bad edges.

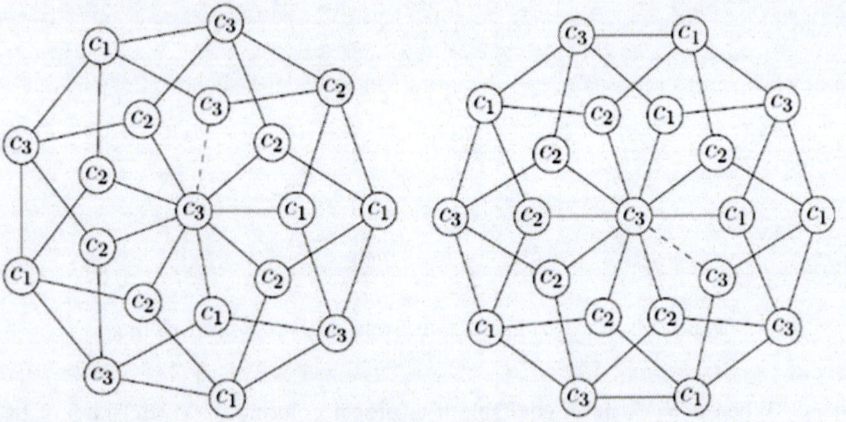

**Fig. 2** Mycielskian of cycles with 3-equitable near proper coloring

*Subcase-1.2*: When $k = 2$ and $n$ is odd, assign the available colors to all $u_i$'s and $v_i$'s as in Subcase-1.1. Here we observe that among the $2n$ vertices, $n + 1$ number of vertices receive color $c_1$ and $n - 1$ number of vertices receive color $c_2$. And assign the vertex $w$ with color $c_2$ to satisfy the equitability condition. And we obtain $\lfloor \frac{n}{2} \rfloor$ bad edges which are incident with $w$. Along with that we get three more $u_i v_j$ bad edges. (That is $v_1 v_n, v_1 u_n, u_1 v_n$). Hence, we obtain $\lfloor \frac{n}{2} \rfloor + 3$ bad edges.

*Case-2*: When $k = 3$, let $\{c_1, c_2, c_3\}$ be the available colors and $\{V_1, V_2, V_3\}$ be the corresponding color classes. In an equitable near-proper coloring of Mycielskian of cycles there is one color class which contains $\lceil \frac{2n+1}{k} \rceil$ vertices and two color classes contain $\lfloor \frac{2n+1}{k} \rfloor$ vertices. Consider the color class $V_3$ with minimum cardinality that is $\lfloor \frac{2n+1}{k} \rfloor$ vertices. We can assign $\lfloor \frac{n}{2} \rfloor$ vertices on the cycle $C_n$ with color $c_3$. Also assign the vertex $w$ with color $c_3$. Since the cardinality of the color class $V_3$ is $\lfloor \frac{2n+1}{k} \rfloor$, we see that $\lfloor \frac{2n+1}{k} \rfloor - \lfloor \frac{n}{2} \rfloor - 1$ number of $u_i$ vertices should receive color $c_3$ in order to satisfy the equitability condition. And since the vertex $w$ is adjacent with all $u_i$'s we see that we obtain $\lfloor \frac{2n+1}{k} \rfloor - \lfloor \frac{n}{2} \rfloor - 1$ bad edges. (See Fig. 2 for illustration.)

Figure 2 depicts a 3-equitable near proper coloring of Mycielskian of cycles.

**Theorem 2.3** *The equitable defective number of Mycielskian of wheels $\mu(W_{1,n})$ is given by*

$$b^k_{\chi_e}(\mu(W_{1,n})) = \begin{cases} 2n & \text{if } k = 2 \\ 2\lfloor \frac{2n+3}{k} \rfloor - 5 & \text{if } k \geq 3. \end{cases}$$

**Proof.** Let $\{v_0, v_1, v_2, \cdots, v_n\}$ be the vertices of the wheel graph $W_{1,n}$. Here $v_0$ denotes the central vertex and $v_1, v_2, \cdots, v_n$ are the rim vertices. Let $\{u_0, u_1, u_2, \cdots, u_n\}$ be the vertices corresponding to $\{v_0, v_1, v_2, \cdots, v_n\}$ and let $w$ be the vertex which is adjacent to all $u_i$'s for $0 \leq i \leq n$. Hence, the Mycielski graph of wheel graph contains $2n + 3$ vertices. In an equitable near-proper coloring we consider the cases as below.

*Case-1*: When $k = 2$, in an equitable near-proper coloring we need to consider two subcases.

*Subcase-1.1*: When $k = 2$ and $n$ is even, we assign the two available colors $c_1$ and $c_2$ to the vertices of $\mu(W_{1,n})$ in the following way. Assign $\{v_0, v_1, v_2, \cdots, v_n\}$ with $c_1$ and $c_2$ alternatively. Hence, both $v_0$ and $v_n$ receive color $c_1$. We observe that we obtain $\frac{n}{2}$ bad edges among the spokes. Now assign all $u_i$'s with $c_1$ and $c_2$ such that both $v_i$ and $u_i$ receive the same color $\forall i$. Since $v_0$ is adjacent with all $u_i$'s where $1 \leq i \leq n$ we get $\frac{n}{2}$ bad edges which are incident with $v_0$. In the same way as $u_0$ is adjacent with all $v_i$'s where $1 \leq i \leq n$ we obtain $\frac{n}{2}$ bad edges which are incident with $u_0$. Now assign $w$ with color $c_2$ to satisfy the equitability condition. And since the vertex $w$ is adjacent with all $u_i$'s $(0 \leq i \leq n)$ we obtain $\frac{n}{2}$ bad edges which are incident with $w$. Hence, we obtain $4\frac{n}{2} = 2n$ bad edges.

*Subcase-1.2*: Let $k = 2$ and $n$ is odd. When we assign the available colors $c_1$ and $c_2$ to the $n+1$ vertices $\{v_0, v_1, v_2, \cdots, v_n\}$ as in Subcase-1.1 we obtain $\lfloor \frac{n}{2} \rfloor$ bad edges among the spokes. Along with that we obtain one bad edge on the cycle since to properly color an odd cycle we require minimum three colors. Now assign all $u_i$'s with $c_1$ and $c_2$ as in Subcase-1.1. Here, we obtain $\lfloor \frac{n}{2} \rfloor$ bad edges which are incident with $v_0$ and $\lfloor \frac{n}{2} \rfloor$ bad edges which are incident with $u_0$. Assign $w$ with color $c_1$ or $c_2$ we obtain $\frac{n+1}{2}$ bad edges. Hence, we obtain $\lceil \frac{n}{2} \rceil + 2\lfloor \frac{n}{2} \rfloor + \frac{n+1}{2} = 2n$ bad edges.

*Case-2*: When $k = 3$, we consider two subcases.

*Subcase-2.1*: When $k = 3$ and $n$ is even, we assign the available three colors to $2n + 3$ vertices. We assign the central vertex $v_0$ of the wheel with color $c_1$ and the remaining $v_i$'s where $1 \leq i \leq n$ can be assigned with colors $c_2$ and $c_3$ alternatively. Since $n$ is even, we can properly color the wheel with three colors. Now we observe that colors $c_2$ and $c_3$ repeated $\frac{n}{2}$ times on the rim. To color the vertices in an equitable manner all the three color classes should contain either $\lfloor \frac{2n+3}{3} \rfloor$ vertices or $\lceil \frac{2n+3}{3} \rceil$ vertices. We assign vertex $u_0$ and vertex $w$ with color $c_1$ and consider that $c_1$ should be repeated $\lfloor \frac{2n+3}{3} \rfloor$ times. Hence, $\lfloor \frac{2n+3}{3} \rfloor - 2$ number of $u_i$'s where $1 \leq i \leq n$ should be colored with $c_1$. Since $v_0$ is adjacent with $n$ number of $u_i$'s where $1 \leq i \leq n$ we obtain $\lfloor \frac{2n+3}{3} \rfloor - 3$ bad edges which are incident with $v_0$. And since $w$ is adjacent with $\lfloor \frac{2n+3}{3} \rfloor - 2$ number of $c_1$ colored vertices there will be $\lfloor \frac{2n+3}{3} \rfloor - 2$ number of $u_iw$ bad edges. Other $u_i$'s can be assigned with $c_2$ and $c_3$ without creating any bad edges. And we obtain $2\lfloor \frac{2n+3}{3} \rfloor - 5$ bad edges in this case.

*Subcase-2.2*: When $k = 3$ and $n$ is odd, we color the $v_i$'s as in Subcase-2.1 for $0 \leq i \leq n-1$ and assign $v_n$ with color $c_1$. And when we assign colors to $u_i$'s, assign vertex $u_n$ also with color $c_1$ and the remaining $u_i$'s can be assigned in an equitable manner. Here we observe that we obtain an extra bad edge on the wheel since both the vertices $v_n$ and $v_0$ are assigned with color $c_1$. Also we obtain $\lfloor \frac{2n+3}{3} \rfloor - 3$ bad edges which are incident with $v_0$ and $u_0$ and $\lfloor \frac{2n+3}{3} \rfloor - 3$ bad edges which are incident with $w$. Hence, the resulting number of bad edges in this case is $2\lfloor \frac{2n+3}{3} \rfloor - 5$.

*Case-3*: When $k \geq 4$, we follow the same procedure as in Subcase-2.1 for all $u_i$'s, $u_0$ and $w$. The $v_i$'s can be properly colored with the available colors. We observe that we obtain $\lfloor \frac{2n+3}{k} \rfloor - 3$ bad edges which are incident with $v_0$ and $\lfloor \frac{2n+3}{k} \rfloor - 2$ bad edges which are incident with $w$. Hence, the equitable defective number is $2\left\lfloor \frac{2n+3}{k} \right\rfloor - 5$.

**Theorem 2.4** *The equitable defective number of Mycielskian of gear graphs $\mu(G_n)$ is given by*

1. If $k = 2$, then $b^k_{\chi_e}(\mu(G_n)) = \lfloor \frac{n}{k} \rfloor$
2. If $k = 3$, then $b^k_{\chi_e}(\mu(G_n)) = \begin{cases} 2n+1-\frac{4n+3}{3} & if \quad n \equiv 0 \pmod 3 \\ 2n+1-\lceil \frac{4n+3}{3} \rceil & if \quad n \equiv 1 \pmod 3 \\ 2n+1-\lceil \frac{4n+3}{3} \rceil & if \quad n \equiv 2 \pmod 3. \end{cases}$

**Proof.** Let $v_0, v_1, v_2, \cdots, v_n$ be the vertices of the gear graph $G_n$ with $2n+1$ vertices. By the construction of Mycielski graph of gear graphs, $\mu(G_n)$ contains $4n+3$ vertices. Let $u_0, u_1, u_2, \cdots, u_n$ be the corresponding vertices of $v_i$'s in $\mu(G_n)$ and let $w$ be the vertex which is adjacent to all $u_i$'s where $0 \leq i \leq n$. We observe that the equitable coloring of Mycielskian of gear graphs contain 3 colors when $n = 3, 4, 5$ and 4 colors when $n \geq 6$. Hence, in an equitable near-proper coloring we need to consider two cases as $k = 2$ for $n \geq 3$ and $k = 3$ for $n \geq 6$.

*Case-1*: When $k = 2$ we have two available colors say $c_1$ and $c_2$. Assign the central vertex $v_0$ with color $c_1$. Let $v_1, v_3, v_5, \cdots, v_{n-1}$ be the vertices which are adjacent with the central vertex $v_0$. Assign these vertices with color $c_2$. Now the remaining vertices $v_2, v_4, v_6, \cdots, v_n$ can be assigned with color $c_1$ again. Hence, we have properly colored the gear graph resulting no bad edges. Now assign the $u_i$'s with the same colors as $v_i$ received $\forall i$. Since $v_i$ is not adjacent with $u_i$ $\forall i$ we end up with no bad edges. Now we assign the vertex $w$ with color $c_2$. Now we observe that we obtain $\lfloor \frac{n}{k} \rfloor$ bad edges since among the $n+1$ number of $u_i$'s, $\lceil \frac{n}{k} \rceil$ vertices receive color $c_1$ and $\lfloor \frac{n}{k} \rfloor$ vertices receive color $c_2$.

*Case-2*: When $k = 3$ and $n \geq 6$ assign all $u_i$'s with $c_1$ and $c_2$ alternatively. Since $k = 3$, either $\lceil \frac{4n+3}{3} \rceil$ or $\lfloor \frac{4n+3}{3} \rfloor$ vertices receive color $c_3$. In order to reduce the number of bad edges we consider $\lfloor \frac{4n+3}{3} \rfloor$ vertices receive color $c_3$. Assign vertex $w$ with color $c_3$. Since all $u_i$'s are assigned with colors $c_1$ and $c_2$, color $c_3$ should be assigned to $\lfloor \frac{4n+3}{3} \rfloor - 1$

number of $v_i$'s. Since $v_i$'s are rim vertices we obtain $2n+1-\frac{4n+3}{3}$ bad edges when $n \equiv 0 \pmod 3$, $2n+1-\lceil\frac{4n+3}{3}\rceil$ bad edges when $n \equiv 1 \pmod 3$ and $2n+1-\lfloor\frac{4n+3}{3}\rfloor$ bad edges when $n \equiv 2 \pmod 3$.

**Theorem 2.5** *The equitable defective number of Mycielskian of helm graphs $\mu(H_{1,n})$ is given by*

1. If $n$ is even, then $b^k_{\chi_e}(\mu(H_{1,n})) = \begin{cases} 2n & if \quad k=2 \\ 1 & if \quad k=3 \end{cases}$

2. If $n$ is odd, then $b^k_{\chi_e}(\mu(H_{1,n})) = \begin{cases} 2(n+1)+\lfloor\frac{n}{2}\rfloor & if \quad k=2 \\ 4 & if \quad k=3 \\ 1 & if \quad k=4\,. \end{cases}$

**Proof.** Let $\{v_0, v_1, v_2, \cdots, v_n, v'_1, v'_2, \cdots, v'_n\}$ be the vertices of the helm graph $H_{1,n}$. Here $v_0$ denotes the central vertex and $v_1, v_2, \cdots, v_n$ are the rim vertices. Let $v'_1, v'_2, \cdots, v'_n$ be the vertices corresponding to $v_1, v_2, \cdots, v_n$ such that $v'_i$ is adjacent with $v_i$ $\forall i$. According to the construction of Mycielski graph $\mu(H_{1,n})$, let $u_0, u_1, u_2, \cdots, u_n, u'_1, u'_2, \cdots, u'_n$ be the vertices corresponding to all $v_i$'s and let $w$ be the vertex which is adjacent to all $u_i$'s. Hence, the Mycielskian of helm graph contains $4n+3$ vertices. The equitable coloring of Mycielskian of helm contain 4 colors when $n$ is even and 5 colors when $n$ is odd. Hence, in an equitable near-proper coloring we consider two cases as $n$ is even and $n$ is odd.

*Case-1*: When $n$ is even, we consider the following subcases.

*Subcase-1.1*: When $k=2$ and $n$ is even, assign the central vertex $v_0$ with color $c_1$ and the rim vertices with $c_1$ and $c_2$ alternatively. Now assign the pendant vertices with the two available colors such that if $v_i$ is assigned with color $c_1$ (or $c_2$) then assign $v'_i$ with color $c_2$ (or $c_1$). Now we observe that we obtain $\frac{n}{2}$ bad edges among the spokes. Now assign all $u_i$'s where $i = 0, 1, 2, \cdots, n, 1', 2', \cdots, n'$ with the same colors which are assigned to $v_i$. And we obtain $\frac{n}{2}$ number of $u_iv_j$ bad edges where $1 \le i \le n$. Now, assign the vertex $w$ with color $c_2$. Since $w$ is adjacent with $2n+1$ vertices and among those vertices, $\lfloor\frac{2n+1}{2}\rfloor$ vertices receive color $c_2$ we obtain $n$ bad edges. Hence, in this case we obtain $\frac{n}{2}+\frac{n}{2}+n = 2n$ bad edges.

*Subcase-1.2*: When $k=3$ and $n$ is even, assign $v_0$ with color $c_3$. Assign all rim vertices with colors $c_1$ and $c_2$ alternatively. Now assign $u_0$ with color $c_3$ and other $u_i$'s with colors $c_1$ and $c_2$. And assign $w$ with color $c_3$. Since $w$ is adjacent with all $u_i$'s and only one $u_i$ received color $c_3$ we obtain only one bad edge in this case. Now color the pendent vertices properly with any of the three colors satisfying the equitability condition and we observe that we get only one bad edge in this case.

*Case-2*: When $n$ is odd, we consider three subcases.

*Subcase-2.1*: When $k = 2$, assign the rim vertices $v_1, v_2, \cdots, v_n$ as in Subcase 1.2. Here we get one bad edge on the cycle since to properly color an odd cycle we require at least three colors. And assign the pendent vertices $v'_1, v'_2, \cdots, v'_n$ with $c_1$ or $c_2$ such that if $v_i$ is assigned with color $c_1$ (or $c_2$) then assign $v'_i$ with color $c_2$ (or $c_1$). Now assign the central vertex $v_0$ with color $c_2$ we obtain $\lfloor \frac{n}{2} \rfloor$ bad edges among the spokes. Now assign the $u_i$'s where $i = 0, 1, 2, \cdots, n, 1', 2', \cdots, n'$ with the same colors as vertex $v_i$ received. Here, we observe that we obtain $\lfloor \frac{n}{2} \rfloor$ bad edges which are incident with $v_0$ and $\lfloor \frac{n}{2} \rfloor$ bad edges which are incident with $u_0$. Along with that we obtain two more bad edges $u_1v_n$ and $u_nv_1$. Now assign the vertex $w$ with color $c_1$. Since $w$ is adjacent with all $u_i$'s which are $2n+1$ in number we get $\lfloor \frac{2n+1}{2} \rfloor$ bad edges. Hence, in this case we obtain $\lceil \frac{n}{2} \rceil + \lfloor \frac{n}{2} \rfloor + \lfloor \frac{n}{2} \rfloor + 2 + \lfloor \frac{2n+1}{2} \rfloor$ bad edges. On simplifying, we have $2(n+1) + \lfloor \frac{n}{2} \rfloor$ bad edges.

*Subcase-2.2*: When $k = 3$ and $n$ is odd, we color the rim vertices $v_1, v_2, \cdots, v_{n-1}$ with two colors $c_1$ and $c_2$ alternatively. Now assign $v_n$ and the central vertex $v_0$ with color $c_3$ which results in only one bad edge among the spokes. Now assign all pendent vertices with color $c_3$ to satisfy the equitability condition without creating any bad edges. Now start assigning colors to the $u_i$'s such that assign $u_0$ with color $c_3$ and all other $u_i$'s can be assigned with $c_1$ and $c_2$ alternatively. Here, we obtain two bad edges $u_0v_n$ and $v_1u_n$ (or $v_0u_n$ and $u_1v_n$). Now assign the vertex $w$ with color $c_3$ creating only one bad edge. Hence, in this case we obtain 4 bad edges.

*Subcase-2.3*: When $k = 4$, Assign the $v_i$'s and $u_i$'s where $2 \leq i \leq n$ with two colors $c_1$ and $c_2$ alternatively. Assign $v'_1, v'_2, \cdots, v'_n$ with colors $c_3$ or $c_4$. Now assign $u'_1, u'_2, \cdots, u'_n$ with any of the three colors $c_1$, $c_2$ and $c_3$ satisfying the equitability condition and without creating any bad edges. Now assign $v_0$, $u_0$ and $w$ with color $c_4$. We observe that we get only one $wu_0$ bad edge in this case.

## 3 Conclusion

In this paper, the equitable near-proper coloring of Mycielskian of paths, cycles, wheel graphs, gear graphs and helm graphs are explored. We determined the equitable defective number by investigating all the possible cases. Other prominent graph classes, derived graphs, graph operations, graph products and graph powers can be studied in this context. A further study on equitable near-proper coloring would result in great contributions in literature and can explore the applications in this area. Hence, a larger scope for strong research is recommended .

## References

1. Harary F (1996) Graph Theory, vol 2. Narosa Publ. House, New Delhi
2. West DB (2001) Introduction to Graph Theory, vol 2. Prentice Hall of India, New Delhi
3. Meyer W (1973) Equitable coloring. Amer Math Monthly 80(8):920–922
4. Kok J, Sudev NK (2021) $\delta^{(k)}$ - colouring of cycle related graphs. Adv Stud Contemp Math (to appear)
5. Sabitha J, Sudev NK On equitable near proper coloring of graphs. Proyecciones J Math Commun (under Review)
6. Fornasiero F, Naduvath S (2019) On J-colorability of certain derived graph classes. Acta Univ Sapientiae Info 11(2):159–173
7. Sudev NK, Susanth C, Kalayathankal SJ (2018) On the rainbow neighbourhood number of Mycielski type graphs. Int J Appl Math 31(6):797–803
8. Vivin V, Kaliraj K (2017) Equitable coloring of Mycielskian of some graphs. J Math Ext 11:1–18

# A Model of Daily Newspaper ATM Using PLC System

Ganesh I. Rathod(✉), Dipali A. Nikam, and Rohit S. Barwade

Dr. J. J. Magdum College of Engineering, Jaysingpur, Kolhapur, Maharashtra, India

**Abstract.** Due to the COVID-19 pandemic, social distancing has become of prior importance in the everyday lifestyle of the people. In this pandemic, though e-newspapers are available, reading print newspaper daily is the habit of many people. The proposed model gives an idea of providing a printed daily newspaper to the people without being in contact with paperboys or newspaper shopkeepers. The proposed model takes ₹05 coin as an input and provides three newspaper options to the user and accordingly delivers the newspaper.

**Keywords:** Programming logic control (PLC) · Online news · Coin detection · Electromagnetic relay · Buzzer · Automated teller machine (ATM)

## 1 Introduction

Globally, as of 12:01 pm CEST, 8 August 2020, there are been 19,131,120 confirmed cases of COVID-19, including 714,873 deaths reported to WHO. In India, itself confirmed cases of COVID-19 has reached 20 lakhs. Since no country has yet found any vaccine for this deadly virus, social distancing is itself one of the main prevention methods for all of us. Luckily with the eruption of technology, smartphones, and cheaper data people can be updated with the latest news anytime and anywhere without going outside. From e-newspaper's to online news, nowadays news is on the palms of the people. But the online news has its disadvantages like there is so much competition around to display the news, that most of the information is not properly checked. Popping ads are so much around that, it distracts the reader from reading. Unlike the printed newspaper there are no front pages in the online news due to which it is not clear that which article is important because they all carry equal weight. According to [12], studies discloses that people are less likely to read long articles (containing more than 1,000 words) online than they are on a printed paper. Research shown in [12], also disclose that people grasp limited information when read on a mobile screen because reading on screen is further substantially and mentally challenging than reading on paper. By [12], volunteers given the task to read information from screen as well as from printed paper. Those who had read the information on the mobile screen trusted more on memorizing, whereas volunteers who read on printed paper trusted on remembering and knowing.

People read more carefully and continue to hold the read information when accessing news via print media. The act of physically grabbing a newspaper and sitting down to


S. Shukla et al. (eds.), *Data Science and Security*, Lecture Notes in Networks and Systems 290,
https://doi.org/10.1007/978-981-16-4486-3_38

read means there is a certain commitment to the news. A print newspaper is undying and stable, whereas after some day's articles can be hidden from an online website. With these positives of the print newspaper, majority of the people read news only in the printed format they are addicted to a reading newspaper in the morning.

While social distancing is important in this pandemic, people still tend to go outside to buy a daily newspaper. Paperboys risk their life and deliver print newspapers daily door to door. Our proposed model, "Daily newspaper ATM using PLC", is the small step to maintain social distancing and yet deliver daily newspaper to the user. As the name indicates proposed model is based on programming, logic, and controlling operation. This model can be used in public places like Airports, Bus Stands, Railway stations, Hospitals, Malls, etc. The inputs to the programming logic controller (PLC) are ₹05 coin and output is the latest printed newspaper. When the user inserts the coin in the coin detection box, the IR sensor senses the coin, and then from the given three options of newspaper, the user has to select the type of newspaper which he wants, once the option is chosen by the user, the programming logic controller turns on the relay and the appropriate newspaper cassette will be switched on and one of the newspaper from the cassette will be dispatched through a dispenser. The buzzer remains on continuously till the newspaper is removed from the dispenser by the user, just like an ATM cash machine.

## 2 Overview of Programming Logic Control

PLCs were invented by Dick Morley in 1964. Since then PLC has upraised the manufacturing and industrial sectors. Programmable Logic Controller (PLC) remains as a microprocessor-based system that operates programmable memory for storing instructions and performs functions such as sequencing, logic, control, counting, and arithmetic as mentioned in [1]. More features and working details can be seen in [1–6]. It is intended to be used in industries that involve machining, home automation, wrapping, material management, automated assemblage, etc. The main feature of the PLC is that it can be easily programmed and it can bear the industrial atmosphere, such as high temperature, moisture, mechanical shocks, and vibrations. The Block diagram of the PLC system is shown in Fig. 1, which contains basic four components, Central Processing Unit, Input and Output module, Power supply, and Programming device.

### 2.1 Central Processing Unit (CPU)

The Central Processing Unit is the core of the PLC, which performs various operations like transmitting, counting, controlling, data assessment, and sequential operations as mentioned in [4]. The magnitude and nature of CPU will regulate the programming functions available, size of the application logic available, amount of memory available, and processing speed [4]. The main operation of the CPU is that it reads input information from numerous sensing/ input devices and performs the client program from memory and sends suitable output commands to output devices. CPU consists of a small area of memory section in which data is stored and retrieved. The memory section can be categorized into two types: Data memory & User memory.

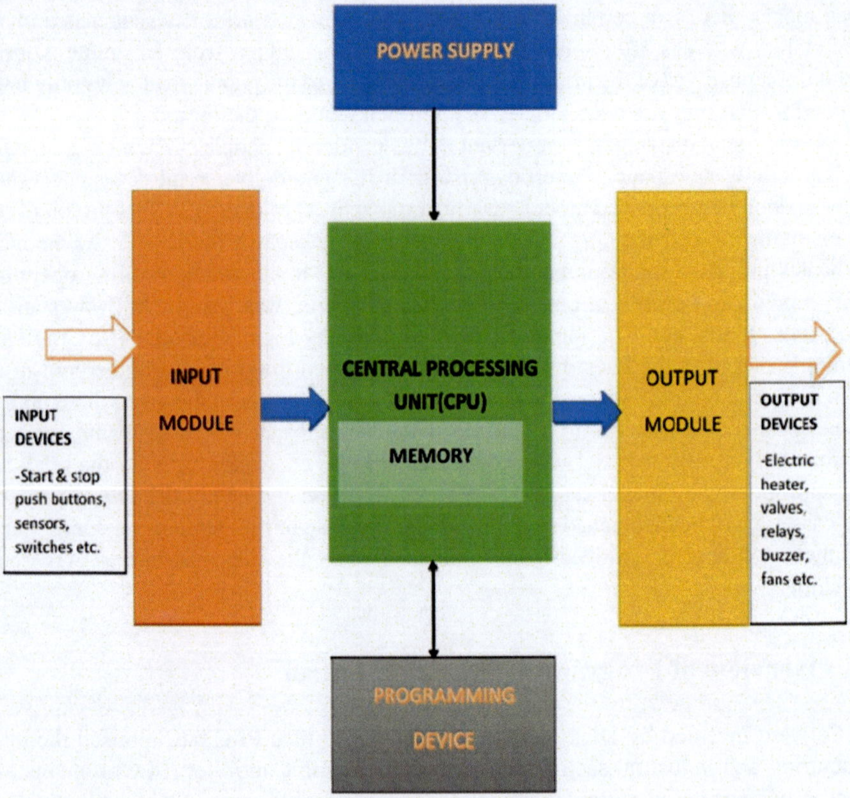


**Fig. 1** Block diagram of PLC

Data Memory is utilized to store numerical data wanted in math calculation, bar code data, etc. and User memory contains the user's application program.

## 2.2 Input and Output Module

The input module is an interface between CPU and input devices and it is used to convert an analog signal into a digital signal which can be used by the CPU. For example, 220 V AC is converted into 5 V DC. More AC-voltage regulator systems and working of electronic devices can be understood from [7–9]. Input devices like sensors, switches, start and stop pushbuttons, etc., mentioned in Fig. 1, are hardwired to terminals on the input modules. The output module functions in the reverse order of the input module, i.e. it is a mediator between output devices and central processing unit (CPU), which converts a digital signal into an analog signal used to control output devices. Output devices like electric heater, buzzer, valves, relays, lights, fans, etc., mentioned in Fig. 1, are hardwired to terminals on the output module.

### 2.3 Power Supply

The Power supply is provided to the CPU, input, and output module unit. The power supply may be separately or integrally mounted unit. Most of the PLC system operates on 0 V DC and 24 V, as can be studied from [10].

### 2.4 Programming Device

Programming devices are dedicated devices like keyboards and monitors, used for programming a PLC. The user program is entered in the PLC processor with the help of a keyboard in the form of a ladder logic program. This ladder logic program can be seen on the display screen. The programmer can communicate with the PLC processor with help of programming devices also they can edit the program and monitor the execution of the program of the PLC. Troubleshooting of the PLC ladder logic program is also done by programming devices.

### 2.5 Working of PLC

PLC has the following scanning operations.

- Through CPU, it begins to read all the data from the input module and checks the status of all the inputs devices that are connected to the PLC.
- Through CPU, it starts executing the program written by the programmer in relay ladder logic.
- It performs all the internal diagnosis and communication tasks (communications with programming terminals).
- Concerning the program results, it writes the data into the output module so that all outputs are updated.

This process will be continued as far as the PLC is in run mode.

## 3 Proposed Methodology

In this paper, Siemens PLC is suggested which the main component of our Daily newspaper ATM system. The operation of our model will take place as per the program entered in the ladder logic program and the time delay is given to it. PLC accepts the input from the different sensors & gives output appropriately and performs the five below operation as follows: & gives output appropriately and performs the five below operation as follows:

- Detects the coin from the coin detector. Only the correct coin (i.e. ₹05) will be accepted and the wrong one will be rejected.
- It accepts the newspaper choice by the user from the input screen. Total three choices of newspaper will be displayed on the screen: "The Hindu", "Loksatta" and "The Asian Age".

- It turns on the newspaper cassette-1 filled with "The Hindu" newspaper if choice 1 ("The Hindu") is pressed.
- It turns on the newspaper cassette-2 filled with "Loksatta" newspaper if choice 2 ("Loksatta") is pressed.
- It turns on the newspaper cassette -3 filled with "The Asian Age" newspaper if choice 3 ("The Asian Age") is pressed.
- Turn-ON/OFF the buzzer when newspaper present/absent respectively in the dispenser.

Block diagram of the proposed model is shown in Fig. 2, it contains basic four modules: Coin Detection, Relay unit, Newspaper cassette and Buzzer.

### 3.1 Coin Detection Module

For the coin detection module, "GD-100F -CPU coin acceptor" is suggested along with Arduino, and the figure for the same is shown in Fig. 3. The advantages of Arduino can be seen in [11]. The coin detection module works by comparing the coin inserted inside the sensor with the coin inserted at the front hole. If the coin is the same, the coin will be accepted, but if a coin is different, the coin will be rejected. So, any coin can be inserted and programmed using Arduino and for this model ₹, 05 coins is been used.

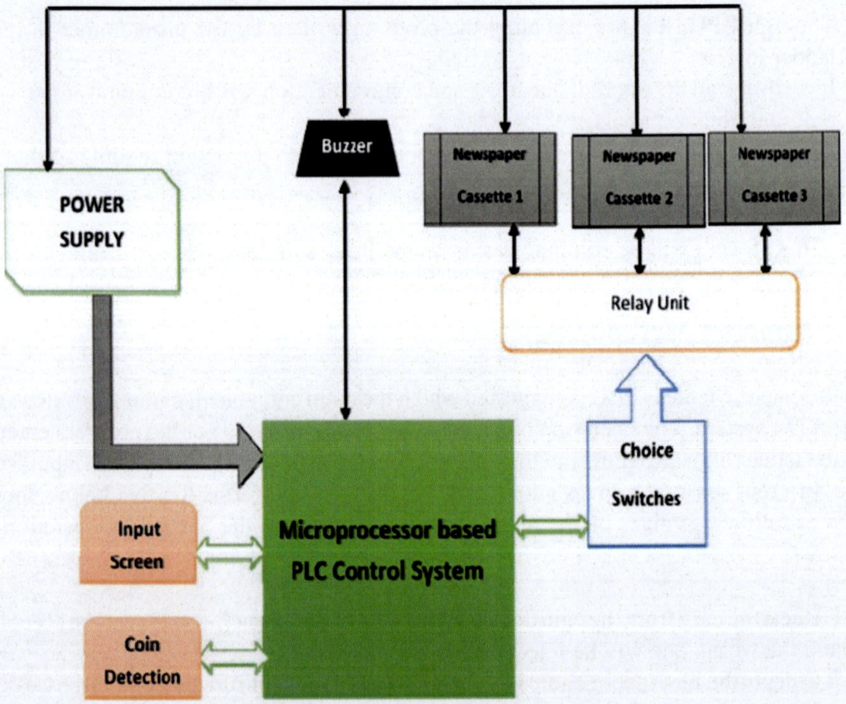


Fig. 2 Block diagram of proposed model

**Fig. 3** GD-100F -CPU coin acceptor

**Fig. 4** Electromagnetic relay

### 3.2 Relay Unit Module

A relay is an electromagnetic switch ran by a fairly small electrical current that can activate or turn off an ample electric current. The key part of a relay is an electromagnet coil. When a current flow through the coil, fans, etc., mentioned in Fig. 1, are hardwired to terminals on the output module electromagnetic field is set up. A sample image of an electromagnetic relay is shown in Fig. 4. In the proposed model, the PLC system takes input from the sensors which produce only small electric currents. But in our model, we need to drive bigger pieces of apparatus like a Newspaper cassette, which uses bigger currents. Relays bridge the gap, making it possible for small currents to activate bigger ones. According to the choice provided by the user, an appropriate newspaper cassette will be turned on to dispatch the individual newspaper.

### 3.3 Newspaper Cassette Module

The proposed model suggests three newspaper cassettes each stacked with different newspapers. The type of newspapers suggested according to their price in India are The Hindu, Loksatta, and The Asian Age. Just like the ATM cash cassette, a newspaper cassette can be developed to stack daily newspapers. In this, COVID-19 pandemic paper boys risk their lives and deliver newspaper door to door, but with this model now paper boy's job is to visit the newspaper ATM daily and fill the cassette.

### 3.4 Buzzer Module

The proposed model suggests a simple piezo buzzer and the image of it is shown in Fig. 5. The buzzer is normally associated with the alarm application making a continuous beep sound when turned ON. A piezo buzzer is a small well-organized component to enhance sound features to our model. It is a very small and compact 2-pin structure hence can be easily used. The long pin identified by the (+) symbol can be powered by 5/6 V DC and

**Fig. 5** Piezo buzzer

the short pin identified by (−) is connected to the ground of the circuit. In the proposed model, the buzzer remains turned on continuously till the newspaper is removed from the dispenser by the user, just like an ATM cash machine. Once the newspaper is removed from the dispenser the buzzer will be turned off.

# 4 Expected Results and Approximate Cost

Below are the two tables providing the details of expected results and approximate cost of important components required (Table 1 and 2).

**Table 1** Expected results

| Sr. no | Use of practice | Expected results |
|---|---|---|
| 1 | Coin detection | Should accept the ₹05 coins only and reject the different coins |
| 2 | Input screen | Once the coin is accepted successfully, three newspaper options should be visible and be ready to take input from a user |
| 3 | Newspaper dispenser | Based on the choice provided by the user, an appropriate newspaper cassette should be turned on and should be dispensed from the dispenser |
| 4 | Buzzer | The Buzzer should be remain turned on continuously till the newspaper is removed from the dispenser, and once the newspaper is removed, the buzzer should be turned off automatically |

**Table 2** Approximate cost

| Sr. no | Name of the component | Approximate cost in Indian rupees |
|---|---|---|
| 1 | Siemens PLC | 75,000 |
| 2 | GD-100F-CPU coin detector | 5,000 |
| 3 | Electromagnetic relay | 70/unit |
| 4 | Piezo buzzer | 50/unit |

## 5 Conclusions

The idea presented in this paper is a blueprint to provide daily newspapers to the user anytime, maintaining total social distancing. Using this model, now the print newspaper readers can come out and grab it without being in contact with any paper boys or newspaper shopkeepers. The job of the paperboys is still intact since they are the ones who fill the cassette daily, but now they don't risk their life, purely because they don't deliver newspapers door to door. Post pandemic too this model can be mounted permanently in public places like Airports, Bus Stands, Railway stations, Hospitals, Malls, etc. Since the model is a blueprint, performance analysis is not done and the future work can be viewed after implementing in real-time.

## References


1. Crispin J (2004) Programmable Logic Controllers and their Engineering Applications, 2nd edn, pp 150–155. McGraw-Hill, New York
2. Nashelsky L (2013) Electronic Devices & Circuit Theory, 10th edn, pp 250–251. Pearson Prentice Hall Publication
3. Johnson CD (2014) PLC Process Instrumentation and Technology, 8th edn, pp 340–345 Tata McGraw Hill
4. Frey G, Litz L (2000) Formal methods in PLC programming. In: IEEE Conference on System Man and Cybernetics SMC, pp 1–4
5. Endi M, Elhalwagy YZ, Hashad A (2010) Three-layer PLC/SCADA system architecture in process automation and data monitoring, pp 420–425. IEEE. (978–1–4244–5586–7/10/C)
6. Hristofovou L, Hatzipetvou K (1998) System with PLC for the control of asynchronous motor. Diploma work, National Tech Univ, Athens
7. Loannides MG, Papadopoulos PJ, Tegopoulos JA (1990) Digital techniques for AC voltage regulation. In: Proceedings of 6th International Conference Power Electronics Motion Control, Budapest, pp 975–979
8. Horowitz P, Hill W (2002) The Art of Electronics, 2nd edn, pp 200–201, Cambridge University Press Publication
9. Loannides MG, Katiniotis IM (2000) Laboratory of Electric Drives. Editions National Tech Univ, Athens
10. Aramaki N, Shimikawa Y, Kuno S, Saitoh T, Hashimoto H (1997) A new architecture for high-performance programmable logic controller. In: Proceedings of 23rd International Conference Industrial Electronics, Control and Instrumentation, vol 1, pp 187–199
11. Gadage R, Mirje S, Yadav P, Makandar Z, Rathod GI (2019) An Approach to Accelerometer Based Controlled Robot. In: International Research Journal of Engineering and Technology (IRJET), vol 6, issue 6, pp 117–119
12. https://www.scientificamerican.com/article/reading-paper-screens/

# The Preservative Technology in the Inventory Model for the Deteriorating Items with Weibull Deterioration Rate

Jitendra Kaushik[1] and Ashish Sharma[2](✉)

[1] Department of Data Science, Christ University, Bangalore, India
[2] School of Sciences, Christ University, Bangalore, India
ashish.sharma@christuniversity.in

**Abstract.** An EOQ model for perishable items is presented in this study. The deterioration rate is controlled by preservative technology. This technology only enhances the life of perishable items. So, retailers invested in this technology to get extra revenue. The Weibull deterioration rate is considered for the ramp type demand. Shortages consider partially backlogged, and discount is provided to loyal customers. The concavity of the profit function is discussed analytically. Numerical examples support the solution procedure; then, Sensitivity analysis is applied to accomplish the most sensitive variable.

**Keywords:** Weibull · Shortage · Retailers

## 1 Introduction

The deterioration refers to the decaying quality of the product. As time increases, the quality of the product deteriorates like in food products taste and smell changes with time. In that situation, customers don't want to purchase these types of items results in a direct loss of the retailers. As per a study by [12], approximately 20% of food waste during the storage period. This financial and social loss motivates us to solve this problem through the inventory model. We are using the concept of preservative technology for the deteriorating items so that the deterioration rate reduces as a consequence shelf life of the product enhances. In this way, retailers get extra revenue by reducing the wastage of food. Retailers invested in the preservative technology to accomplish extra revenue up to the second phase only because after that, food quality is not acceptable.

In this study, we are taking a non-linear price and time-dependent ramp type demand with partial backlog shortage. The concavity of the profit function is discussed analytically. The numerical examples support our study. 3D graphs show the concavity of profit function based on decision variables. The sensitivity analysis is applied to find the most sensible variable. In the last, we conclude our study with future aspects.


S. Shukla et al. (eds.), *Data Science and Security*, Lecture Notes in Networks and Systems 290,
https://doi.org/10.1007/978-981-16-4486-3_39

## 2 Literature Review

The inventory model was pioneered by [5]. [10] first considered ramp-type demand with a constant rate of deterioration. The shortage was allowed as fully backlogged. [3] presented a study of ramp-type demand patterns for perishable items. The demand function was time-dependent, considered with allow shortage then compare without shortage in their study. [7, 13, 16] worked ramp type demand of inventory models. [2] presented a ramp-type demand pattern that is dependent on time. The Weibull deterioration was considered completely backlogged. [14] worked on ramp-type time-dependent demand function. The complete backlog was allowed in their study, which was further extended by [15], allowing Weibull deterioration. [15] presented an inventory model for perishable items. Weibull deterioration rate was considered for quadratic demand function of time.

Preservative technology has also emerged as a new topic of research in inventory modeling. The concept of preservation technology introduced by [6] a time-dependent linear demand function was adopted for the deteriorating item. Then [11] applied the preservative technology concept into a trapezoidal model with the assumption that retailer will stop their investment into the third phase of the deteriorating items with allowing shortage as completely backlogged. The preservative technology controls the deterioration rate, but after a specific period of time, the quality of the products gets worse. Then, [1] also worked on trapezoidal type demand function with Weibull as deterioration rate with allowing shortage as partial backlog. [8, 9] presented a trapezoidal type demand model in their study, which shall be extended in this paper by considering preservation technology. The ramp-type demand pattern was adopted for our study,the two parameters Weibull distribution is considered. The shortage is allowed with partial backlogging.

## 3 Notations and Assumptions

### 3.1 Notations

$T_2$: Interval Length
$T_1$: Time up to which Demand Increases
$T_{L_1}$: Shortage Starting after Phase I
$T_{L_2}$: Shortage Starting after phase II
I: Initial Inventory Level
$I_b$: Backlogged Shortage
$\theta$: Deterioration with Preservation
$c_0$: Ordering Cost
$c_1$: Holding Cost Per Unit of Time
c: Purchasing Price of Items Per Unit
P: selling price as per Unit of the Item
$\lambda P$: The Backlogged Price; $(0 < \lambda < 1)$

In this paper we consider $gh(1-t)^{h-1}$ (two parameter weibull distribution function)

### 3.2 Demand Function

We consider a ramp-type demand function that increases first then becomes constant. Time-dependent Weibull deterioration rate is considered when preservative technology is allowed.

For phase I: $f_1(P, t) = \frac{a+bt}{Pj}; 0 \leq \mathrm{t} \leq \mathrm{T}_1$.

For phase II: $f_2(P, t) = \frac{a+bT_1}{Pj}; \mathrm{T}_1 \leq \mathrm{t} \leq \mathrm{T}_2$.

Where P and T are decision variables and others like a,b and j are demand parameters.

### 3.3 Assumption

(1) Discount to loyal customers

$$\gamma(\eta) = 1 - (\eta/T_2)\,;\ 0 \leq \eta \leq T_2$$

(2) Backorder Price is $\lambda P$ such that $c < \lambda P <$ P.
(3) Positive demand function considered; $f_F(P, T) > 0$; $F = 1, 2$.
(4) $P >$ c + T$c_1$. It is considered a price floor.
(5) Infinite replenishment rate is considered.
(6) Weibull deterioration rate is considered.
(7) The preservative technology enhances the shelf life of perishable items.

## 4 Analysis

**Case 1: Inventory Depletes in the Growth Phase** ($0 \leq T_{L_1} \leq T_1$)
The Holding cost is given by

$$H_1 = c_1[\int_0^{T_{L_1}} \{(\mathrm{T}_{\mathrm{L}_1} - 0)\mathrm{f}_1(\mathrm{P, t}) + \theta t\}\,\mathrm{dt}]$$

The revenue earned

$$R_1 = \int_0^{T_{L_1}} Pf_1(P, t)dt + \lambda P I_{b_1}$$

Here $I_{b_1}$ is the amount of backlogging given by

$$I_{\mathbf{b}_1} = \int_{T_{L_1}}^{T_1} f_1(\mathrm{P, t}) * \gamma(\mathrm{T}_1 - \mathrm{t})\mathrm{dt} + \int_{T_1}^{T_2} f_2(\mathrm{P, t}) * \gamma(\mathrm{T}_2 - \mathrm{t})\mathrm{dt}$$

The initial inventory level is

$$I_1 = \int_0^{T_{L_1}} \{f_1(p,t) + \theta t\} \, dt$$

The profit of the first stage for per unit of time

$$Net_1 = (R_1 - H_1 - c_0 - c(I_1 + I_{b_1})) \tag{1}$$

**Case 2: Inventory Depletes in a Constant Phase** $(T_1 \le T_{L_2} \le T_2)$
The cost of the holding is given by

$$H_2 = c_1[\int_0^{T_1} \{(\mathrm{T_1} - 0)\mathrm{f_1(P, t)} + \theta t\} \, \mathrm{dt} + \int_{T_1}^{T_{L_2}} \{(\mathrm{T_{L2}} - 0)\mathrm{f_2(P, t)} + \theta t\}\mathrm{dt}]$$

$R_2$ is the earned revenue

$$R_2 = \int_0^{T_1} \mathrm{Pf_1(P, t)dt} + \int_{T_1}^{T_{L_2}} \mathrm{Pf_2(P, t)dt} + \lambda \mathrm{PI_{b_2}}$$

Here backlogged amount is

$$I_{b_2} = \int_{T_{L_2}}^{T_2} f_2(\mathrm{P, t}) * \gamma(\mathrm{T_2 - t})\mathrm{dt}$$

In this way, the initial level of inventory is

$$I_2 = \int_0^{T_1} \{f_1(p,t) + \theta t\}dt + \int_{T_1}^{T_{L_2}} \{f_2(p,t) + \theta t\}dt$$

For the second phase, profit per unit of time is

$$Net_2 = (R_2 - H_2 - c_0 - c(I_2 + I_{b_2}))/T_2 \tag{2}$$

**Theorem 1:** (a) $Net_1$ is the concave function $T_{L_1}$ iff.

$$T_1 ghP^j[c(1 - T_{L_1})^h + 1] + bcT_{L_1}[T_1 - T_{L_1}(T_1 + bcT_{L_1} - 1)] + bPT_{L_1}[T_1\{T_{L_1} - 1\}$$
$$+ \lambda T_{L_1}\{1 - T_{L_1}\}] + cT_{L_1}(P - c) + a(1 - T_{L_1}) + c_1 T_{L_1}(T_1 - c_1 T_{L_1})$$
$$+ T_1(c - P) + PT_{L_1}\lambda \le 0 \tag{3}$$

(b) $Net_2$ is the conditionally concave function in $T_{L_2}$ iff.

$$P^{-j}(T_2ghP^j(1-T_{L_2})^h(c+c_1(-T_1+T_{L_2}))+a\mathrm{T}_2[c-P+T_{L_2}(P-c+c_1(1-T_{L_2})]$$
$$+T_1T_2[ac_1T_{L_2}+bc(1-T_{L_2})+bT(T_{L_2}-1)] \leq 2\mathrm{T}_1^2T_2c_1(P+c_1T_{L_2})-2T_{L_2}(c-P\lambda) \quad (4)$$

**Proof:** (a) we partially differentiate $Net_1$ from expression (1) with respect to $T_{L_1}$, we get

$$\frac{\partial Net_1}{\partial T_{L_1}} = \frac{1}{T_1T_2(-1+T_{L_1})}P^{-j}(T_1(cgP^j(1-T_{L_1})^h + c_1ghP^j(1-T_{L_1})^h - bc(-1+T_{L_1})^hT_{L_1}$$
$$+bP(-1+T_{L_1})^hT_{L_1} - bc_1(-1+T_{L_1})^hT_{L_1}^2) + b(-1+T_{L_1})^hT_{L_1}^2(c-P\lambda)$$
$$-a(-1+T_{L_1})(-cT_{L_1}+T_1(c-P+c_1T_{L_1})+PT_{L_1}\lambda)) \quad (5)$$

**(b)** then partial differentiation of profit function $\mathrm{Net}_2$ from the stated above expression (2) with respect to $T_{L_2}$, then we get

$$\frac{\partial Net_2}{\partial T_{L_2}} = \frac{1}{\mathrm{T}_2^2(-1+T_{L_2})}P^{-j}(T_2ghP^j(1-T_{L_2})^h(c+c_1(-T_1+T_{L_2}))$$
$$+a(-1+T_{L_2})(T_2(-c+T_1c_1+P-c_1T_{L_2})+T_{L_2}(c-P\lambda))$$
$$+bT_1(-1+T_{L_2})(\mathrm{T}_2(-c+T_1c_1+P-c_1T_{L_2})+T_{L_2}(c-P\lambda))) \quad (6)$$

The profit function is concave because the above condition gets satisfied w.r.t $T_{L_2}$.Therefore, we have to check either this profit function is concave in terms of P.

$$\frac{\partial Net_1}{\partial p} = 0 \quad (7)$$

The concavity can't be check manually of profit function regarding P; then we use Graphical methods as shown in Fig. 1.

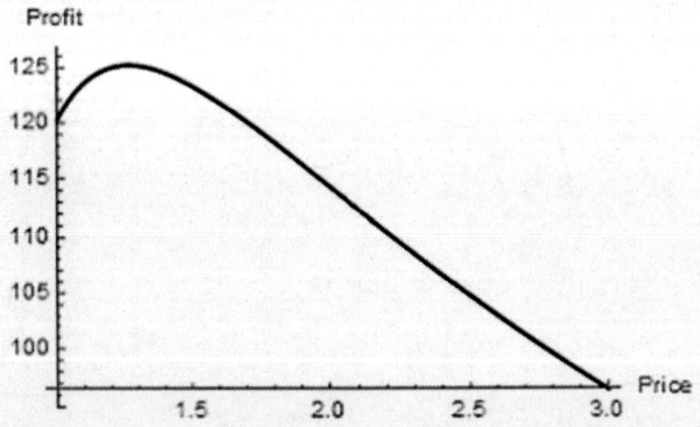


**Fig. 1** Shows concavity of profit function based on P

# 5 Solution Procedure

For F = 1.

Step 1: Solve $\frac{\partial Net_1}{\partial T_{L_1}} = 0$ from expression (5) and $\frac{\partial Net_1}{\partial P} = 0$ from expression (7) to get the value of $T^*_{L_1}$ and $P^*$.þ
Step 2. $Check(0 < T_{L_1} < T_1)$&P* > Pr *ice* floor. The initial test parameter is applied if a satisfactory result is found in this stage.
Step 3. For this set $(T^*_{L_1}, P^*)$, receive the value of $Net_1$ from expression (1).
Step 4. We may repeat the steps from 1 to 3 for F = 2. From $Net_1$ & $Net_2$, select the maximum one.

## 5.1 Numerical Example

The Numerical Examples results may find when $g = 0.2, h = 2, t = 2, T_1 = 50, T_2 = 75, a = 20, b = 7.5, j = 1.7, c_0 = 100, \lambda = .99, c_1 = .001, c = 0.5$ (Tables 1, 2 and Figs. 2, 3).

**Table 1** Result of phase I $(0 < T_{L_1} < T_1)$

| $T^*_{L_1}$ | $P^*$ | $Net^*_1$ | $I^*_1$ | $I^*_{b_1}$ |
|---|---|---|---|---|
| 49.9784 | 1.26388 | 125.174 | 12014.7 | 5532.3 |

**Table 2** Result of phase II, $(T_1 < T_{L_2} < T_2)$

| $T^*_{L_2}$ | $P^*$ | $Net^*_2$ | $I^*_2$ | $I^*_{b_2}$ |
|---|---|---|---|---|
| 75 | 1.27451 | **140.677** | 12312.3 | 5923.13 |

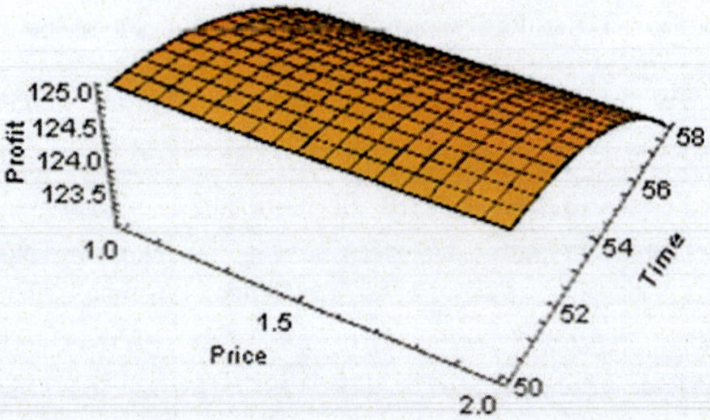


**Fig. 2** Showing joint concavity of $Net_1$ based on $T_{L_1}$ and P when preservation allowed

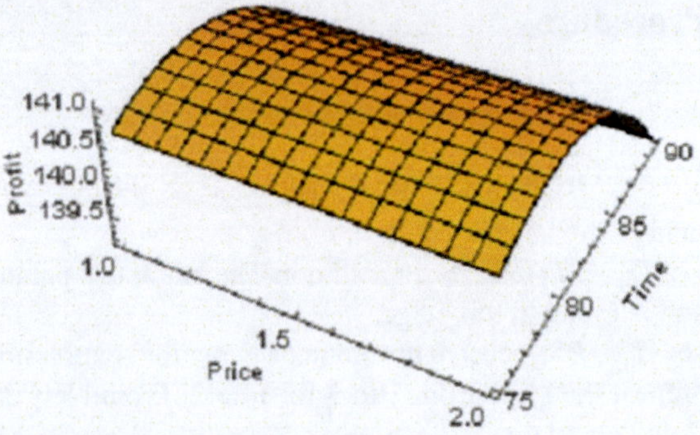


**Fig. 3** Showing joint concavity of $Net_2$ based on $T_{L_2}$ and P. when preservation allowed

Results: When preservative technology is applied, the profit reaches the first stage $Net_1 = 112.728$ then increases into the second stage $Net_2 = 131.586$. We restrict the value of $T_{L_2} = 75$ then we find maximum Profit **131.586**, Maximum inventory = 13,213.2; Maximum backlogged = 7276.65 and price = 1.17812 into the second stage.

# 6 Sensitivity Analysis

Sensitivity analysis is applied on phase II, where we received maximum profit $Net_2 = 140.677$ to find the most sensitive variable among the parameters (Table 3).

**Results:** P and T are decision variables. a, b, and j are demand parameters. We increase their values by 25% at each stage and reduce vice versa. We found maximum change into decision variable P when we reduced value as –75% we receive 4671.53% change in variable P, which is the most sensitive variable.

We are finding percentage loss so which value occurs in minus showing profit percentage. The maximum profit percentage found –590.85% when we increase the value of 'b' by 75%.

**Table 3** Results of sensitivity analysis as loss percentage

| | −100% | −75% | −50% | −25% | 0% | 25% | 50% | 75% | 100% |
|---|---|---|---|---|---|---|---|---|---|
| a | 7.45 | 5.59 | 3.73 | 1.86 | 0 | −1.86 | −3.73 | −5.59 | −7.45 |
| b | −49.08 | −42.34 | −133.76 | −225.18 | 0 | −408.02 | −499.44 | **−590.85** | −682.27 |
| j | −452.73 | −414.9 | −379.66 | −346.97 | 0 | Not Define | Not Define | Not Define | Not Define |
| T | 3.75* | 130.3 | −263.13 | −290.25 | 0 | −307.48 | −263.14 | −183.78 | −69.63 |
| P | Not valid | 4671.53 | 2437.59 | 1697.7 | 0 | 1087.9 | 929.562 | 814.01 | 725.62 |

# 7 Conclusion

In this study, profit may increase in the first phase and become constant in the second phase as ramp type demand. We received $Net_1 = 125.174$ then maximum profit $Net_2 = 140.677$ when preservative technology is applied. The retailer may invest in preservative technology to enhance the profit till Phase II.

# References


1. Das SC, Zidan AM, Manna AK, Shaikh AA, Bhunia AK (2020) An application of preservation technology in inventory control system with price dependent demand and partial backlogging. Alexandria Eng J 59(3):1359–1369
2. Deng PS (2005) Improved inventory models with ramp type demand and Weibull deterioration. Int J Inf Manag Sci 16(4):79
3. Deng PS, Lin RHJ, Chu P (2007) A note on the inventory models for deteriorating items with ramp type demand rate. Eur J Oper Res 178(1):112–120
4. Erlenkotter D (1990) Ford Whitman Harris and the economic order quantity model. Oper Res 38(6):937–946
5. Harris FW (1913) How much stock to keep on hand. Factory Mag Manag 10:240–241
6. Hsu PH, Wee HM, Teng HM (2010) Preservation technology investment for deteriorating inventory. Int J Prod Econ 124(2):388–394
7. Jain S, Kumar M, Advani P (2009) An optimal replenishment policy for deteriorating items with ramp type demand under permissible delay in payments. Pak J Stat Oper Res 5(2):107–114
8. Kaushik J, Sharma A (2019) Procurement and pricing decision for trapezoidal demand rate and time-dependent deterioration. Int J Innov Technol Explor Eng 8(12)
9. Kaushik J, Sharma A (2020) Inventory model for the deteriorating items with price and time-dependent trapezoidal type demand rate. Int J Adv Sci Technol 29(1):1617–1629
10. Mandal B, Pal AK (1998) Order level inventory system with ramp type demand rate for deteriorating items. J Interdiscip Math 1(1):49–66
11. Mishra U (2015) An inventory model for deteriorating items under trapezoidal type demand and controllable deterioration rate. Prod Eng Res Devel 9(3):351–365
12. Sethi V, Sethi S (2006) Processing of fruits and vegetables for value addition. Indus Publishing
13. Sharma A, Kaushik J (2020) Inventory model for deteriorating items with ramp type demand under permissible delay in payment. Int J Procurement Manag. https://doi.org/10.1504/IJPM.2020.1003328
14. Singh T, Mishra PJ, Pattanayak H (2018) An EOQ inventory model for deteriorating items with time-dependent deterioration rate, ramp-type demand rate, and shortages. Int J Math Oper Res 12(4):423–437
15. Singh T, Muduly MM, Asmita N, Mallick C, Pattanayak H (2020) A note on an economic order quantity model with time-dependent demand, three-parameter Weibull distribution deterioration, and permissible delay in payment. J Stat Manag Syst 23(3):643–662
16. Wu JW, Lin C, Tan B, Lee WC (1999) An EOQ inventory model with ramp type demand rate for items with Weibull deterioration. Int J Inf Manag Sci 10(3):41–51

# Lifestyle Diseases Prevalent in Urban Slums of South India

Abhay Augustine Joseph(✉), Hemlata Joshi, Matthew V. Vanlalchunga, and Sohan Ray

Department of Statistics, CHRIST (Deemed To Be University), Bangalore, India
abhay.joseph@science.christuniversity.in

**Abstract.** Lifestyle diseases have always been considered to be a malady of the middle and upper classes of society. Recent findings indicate that these chronic non-communicable diseases are common among the lower socioeconomic classes as well. The objective of this study was to assess the prevalence of lifestyle diseases in three cohorts of urban slums, namely, waste pickers living in non-notified slums, communities living in notified slums, and BBMP Pourakarmikas, and to identify the risk factors among the three cohorts contributing to the common lifestyle diseases including hypertension, diabetes, and cardiovascular diseases. In this study, the data was collected by conducting health camps, followed by analysis of the data using logistic regression, Hosmer–Lemeshow test and ROC Curve Analysis. The prevalence of hypertension was found 13.35%, diabetes-8.53% and cardiovascular disease-3.59%. These were significantly associated with substance abuse, high BMI, and age.

**Keywords:** Diabetes · Health · Heart disease · Hypertension · Slums · Urban health · Logistic regression

## 1 Introduction

Lifestyle diseases, referred to as chronic non-communicable diseases (CNCDs), have long been thought of as 'diseases of affluence'. However, recent studies have shown that these diseases are prevalent in urban slums as well. Lower socioeconomic countries have provided the best models in studying the effect of chronic non-communicable diseases [1]. CNCDs are growing in numbers, including among the urban slums. These diseases lead to an economic burden on the families and the society, as they reduce the productivity and add on to the expenses.

CNCDs, such as diabetes, obesity, coronary disorders, asthma, mental illness and hypertension, have been proven to be the foremost cause of death the world over [1]. A study by the World Health Organization (WHO), in 2001 showed that chronic diseases were the cause for about 60% of global deaths and around 46% of the global burden of disease [2]. In India, these diseases account for 53% of all deaths and about 44% of all disability-adjusted life years (DALYs) [3].

Hypertension continues to be one of the world's biggest public health challenges. High blood pressure or Hypertension accounts for about 7.5 million deaths worldwide


S. Shukla et al. (eds.), *Data Science and Security*, Lecture Notes in Networks and Systems 290,
https://doi.org/10.1007/978-981-16-4486-3_40

[4]. Diabetes has been another concerning problem in the country. India produces a sixth of the total number of diabetic cases worldwide. It is distributed to 9% of the total population in urban India and 3% in rural areas [5]. It has been observed that the urban population are more vulnerable to chronic non-communicable diseases as opposed to the rural population. There is very little information about the spectrum and burden of CNCDs in urban slums [6]. Studies done in Indian slum population have shown significant risk factors of CNCDs [7, 8].

The purpose of the present study is to assess the prevalence of CNCDs in the urban slums of South India, and determine the variables that are significantly associated with the lifestyle diseases under study, namely, diabetes mellitus, hypertension and cardiovascular diseases. For this, logistic regression models have been constructed, using the primary data collected. Hosmer Lemeshow test and ROC Curve analysis have been used to determine the goodness of fit and the predictive power of the models, and the results have been interpreted.

## 2 Methodology

A cross-sectional study was conducted in 44 camps, organised by Anahat Foundation, from August 2019 to September 2020. The study sample included 4539 people of all age groups and genders, residing in the urban slum areas of Karnataka. The individuals in the study were classified into three cohorts based on the types of communities- waste pickers living in non-notified slums, communities living in notified slums, and BBMP Pourakarmikas.

Community-based cross-sectional survey was done to collate each person's data, based on their lifestyle habits and medical reports. The surveys were conducted at health camps established at the various slum communities across the state, and the demographic parameters and medical conditions were studied at the individual level. The patient data was anonymized to ensure their personal information is not shared.

### 2.1 Variables Used

This study was done to find the prevalence of the lifestyle diseases, namely Diabetes, Hypertension and Cardiovascular Diseases and to study the correlation between these diseases and health factors. The dependent variables, with respect to each lifestyle disease are categorical and tests tells us whether the disease is found to be present in an individual or not. This implies the dependent variables are dichotomous variables. Hence, the logistic regression is used as the primary method to determine factors affecting lifestyle diseases.

The independent variables considered in this study include—

**Age.** This tells us the age of the individual under study and has been categorised into four categories namely—$< 18$, 18–29, 29–39 and $>39$.

**BMI.** The Body Mass Index measures weight with respect to height of an individual. In this study, the BMI has been categorised into four categories—Underweight, Normal, Overweight and Obese.

**Gender.** This variable refers to the gender of the individual and is categorised as either Male or Female.

**Oral Health.** This is a binomial variable stating the presence or absence of Poor Oral Hygiene (POH) or Oral Lesions in individuals.

**Substance Abuse.** This binomial refers to self-reported results of presence or absence of alcoholism or usage of tobacco among individuals.

## 2.2 Methods

**Logistic Regression.** Logistic regression is used to estimate a good explanatory model in predicting values of a response variable, when we have only binary outcomes or two values, such as the presence or absence of a disease. In logistic regression we are generally most interested in modelling the proportion of individuals with the outcome of interest using a link function which describes the relationship that links the response variable to the variable(s) that we use in the regression model.

Mathematically, a logistic regression model estimates a multiple regression model defined as:

$$\log(\text{odds}) = \text{logit}(P) = \ln\left(\frac{P}{1-P}\right) = \beta_0 + \beta_1 X_1 + \ldots + \beta_P X_P \quad (1)$$

where P is the probability of the outcome of interest. Here, it is the probability of presence of a disease. The odds of having a diseases is given by:

$$\text{odd} = \frac{P}{1-P} \quad (2)$$

The intercept term, on exponentiation, tells us the log odds of a person having a disease in the absence of any other factor affecting it. The coefficients of each predictor variable tells us by how much a change in the predictor variable affects the log odds of a person having the disease.

The output variable should be dichotomous in nature and ideally there should be no outliers in the data. Larger number of variables will yield better and accurate results, but it must be taken care there is no multicollinearity between the predictor variables.

**Hosmer–Lemeshow Test.** The Hosmer–Lemeshow test is statistical method used to test the goodness of fit of the estimated regression model. It is essentially used to test how actual and predicted event rates in a model match. This test works by dividing the sample based on their predicted probabilities. In this study, the logistic regression model estimates the probability of a person possessing a disease as a success. The Hosmer–Lemeshow test groups the observations, ordering them from lowest to highest, splitting them into deciles based on increasing order of their estimated risks. The following formula is used to compute the Hosmer–Lemeshow test statistic:

$$G^2_{HL} = \sum_{j=1}^{10} \frac{(O_j - E_j)^2}{E_j\left(1 - \frac{E_j}{n_j}\right)} \sim \chi^2 \quad (3)$$

where, $\chi^2$ is the chi-squared value, and $n_j$, $O_j$ and $E_j$ are the total number of observations, observed cases and expected cases in the jth group respectively. This statistic follows Chi-squared distribution with $j-2$ degree of freedom. Pearson's chi-square test is applied to compare observed counts with expected counts. A significant value of the statistic indicates that the model is not good fitted to the data and a non-significant test (large p-value) indicates a good fit. Here, the graph of observed presence of the disease is plotted against the expected presence of the disease.

**ROC Curve Analysis.** The receiver operating characteristic (ROC) curve is one of the measures used to tell us the "predictive power" or "explanatory power" of the constructed regression model. This indicates how well the model can predict the dependent variable based on the independent variables. The ROC curve is a plot of sensitivity of the model, which is probability of a positive result among the cases, or in this study the probability of the presence of a disease among those who were actually diagnosed positive, against 1-specificity of the model, which is the probability of a negative result among the non-cases, or probability of no disease found among those who did not test positive for any of them. Thus, it plots the "true positive rate" against the "false positive rate".

The area under the curve called c-statistic, which is calculated by integration, indicates the magnitude of discrimination in the model. Here, discrimination refers to how well a model can separate those who do and do not have the outcome of interest i.e. distinguishing between people with or without a disease. This can be used to see whether the model can predict a higher risk score for those who have been diagnosed to have a diseases against those who haven't been diagnosed.

## 3 Empirical Analysis

The data collected comprised of individuals of all age groups, genders, occupations and educational backgrounds. Out of these, 39.57% of them were male subjects and 60.43% were female. When considering the age distribution, 35% of the subjects were children (<18), 19% were 18 to 29 years old, 19% were 30 to 39 years old and 27% were above 40.

The observations in the study were taken from lower socioeconomic areas and consisted largely of uneducated (31%) or less educated individuals. Figure 1a displays the frequencies of the level of education of the population. The majority of the population work as BBMP Pourakarmikas (13%) and waste pickers (18%) indicating a low socioeconomic status and poverty stricken population. Figure 1b highlights the occupation distribution of the population:

### 3.1 Results

The prevalence of hypertension in the whole study population was 13.35%, diabetes 8.53% and cardiovascular disease 3.59%.

**Cohort-Wise Comparison.** Lifestyle diseases were highly prevalent in the urban slum population under study. From the cohort-wise comparison in Table 1, it is seen that

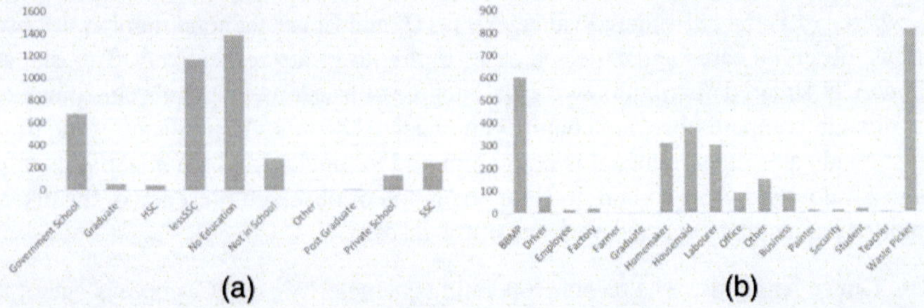


**Fig. 1** **(a), (b)** Bar diagrams showing the distribution of education levels and occupations among the people respectively

**Table 1** Prevalence of non-communicable diseases in urban slum communities

| Table 1 | Cohort-wise comparison | | |
|---|---|---|---|
| Cohort | Diabetes mellitus | Hypertension | Cardiovascular diseases |
| BBMP | 13.48 | 40.07 | 10.64 |
| Migrants | 6.09 | 7.51 | 3.38 |
| Non-migrants | 8.85 | 10.78 | 2.12 |

extremely high percentage of hypertension and moderately high percentage of diabetes and cardiovascular diseases were observed among the BBMP communities. This can be attributed to their sedentary lifestyle, lack of physical activity and dietary habits. Most of them are involved in clearing waste using vehicles. Lowest prevalence was seen among the migrant workers.

**Diabetes Mellitus.** The prediction model in the case of diabetes mellitus included the health factors—age, BMI, gender, cardiovascular diseases, hypertension and substance abuse. Table 2 shows the association of these factors with diabetes based on cross-tabulation and a logistic regression model. Results indicate that the probability of having diabetes significantly increases with increasing age, with most cases above 40 years of age. It was also found that higher the Body Mass Index (BMI), higher was the probability of having diabetes, the risk being significantly higher among obese individuals (BMI > 30). Hypertension and substance abuse were also found to be significantly associated with diabetes. However, contrary to past studies, gender did not show significant correlation with diabetes (p-value = 0.746). Similarly, there was no significant association between diabetes and cardiovascular diseases in this population.

Figure 2a gives the plot for the Hosmer–Lemeshow test of observed rate of diabetes versus expected rate of diabetes. The p-value of the Hosmer–Lemeshow test was found to be 0.99. This tells us that for the logistic regression model to predict presence of diabetes, the model's predicted values were a good match for the observed values, indicating a good fit. Figure 2b shows the ROC Curve for the prediction model of diabetes mellitus. The value of the c-statistic is 0.877. This implies that the model had a high predictive power, deeming highly accurate results.

**Table 2** Factors affecting the probability of having diabetes mellitus

| Table 2 | Diabetes mellitus | | | |
|---|---|---|---|---|
| | Percentage | Odds ratio (95% CI) | p-value | Std. error |
| **Gender** | | | | |
| Male | 7.91 | 1.05 | 0.226345 | 0.1289 |
| Female | 8.93 | 1.0 (reference) | | |
| **Age groups** | | | | |
| <18 | 0.06 | 0.06 | 0.006195* | 1.0431 |
| 18–29 | 1.61 | 1.0 (reference) | | |
| 30–39 | 6.79 | 3.72 | 1.68e−05* | 0.3054 |
| 40+ | 25.55 | 16.32 | < 2e−16* | 0.2822 |
| **BMI Categories** | | | | |
| Underweight | 2.23 | 0.66 | 0.0552 | 0.2143 |
| Normal | 10.05 | 1.0 (reference) | | |
| Overweight | 15.88 | 1.23 | 0.1587 | 0.1469 |
| Obese | 23.18 | 1.77 | 0.000538* | 0.1652 |
| **Cardiovascular Diseases** | 28.22 | 1.36 | 0.127936 | 0.2033 |
| **Hypertension** | 28.88 | 2.02 | 3.05e−08* | 0.1268 |
| **Substance Abuse** | 12.79 | 0.67 | 0.018* | 0.1592 |

*p-value significant (<0.05)

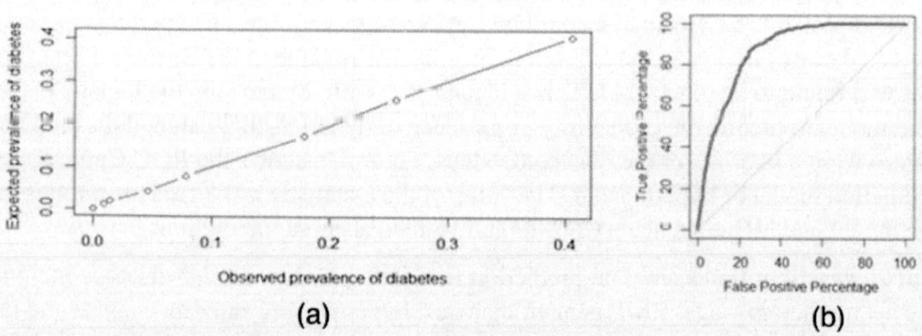


**Fig. 2** **(a), (b)** Hosmer–Lemeshow plot and ROC curve of predictive model for diabetes mellitus

**Hypertension.** The prediction model used to determine presence of Hypertension included the health factors—age, BMI, gender and substance abuse. Table 3 shows the association of these factors with hypertension based on cross-tabulation and a logistic regression model. In correlation with the previous studies, increasing age, higher BMI, and substance abuse were all significant in determining the probability of having hypertension. Males were less prone to hypertension as compared to females, though this was not statistically significant.

**Table 3** Factors affecting the probability of having hypertension

| Table 3 | Hypertension | | | |
|---|---|---|---|---|
| | Percentage | Odds ratio (95% CI) | p-value | Std. error |
| **Gender** | | | | |
| Male | 11.36 | 0.90 | 0.3358 | 0.1071 |
| Female | 14.66 | 1.0 (reference) | | |
| **Age groups** | | | | |
| <18 | 0.31 | 0.08 | 8.42e−08* | 0.4812 |
| 18–29 | 5.88 | 1.0 (reference) | | |
| 30–39 | 16.16 | 2.71 | 1.23e−08* | 0.1749 |
| 40+ | 33.52 | 7.62 | <2e−16* | 0.1590 |
| **BMI Categories** | | | | |
| Underweight | 4.25 | 0.78 | 0.1307 | 0.1648 |
| Normal | 15.18 | 1.0 (reference) | | |
| Overweight | 25.04 | 1.46 | 0.0017* | 0.1207 |
| Obese | 34.90 | 2.20 | 1.44e−08* | 0.1393 |
| **Substance Abuse** | 26.34 | 1.29 | 0.0382* | 0.1235 |

*p-value significant (<0.05)

The plot for the Hosmer–Lemeshow test of observed rate of hypertension versus expected rate of hypertension is given in Fig. 3a. The p-value of the Hosmer–Lemeshow test was found to be 0.0001047. This indicates that with reference to the logistic regression model to predict the probability of presence of hypertension, the model's predicted values were a poor fit for the observed values. Figure 3b shows the ROC Curve for the prediction model of hypertension. The value of the c-statistic is 0.842. This implies that the model had high explanatory power and yielded accurate predictions.

**Cardiovascular Diseases.** The prediction model for cardiovascular diseases included the health factors—age, BMI, gender, diabetes, hypertension, substance abuse and oral health. Table 4 shows the association of these factors with cardiovascular diseases based on cross-tabulation and a logistic regression model.

Cardiovascular diseases have been known to be associated with poor oral hygiene (POH) and oral health. However, in this study, it was observed that POH and oral lesions are not significant in determining the presence or absence of cardiovascular diseases. In correspondence with the previous results, it was seen that cardiovascular diseases significantly correlated with the factors age, gender and substance abuse. It was seen that males were more susceptible to cardiovascular diseases. This can be correlated to their substance abuse, which was another variable that significantly correlated with

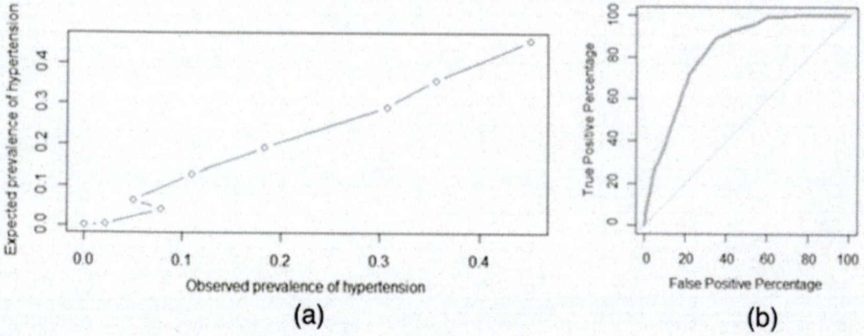


**Fig. 3** **(a), (b)** Hosmer–Lemeshow plot and ROC curve of predictive model for hypertension

**Table 4** Significant factors affecting probability of having cardiovascular diseases

| Table 4 | Cardiovascular diseases | | | |
|---|---|---|---|---|
| | Percentage | Odds ratio (95%CI) | p-value | Std. error |
| **Gender** | | | | |
| Male | 4.40 | 1.80 | 0.001* | 0.1785 |
| Female | 3.06 | 1.0 (reference) | | |
| **Age groups** | | | | |
| <18 | 0.57 | 0.51 | 0.16147 | 0.4751 |
| 18–29 | 1.73 | 1.0 (reference) | | |
| 30–39 | 2.46 | 1.07 | 0.843732 | 0.3485 |
| 40+ | 9.60 | 3.61 | 1.52e-05* | 0.2965 |
| **BMI Categories** | | | | |
| Underweight | 1.21 | 0.57 | 0.061743 | 0.3005 |
| Normal | 4.32 | 1.0 (reference) | | |
| Overweight | 6.58 | 1.24 | 0.291340 | 0.2075 |
| Obese | 8.33 | 1.50 | 0.090522 | 0.2396 |
| **Hypertension** | 12.71 | 2.28 | 6.91e-06* | 0.1835 |
| **Diabetes** | 11.89 | 1.39 | 0.100575 | 0.2019 |
| **Substance Abuse** | 9.54 | 1.68 | 0.011832* | 0.2064 |
| **Oral Health** | 6.68 | 1.23 | 0.383081 | 0.2391 |

*p-value significant (<0.05)

the presence of cardiovascular diseases, as substance abuse is more common in men (16.54%) as compared to women (8.28%). Diabetes and BMI were not found to be significant in determining the presence of cardiovascular diseases in this population, but hypertension was highly significant.

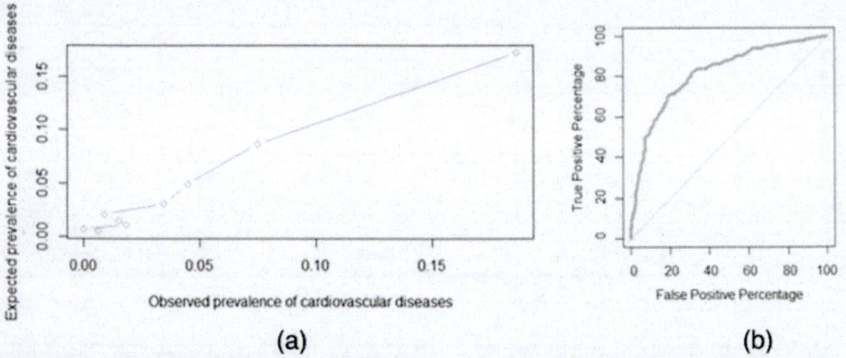


**Fig. 4** **(a), (b)** Hosmer–Lemeshow plot and ROC curve of predictive model for cardiovascular diseases

Figure 4a gives the plot for the Hosmer–Lemeshow test of observed rate of cardiovascular diseases versus expected rate of cardiovascular diseases.

The p-value of the Hosmer–Lemeshow test was found to be 0.5275. This indicates that the model has a good fit, as the predicted values are a good match for the observed values. The ROC Curve of the prediction model for cardiovascular diseases is given in Fig. 4b. The value of the c-statistic is 0.817. This implies that the model had a high predictive power.

## 4 Conclusion

In conclusion, lifestyle diseases were common among the urban slum population of Karnataka. The factors found to be significantly associated with diabetes, hypertension and cardiovascular diseases were age, gender, BMI and substance abuse. Lifestyle diseases pose a big threat to productivity and add to the economic burden of families and the society. Increasing awareness about the preventable risk factors will go a long way in reducing the incidence of these diseases. Provision of adequate healthcare facilities in these communities will ensure early detection and timely treatment of such diseases, minimising the health and financial consequences.

## References

1. Chadee D, Seemungal T, Pereira LM, Chadee M, Maharaj R, Teelucksingh S (2013) Prevalence of self-reported diabetes, hypertension and heart disease in individuals seeking State funding in Trinidad and Tobago, West Indies. Journal of Epidemiology and Global Health
2. World Health Organization (2007, June 22). Retrieved from https://www.who.int/nutrition/topics/2_background/en/
3. Deepadarshan H, Hiremath SD (2017) Lifestyle factors and lifestyle diseases among rural population of Bengaluru rural district. International Journal Of Community Medicine And Public Health

4. Singh S, Shankar R, Singh GP (2017) Prevalence and Associated Risk Factors of Hypertension: A Cross-Sectional Study in Urban Varanasi. Int J Hypertens
5. Kannan R (2019, November 15) India is home to 77 million diabetics, second highest in the world. Retrieved from https://www.thehindu.com/sci-tech/health/india-has-second-largest-number-of-people-with-diabetes/article29975027.ece?utm_source=admitad
6. Riley LW, Ko AI, Unger A, Reis MG (2007) Slum health: Diseases of neglected populations. BMC International Health and Human Rights
7. Anand K, Shah B, Yadav K, Singh R, Mathur P, Paul E, Kapoor SK (2007) Are the urban poor vulnerable to non-communicable diseases? a survey of risk factors for non-communicable diseases in urban slums of Faridabad. Natl Med J India 20(3):115–120
8. Shanmugam J, Gurupatham D, Arumugam A (2018) Prevalence of risk factors for non-communicable diseases in urban slum of Salem, Tamil Nadu. Int J Commun Med Public Health 5(5):1863–1868

# Eccentric Graph of Join of Graphs

M. Rohith Raja(✉), M. Tabitha Agnes, and Sudev Naduvath

Department of Mathematics, CHRIST (Deemed to be University), Bangalore 560029, Karnataka, India

rohith.m@res.christuniversity.in

**Abstract.** The eccentric graph $G_e$ corresponding to a graph $G$ is a derived graph with the same vertex set of $G$ and two vertices in $G_e$ are neighbours if one of them is the eccentric vertex of the other. Motivated by the studies on derived graphs and graph operations, in this article, the eccentric graph of the join of two graphs is analysed based on the variations in the radius. The notion of eccentric join of two graphs with at least one of them having radius 1, is introduced. The eccentric graph of eccentric join of graphs is also examined. Finally, the concept of r-eccentric join of graphs is also introduced. This study is analytical in nature, which involves deductive and logical reasoning.

**Keywords:** Eccentric graphs · Eccentricity · Eccentric join

## 1 Introduction

Consider a graph $G$ with $V(G)$ and $E(G)$ as the vertex set and edge set of $G$ respectively. All graphs discussed in this article are simple undirected graphs. For terminology and the notations in graph theory, we refer to [2, 8]. The *eccentricity* of a vertex $v \in V(G)$, denoted by $e(v)$, is the length of the shortest path between $v$ and the vertex at its maximum distance. Let $u$ be the vertex at maximum distance of $v$. Then $u$ is called the *eccentric vertex* of $v$. The *radius* of a connected graph $G$, denoted by $rad(G)$ is the minimum of the eccentricities of all the vertices and the *diameter* of $G$, denoted by $diam(G)$, is the maximum of the eccentricities of all the vertices in $G$. The eccentricities of all the vertices of a disconnected graph are taken to be $+\infty$ which will also be the radius as well as the diameter of the graph. The *join* of two graphs $G$ and $H$, denoted by $G + H$, is obtained by considering disjoint copies of $G$ and $H$ and making each vertex of $G$ adjacent to each vertex of $H$.

The notion of eccentric graph of a graph was introduced in [1] as follows: An *eccentric graph* of a graph $G$, denoted by $G_e$, is a derived graph with the vertex set same as that of the graph $G$ and two vertices in $G_e$ are adjacent if one of them is the eccentric vertex of the other in $G$. In [3], the eccentric graphs are defined differently. The eccentricity $e(G)$ of a connected graph $G$ is the smallest integer $k$ satisfying the condition that eac vertex of $G$ having eccentricity $k$ or more is an eccentric vertex. In [3], any graph $G$ with every vertex being an eccentric vertex, or equivalently, the radius of $G$ equals $e(G)$, is called an eccentric graph. Further, in [5], $l$-eccentric graph is defined as a graph with


S. Shukla et al. (eds.), *Data Science and Security*, Lecture Notes in Networks and Systems 290,
https://doi.org/10.1007/978-981-16-4486-3_41

each of the vertices having $l$ eccentric vertices. The definition in [1], gives a derived graph class, whereas that in [3], gives a new graph class. In this paper, we follow the definition in [1]. In [1], the authors have characterized the graphs with $G_e = K_p$ as well as $G_e = pK2$. The equation $G_e = \bar{G}$ has also been solved in [1]. The following results are significant and relevant for the current study:

**Result 1.1.** [1] *If Gis a graph with radius greater than 2, then* $(G + K_n)_e = \bar{G} + K_n$, $n \geq 1$.

**Result 1.2.** [1] *Corresponding to any graph Gof order k, there exists an eccentric graph Hof order k + 2such that Gis an induced sub-graph of H.*

Further studies have been carried out in [4, 7] and [6] which deals with the eccentric graphs of particular classes of graphs. Motivated by the studies and results mentioned above, in this article, the eccentric graph of join of graphs is examined. The notion of eccentric join and $r$-eccentric join of graphs are also introduced.

## 2 Eccentric Graph of Join of Graphs

We classify connected graphs into two categories based on their radius classification: the graphs with radius 1 and those with radius greater than or equal to 2. Since the radius of the graph is the least value of the eccentricities of all its vertices, graphs having radius 1 will have at least one vertex having eccentricity 1 which will be adjacent to all the other vertices. It is evident that, for a graph $G$ with $rad(G) = 1$, the maximum eccentricity that can be attained, which is the diameter of the graph, will be utmost 2. If the diameter is 1, each vertex is adjacent to all the other vertices and such types of graphs are complete graphs. If the diameter is 2, those vertices not having eccentricity 1, will be having eccentricity 2. Star graphs and wheel graphs are examples of such graphs. The second category of connected graphs involves the graphs having radius at least 2. In these types of graphs, there will be no vertex neighbouring all the remaining vertices. Cycles $C_n, n \geq 4$, paths $P_n, n \geq 4$ etc. are examples of such types of graphs.

For disconnected graphs, each vertex will have eccentricity $+\infty$. Therefore, the radius and the diameter of a disconnected graph are $+\infty$. In this article, we initiate a study of the eccentric graph of the join of two graphs with reference to the above classifications.

**Theorem 2.1.** *Let Gbe a graph with* $rad(G) = 1$. *Then,* $(G + K_n)_e = G_e + K_n$.

**Proof.** Consider a graph $G$ with $rad(G) = 1$. Then, $G$ has at least one vertex $u$ having all the remaining vertices of $G$ as its neighbours. All such vertices will have the remaining vertices of $G + K_n$ and $(G + K_n)_e$ as their neighbours. Rest of the vertices of $G$ have eccentricity 2 in $G$ and $G + K_n$. Therefore, in $(G + K_n)_e$, those vertices will form adjacency with their eccentric vertices in $G$. Now, for every $u \in V(K_n)$, $u$ is adjacent to all the other vertices in $K_n$ and therefore in $G + K_n$ and $(G + K_n)_e$. Therefore, for $rad(G) = 1$, $(G + K_n)_e = G_e + K_n$. □

The next theorem is an extension of the Result 1.1 stated in [1].

**Theorem 2.2.** *For any graph G,* $(G + K_n)_e = \bar{G} + K_n$ *if and only if* $rad(G) \geq 2$.

**Proof.** The case when $rad(G) > 2$ has already been settled in [1]. Suppose that $rad(G) = 2$. Let $u \in V(K_n)$. Then all the vertices of $K_n$ are neighbours of $u$ and therefore $u$ is adjacent to all the vertices of $G + K_n$ and $(G + K_n)_e$. Consider $v \in V(G)$. Since $rad(G) = 2$, for all $v \in V(G)$, there exists at least one vertex $w \in V(G)$ such that $v$ is not adjacent to $w$ which implies all the vertices of $G + K_n$ have eccentricity 2. Therefore, in $(G + K_n)_e$, the vertices of $G$ will be adjacent only to those vertices to which they are not adjacent in $G$ and therefore, $(G + K_n)_e = \bar{G} + K_n$.

Consider the case $rad(G) < 2$. Then, either $rad(G) = 0$ or $rad(G) = 1$. When $rad(G) = 0$, the graph is trivial. Suppose $rad(G) = 1$. Then, $V(G)$ consists of at least one vertex $u$ as discussed while proving Theorem 2.1. But in $\bar{G}$, $u$ is adjacent to none of the vertices that are adjacent in $G$. Therefore, $(G + K_n)_e \neq \bar{G} + K_n$, completing the proof.

Now, we give an extension of the Result 1.2.

**Corollary 2.3.** *Let G be any graph of order k such that* $rad(G) \geq 2$. *Then, there exists an eccentric graph H of all higher orders,* $k + n$ *for any positive integer n, such that G is an induced sub-graph of H.*

**Proof.** Take $H = G + K_n$, where $n$ is any positive integer. Now, $G$ contains all edges present in $H$ excluding those adjacent with $K_n$. Therefore, $G$ is an induced sub-graph of $H$. $H$ will be the eccentric graph of $\bar{G} + K_n$ by Theorem 2.1. The order of $H$ is $k + n$ where $n$ is any positive integer. Hence, the proof is complete. □

Now, we extend the above results to the general graphs $G$ and $H$. First, we consider the case where both the graphs are of radius at least 2.

**Theorem 2.4.** *Let G and H be two graphs such that* $rad(G) \geq 2$ *and* $rad(H) \geq 2$. *Then,* $(G + H)_e = \bar{G} \cup \bar{H}$.

**Proof.** When the radius of a graph is at least 2, the graph can be either connected or disconnected. If the radius is finite, the graph is connected and if it is infinite, the graph is disconnected. For a disconnected graph, each vertex will have eccentricity $\infty$.

Here, the graphs $G$ and $H$ can either be connected or disconnected. Since $rad(G) \geq 2$ and $rad(H) \geq 2$, there exists no vertex of eccentricity 1 in both $G$ and $H$, which means no vertex is adjacent to all the other vertices. Therefore, in $G + H$, the eccentricity of all the vertices is 2. Now, in $(G + H)_e$, each vertex will be adjacent only to those vertices to which it is not adjacent in the original graphs, respectively. Therefore, in the eccentric graph, there is no adjacency between the vertices in $G$ and $H$. Hence, $(G + H)_e = \bar{G} \cup \bar{H}$. □

The eccentric graph of join of graphs where at least one of them is of radius one, can be discussed only after introducing a new operation on graphs namely, the eccentric join of graphs.

(a) *Graphs $G = K_{1,4}$ and $H = P_3$.* (b) *The eccentric join of G and H.*

**Fig. 1** The eccentric join of $G$ and $H$

**Definition 2.5.** *Let Gand Hbe two graphs such that at least one of them has vertices of eccentricity 1. The eccentric join of Gand H, denoted by $G(+)_eH$ is a graph with the vertex set consisting of all vertices of Gand Hand the edge set consisting of all edges in Gand Hand additional edges between vertices having eccentricity 1and all the remaining vertices in Hand Grespectively.*

From Definition 2.5, it can be observed that, eccentric join can be defined only when at least one of them has eccentricity 1. Let $S_G = \{u \in G : e(u) = 1\}$ and $S_H = \{v \in H : e(v) = 1\}$. Then the number of new edges in $G(+)_eH = |S_G||H| + |S_H||G| - |G||H|$. We now look at an example.

**Example 2.1.** Consider a star graph $G$ with 4 pendant vertices, denoted by $K_{1,4}$ and $H$ be a path on 3 vertices. In this case, both the graphs have radius 1. Therefore, we can find the eccentric join of $G$ and $H$. The graphs $K = S_{1,4}$ and $H = P_3$ are represented in Fig. 1 and their eccentric join $G(+)_eH$ is represented in Fig. 1.

Both the graphs are of radius 1 and have one vertex each of eccentricity 1. Therefore, in the eccentric join of $G$ and $H$, the neighbourhood of these two vertices will comprise all all the other vertices of $H$ and $G$ respectively, as shown in Fig. 1.

**Theorem 2.6.** *Let Gand Hbe two graphs such that $rad(G) = rad(H) = 1$. Then $(G + H)_e = G_e(+)_eH_e$where $(+)_e$stands for the eccentric join of graphs.*

**Proof.** Since $rad(G) = rad(H) = 1$, both $G$ and $H$ has at least one vertex adjacent to the remaining vertices of $G$ and $H$ respectively. Also, the remaining vertices of $G$ and $H$ will have eccentricity 2. In $G + H$, each vertex of $G$ will have the vertices of $H$ as its neighbours in addition to the existing neighbours in $G$. Therefore, in $(G + H)_e$, the vertices having eccentricity 1 in $G+H$ will have adjacency with all the other vertices and the remaining vertices having eccentricity 2 will be adjacent to their eccentric vertices in the original graphs. By the definition of eccentric join given above, the eccentric graph of the join of graphs, both having radius 1 will be the eccentric join of the eccentric graphs $G_e$ and $H_e$. Therefore, $(G + H)_e = G_e(+)_eH_e$. □

Now let us consider the case when only one of them is of radius 1. We assume that $rad(G) = 1$.

**Theorem 2.7.** *Let Gand Hbe two graphs such that $rad(G) = 1$and $rad(H) \geq 2$. Then $(G + H)_e = G_e(+)_e\bar{H}$.*

**Proof.** Since $rad(G) = 1$, there will be at least one vertex having adjacency with the remaining vertices of $G$. All such vertices of $G$ will be adjacent to all the vertices, both in $G + H$ as well as $(G + H)_e$. The remaining vertices of $G$ will have eccentricity 2 in $G$ as well as in $G + H$. Now, since $rad(H) \geq 2$, $H$ can either be a connected or a disconnected graph. In either case, each vertex of $H$ will have eccentricity 2 in $G + H$. Therefore, in $(G + H)_e$ the vertices of $H$ will be adjacent to those vertices to which they were not adjacent initially in $H$. Therefore, the eccentric graph of the join will be the eccentric join of the eccentric graph of G and the complement of $H$. Therefore, $(G + H)_e = G_e(+)_e\bar{H}$. □

Here, Theorem 2.2 is a special case of Theorem 2.6. In Theorem 2.1, one of the graphs is a complete graph which has radius 1 and the other is a graph having radius greater than or equal to 2.

With these above results, we have obtained the eccentric graph of the join of graphs for all possible cases. Now, we discuss the nature of eccentric graph of the eccentric join of graphs. The eccentric join of graphs are defined only when either of the graphs have radius 1. Thus, we have following two cases:

**Theorem 2.8.** *Consider two graphs Gand Hsuch that* $rad(G) = rad(H) = 1$. *Then,*

$$(G(+)_eH)_e = G_e + H_e.$$

**Proof.** Since both the graphs $G$ and $H$ have radius 1, the eccentric join of $G$ and $H$, namely $G(+)_eH$, will have all existing edges in both the graphs and additional edges between the vertices having radius 1 and all the other vertices. Vertices having eccentricity 1 in $G(+)_eH$ will have adjacency with all the vertices of $(G(+)_eH)_e$. All the other vertices of $G(+)_eH$ will have eccentricity 2. These vertices will have adjacency with all the remaining vertices of $G(+)_eH$ to which they are not adjacent, which yields the join of the corresponding eccentric graphs. Hence, the result. □

Now, we illustrate this in the next figure. Consider the graphs $G$ and $H$ as given in Fig. 1(a). Figure 1(b) shows the eccentric join of the graphs. Now, the eccentric graph of eccentric join of $G = K_{1,4}$ and $H = P_3$ is given in Fig. 2.

Next, we consider the case where one of the graphs has radius 1 and the other may be a connected or disconnected graph.

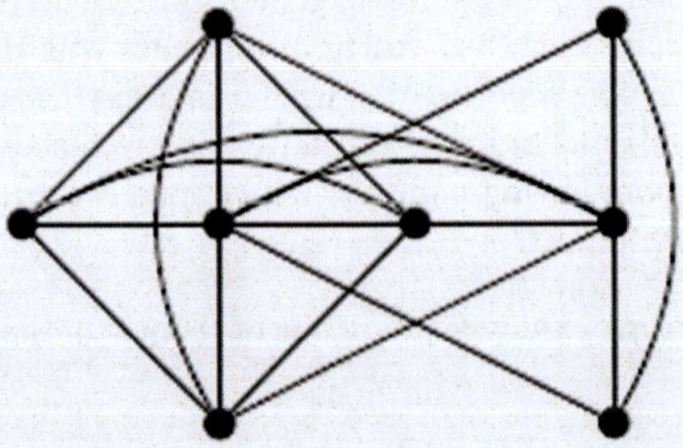

**Fig. 2** The eccentric graph of eccentric join of graphs $G = K_{1,4}$ and $H = P_3$, both having radius 1

**Theorem 2.9.** *Let G and H be two graphs such that rad(G) = 1 and rad(H) ≥ 2. Then,* $(G(+)_e H)_e = G_e + \bar{H}$.

**Proof.** Since $rad(G) = 1$, the vertices having eccentricity 1 in $G$ will have adjacency with all the vertices of $G(+)_e H$ and $(G(+)_e H)_e$. The remaining vertices of $G$ will have eccentricity 2 in $G(+)_e H$, and therefore, in $(G(+)_e H)_e$, they will be adjacent to the eccentric vertices of $G$ and all the vertices of $H$. Now, since $rad(H) \geq 2$, $H$ can be either connected or disconnected. In either case, the remaining vertices will have eccentricity 2 in $G(+)_e H$. In $(G(+)_e H)_e$, the vertices of $H$ will have adjacency with those vertices to which they are not initially adjacent. Thus, $(G(+)_e H)_e = G_e + \bar{H}$. □

Now, we generalize the definition of eccentric join given above. The above definition holds only for two graphs with at least one of them having radius 1. Here, we define an operation on two graphs, namely the radius eccentric join or the r-eccentric join of two graphs with any given radius. The graphs may be either connected or disconnected.

**Definition 2.10.** *Consider two graphs G and H with any radius. Let* $S = \{v : e(v) = min\{rad(G), rad(H)\}\}$. *The radius eccentric join or r-eccentric join of G and H, denoted by* $G(+)_e^r H$ *is a graph with the vertex set having all vertices in G and H and the edge set consisting of all edges in G and H and additional edges between vertices in S and all the other vertices in G and H.*

We now see an example.

**Example 2.2.** Consider two graphs $G$ and $H$ as given in Fig. 3 such that $rad(G) = 2$ and $rad(H) = 3$. $G$ has minimum radius and therefore the set $S$ described in Definition 2.10 consists of the single central vertex in $G$. The r-eccentric join of $G$ and $H$, given in Fig. 3 consists of all the edges in $G$ and $H$, and additional edges between the central vertex of $G$ and all the other vertices in $G$ and $H$. The obtained graph has radius 1.

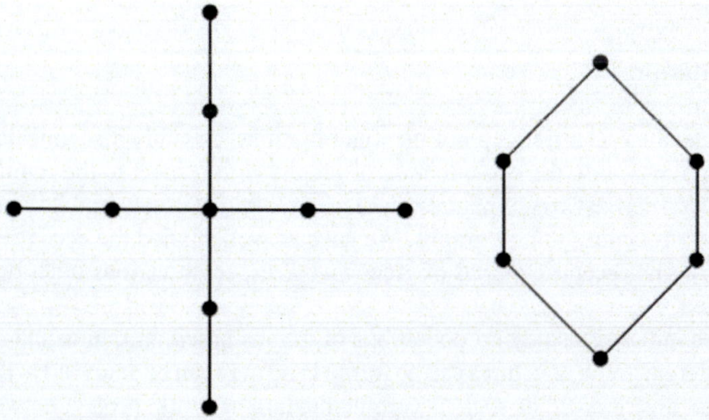

**Fig. 3** The graphs $G$ and $H$ with $rad(G) = 2$ and $rad(H) = 3$

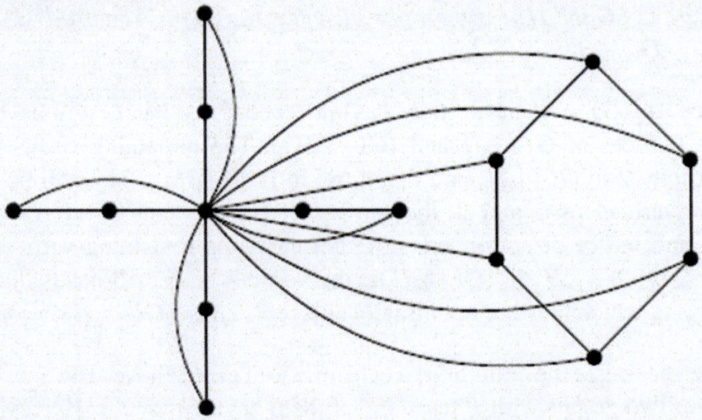

**Fig. 4** The 2-eccentric join of graphs $G$ and $H$ with $rad(G) = 2$ and $rad(H) = 3$

**Theorem 2.11.** *Let Gand Hbe two graphs with* $\min\{rad(G), rad(H)\} = r$. *Then the r-eccentric join of Gand H, denoted by* $G(+)_e^r H$, *is a complete graph if and only if both Gand Hare r self-centered.*

**Proof.** Suppose that $G$ and $H$ are $r$ self-centered graphs. Then, $rad(G) = rad(H) = r$ and each of the vertices has eccentricity $r$. By Definition 2.10, the set $S$ will contain all the vertices in both $G$ and $H$. Therefore, in $G(+)_e^r H$, each vertex will be adjacent to all other vertices thus giving a complete graph.

Conversely, suppose that $G$ and $H$ are two graphs with $\min\{rad(G), rad(H)\} = r$ with at least one of them not being self-centered of radius $r$. It follows that, there will be at least one pair of vertices in $G$ or $H$ or both having eccentricity greater than $r$. Thus, in $G(+)_e^r H$, there will be no adjacency between such pair of eccentric vertices. Therefore, the graph $G(+)_e^r H$ is not complete. □

## 3 Conclusion

In this article, the eccentric graph of the join of graphs, possibly disconnected also, has been explored based on the classification of graphs with respect to the radius. We have also introduced derived graphs namely, the eccentric join of graphs and its generalization which is the r-eccentric join of graphs. We have as well studied the eccentric graphs of these graphs. The eccentric graphs of graphs with respect to various other operators are being studied.

We have initiated a study on operations of graphs based on radius and diameter of graphs. Different graph structures can be formed and they can be analysed in detail to find applications in various fields such as communication network, project networks, social networks etc. The $r$-eccentric join of graphs are radius 1 graphs and they can be applied in cases where a single vertex contact is required to deal with two different graphs. Resultant graphs based on graph operations can be defined with the radius or diameter equal to the minimum or maximum of the two graphs. A study can also be initiated on minimising the number of additional edges such that the radius or the diameter of the resulting graph obeys a particular condition. Such graphs can have numerous applications and can be studied at detail.

**Acknowledgements.** The first author would like to thank Mr. Libin Chacko Samuel for his valuable suggestions which were instrumental in getting the paper into this format.

## References

1. Akiyama J, Ando K, Avis D (1985) Eccentric graphs. Discrete Math 1:1–6
2. Buckley F, Harary F (1990) Distance in graphs. Addison-Wesley- California
3. Chartrand G, Gu W, Schultz M, Winters SJ (1999) Eccentric Graphs. Networks 2:115–121
4. Kaspar S, Gayathri B, Kulandaivel MP, Shobhanadevi N (2018) Eccentric graphs of some particular classes of graphs. Int J Pure Appl Math 16:145–152
5. Narayankar KP, Lokesh SP (2018) On l-eccentric graphs. Energy 5:1–15
6. Sriram S, Ranganayakulu D, Sarmin NH, Venkat I, Subramaniam KG (2014) On eccentric graphs of unique eccentric point graphs and diameter maximal graphs. Appl Math Comput Intell 1:283–291
7. Sriram S, Ranganayakulu D, Sarmin NH, Venkat I, Subramaniam KG (2014) On eccentric graphs of broom graphs. Annal Pure Appl Math 2:146–152
8. West DB (2001) Introduction to graph theory. 2nd edn. Prentice Hall of India- New Delhi

# Differential Privacy in NoSQL Systems

Navraj Singh(✉), Abhishek Shyam, Samatha R. Swamy, and Prasad B. Honnavalli

PES University, Bengaluru, India

**Abstract.** With the advent of the volume, velocity and variety of data in today's day and age, big data has become the norm in many places. Data at rest, the same data that is fed to the analyst of a company is at most times, not encrypted and depends on the fact that a person can be trusted by the company. Our contribution is implementing differential privacy in a NoSQL database while combating the few drawbacks in differential privacy with techniques such as cluffering (clustering and shuffling) and k-Anonymization. We implement an end to end implementation in a NoSQL database, MongoDB to promote privacy by perturbation using differential privacy which has been shown to be one of the best ways to do so. A generic system gets formed where local differential privacy is enforced before sending the sensitive database for analysis and enforcing differential privacy concepts on aggregate queries ensuring that no particular individual gets exposed in the process. We further provide the basis for implementing other aggregates apart from the ones we implemented and for complex queries that are not yet supported while the core concepts remain the same.

**Keywords:** Privacy preservation · Cluffering · Differential-privacy · Sensitivity · Distance · NoSQL · MongoDB

## 1 Introduction

With the amount of varied data, relational tables become too restricted for data storage. NoSQL databases store data differently than the traditional relational tables. The design for such tables is simple and it is extremely flexible. NoSQL tables have become popular not only due to the fact that they are easy to develop but they are also horizontally scalable. This takes place as data is split into shards which may be of fixed or variable length. Simply adding more common hardware will suffice for more data storage while in relational SQL, we would have to increase the storage of the system and be continuously upgrading vertically. A lot of companies from big giants like Amazon to startups have adopted NoSQL starting the 'NoSQL movement' and they have proclaimed that with the nature of today's data, this is the way to go.

## 2 Related Work

### 2.1 Privacy

There are two main issues we tackle in NoSQL systems. Firstly, data at rest is completely unencrypted by default in most cases. If the data is not encrypted the companies, by


S. Shukla et al. (eds.), *Data Science and Security*, Lecture Notes in Networks and Systems 290,
https://doi.org/10.1007/978-981-16-4486-3_42

default, have to trust their analysts not to go through the data at hand and use it in a judicial way. Secondly, these systems provide a way to communicate with the data using query languages, which are always susceptible to injection attacks.

Encryption is the most widely used method to guarantee privacy, but if data at rest is encrypted, with each query all the data would be decrypted and output would be displayed. This fares well with a small amount of data, but as data size increases, the decryption time increases and speed for query processing decreases and is not desirable. Fuller, Benjamin et al. [3] is an example of a database that is cryptographically protected.

Nalini C. and A.R. Arunachalam [1] talks about Data Perturbation, a technique to increase individuals' confidentiality is implemented by adding noise to the database. Even with all the ways to achieve privacy, data perturbation has been proven to be the most effective technique to deal with securing the data at hand.

Shrivastva, Krishna Mohan Pd, et al. [2] comes to the conclusion that of all the methods based on privacy, differential privacy completes the check of true privacy in systems. At the same time, the paper describes that this technique of privacy has shown a unique compatibility to the volume, velocity, variety and veracity of the data.

Johnson, Noah, Joseph P. Near, and Dawn Song [5] is the very first practical way of preserving privacy using differential privacy in a SQL based, real world system using FLEX, which is an end to end implementation in SQL enforcing almost no performance overhead. The drawbacks of this paper are discussed thoroughly and solved in one of the future papers we are about to discuss ([4]).

Our implementation is based on the idea of differential privacy based on bounded user contribution. Wilson, Royce J. et al. [4] have implemented the same in PostgreSQL with a full end-to-end implementation. They propose methods to provide differentially private aggregations on the databases. The most important distinction is that a particular individual can spam as many rows as possible in this system. The other papers limited this fact by repressing each unique user to one row being highly unrealistic and impractical in nature. It limits the implementation of differential privacy to SQL systems. The paper concludes by stating that the concept of contribution bounds can be theoretically applied to NoSQL based systems.

Ye, Qingqing et al. [6] introduces local differential privacy to a key-value data collection scenario and develops three solutions PrivKV, PrivKV M and PrivKV M+ as a decentralized privacy-preserving mechanism while he paper deals only with local differential privacy on key-value based models while we address the issue of privacy with a more end to end implementation.

Barthe, Gilles et al. [9] talks about verifying differential privacy by proving that if we have a program and it satisfies a Hoare specification, then the program being probabilistic is differentially private. With all the Local Differential Privacy protocols deployed in real world scenarios Wang, Tianhao et al. [8] introduces a framework generalizing these protocols for a fast aggregation algorithm.

Mironov, Ilya [10] technique introduces the concept of Renyi divergence. It seems to be a natural relaxation to differential privacy. It extends the properties of the standard differential privacy getting to be termed as Concentrated Differential Privacy. Bun, Mark, and Thomas Steinke [11] helps in establishing lower bounds in the said technique. It also achieves a unification by defining approximate concentrated differential privacy.

### 2.2 MongoDB

We have chosen to implement Differential Privacy in MongoDB, which uses a document data model for its service. Kumar, Jitender, and Varsha Garg [7] talks about the issue of security in NoSQL databases in general, MongoDB in particular and that MongoDB does not support encryption or decryption by default and the vulnerability it has to injection attacks due to its query language. This further justifies the need of differential privacy in a NoSQL environment and on MongoDB in particular.

## 3 Techniques Used

### 3.1 Cluffering

Before the Analyst receives the data from a source, we apply cluffering to our dataset. This ensures that even if the analyst gets hold of the raw dataset, he would not be able to get the exact value of any individual row. Since we have to preserve results of the aggregate functions, the aggregate values like mean, sum, median and the rest remain untouched. Cluffering is a term formed by Clustering and Shuffling.

When a specific numerical column is to be clustered, we take the column and find out the relative error of each row in the column. The relative error of each row is calculated using Eq. 1.

$$Relative\ Error = \frac{|Value - Mean|}{Standard\ Deviation} \tag{1}$$

We calculate the relative error of the particular column, along with another value and if a given id is available we can make a new clustered collection and store it back.

Based on the number of values, we can round off all the relative error values to x decimal places. We group the values which have the same altered relative values together and we do shuffling.

In shuffling we take each such group of values having the same relative errors and shuffle them. We group both the index and the values according to the relative error and shuffle only the values. The shuffled values then go to the respective indexes. Cluffering is a fast technique and a pre-processing measure to avoid the exact value of an individual to be found out. However, it does not provide for non numerical values in the database and they are exposed if the whole dataset is exposed.

### 3.2 Differential Privacy

Differential privacy is a concept of privacy preservation that provides a guarantee to the user that it is impossible to get too much information about any individual from any query of the database. The criteria for a function to satisfy differential privacy for all possible outputs S and all neighbouring datasets x and x' is in Eq. 2.

$$Pr[F(x) = S] >= \left(e^{\varepsilon} * Pr\left[F\left(x'\right) = S\right]\right) \tag{2}$$

### 3.2.1 The Laplace Mechanism

The easiest way to achieve privacy preservation is to add noise to the values present in the database. This concept of adding just noise is known as the laplace mechanism of privacy preservation. The point to understand here is the amount of noise necessary to satisfy the concepts of differential privacy.

$$F(x) = f(x) + \mathrm{Lap}(\mathrm{s}\,\varepsilon) \tag{3}$$

where ε is the sensitivity of $f$, and Lap($S$) denotes sampling from the Laplace distribution with center 0 and scale $S$.

## 3.3 Sensitivity

As we have shown above the amount of noise necessary to ensure differential privacy using the laplace mechanism for a given query depends on the sensitivity of the query. We define sensitivity of a function as the amount of change in the output of a function by a unit change in the input. For a function $f$: D→ R mapping datasets (D) to real numbers, the global sensitivity of $f$ is defined by Eq. 4.

$$GS(\mathrm{f}) = maxx,\ x : d\left(\mathrm{x}, \mathrm{x}^{j}\right) <= 1\left|f(\mathrm{x})f\left(\mathrm{x}^{j}\right)\right| \tag{4}$$

In Eq. 4, $d(x, x^{J})$ represents the distance between two datasets $x$ and $x^{J}$, we define two datasets as neighbors if their distance is 1 or less.

# 4 Methodology

In this paper we propose a mechanism for privacy preservation in NoSQL databases using the core concepts of differential privacy. The image below shows the important stages we go through to make our database differential private. It consists of 3 key concept:

1) Partitions and contributions
2) Bounding the number of contributed partitions
3) Clamping the values of each partition

## 4.1 Partition of Contributions

In this step we split the result of our query into multiple partitions. A single partition in the whole set represents a subset of aggregated data corresponding to a given value of the aggregation. Graphically, we represent this step in a bar chart with each partition combined in a bar on a bar chart.

Criteria for making the partitions are as follows:

1) Maximum/Minimum partitions contributed
2) Maximum/Minimum contributed value
3) Maximum/Minimum contributions per partition

Imagine the case of an online food delivery website where the owner wants to run the queries on his business. In this case the contributions will be made as follows:

1) Maximum partitions contributed: to how many partitions can a user ordering food contribute? Here a user ordering food can contribute up to one partition. In other words, there is at most one time-slot when a user with a given id can order food from the website.
2) Maximum contributed value: what is the maximum value that can be contributed by a user ordering food to a partition? Here we count the number of orders.
3) Maximum contributions per partition: how many times can a user ordering food can contribute to a partition? In our example, a user can contribute to a partition at most once. In other words, a user can order food only once at a given hour. This is important because differential Privacy adjusts the amount of noise to mask contributions of each visitor. More contributions require more noise.

### 4.2 Bounding the Number of Contributed Partitions

Larger noise affects the utility of the data. In order to preserve the data utility, we made an approximate estimate of how many times a week a person may visit a restaurant on average, and assumed that the value is around 3 instead of scaling the noise by the factor of 7.

### 4.3 Clamping Individual Contributions

Once we have the upper and the lower bounds every input value will be automatically clamped to the specified bounds. This is needed for calculating the sensitivity of the aggregation, and to scale the noise that will be added to the sum.

### 4.4 Laplace Noise

Once the values are clipped we apply Laplace's noise to add another level of security and output the result.

### 4.5 Approach

Johnson, Noah Near, Joseph Hellerstein, et al. [12] presents Chorus, it is a system which provides differential privacy for SQL queries. In this paper we have taken reference from the above CHORUS mechanism and designed and implemented 3 methods for applying the concepts of differential privacy in No sql database.

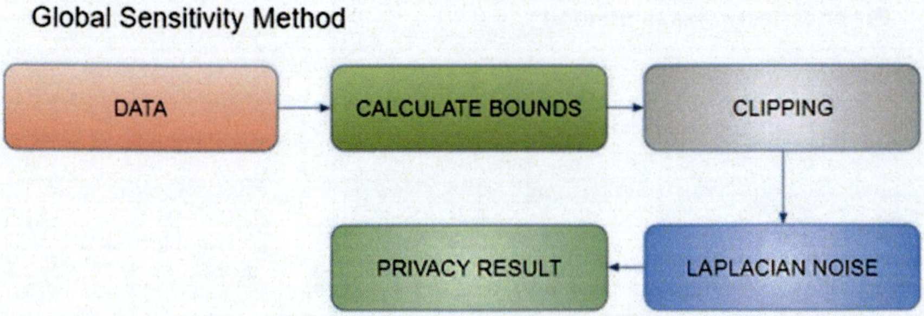


**Fig. 1** Global sensitivity method

### 4.5.1 Global Sensitivity Method

According to the concept of Laplace Mechanism and Global Sensitivity, the amount of noise necessary to ensure differential privacy for a given query depends on the sensitivity of the query. The definition of global sensitivity says that for any two neighboring datasets $x$ and $x^J$, the difference between $f(x)$ and $f(x^J)$ should not be greater than $GS(f)$ to satisfy the concept of differential privacy (Fig. 1).

### 4.5.2 Propose-Test-Release Method

The primary problem with local sensitivity is that it is a function of both the query and the data set which is being queried. The problem with the concept of Local Sensitivity is that it itself reveals some important information about the data. In this method of Propose-test release we aim to make the sensitivity itself differentially private. The framework first asks the analyst to propose an upper bound on the local sensitivity of the function being applied. Then, the framework runs a differentially private test to check that the dataset being queried is "far from" a dataset where local sensitivity is higher than the proposed bound. If the test passes, the framework releases a noisy result, with the noise calibrated to the proposed bound. We define the notion of local sensitivity at distance $k$ and $A(f, x, k)$ to denote the maximum local sensitivity achievable for $f$ by taking $k$ steps away from the dataset $x$ as shown in the Eq. 7.

$$A(f, x, k) = max_{y:d(x,y)<=k} LS(f, y) \tag{7}$$

To define a query to answer the question: "how many steps are needed to achieve a local sensitivity greater than a given upper bound $b$?"

$$D(f, x, b) = argmin_k A(f, x, k) > b \tag{8}$$

The 3 key steps in propose-test-release method are as follows:

1) Propose a target bound $b$ on local sensitivity.
2) If $D(f, x, b) + Lap(\frac{1}{\varepsilon}) < \frac{log(2/\delta)}{2\varepsilon}$, return 0
3) else Return $f(x) + Lap(b/\varepsilon)$

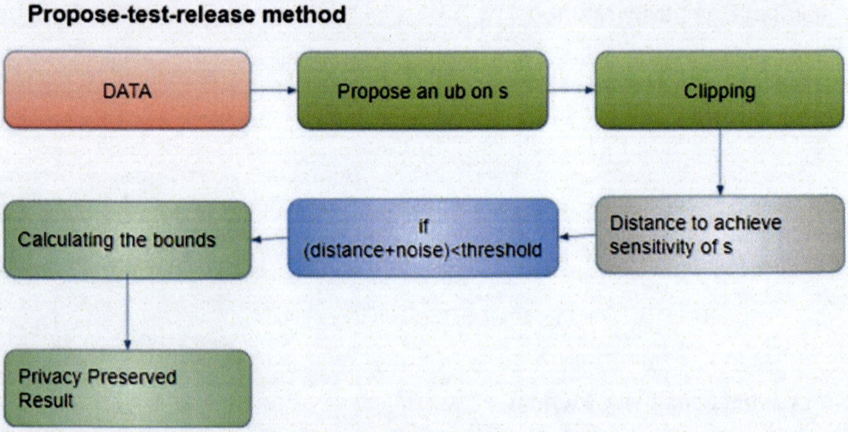


**Fig. 2** Propose-test-release method

We notice that *D(f, x, b)* has a global sensitivity of 1: adding or removing a row in *x* might change the distance to "high" local sensitivity by 1. Thus, adding Laplace noise scaled to yields a differentially private way to measure local sensitivity.

The drawback of this approach is that it does not satisfy the concepts of pure differential privacy. It's because there's a 50% chance of passing the test in step 2 by accident. The noise added in step 2 might be large enough to pass the test, even though the value of *D(f, x, b)* is actually less than the minimum distance required to satisfy differential privacy (Fig. 2).

### 4.5.3 Sample and Aggregate Method

The sample and Aggregate method satisfies the concepts of both global and local sensitivity.

1) Split the dataset $X \in D$ into k disjoint chunks $x_1, \ldots, x_k$
2) Compute a clipped answer for each chunk: $a_i = \max(l, \min(u, f(x_i)))$
3) Compute a noisy average of the answers: $A = (\frac{1}{k} \sum_{i=1}^{k} a_i) + \text{Lap}(\frac{u-l}{k\varepsilon})$

The framework can be shown to satisfy differential privacy just by global sensitivity and parallel composition. We split the dataset into $k$ chunks, so each individual appears in exactly one chunk. We don't know the sensitivity of $f$, but we clip its output to lie between $u$ and $l$, so the sensitivity of each clipped answer $f(x_i)$ is $u - l$. Since we take the mean of $k$ invocations of $f$, the global sensitivity of the mean is $(\frac{u-l}{k})$. This framework satisfies pure ε-differential privacy, and it actually works without the use of local sensitivity. In fact, we don't need to know anything about the sensitivity of $f$ (global or local). We also don't need to know anything about the chunks $x_i$, except that they're disjoint (Fig. 3).

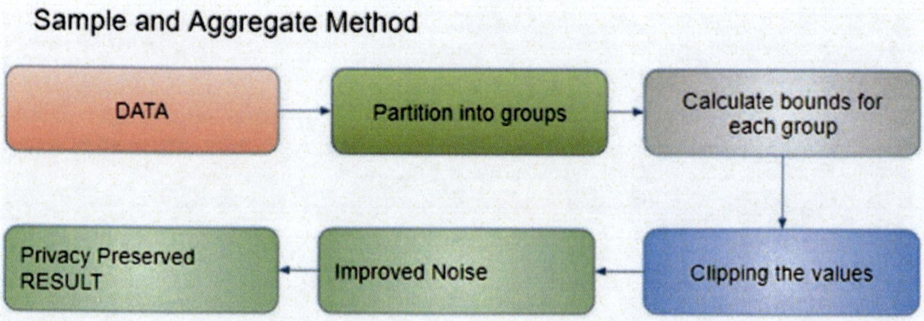


**Fig. 3** Sample aggregate method

# 5 Results

Some important results:

1) The graph in Fig. 4 shows the relative error of No sql Average queries. All errors are within 0.2 (20%) error limit and satisfy the concepts of differential privacy.
2) The graph Fig. 5 shows the average execution time of an average aggregate query in normal nosql and the 3 differential privacy methods implemented.

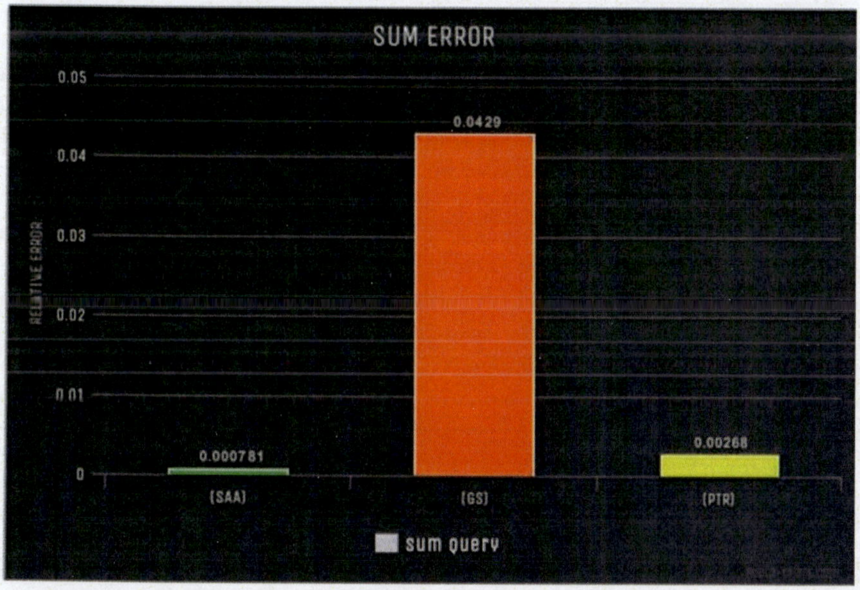


**Fig. 4** Relative error

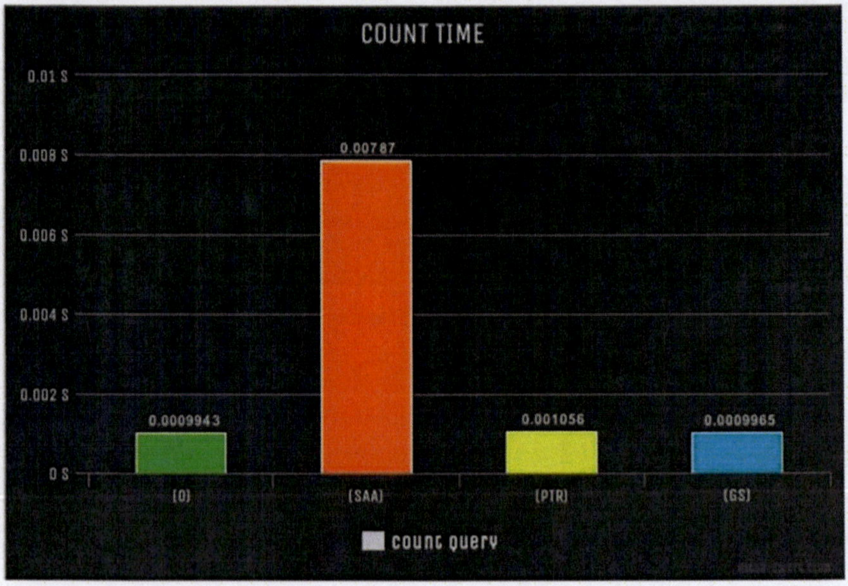


**Fig. 5** Average execution time

# 6 Comparison

The approach which we have proposed in this paper is a very algorithm based approach for privacy preservation. Most of the current existing methods which are in use in the industry are mostly using cryptography for privacy preservation. The drawback with this approach is that it takes a lot of time for query execution. As the result of the query has to be both encrypted and decrypted at the sender and receiver end respectively. Our approach is comparatively very fast and the execution time of our differential-privacy approach is very much comparable to the actual nosql query execution time.

The graph below shows the execution time of min and max aggregate queries using normal no-sql and simple-aggregate mechanisms (Fig. 6).

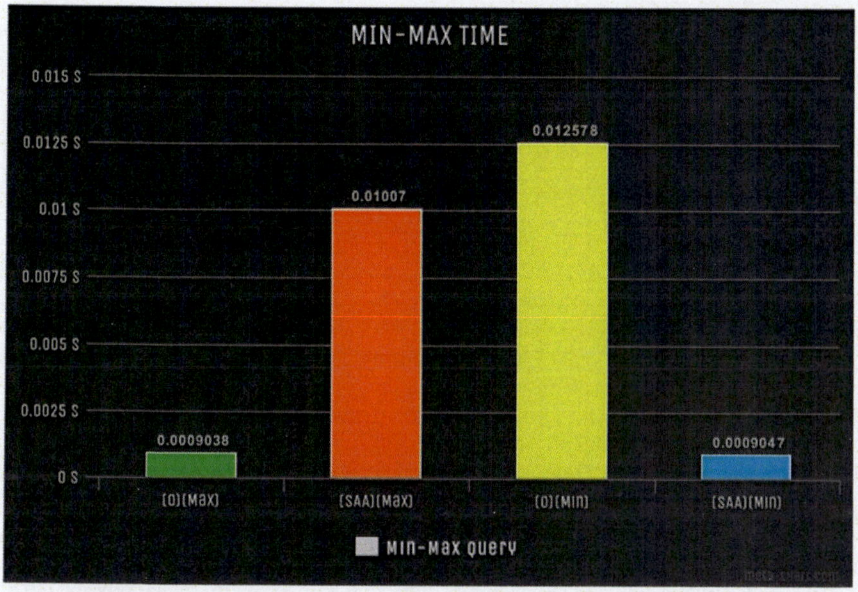


**Fig. 6** Comparison of SAA-No-SQL vs normal No-SQL (execution time)

## 7 Conclusion

In this paper, we have presented a generic system which answers NoSQL queries using user-level differential-privacy. We know that most data analysis tasks are based on certain aggregations including summation, counting, average, maximum and minimum and the method by our paper is able to capture most of the above tasks. The systems developed by us works well even for typical use cases while providing a mechanism to deduce privacy parameters from accuracy requirements, and also preserving the utility of the result so that the privacy preserved result is not very much deviated from the actual value and the noise added is also within a certain range so that the data result is useful in all respects. The algorithms presented in this work are relatively simple, but Real world use cases and evidence show that this approach is useful, robust and scalable.

## 8 Future Enhancements

There is a wide array of prospects that this paper can be taken forward by: The number of aggregation functions supported by NoSQL databases in general, MongoDB in particular is very high and a combination of the existing methods can be reused to support all the other functions as well.

1) Parsing the query for complicated and nested cases is not supported and support can be added by finding the nested queries, getting the result of the said queries in the form of objects and applying the differential privacy step in the final step which is already supported by our model. Joins can be supported the same way.

2) Aggregation can alternatively be done with map-reduce jobs which are not yet supported by the script. The core concepts and the functions for differential privacy, however, remain the same

## References


1. Nalini C, Arunachalam AR (2017) A study on privacy preserving techniques in big data analytics. Int J Pure Appl Math 116(10):281–286
2. Shrivastva, KMP, Rizvi MA, Singh S (2014) Big data privacy based on differential privacy is a hope for big data. In: 2014 international conference on computational intelligence and communication networks. IEEE
3. Fuller B et al (2017) Sok: cryptographically protected database search. In: 2017 IEEE symposium on security and privacy (SP). IEEE
4. Wilson RJ et al (2020) Differentially private SQL with bounded user contribution. In: Proceedings on privacy enhancing technologies 2020(2):230–250
5. Johnson N, Near JP, Song D (2018) Towards practical differential privacy for SQL queries. Proc VLDB Endow 11(5):526–539
6. Ye Q et al (2019) PrivKV: Key-value data collection with local differential privacy. In: 2019 IEEE symposium on security and privacy (SP). IEEE
7. Kumar J, Garg V (2017) Security analysis of unstructured data in NOSQL MongoDB database. In: 2017 International conference on computing and communication technologies for smart nation (IC3TSN). IEEE
8. Wang T et al (2017) Optimizing locally differentially private protocols. arXiv preprint arXiv: 1705.04421
9. Barthe G et al (2014) Proving differential privacy in Hoare logic. In: 2014 IEEE 27th computer security foundations symposium. IEEE
10. Mironov I (2017) Rényi differential privacy. In: 2017 IEEE 30th computer security foundations symposium (CSF). IEEE
11. Bun M, Steinke T (2016) Concentrated differential privacy: simplifications, extensions, and lower bounds. In: Hirt M, Smith A (eds) Theory of cryptography. TCC 2016. LNCS, vol 9985. Springer, Heidelberg
12. Johnson N, Near JP, Hellerstein J, Song D (2018) Chorus: differential privacy via query rewriting

# Movie Success Prediction from Movie Trailer Engagement and Sentiment Analysis

Abin Emmanuvel(✉), Vandana Bhagat, and Lija Jacob

CHRIST (Deemed to be University), Lavasa, Pune, India
abin.emmanuvel@science.christuniversity.in

**Abstract.** The diverse movie industry faces many challenges in the promotion of the product across different demographics. Movie trailer engagements provide valuable information about how the audience perceives the movie. This information can be used to predict the success of the upcoming movie before it gets released. The previous research works were mainly concentrating on Hindi language movies to predict success. The current research paper includes the success prediction of movies other than Hindi. This paper aims to analyze various Machine Learning models' performance and select the best performing model to predict movie success. The developed model can efficiently classify successful and unsuccessful movies. For the current research, the data is collected from various sources through web scrapping and API calls in Sacnilk, The Movie Database (TMDB), YouTube, and Twitter. Different machine learning classification models such as Random Forest, Logistic Regression, KNN, and Gaussian Naïve Bayes are tested to develop the best-performing prediction model. This research can help moviemakers to understand the popularity of the movie among the viewers and decide on an efficient promotional strategy to make the movie more successful.

**Keywords:** E-word of mouth · Machine learning · Movie trailer engagement · Predictive analysis · Sentiment analysis · Web scraping

## 1 Introduction

The movie Industry is prevalent in India and India is a diverse country with various languages, each with different movie industries catering to those languages. Generally, the promotional budget takes up a fair amount of the total production budget. Mega budget movies are usually associated with a renowned director, well-known cast, and high marketing budget. In addition to this, high budget movies' promotional activities are much higher than movies with a low production budget. A countless number of low-budget movies have disappeared without ever reaching the audience due to fewer promotional activities. However, the producers and studios, making low-budget movies must look into cost-effective promotional activities.

Various research works have targeted the analysis of Hindi movies in the research. The current research aims to find the best machine learning model to predict the movie's success based on WOM (Word of Mouth), movie trailer engagement, and movie release


S. Shukla et al. (eds.), *Data Science and Security*, Lecture Notes in Networks and Systems 290,
https://doi.org/10.1007/978-981-16-4486-3_43

details. The prior knowledge of movie reception by the audience can help the moviemaker to improve the promotional activities.

Movie Trailer Engagement and Tweets before the movie's release can be used to understand, how well the audience perceives the movies before the release. While the trailer statistics provide information about popularity, the sentimental analysis from tweets can provide insight into a movie's perception from the audience.

In the current research, movie data is collected through web scrapping of Sacnilk [1] and using the API of TMDB [2], YouTube [3] and Twitter [4]. For the prediction of movies, previous researchers only considered Hindi movies in their data, while the current research includes Hindi, Tamil, Telugu, Malayalam, Kannada, and English movies in the data. WOM about the movie is not used in many research. Machine learning classification models such as Random Forest, Logistic Regression, KNN, and Gaussian Naïve Bayes are build using the collected data. It is observed that Random Forest provides the best performance among all the selected models.

The paper is organized as follows, Sect. 1 explains the notion of Movie success prediction and related work about the study is discussed in Sect. 2. Section 3 describes the architecture of the proposed model and explains each step on the architecture in detail. Observations and results of the models are discussed in Sects. 4 and 5 concludes the findings of the research.

## 2 Related Work

The literature review focused on articles on predicting a movie's success, finding the sentiment analysis from reviews or social media activities on Twitter or YouTube. Ten papers from 2017 to 2020 were collected mainly from IEEE to review.

Shamsur et al. [5] work was to find the best possible method for predicting the Box office success from YouTube data. Their Multiple Linear Regression and Random Forest model has about 90% accuracy. Sameer et al. [6] concentrates on Bollywood and has a dataset that includes reviews, YouTube statistics, music score, movie details, and other statistics. The dataset was used in 5 different classifier models, and Bagging had the best accuracy of 82.010%. They also found some variables with high contribution to the model: the number of screens, quality of the songs, actors, and actresses, and timing of the release. Javaria et al. [7] work uses mathematical models on the Bollywood dataset. Movie genres, actors, directors, music directors, producers, and language are used as factors to find the success of movies. It was found that the success of a movie is highly dependent on its cast and there was a strong correlation found between actors and some genres, indicating that actors tend to work in certain genres. Ashutosh's [8] work uses various algorithms to predict Bollywood movies' success and the better algorithm was KNN with 49.33% of accuracy. Rijul et al. [9] and Hemraj et al. [10] also in this work uses different models, and Random Forest was found to be better in both of them with 61 and 92% accuracy, respectively. While the former uses IMDb as a metric of success, the latter tries to predict the box office collection. IMDb is an online database of movies collected, and the rating here considered is the evaluation of that movie by its users.

From the literature survey, it was observed that only English/Hindi movies were considered in a majority of works. Various researches have focused only on Hindi movies in India and did not consider WOM for the prediction.

# 3 Methodology

This section describes the source of data and how are they extracted, pre-processed and transformed to train classification models.

## 3.1 Proposed Methodology

Building a machine learning model requires specific steps like data extraction and data pre-processing. The proposed architecture of the model is shown in Fig. 1 and explained in Sects. 3.2, 3.3, 3.4, 3.5 and 3.6.

## 3.2 Data Extraction

The data for the experiment is scrapped through various websites. Box Office data of Hindi, Tamil, Telugu, Malayalam, Kannada and English movies are collected from the Sacnilk website through web scrapping and movie details, such as genre, release date, etc., from TMDB are collected through the TMDB API. YouTube API is used to collect movietrailer statistics of the movie trailer/song and Twitter API to collect tweets about the movie.

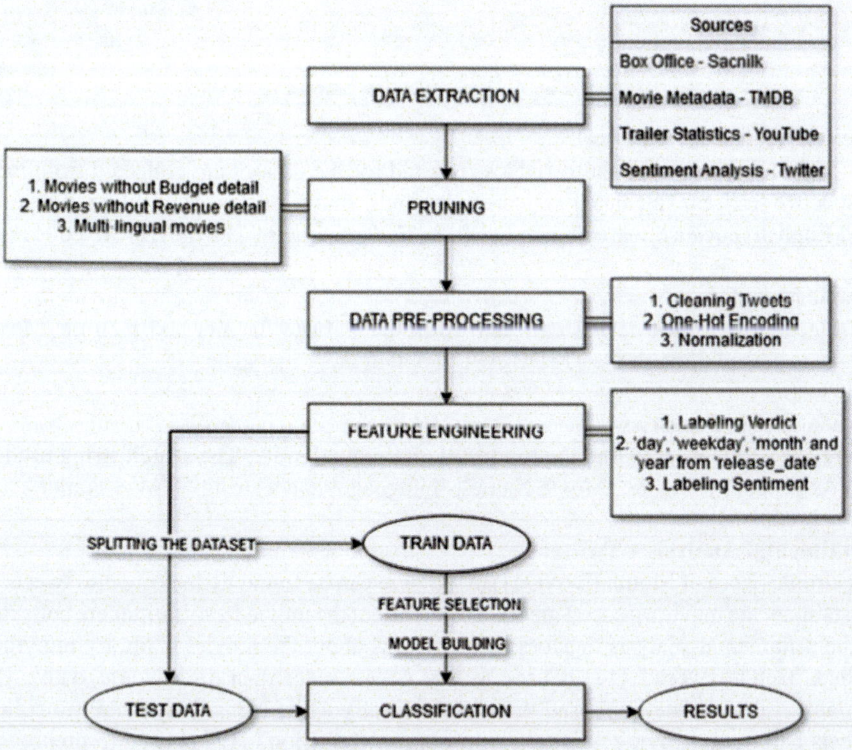


**Fig. 1** Model architecture

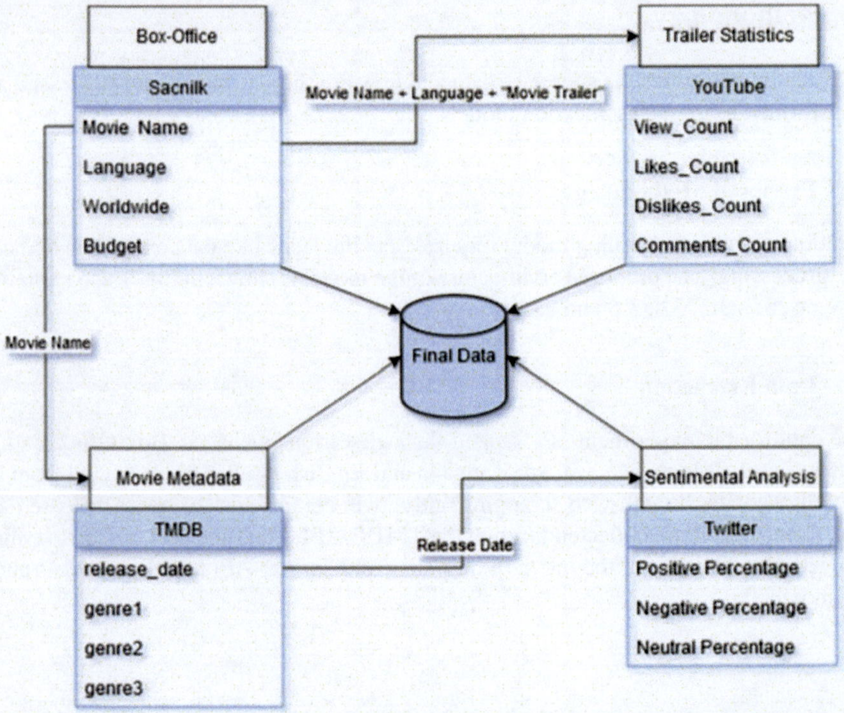


**Fig. 2** Data collection process

Figure 2 Shows how the variables required for the dataset are collected from their respective sources. The data is collected for Hindi, Tamil, Telugu, Malayalam, Kannada, and English movies released between 2018–2020. A total of 275 movies are collected.

**Movie Metadata**
The movie box office data is collected from the website Sacnilk and then the Movie Name is passed through the TMDB API to obtain more metadata.

**Movie Trailer Statistics**
YouTube API is used to find the top result of a movie trailer. The search string used in the API is built using the 'Movie Name', 'Language' and a string "Movie Trailer".

**Sentimental Analysis – Twitter**
Electronic-Word of Mouth (E-WOM) of a movie can be found by looking into tweets or posts about the movie on social media. Tweets about the movie from Twitter are collected to do sentimental analysis to access the E-WOM about the movie. A library in python called "GetOldTweets" [11] is used to find tweets older than the 'release_date'. The variable 'release_date' is passed through the library while using Twitter API to collect tweets before the movie release. Python library "TextBlob" [12] is used to preprocess the collected tweets and calculate sentiment polarity.

### 3.3 Pruning

For efficient model building, relevant data is required. The pruning stage removes irrelevant data. The Performance of the movie in Box-Office is calculated from Budget and Revenue details. Budget and revenue details are missing for some movies, such records are considered irrelevant for the study and are removed from the dataset.

### 3.4 Data Pre-processing

The collected data is converted to a suitable form for model building through the various pre-processing tasks. The collected tweets about movies need to be pre-processed by removing the '@' mentions, '#' hashtag, 'RT' retweet and hyperlink.

Nominal variables like 'genre' and 'verdict' are converted into binary variables through the 'One Hot Encoding technique' because machine learning algorithms need numerical data.

Min–Max Normalization is used to scale the data to bring the values of data between 0 and 1. Scaling will not change the performance of the model Random Forest, Logistic Regression and Naïve Bayes. But KNN being a distance-based model improves the performance when the data is normalized.

### 3.5 Feature Engineering

New features are engineered from the collected data for building better models. The movies are divided into two classes of successful and non-successful movies. If revenue is more than budget then the movie is labelled successful otherwise it's not.

The 'release_date' variable is used to create new variables day, weekday, month and year. The pre-processed tweets of a movie are used to generate public sentiment value. Among 'Positive_Percentage', 'Negative_Percentage' and 'Neutral_Percentage', the highest is found and a new variable 'Sentiment' is created with labels 'Positive', 'Negative' or 'Neutral' respectively.

### 3.6 Classification Models

#### Feature Selection

'ExtraTreesClassifier' is used to find essential features for model building. During the feature selection, the variables 'genre2' and 'genre3' had the least importance. The removal of these variables did not harm the performance of the model. Therefore, these variables are removed from the dataset.

#### Model Building

The data is split into a 70:30 ratio for training and testing the model. Hyperparameter tuning such as 'Random Search' for Random Forest and 'Grid Search' for other models is used on training data to find the best parameters of that model.

Prediction of Movie Success is a classification problem. This dataset contains labelled data. The objective is to better classify the 'Verdict' Column in the dataset.

The 'Verdict' column contains a binary value of either 0 or 1. Therefore, supervised classification algorithms are used to train the dataset and predict the outcome. Classification models Random Forest, Gaussian Naïve Bayes, Logistic Regression and KNN are applied to the modified dataset.

## 4 Experimental Results and Discussion

This section contains the results obtained from the experiments performed during the research. Box Office data and other metadata of 275 movies are obtained from Sacnilk and TMDB, respectively, while trailer statistics and sentiment of users before the release of the movie are obtained from YouTube and Twitter respectively.

The experiment is done using the python programming language. The experiment included Random Forest, Naïve Bayes, Logistic Regression and KNN models made using the "Scikit-learn" library [13]. These models are selected as it's better for datasets with fewer observations.

The training accuracy and testing accuracy indicate the percentage of correct predictions in training data and testing data respectively by the model. The F1-score combines the precision and recall of the model.

**Table 1** Evaluation of classifiers

| Algorithms | Training accuracy | Testing accuracy | F1 score |
|---|---|---|---|
| Random forest | 0.802 | 0.771 | 0.77 |
| Logistic regression | 0.531 | 0.530 | 0.48 |
| KNN | 0.73 | 0.73 | 0.73 |
| Gaussian Naïve Bayes | 0.791 | 0.614 | 0.58 |

As shown in Table 1, Random Forest makes 80% correct predictions on the training data and 77% correct predictions of testing data. This is the better model in comparison to other models in classifying successful and non-successful movies. Random Forest is also a better model in F1 score metrics. F1 Score is a better metric than Precision and Recall here as it gives a balance between Precision and Recall.

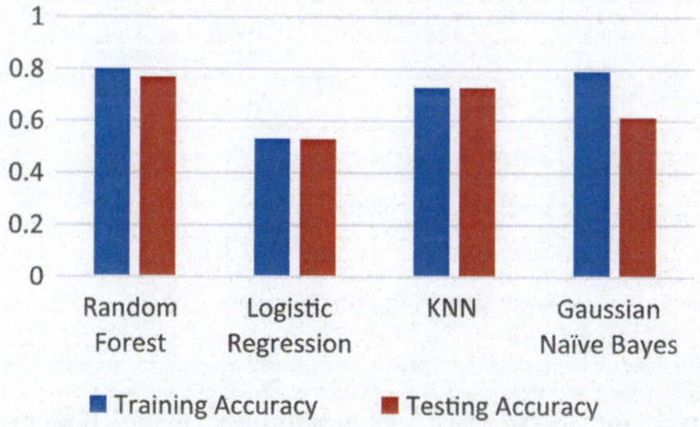


**Fig. 3** Comparison of models

From Fig. 3 it is evident that Random Forest is the better model with 77% accuracy followed by KNN with 73% of accuracy and Logistic Regression with 53% accuracy while Gaussian Naïve Bayes overfits.

Receiver Operator Characteristic (ROC) curve is an evaluation metric for binary classification problems. It is a probability curve that plots the True positive rate against the false positive rate. The ROC curves of all models are shown in Fig. 4.

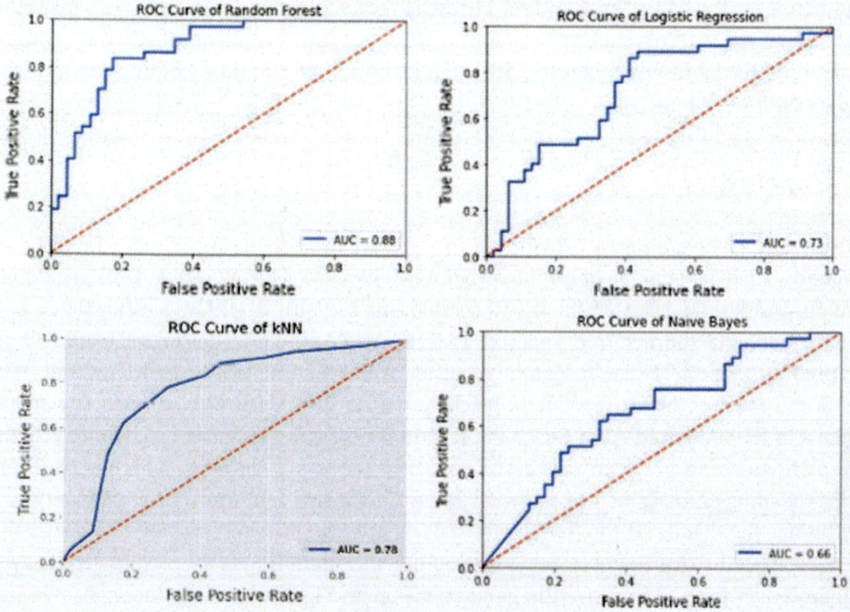


**Fig. 4** ROC curve of all models

**Table 2** AUC score

| Models | AUC score |
|---|---|
| Random forest | 0.88 |
| Logistic regression | 0.73 |
| KNN | 0.78 |
| Gaussian Naïve Bayes | 0.66 |

The Area Under the Curve (AUC) is a summary of the ROC curve. AUC Score measures the ability of a classifier to distinguish between classes i.e. higher the AUC, the better the model is at distinguishing between classes.

From Fig. 4, The ROC curve of the Random Forest model is the bigger one i.e., closest to the left corner of the plot and thus creates more Area Under the Curve (AUC). From Table 2, Random Forest has the highest score of 88 among the AUC Scores of the classifier models; therefore, Random Forest can better distinguish between the two classes. The Clients of the model built using Random Forest can now use it to predict whether the movies are successful or non-successful.

Even in previous works [9] and [10], Random Forest was the better model with 61 and 92% of accuracy respectively. The Random Forest model contains decision trees. The better performance by Random Forest is indicating the decisions taken inside the trees. Here, the features such as Trailer Statistics, year and month of release have a higher correlation to the movie's verdict. So, such information could be helping the model to make better decisions.

## 5 Conclusion

Movie trailer engagements provide an idea about the perceiving of the movie by the audience. This data can be helpful to predict the succusses of the movie well in advance. The proposed model has efficiently classified that the movie will be successful or not.

Various data sources like Sacnilk, TMDB, YouTube and Twitter are used to build the prediction models. The research has tested four machine learning models and found Random Forest as the most efficient prediction algorithm with 77% of accuracy and the highest in F1 Score and AUC Score i.e. Random Forest is the better model in predicting and distinguishing between successful and non-successful movies. This research provides meaningful insights to producers hoping to invest late during the production. In the future, the details of moviemakers, specifically the director, writer, actors etc. may be used to add extra features for constructing the model, and more statistics may be gathered over time to improve the generalization and learning of the models.

## References

1. Sacnilk, https://www.sacnilk.com/. Accessed 25 Oct 2020
2. TMDB API, https://developers.themoviedb.org/3. Accessed 25 Oct 2020

3. YouTube API, https://developers.google.com/youtube/v3. Accessed 25 Oct 2020
4. Twitter API, https://developer.twitter.com/en/docs. Accessed 25 Oct 2020
5. Rahim MS, Chowdhury AZME, Islam MA, Islam MR (2018) Mining trailers data from youtube for predicting gross income of movies. In: 5th IEEE region 10 humanitarian technology conference 2017, R10-HTC 2017, pp 551–554, January 2018. https://doi.org/10.1109/R10-HTC.2017.8289020
6. Jaiswal SR, Sharma D (2017) Predicting success of bollywood movies using machine learning techniques. In: ACM international conference proceeding series, pp 121–124. https://doi.org/10.1145/3140107.3140126
7. Ahmad J, Duraisamy P, Yousef A, Buckles B (2017) Movie success prediction using data mining. In: 8th international conference on computing communication and networking technologies, ICCCNT 2017, pp 2015–2018. https://doi.org/10.1109/ICCCNT.2017.8204173
8. Kanitkar A (2018) Bollywood movie success prediction using machine learning algorithms. In: 2018 IEEE 3rd international conference on circuits, control, communication and computing, I4C 2018. https://doi.org/10.1109/CIMCA.2018.8739693
9. Dhir R, Raj A (2018) Movie success prediction using machine learning algorithms and their comparison. In: ICSCCC 2018 - 1st international conference on secure cyber computing and communication (ICSCCC), pp 385–390. https://doi.org/10.1109/ICSCCC.2018.8703320
10. Verma H, Verma G (2020) Prediction model for bollywood movie success: a comparative analysis of performance of supervised machine learning algorithms. Rev Socionetw Strateg 14:1–17. https://doi.org/10.1007/s12626-019-00040-6
11. GetOldTweets3, https://pypi.org/project/GetOldTweets3/. Accessed 25 Oct 2020
12. TextBlob: Simplified text processing. https://textblob.readthedocs.io/en/dev/. Accessed 25 Oct 2020
13. scikit-learn: machine learning in Python. https://scikit-learn.org. Accessed 25 Oct 2020

# On the *k*-Forcing Number of Some DS-Graphs

M. R. Raksha(✉) and Charles Dominic

Department of Mathematics, CHRIST(Deemed to be University), Bangalore, India
raksha.mr@res.christuniversity.in

**Abstract.** Amos et al. introduced the notion of $k$-forcing number as a generalization of Zero forcing number and is denoted by $F_k(G)$ where $k > 0$ is any positive integer, the $k$-forcing number of a graph is the minimum cardinality among all $k$-forcing sets of a graph $G$. In this paper, many bounds for $k$-forcing number of degree splitting graph $DS(G)$ for different graph classes are found. We evaluate the value of $k$-forcing number of degree splitting graph of some of the Cartesian product graph for different values of $k$. Also we observed that for Turán graph $T_{n,t}$, upper and lower bound is given by, $F_k(T_{n,t}) \leq F_k(DS(T_{n,t})) \leq F_k(T_{n,t}) + 1$.

**Keywords:** $k$-forcing number · Degree splitting graph · Turán graph · Regular graph

## 1 Introduction

In the following manuscript, we examine only simple graphs. That is finite graphs without any loops or multiple edges. Let $G = (V, E)$ be a graph. For a vertex $v \in V(G)$, the set of vertices in $G$ which are adjacent to it are known as the neighborhood $N(v)$ of $v$ and the number of edges incident to the vertex $v$ is called degree of the vertex $v$. The minimum and maximum degree of a graph $G$ can be denoted by $\delta(G)$ and $\Delta(G)$ respectively. All other notation and terminology which are not defined here is referred from [1]. David Amos et al. in [2] introduced $k$-forcing number of graphs as a generalization of zero forcing number. The conjecture state in [2] is proved in [4]. When $k = 1$, the definition 1-forcing number is identical to the definition of zero forcing number.

The main motivation behind the study of $k$-forcing number is the wide range of application that zero forcing number holds, from building logic circuits [8], analysing controlability of quantum sytems [7] to network analysis of active propagation model, where any person with an idea in the network is considered as infected and the spreading is based on zero forcing [9]. In a similar manner one can study the spreading based on $k$-forcing sets in graphs. In this article, we consider the underline network as the degree splitting graph of the given graph. We use the underlying definitions for the furture outcomes of this manuscript.

Consider a graph $G$ in which all verices are either colored white or black. Then the $k$-forcing number of $G$ can be defined as follows:

- $k$-Color Change Rule: Any black vertex can force at most $k$ white vertices provided it has at most $k$ white neighbours.


S. Shukla et al. (eds.), *Data Science and Security*, Lecture Notes in Networks and Systems 290,
https://doi.org/10.1007/978-981-16-4486-3_44

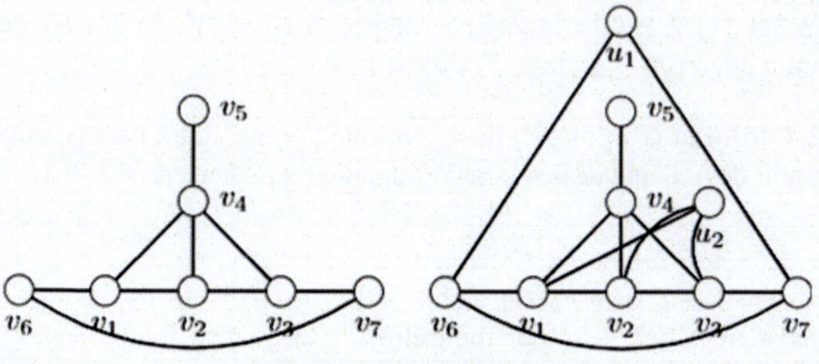


**Fig. 1** Graph G and DS(G)

- For a black and white coloring of the given graph $G$, the $k$-derived coloring of the given graph $G$ is the result of applying the $k$-color-change rule until no more changes are possible.
- If $S \in V(G)$ is a forcing set, which consist of those vertices which helps to force the entire graph black by applying $k$-color changing rule, then $k$-forcing number is defined as the minimum cardinality of the $k$-forcing sets.

Zero forcing number can be obtained when $k = 1$. Zero forcing number of degree splitting and complete degree splitting graph are studied in [3]. In this paper, we only focus on $k > 1$.

The degree splitting graph was introduced by R. Ponraj in [5]. Let G be a graph with $V(G) = U_1 \cup U_2 \cup \ldots U_t \cup B$ where each $U_i$ is a set containing at least two vertices having same degree and $B = V(G) \setminus \cup_{i=1}^{t} U_i$. The degree splitting graph of G, denoted by DS(G), is obtained from G by adding vertices $u_1, u_2, \ldots, u_t$ and joining the vertex $u_i$ to each vertex of $U_i$ for $1 \leq i \leq t$ (Fig. 1).

In Sect. 2 we give some of the prerequisites which are necessary to understand the current findings. Section 3 deals with some new results on the $k$-forcing number of degree splitting graph.

## 2 Preliminaries

In the current section we quote few previous findings which will be used in the illustration of the obtained results.

**Proposition 2.1** [2]. Let $G = (V, E)$ be a simple graph where V is the set of all vertices and E is set of all edges in $G$. (1) If G is connected with

i) $|V(G)| \geq 2$, then $F_{\Delta}(G) = 1$.
ii) $\Delta \geq 2$, then $F_{\Delta-1}(G) \leq 2$ with equality holding if and only if G is $\Delta$-regular.

(2) If $k$ is a positive integer, then $F_k(G) \geq F_{k+1}(G)$.

**Proposition 2.2** [2]. If $G = (V, E)$ be a graph and let $k$ be a positive integer, then $F_k(G) \geq \delta - k + 1$. Where $\delta$ is the minimum degree of $G$.

**Proposition 2.3** [6]. Let $k$ be a positive integer. If $G = (V, E)$ is a connected graph with maximum degree $\Delta \leq k$, then $F_k(G) = 1$.

**Theorem 2.4** [6]. Let $G = (V, E)$ be a connected graph and $k$ be a positive integer, $\delta$ is the minimum degree and $\Delta$ is the maximum degree, where $\Delta \geq k + 1$.

(i) If $\delta < \Delta = k + 1$, then $F_k(G) = 1$.
(ii) If $\delta = \Delta = k + 1$, then $F_k(G) = 2$.
(iii) Otherwise, if $\Delta \geq k + 2$, then the following inequality holds,

$$F_k(G) \leq \frac{(\Delta - k - 1)|V(G)| + max\{\delta(k + 1 - \Delta) + k, k(\delta - \Delta + 2)\}}{\Delta - 1}.$$

## 3 New Results

In this section we give new results which are obtained for degree splitting graph of a graph. We also try to find the bound for few specific classes of graphs. In this section, we try to bring a new relation between $k$-forcing number and $\Delta-$ forcing number of a graph. We have also tried to improvise the bound and in few cases we are trying to define upper and lower bounds specifically for degree splitting graph, rather than giving a wide range of bound for generalised graph classes.

**Theorem 3.1.** For any graph $G$, $F_k(G) \geq F_\Delta(G)$ where $k < \Delta$.

The proof of the above theorem follows from Proposition 2.1. Note that $F_\Delta(G) = 1$. For any value of $k$, the cardinality of the $k$-forcing set will be either 1 or more. Hence $F_k(G) \geq F_\Delta(G)$.

In the following few results we will try to understand the behaviour of $k$-forcing number of a $DS(G)$ when a graph $G$ has a vertex of degree $n - 1$ and $n - 2$, where $n$ is the order of graph $G$.

**Theorem 3.2.** Let $G$ be a graph and $DS(G)$ is the degree splitting graph of $G$. If $\Delta(G) = n - 1$ where order of the graph $G$ is n, then $F_k(DS(G)) \leq n + 1 - k$.

**Proof.** Since $\Delta(G) = n - 1$, any vertex in $G$ has degree less than or equal to $n - 1$. Therefore in $DS(G)$, degree of each of the vertex will be at most one more than that of degree of the same vertex in $G$. Let $v$ be the vertex of degree $\Delta(G) + 1$ in $DS(G)$. If we choose this vertex to be black, then it is clear that at least $n - k$ of its neighbour should be black in order for the forcing process to occur. Once $n - k$ neighbours of $v$ are black remaining all its white neighbours are forced black. Hence forcing the entire graph black. This implies that $F_k(DS(G)) \leq n + 1 - k$. ■

**Theorem 3.3.** Let $G$ be a graph and $DS(G)$ is the degree splitting graph of $G$. If $\Delta(G) = n - 2$ where order of the graph $G$ is n, then $F_k(DS(G)) \leq n - k$.

**Proof.** Since $\Delta(G) = n - 2$, any vertex in $G$ has degree less than or equal to $n - 2$. Therefore in $DS(G)$, degree of each of the vertex will be at most one more than that of degree of the same vertex in $G$. Let $v$ be the vertex of degree $\Delta(G) + 1$ in $DS(G)$. If we choose this vertex to be black, then it is clear that at least $n - 1 - k$ of its neighbour should be black in order for the forcing process to occur. Once $n - 1 - k$ neighbours of $v$ are black remaining all its white neighbours are forced black. The only white vertex not adjacent to $v$ is forced by the black neighbour of it. Hence forcing the entire graph black. This implies that $F_k(DS(G)) \leq n - k$. ■

**Observation 3.4.** When $\delta = t$ and $\Delta = t+1$, $F_k(G) = 1$ for $k > t$ and $F_k(DS(G)) = 1$ for $k \geq t + 1$.

The above result follows from Theorem 2.4.

**Lemma 3.5.** Let $G$ be a graph of order $n$ and $k$-forcing number $F_k(G) = t$ where $k < \delta$. If $G$ has a vertex of degree $n - 1$, then by removing such a vertex the $k$-forcing number will be reduced by one.

**Proof.** Here in graph $G$ any vertex will have degree at least $\delta$ also $k < \delta$. Let us consider a vertex of $G$ to be black in order for this vertex to force its white neighbours at least $\delta - k$ of its neighbours should be black. Now when we remove a vertex of degree $n - 1$, degree of this vertex will reduce to at least $\delta - 1$. Thereby number of neighbours which should be initially black also reduces by one. ■

**Theorem 3.6.** $F_k(DS(P_n)) = 1$, for $k \geq 2$.

In the graph $DS(P_n)$ we have $n+2$ vertices of which 2 vertices are of degree 2 and $n-2$ of degree 3 and one of degree $n - 2$ and one of degree 2. Starting the forcing process with a vertex of degree 2 can lead to forcing of entire graph black.

A r-regular graph: a graph of order n, where all the n vertices have degree r. In a regular graph $G$ $\delta(G) = \Delta(G) = r$.

**Theorem 3.7.** Let $G$ be a $r$-regular graph. Then, $F_k(DS(G)) \geq r + 2 - k$ where $k \geq 2$.

**Proof.** Assume that $G$ is a regular graph. Let the order of $G$ be $n$. All the vertices in $G$ have degree $r$, hence there will be one vertex adjacent to all the $n$ vertices of $G$ in $DS(G)$.

**Claim:** $k$-forcing number of regular graph $G$ is greater than $r + 1 - k$.
Consider any arbitrary vertex $v$ in $G$ to be initially black, degree of $v$ is r. Hence it requires $r - k$ of its neighbour to be black. This shows that $F_k(G) \geq r + 1 - k$. Clearly from the Lemma 3.5 we can say that $F_k(DS(G))$ is at least one more than $F_k(G)$. Therefore the result follows. ■

**Observation 3.8.** For every regular graph $G$ there exist some value $k$ such that the above result is sharp. For example consider the cycle $C_n$, which is a 2 regular graph. When $k = 2$, $F_2(DS(C_n)) = r + 2 - k = 2 + 2 - 2 = 2$ (Fig. 2).

**Theorem 3.9.** For a $r$-regular graph $G$, if $k \geq r - 1$, then $F_k(DS(G)) \leq 3$.

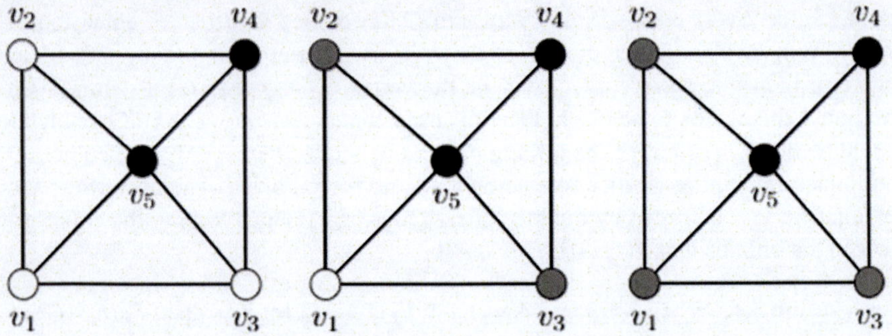


**Fig. 2** Steps involved in 2 forcing of DS($C_4$)

**Proof.** Clearly from Proposition 2.3 we know that for $k \geq r$, $F_k(G) = 1$ and from Proposition 2.1 we can see that for $k = r - 1$, $F_k(G) \leq 2$. From Lemma 3.5 we can conclude the above result. ■

Next we show that the above bound is sharp.

**Corollary 3.10.** Let $P(n, m)$ be the Generalised Petersen graph. Then, the $k$-forcing number of $P(n, m)$ is stated below.

$$F_k(DS(P(n, m))) = \begin{cases} 3 \; k = 2 \\ 2 \; k = 3 \\ 1 \; k > 3 \end{cases}$$

**Proof.** Generalised Petersen graph is 3-regular. From the above Theorem 3.9 we can say that $F_k(DS(P(n, m))) \leq 3$ when $k \geq 3 - 1$.

**Case 1.** When $k = 2$, since $P(n, m)$ is a 3-regular graph $DS(P(n, m))$ has minimum degree 4. Hence initially a vertex $v$ of degree 4 and two of its neighbour has to be colored black. After which any black vertex will have at most 2 white vertices hence forcing the entire graph black with a total of 3 initially colored black vertices.

**Case 2.** When $k = 3$ this follows immediately from the Proposition 2.3 and Lemma 3.5.

**Case 3.** When $k > 3$, or $k \geq \delta(DS(G)) = 4$. Exactly one vertex will be of degree $n$ say $u$ and all the other vertices will be of degree 4 in $DS(G)$. Here by taking a vertex of degree 4 to be initially black can force all its 4 neighbours including $u$ to be black. After which any black vertex except $u$ will have at most 3 white neighbour. Using which we can force the entire neighbour black. ■

Turán *graph* $T_{n,t}$ is the complete r partite graph with n vertices whose partite set differ in size by at most 1. According to pigeonhole principle all the partite set will have either $\lfloor \frac{n}{t} \rfloor$ or $\lceil \frac{n}{t} \rceil$ as their cardinality. There are b partite sets having cardinality $\lceil \frac{n}{t} \rceil$ and $n - b$ partite sets having cardinality $\lfloor \frac{n}{t} \rfloor$.

**Theorem 3.11.** Let $T_{n,t}$ be the Turán's graph, then $F_k(T_{n,t}) \leq F_k(DS(T_{n,t})) \leq F_k(T_{n,t}) + 1$.

**Proof.** The degree splitting graph is obtained by adding few vertices to the original graph $G$. In course of which degree of few of the vertices of $G$ might increase in $DS(T_{n,t})$. Hence $F_k(DS(T_{n,t}))$ can never be less than $F_k(T_{n,t})$. Hence the first part of the proof is evident.

In a Turán's graph cardinality of any two partite set differ by at most one meaning there are at most 2 different degrees in the degree sequence of the graph. So in the degree splitting graph of Turán's graph there will be at most 2 newly added vertices.

**Case 1.** When only one vertex is added to the graph $G$ to get $DS(T_{n,t})$.

If the added vertex is adjacent to initially forcing vertex and its degree increases more than $k$. Then we need to consider this vertex to be black. There by $F_k(DS(T_{n,t})) = F_k(T_{n,t}) + 1$. If the added vertex doesn't increase the degree of the initial forcing vertex more than $k$, or if it not adjacent to the initial forcing vertex then $k$-forcing number of such a degree splitting graph will be same as the $k$-forcing number of the original graph.

**Case 2.** When 2 vertices are added to the graph $G$ to get $DS(T_{n,t})$.

i) When the added vertex doesn't increase the degree of any vertex in the graph $DS(T_{n,t})$ beyond $k$, then $F_k(DS(T_{n,t})) = F_k(T_{n,t})$.
ii) When the added vertex increases the degree of some or all the vertex beyond$k$. Let $v$ be the vertex which was taken initial black in order to obtain $F_k(T_{n,t})$ and let $u_1$ be the vertex added in $DS(T_{n,t})$ corresponding to its degree and $u_2$ be the other vertex which is added in $DS(T_{n,t})$. Then by taking $u_2$ as black, all the vertices in other partite set are forced black using the same approach as $F_k(T_{n,t})$, the remaining vertices in the partite set of $v$ can be forced by the vertices which have same degree as $v$ and the remaining one white vertex $u_1$ can be forced by any one of its adjacent vertex as all the adjacent vertex are black. Suppose there are no partite set of degree $v$. Then take the vertex $u_2$ instead of $v$ as the initially colored black vertex. This step resembles the initial step of $k$-forcing number of $T_{n,t}$ by employing color changing rule we can make all the vertex adjacent to $u_2$ as either forced black or initially colored black. Now we need to force the vertices in the partite set of $v$, we can observe that this process is similar to the forcing patter of $T_{n,t}$ except for the fact that $v$ is white here where as in the forcing process of $G$ it was black. If the cardinality of the partite set containing $v$ is less than $k$ then we are done. If not, by taking $v$ to be black we can perform the same process as that of $k$- forcing in $T_{n,t}$. Finally $u_1$ can be forced by $v$. There by proving that $F_k(T_{n,t}) \leq F_k(DS(T_{n,t})) \leq F_k(T_{n,t}) + 1$. ■

**Theorem 3.12.** Let $T_{n,t}$ be a Turán's graph with $t-b = 1$ and $\lfloor \frac{n}{t} \rfloor$, then $F_k(DS(T_{n,t})) = 2t - (k+1)$.

**Proof.** In the Turán's graph $G$, t is the number of partite set and since $t - b = 1$ meaning there is exactly one set with $\lfloor \frac{n}{t} \rfloor = 1$ vertex. Now degree splitting graph of such a graph becomes $2(t-1)$-regular graph. According to the Theorem 3.7 $F_k(DS(T_{n,t})) \geq 2(t-1) + 2 - k = 2t - (k+1)$.

Now it is enough if we can show that $F_k(DS(T_{n,t})) \leq 2t - (k+1)$. Since $DS(T_{n,t})$ is a $2(t-1)$-regular graph, let us consider any vertex $v$ to be initially colored black. In order for this vertex to force its white neighbour, $2(t-1) - k$ of its neighbours have to be initially colored black. After which the vertex $v$ is left with exactly $k$ white vertices. Now in the entire graph we can see that there is exactly one white vertex left which can be forced by any of its black neighbour. Thereby forcing the entire graph black with $1 + 2(t-1) - k = 2t - (k+1)$ black vertices. Hence $F_k(DS(T_{n,t})) \leq 2t - (k+1)$. Hence the proof. ■

The Cartesian product $G \Box H$ of two graphs G and H is the graph with vertex set equal to the Cartesian product $V(G)\ X\ V(H)$ and where two vertices $(g_1, h_1)$ and $(g_2, h_2)$ are adjacent in $G \Box H$ if and only if $g_1 = g_2$ and $h_1$ is adjacent to $h_2$ or $h_1 = h_2$ and $g_1$ is adjacent to $g_2$.

**Theorem 3.13.** Let $P_n$ be a path and $C_n$ be a cycle of order n. Then,

i) For $n, m > 2$ $F_k(DS(P_n \Box P_m)) = \begin{cases} 3 & k = 2 \\ 1 & k > 2 \end{cases}$

ii) $F_k(DS(P_n \Box C_m)) = \begin{cases} 3 & k = 2 \\ 2 & k = 3 \\ 1 & k > 3 \end{cases}$

**Proof.** i) $P_n \Box P_m$ is a grid graph with corner vertices having degree 2, inner vertices with degree 4 and outer vertices except corner vertices having degree 3.

**Case 1.** Let us first consider when k = 2. Subcase 1: When at least one of $n, m$ is greater than 3. Then 3 extra vertices say $u_1, u_2, u_3$ will be added in the degree splitting graph of $P_n \Box P_m$ and made adjacent to all the vertices of degree 2, 3, 4 respectively.

1) Consider a vertex of degree 3 in the $DS(P_n \Box P_m)$ say $v_{1,1}$ to be initially black. Now it need at least one of the 3 white neighbour to be black in order to continue the process. Suppose $u_1$ is taken to be black then, the remaining two white neighbours $v_{2,1}, v_{1,2}$ which are of degree 4 will be forced black. At this stage any black vertex $u_1, v_{2,1}, v_{1,2}$ other than $v_{1,1}$ will have 3 white neighbours. consider $u_1$ then by taking one of the corner vertex as black remaining corner vertex will be forced black. But if *nandm* are both more than 3, then we with 3 initial black vertices we cannot force the entire graph black. Suppose we take $v_{1,2}$ or $v_{2,1}$ we need to take one of its 3 white neighbours say, $u_2, v_{3,1}, v_{2,2}$ or $u_2, v_{1,3}, v_{2,2}$ black. In both the cases after thw forcing the vertex with degree 5 vertex $(v_{2,2})$ and $u_2$ will be black. Also $v_{3,1}$ or $v_{1,3}$ now has at most 2 white neighbours hence they can force their neighbours. Hence $v_{2,2}$ will have exactly 2 white neighbours. After this forcing any black vertex will now have at most 2 white vertices. Hence we need at least 3 black vertices.

2) Consider a vertex of degree 4 in the $DS(P_n \Box P_m)$ say $v_{1,2}$ to be initially black. In the very first step itself we can see that we need to take 2 of white neighbour as initially black. Hence we need at least 3 black vertices. Now since this $v_2$ forces vertex of degree 3 and degree 4 both of them have exactly 2 white neighbours. The process continues till the entire graph is forced black.

3) Consider a vertex of degree 5 in the $DS(P_n \Box P_m)$ say $v_i$ to be initially black. Then we need 3 of its vertices to be black. But this forcing will not give us minimum as by 1) or 2) we can force the entire graph black with 3 black vertices.

4) By taking $u_1$ *or* $u_2$ *or* $u_3$ as initially black we will get $F_k(DS(G))$ to be at least 3 as degree of each of them is atSleast 4. Also all the possible forcing is writen in 4 cases and it is clear that with 2 vertices it is not possible to force the entire graph black.

**Case 2.** Let us consider when k = 2. When $n, m = 3$. Then there will be 2 extra vertices say $u_1, u_2$, are added in the degree splitting graph of $P_n \square P_m$ and made adjacent to all the vertices of degree 2, 3 respectively.

1) Consider a vertex of degree 3 in the $DS(P_n \square P_m)$ say $v_1$ to be initially black: Follows from the above proof.
2) Consider a vertex of degree 4 adjacent to $u_2$ in the $DS(P_n \square P_m)$ say $v_2$ to be initially black. Follows from the above proof.
3) Consider a vertex of degree 4 not adjacent to $u_2$ in the $DS(P_n \square P_m)$ say $v_i$ to be initially black. Then 2 of its neighbours should be consider black so that the vertex $v_i$ forces its white neighbour. Once $v_i$ forces its neighbour. All the degree 4 becomes black. But they have 3 white neighbours hence we need at least 4 but since its not minimum we can adapt 1) or 2). Also with 2 black vertices we will not be able to force the entire graph black.

**Case 3.** When $k > 2$. By Propositions 2.2 and 2.3 we can see that $F_k(DS(P_n \square P_m)) \geq 1$. Now we need to show the reverse inequality.

Let us consider one of the corner vertex whose degree is 3. since $k \geq 3$. this vertex can easily force its white neighbours black. At this stage any black vertex will have exactly 3 white neighbours this process continues till the entire graph is forced black. Hence $F_k(DS(P_n \square P_m)) \leq 1$. Hence the proof.

ii) $P_n \square C_m$ has two different degrees and the degree splitting graph will have 2 vertices $u_1, u_2$ added and made adjacent to all the vertices having degree 3, 4 respectively.

**Case 1.** When k = 2. This is similar to the above argument for $DS(P_n \square P_m)$ when k = 2. except for the fact that all the corner vertices are of degree 3 in $G$.

**Case 2.** When $k = 3$, According to Proposition 2.2. $F_k(DS(P_n \square c_m)) \geq 2$.

Now consider a vertex $v_1$ of degree 4 in $DS(P_n \square P_m)$ to be initially black. Now by taking 1 of its white vertex black $v_1$ can force remaining vertices black. After which any black vertex will have at most 3 white neighbour. Hence we need at least 2 black vertices to force the entire graph black.

**Case 3.** When $k > 3$, the results follows from the Proposition 2.3. ■

## 4 Conclusion and Scope

This manuscript deals with the problem of finding the $k$-forcing number of degree splitting graphs of few classes of graphs such as regular graph, generalised petersen graph, Turán's graph, $P_n \square P_m$, $P_n \square C_m$.

Characterization of graph classes for which $F_k(DS(G)) = F_k(G)$ is an open problem. The scope of the $k$-forcing number of graphs extends to studies on disease spreading patterns and their control, analysing the logical circuits and studies on social connectivity networks.

**Acknowledgements.** We, the authors, expressing our sincere gratitude towards all reviewers for their valuable time and prestigious comments.

## References

1. West DB (2001) Introduction to graph theory. 2nd edn. Prentice Hall of India, New Delhi
2. Amos D, Caro Y, Davila R, Pepper R (2015) Upper bounds on the k-forcing number of a graph. Discret Appl Math 181:1–10
3. Charles D (2019) Zero forcing number of degree splitting graphs and complete degree splitting graphs. Acta Univ Sapientiae Math 11(1):40–53
4. Yan Z, Chen L, Li H (2019) On tight bounds for the k-forcing number of a graph. Bull Malays Math Sci Soc 42(2): 743–749 (2019)
5. Raja P, Somasundaram S (2004) On the degree splitting graph of a graph. Nat Acad Sci Lett 27(7–8):275–278
6. Yair C, Pepper R (2014) Dynamic approach to k-forcing. arXiv preprint arXiv:1405.7573
7. Burgarth D, Giovannetti V (2007) Full control by locally induced relaxation. Phys Rev Lett 99(10):100501
8. Burgarth D, Giovannetti V, Hogben L, Severini S, Young M (2015) Logic circuits from zero forcing. Nat Comput 14(3):485–490
9. Kim IJ et al (2014) Network analysis foractive and passive propagation models. Networks 63(2):160–169

# Document Classification for Recommender Systems Using Graph Convolutional Networks

Akhil M. Nair(✉) and Jossy George

Christ University, Bangalore, India
akhil.nair@christuniversity.in

**Abstract.** Graph based recommender systems have time and time again proven their efficacy in the recommendation of scientific articles. But it is not without its challenges, one of the major ones being that these models consider the network for recommending while the class and domain of the article go unnoticed. The networks that embed the metadata and the network have highly scalable issues. Hence the identification of an architecture that is scalable and which operates directly on the graph structure is crucial to its amelioration. This study analyses the accuracy and efficiency of the Graph Convolutional Networks (GCN) on Cora Dataset in classifying the articles based on the citations and class of the article. It aims to show that GCN based networks provide a remarkable accuracy in classifying the articles.

**Keywords:** Convolutional graph neural networks · Recommender systems · Cora dataset · Classification

## 1 Introduction

The world revolves around the importance of data and finding out ways to manage it in a wise and organized way. Data has been quoted as the oil of the new era. The amount of data being created in a day is enormous and thereby leading to a problem of information overload. Information overload is a term used by the domain experts to explain a situation when the resources available to organize the data is less than the amount of data being generated. Hence a technology termed as Information Filtering System is used to retrieve a piece of information from the mass overload of data produced. One of the most commonly used systems for retrieving the required data is the Recommender Systems, a subclass of Information Filtration Systems. Research scholars find a difficult time in retrieving relevant over prestigious academic articles for their studies. The study of citation recommender systems is not new. It has gained the attention of the research community in the past decade and has evolved to a state-of-the-art system.

The process of literature review is the task that takes in 70% of the time in a study. The reason for such time consumption is the vast availability of research articles and resources to search for. The advancements in the area of machine learning and deep learning has brought in a wide range of evolution in the domain of Recommender Systems. The area of citation recommender system has been a major topic of research. These recommender


S. Shukla et al. (eds.), *Data Science and Security*, Lecture Notes in Networks and Systems 290,
https://doi.org/10.1007/978-981-16-4486-3_45

systems predict the preference of the authors and other research scholars based on their past readings. The research community has been eager to understand and utilize the unstructured data whereas the structured data has not been given required attention. Most of the data available currently are structured in the form of graphs and networks with nodes and networks. Some of the major real-world datasets occur in the form of graphs or networks: social networks, knowledge graphs, protein-interaction networks, the World Wide Web (WWW), etc.

The research community has a lot of importance for the structured data like the paper citations, specifically in-citations and out-citations, author-venue of publishing relation, bibliographic networks etc. Yet very little attention has been given in the utilization of such networks. The graph-based recommender system yields better efficiency on the feedback-based datasets like 10 M movie dataset, FM dataset etc. The major challenge in imparting the graph-based recommenders in the field of citations is the unavailability of implicit and explicit feedback. Also, the content citation and bibliographic datasets bare is too high and is not scalable [1, 2].

Recently the concept of graph convolutional network was introduced by Kipf, Thomas N. and Welling. Max [3]. Graph convolutional network classification is a method to classify documents based on the citation network and the feature associated with the document. The proposed method analyzes the classification accuracy of the Graph Convolutional Networks (GCN) on the Cora Dataset and understands the challenges and drawbacks of the GCN. The compatibility of the existing model with the bibliographic dataset is one of the major concerns.

Section 1 includes brief introduction of the overall study, Sect. 2 contains related research work of other authors, Sect. 3 contains the framework of GCN, Sect. 4 focuses on Experimental analysis, Sect. 5 contains Results and Discussions and finally, Sect. 6 concludes the study.

## 2 Related Work

To generalize the architecture of Convolutional Neural Networks (CNN) and Recurrent Neural Networks (RNN) to a structured graph is a challenging task. So the research community tried finding out some problem-specific architectures [4–6] Michaël Defferrard et al. [7] have brought up the usage of GCN from a lower grid, like image or audio representation to an advanced grid of social networks and knowledge graphs. They formulated CNN in the format of a spectral graph theory keeping the same linear computational complexity where the scalability was challenged. A scalable approach to semi supervised classification on structured data was introduced by Thomas N. Kipf, Max Welling [3]. The efficiency of this classification technique is based on an efficient variant of neural networks that directly operates on the graph structure. This method was implemented on several structured data like citation networks and knowledge graphs and it has outperformed the other state of the art methods with a significant margin. One of the major advantages of this method is the scalability of its computation on large data. Embedding the learnings from the meta path to assign the weights to the edges on its own was developed by Fengli Xu et al. [8]. The model was called the RElation-aware CO-attentive GCN (RecoGCN) that cover up the limitations of the GCN of modelling the heterogenous relations and identifying and including the distant neighbours for the predictions.

Yu Zheng et al. [9] modelled a transitive relationship between user-item interactions and item-price interaction which bridges the link between price on user with the items so as to make the user representation price aware. Further analysis on the data proved that the awareness of price of the items is useful for making user preferences. Bowen Jin et al. [10] constructed a unified graph in order to represent a multi-behavior data. This model is equally usable in cases of the cold-start problems. A recommendation model with BERT and GCN for context-aware citation was then developed motivated from the GCN networks. This model integrates a document encoder and a context encoder [11]. Bidirectional encoders from transformers and GCN layers are used with a pretrained model. This model generates a significant accuracy as compared to the state-of the art graph-based recommender systems. Text classification from the documents remained a challenge while using GCN which was solved with the Heterogenous Graph Convolutional Networks (HeteGCN) by Rahul Ragesh et al. [12]. The shortcomings of the existing methods like scalability and prediction accuracy were overcome by integrating the best aspects of predictive text embedding (PTE) and TextGCN.

## 3 GCN Framework

There is a universal architecture for any graph based neural network. These architectures can be termed Graph Convolutional Networks (GCNs). The major difference between CNNs and GCNs are that the CNN are built for Euclidean structured data and GCN for irregular or unstructured data. The major objective of these structures is to learn a function of a feature on a graph represented by $G = (V, E)$ where the V and E stands for vertices and edges respectively. This graph structure takes in the inputs as:

- A feature matrix represented by $X = N X D$ (where N is the number of nodes and D is the number of features).
- A graph structure represented with an adjacency matrix A.

The GCN model has its advantages and its limitations as well. Figure 1 shows the architecture of the GCN based model for semi-supervised classification and Fig. 2 shows the hidden layers of the model. One of the major limitations is the multiplication of the structure with the adjacency matrix A. The feature vectors of all the neighboring nodes are summed up except for the node itself.

The challenge of adjacency multiplication is overcome by enforcing an identity matrix with A for self-loops. Normalization of the adjacency matrix A such that summation of all the rows equals to 1 i.e., $D^{-1} X A$, where D represents diagonal node degree matrix deals with the second limitation of adjacency matrix A not being normalized.

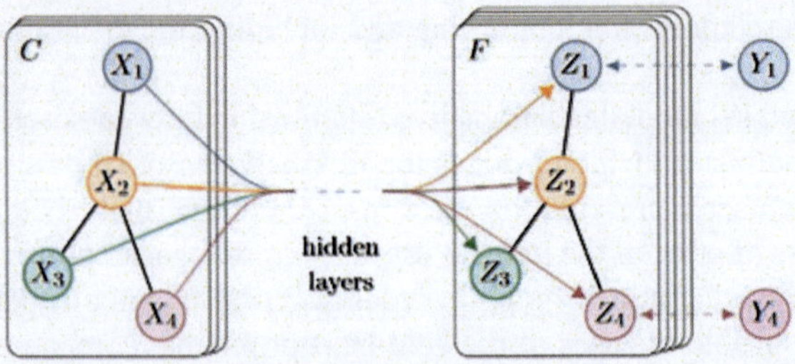


**Fig. 1** Graph convolutional network layers

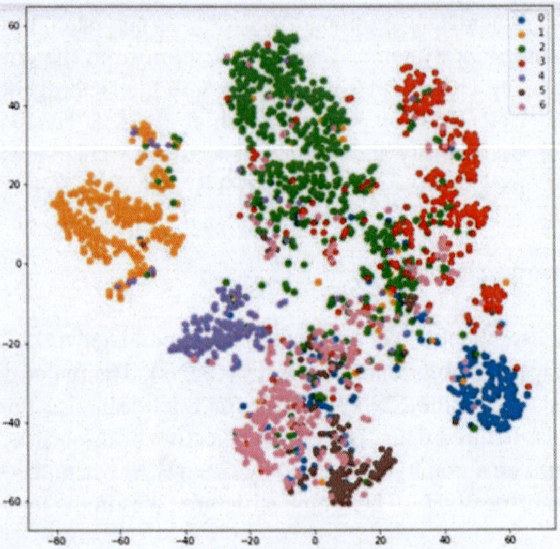


**Fig. 2** Hidden layers

# 4 Experimental Analysis

The study has followed the methodology as shown in Fig. 3. Dataset used for the experimental analysis is the Cora Dataset with citation graphs and features. The dataset contains 2708 nodes. Each node in the citation graph is a technical paper. The node features are represented by a bag-of-words with a binary value of "0" or "1" which represents the presence of the word in the line.

The dataset for the analysis is taken as undirected graph that shows if a document cites another document or vice-versa. Once the data is loaded a node matrix X is generated along with a list of tuples with the adjacent nodes. Later an adjacency matric A is created from the list of tuples with adjacent nodes. The node features contains 1433 words as shown in Fig. 4.

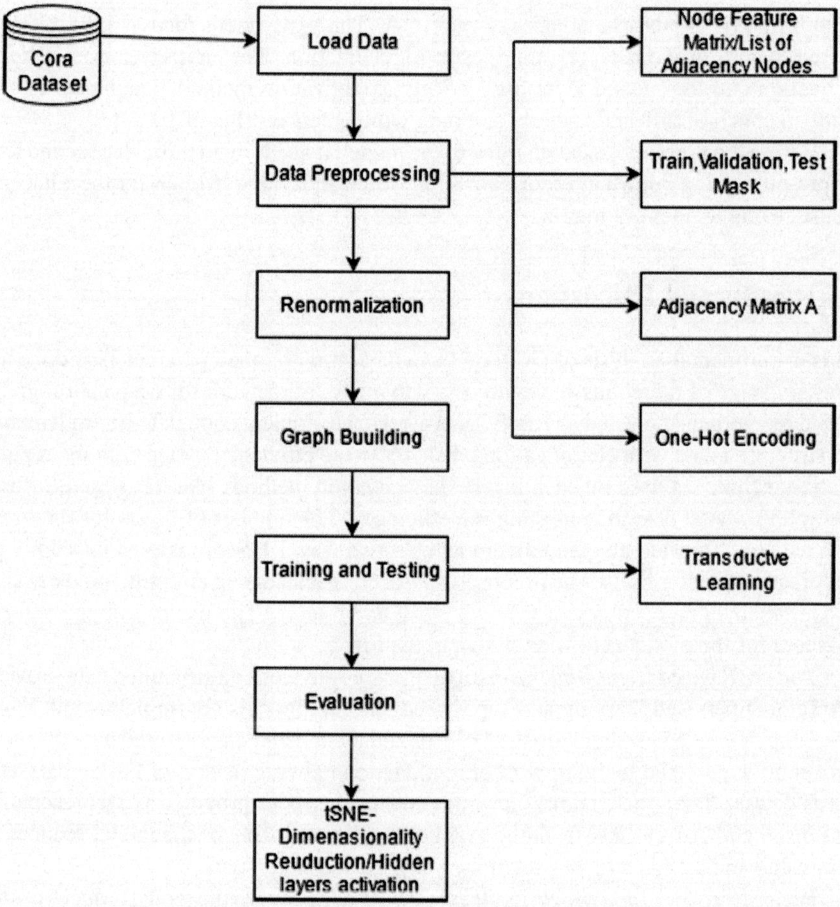


**Fig. 3** Workflow of the GCN analysis

```
X shape:  (2708, 1433)

Number of nodes (N):  2708

Number of features (F) of each node:  1433

Categories:  {'Rule_Learning', 'Theory', 'Genetic_Algorithms', 'Neural_Networks', 'Reinforcement_Learning', 'Case_Bas
ed', 'Probabilistic_Methods'}

Number of classes:  7
```

**Fig. 4** Node feature and category description

During the preprocessing phase, a test, train and a validation mask is generated so that the respective data is available to the corresponding phase. The next step of the preprocessing is to generate an adjacency matrix A. This adjacency matrix A is used to initialize a graph with nodes and edges. The graph G has 2708 and 5728 edges with an average degree of 3.89. The labels for the categories are encoded to numerical categories from 0 to 7 using one-hot encoding. The model will take in 2 inputs namely the node feature matrix N and the adjacency matrix A. The experiment focuses on a GCN of 2 layers with dropout layers and elastic regularization. The first layer uses a ReLU activation and the second layer uses a Softmax activation method. The model is built with an epoch of 200 and an early stopping with patience value of 10.

Before the preprocessed data is fed to the model, it needs to go through a second level of pre-processing known as renormalization which adds the self-loops for the adjacency matrix using an identity matrix.

## 5 Results and Discussions

This experimental analysis of GCN on Cora dataset with tuned parameters is done as a primary stage of recreating the entire work to media framework for deep neural graph-based recommender systems. The GCN proved to be efficient enough to be implemented on structured data with classification labels. Also the efficiency varied with the regularization techniques used on each layer. The activation methods like ReLU and Softmax also plays a vital role in increasing the efficiency. The number of layers for the model is fixed at 2. To evaluate the model in an effective way, F1-Score is used instead of the accuracy and loss metrics.The proposed approach is scalable on any structured data.The model attained an accuracy of 0.75 i.e. 75% on f1-score. The macro average value of f1-score for the model is 0.74 as shown in the Fig. 5.

The GCN based framework has a lot of efficiency to work on structured data provided the semi-supervised learning and supervised learning models are implemented. Future work may include working with the GCN on unlabelled and unstructured data like AAN or a semi-supervised learning with only a citation network as that of DBLP dataset or AAN dataset. The epoch training accuracy and loss has been provided as a representation in Figs. 6 and 7. It is clear from the graph that with each epochs the model reduces the loss value and attains a higher accuracy value on validation data.

Hence from the graph above, its clear that with each epochs the model reduces the loss value and attains a higher accuracy value on validation data. The GCN model increases the accuracy with training by not only incorporating the interconnection between the nodes but also based on the node features.

GCN Classification Report:

| | precision | recall | f1-score | support |
|---|---|---|---|---|
| Case_Based | 0.75 | 0.69 | 0.72 | 114 |
| Genetic_Algorithms | 0.90 | 0.85 | 0.87 | 156 |
| Neural_Networks | 0.82 | 0.70 | 0.76 | 290 |
| Probabilistic_Methods | 0.80 | 0.76 | 0.78 | 172 |
| Reinforcement_Learning | 0.70 | 0.82 | 0.76 | 85 |
| Rule_Learning | 0.55 | 0.77 | 0.64 | 60 |
| Theory | 0.56 | 0.70 | 0.62 | 123 |
| accuracy | | | 0.75 | 1000 |
| macro avg | 0.73 | 0.76 | 0.74 | 1000 |
| weighted avg | 0.76 | 0.75 | 0.75 | 1000 |

**Fig. 5** GCN classification report

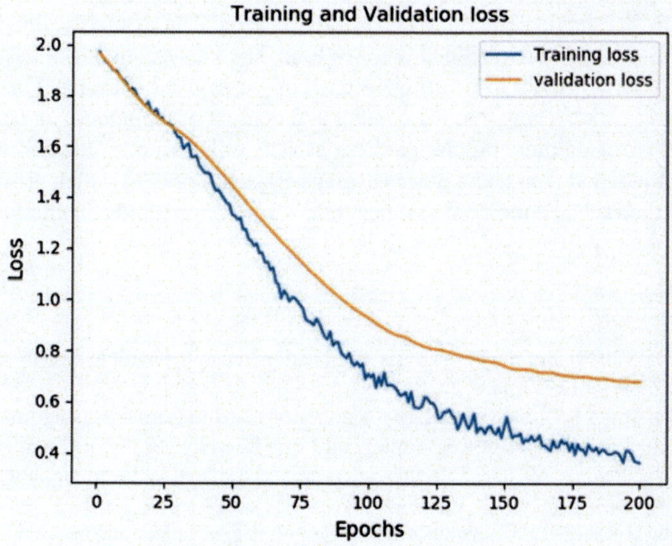


**Fig. 6** Training a validation loss

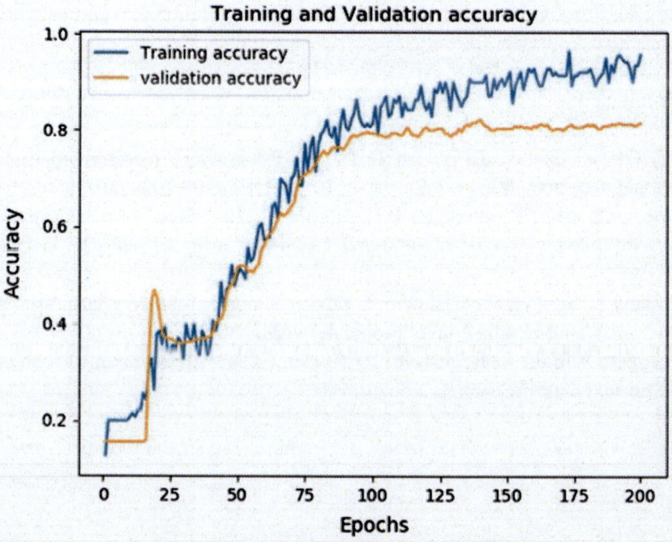


**Fig. 7** Training and validation accuracy

# 6 Conclusion

There have been a numerous amount of advancements in the field of recommender systems especially citation recommenders. The citations mostly depend on the networks and relations of each author and paper. The major challenge with such data is that sometimes the network is enough to tell the relations between two nodes or papers.

GCN directly acts on the structured and semi-structured data to analyze the relationship between each node with embedded features to it. The analysis on GCN based network on Cora dataset is proven to be efficient on semi-supervised learning. The model can be fine tuned in such a way that the content and context of the nodes can be embedded as features to provide more precise predictions. The trained model attained an accuracy of 75% on a labelled data and a macro average F1-score of 0.74 which proves to be an efficient method when compared to other state-of-the-art methods on citation network.

## References

1. George JP (2020) Similarity analysis for citation recommendation system using binary encoded data, pp 12–13, June 2020
2. Nair AM, Wagh RS (2018) Similarity analysis of court judgements using association rule mining on case citation data-a case study. Int J Eng Res Technol 11(3):373–381
3. Kipf TN, Welling M (2017) Semi-supervised classification with graph convolutional networks, arxiv
4. Duvenaud D et al (2015) Convolutional networks on graphs for learning molecular fingerprints
5. Li Y, Zemel R, Brockschmidt M, Tarlow D (2016) Gated graph sequence neural networks
6. Jain A, Zamir AR, Savarese S, Saxena A (2016) Structural-RNN: deep learning on spatio-temporal graphs. https://doi.org/10.1109/CVPR.2016.573
7. Defferrard M, Bresson X, Vandergheynst P (2016) Convolutional neural networks on graphs with fast localized spectral filtering
8. Xu F, Lian J, Han Z, Li Y, Xu Y, Xie X (2019) Relation-aware graph convolutional networks for agent-initiated social e-commerce recommendation. https://doi.org/10.1145/3357384.3357924
9. Zheng Y, Gao C, He X, Li Y, Jin D (2020) Price-aware recommendation with graph convolutional networks. https://doi.org/10.1109/ICDE48307.2020.00019
10. Ying R, He R, Chen K, Eksombatchai P, Hamilton WL, Leskovec J (2018) Graph convolutional neural networks for web-scale recommender systems. https://doi.org/10.1145/3219819.3219890
11. Jeong C, Jang S, Shin H, Park E, Choi S (2019) A context-aware citation recommendation model with BERT and graph convolutional networks. http://arxiv.org/abs/1903.06464
12. Iyer A, Lingam V, Bairi R, Ragesh R (2020) HeteGCN: heterogeneous graph convolutional networks for text classification, arXiv

# A Study on the Influence of Geometric Transformations on Image Classification: A Case Study

Shashikumar D. Nellisara[1](✉), Jyotirmoy Dutta[2], and Sibu Cyriac[2]

[1] Center for Advanced Research and Training, CHRIST (Deemed To Be University), Bangalore, India
shashikumar.nd@christuniversity.in

[2] Center for Digital Innovation, CHRIST (Deemed To Be University), Bangalore, India

**Abstract.** The present research work involves the study of the geometrical transformations which influences the training and validation accuracies of machine learning models. For the study, rice plant leaf disease dataset of 2096 images consisting of 4 classes with 523 images per class were used. The dataset subjected to 24 models out of which three models namely - DenseNet201, Densenet169 and InceptionResNetV2 are selected based on highest training accuracy and less difference between training and validation accuracy. To evaluate the performance of the selected three models, loss functions and accuracies have been computed.

**Keywords:** Augmentation · Rice plant leaf disease · DenseNet · InceptionResNet

## 1 Introduction

Deep Convolutional Neural Network (CNN) have been found to perform good for classification of images. However, the performance of the networks is dependent on the data size as well, higher the data lesser the error. Practically getting valid big data in medical and agricultural related sectors [1–3] is challenging and with less data overfitting is an issue [4]. Overfitting issue gave wide scope for the researchers to develop new techniques and, image augmentation is one of the solutions [5]. There are many augmentation techniques of which some are geometric transformations, color space transformations, kernel filters, mixing images, random erasing, feature space augmentation, adversarial training, GAN-based augmentation, neural style transfer, and meta-learning schemes [6].

Geometric transformation is a traditional technique which is commonly used to increase the number of samples for training the deep neural models, to balance the size of datasets as well for their efficiency improvement and widely used as affine transformations for data augmentation but it is still the subject of research [7]. The current practice for data augmentation is to perform a combination of the different image transformations and colour modification. The focused affine transformations are rotation, scaling (zoom in/out) and shearing and their combinations.


S. Shukla et al. (eds.), *Data Science and Security*, Lecture Notes in Networks and Systems 290,
https://doi.org/10.1007/978-981-16-4486-3_46

Rice is the basic food crop and India is one of the world's largest producers of rice. Rice being a widely used grain of India, the researchers from multiple domains are trying to increase the yield with early detection of diseases. [8] The development in computer vision research gives way to promising area in increasing accuracy in detection of rice plant disease [1, 9–11]. Current study focusses on the impact of transformation in classification of image using neural networks with rice disease image dataset.

## 2 Methodology

The open-source dataset of rice plant leaf disease is used for the study with 2096 images [12]. The image dataset consists of 3 classes of rice leaf disease, i) Hispa, ii) Brown spot and iii) leaf blast along with healthy leaf images as a fourth class [9] as shown in the Fig. 1. With each group having 524 images, the images are split as 400 images (76.3%) for training and 124 images (23.6%) for validation.

**Fig. 1** Rice leaf images of BrownSpot, Hispa, Leaf blast and Healthy (Left to right)

The twenty-four KERAS applications are used as standard models for comparison to select the good performing models [13].

The image size set to 256 pixels by default and the background of all the images is set to automatically converted to white before training the model. Three types of transformations were implemented independently as well as the combinations.

1. Shear – deformation values of 0.1, 0.2 and 0.3 are used
2. Rotation – rotation was done for the values 30° and 45°
3. Zoom – zoom range of [0,2.5] is used

## 3 Results and Discussion

The rice leaf dataset with 2096 images is first tried to classify with twenty-four Keras applications which are deep learning models. The results (Table 1) showed that the majority of the models are overfitting for the selected dataset. Densenet201 and InceptionResNetV2 are considered to be promising models with high training and validation accuracy. DenseNet169 also considered for further investigation as the difference between training and validation accuracy is less compared to other models. The three above selected models were patterned with geometrical transformations and checked the effect on accuracy was checked.

**Table 1** Accuracy and loss for training and validation of Keras models

| Model | Training accuracy (in %) | Validation accuracy (in %) | Training loss | Validation loss |
|---|---|---|---|---|
| Xception | 75 | 50 | 0.46 | 1.58 |
| VGG16 | 96 | 64 | 0.12 | 1.58 |
| VGG19 | 24 | 25 | 1.38 | 1.39 |
| ResNet50 | 95 | 60 | 0.43 | 1.23 |
| ResNet101 | 70 | 60 | 0.76 | 1.22 |
| ResNet152 | 90 | 60 | 0.42 | 1.15 |
| ResNet50V2 | 99 | 60 | 0.13 | 1.53 |
| ResNet101V2 | 98 | 60 | 0.12 | 1.59 |
| ResNet152V2 | 75 | 55 | 0.46 | 1.38 |
| InceptionV3 | 94 | 63 | 0.12 | 1.51 |
| InceptionResNetV2 | 99 | 70 | 0.17 | 1.56 |
| MobileNet | 96 | 60 | 0.02 | 1.86 |
| MobileNetV2 | 98 | 50 | 0.04 | 2.11 |
| DenseNet121 | 75 | 60 | 0.43 | 1.23 |
| DenseNet169 | 93 | 73 | 0.12 | 1.12 |
| DenseNet201 | 99 | 70 | 0.13 | 1.18 |
| NASNetMobile | 98 | 55 | 0.14 | 1.55 |
| NASNetLarge | 98 | 68 | 0.12 | 1.25 |
| EfficientNetB0 | 45 | 25 | 1.13 | 1.63 |
| EfficientNetB1 | 45 | 30 | 1.15 | 1.45 |
| EfficientNetB2 | 46 | 30 | 1.16 | 1.45 |
| EfficientNetB3 | 50 | 25 | 1.04 | 1.76 |
| EfficientNetB4 | 45 | 25 | 1.15 | 1.59 |
| EfficientNetB5 | 50 | 30 | 1.09 | 1.51 |

The graph of Accuracy versus Epoch of DenseNet201 and Loss versus Epoch of InceptionResNetV2 are shown in Fig. 2. As It has been reported in the literature that model accuracy will increase after the geometrical augmentations. Conversely, current work showed with InceptionResNetV2 model performance reduced drastically from 99 to 24% (Table 2) indicating the adverse effect of rotation and shear transformation. The accuracy has not changed for the zoom transformation alone, but when it is combined with rotation, shear or both has reduced the accuracy by 24%.

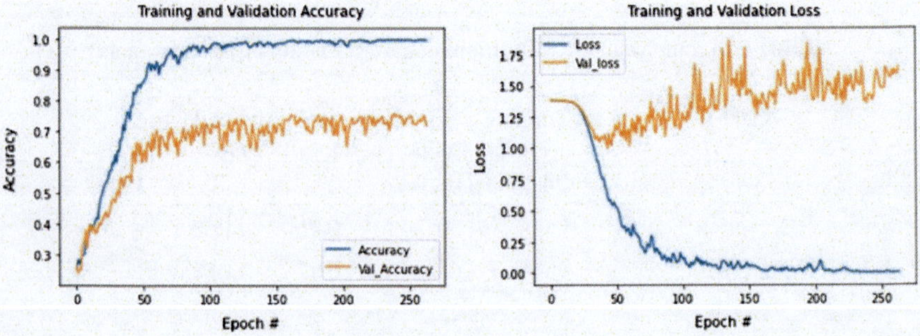


**Fig. 2** Accuracy graph of DenseNet201 (left side) and loss graph of InceptionResNetV2 (right side)

**Table 2** Sample augmented results with InceptionResNetV2 model

| Model | Training accuracy (in %) | Validation accuracy (in %) |
|---|---|---|
| Without augmentation | 99 | 70 |
| Rotation | 24.1 | 24.7 |
| Shear | 24.2 | 25.0 |
| Rotation + shear | 24.8 | 25.2 |

DenseNet169 model also follows the trend of InceptionResNetV2 model in performance, but the quantity of reduction in accuracy is less. DenseNet169 has shown an accuracy of 93% and 73% for training and validation (Table 3). The augmented models showed less than 75% and 63% for training and validation accuracies respectively. Same trend appeared for zoom transformation and has no influence in enhancing the accuracy.

**Table 3** Sample augmented results with DenseNet169 model

| Model | Training accuracy (in %) | Validation accuracy (in %) |
|---|---|---|
| Without augmentation | 93 | 73 |
| Rotation | 75 | 62 |
| Shear | 73 | 59 |
| Rotation + shear | 74 | 63 |

The performance of the DenseNet201 model remains almost the same and there is no influence of transformations unlike DenseNet169 and InceptionResNetV2 models. The base model without any transformations has 99% to 73% training and validation accuracy (Table 4). All the transformations (shear, zoom and rotation) and their combination failed to overcome the overfitting of the model for the rice leaf image dataset.

**Table 4** Sample augmented results with DenseNet201 model

| Model | Training accuracy (in %) | Validation accuracy (in %) |
|---|---|---|
| Without augmentation | 99 | 70 |
| Rotation | 99 | 73 |
| Shear | 99 | 76 |
| Rotation + shear | 97 | 75 |

## 4 Conclusion

The DenseNet201 model with shear transformation gave the good results with 99% training accuracy and 76% validation accuracy.

In general, the geometrical transformation should increase the performance of the model with the reduction of overfitting in terms of accuracy and loss. The training and validation accuracies has not changed with the used dataset for zoom transformation, but for shear and rotation transformations (with or without combinations) has decreased the performance indicating there is an adverse effect on the performance.

**Acknowledgements.** We herewith acknowledge Dr Guydeuk Yeon, Director, Innovation Center, Central Campus, Banglore, for his kind support and advise in pursuing this research work.

## References

1. Prakash K, Saravanamoorthi P, Sathishkumar R, Parimala M (2018) A study of image processing in agriculture. Int J Adv Netw Appl 09(01):3311–3315
2. Guo Y et al. (2020) Plant disease identification based on deep learning algorithm in smart farming. Discret Dyn Nat Soc 2020
3. Chen J, Chen J, Zhang D, Sun Y, Nanehkaran YA (2020) Using deep transfer learning for image-based plant disease identification. Comput Electron Agric 173:105393
4. Shorten C, Khoshgoftaar TM (2019) A survey on image data augmentation for deep learning. J. Big Data 6(1):1–48
5. Poojary R, Raina R, Mondal AK (2021) Effect of data-augmentation on fine-tuned CNN model performance. IAES Int J Artif Intell (IJ-AI) 10(1):84
6. Zheng Q, Yang M, Tian X, Jiang N, Wang D (2020) A full stage data augmentation method in deep convolutional neural network for natural image classification. Discrete Dyn Nat Soc 2020:1–11. https://doi.org/10.1155/2020/4706576

7. Mikołajczyk A, Grochowski M (2018) Data augmentation for improving deep learning in image classification problem. In: 2018 International Interdisciplinary PhD Work. IIPhDW 2018, pp 117–122
8. Review M, Mohiuddin K, Alam MM (2019) A short review on agriculture based on machine learning and image processing. Acta Sci Agric 3(5):55–59
9. Sarma S, Singh K, Singh A (2010) An expert system for diagnosis of diseases in rice plant. Int J Artif Intell 1(1):26–31
10. Kaur H (2019) Applications of machine learning in plant disease detection. (17):3100–3106
11. Prakash K, Saravanamoorthi P, Sathishkumar R, Parimala M (2017) A Study of image processing in agriculture 3315:3311–3315
12. Masood M.H, Saim H, Taj M, Awais MM (April 2020) Early Disease Diagnosis for Rice Crop, arXiv
13. Chollet F et al. (2015) Keras

# Asynchronous Method of Oracle: A Cost-Effective and Reliable Model for Cloud Migration Using Incremental Backups

Vaheedbasha Shaik(✉) and K. Natarajan

Christ (Deemed to be University), Bengaluru, Karnataka, India
vaheedbasha.shaik@res.christuniversity.in

**Abstract.** Cloud Computing has reached a new level in flexibility to provide infrastructure. The proper migration method should be chosen for better cost management and to avoid overpayments to unused resources. So, the migrations from On-Premises to cloud infrastructure is a challenge. The migration can be done in synchronous or asynchronous modes. The synchronous method is mostly used to minimize downtime while doing the cloud migrations. The asynchronous methods can do the migrations in offline mode and very consistently. This paper addresses various issues related to the synchronous mode of Oracle while doing highly transactional database migrations. The proposed methodology provides a solution with a combination of asynchronous and incremental backups for highly transactional databases. This proposed method will be a more cost-effective and reliable model without compromising consistency and integrity.

**Keywords:** Database cloud migrations · Efficient cloud migration · Short-way for migration · Oracle cloud migrations · Cloud migrations

## 1 Introduction

In recent times, cloud migrations from on-premises environments to the cloud has gone up tremendously. There are different models are available currently in the cloud as Infrastructure as a Service (IaaS), Software as a Service (SaaS), and Platform as a Service (PaaS) [1]. The Cloud users will prefer the combination of these services according to their requirements and business model. During the migration of databases, the right strategy model should be picked up. Otherwise, the bad strategy model might put overburden on cost and management [2]. The cloud providers also started the Database as a Service (DaaS) for their customer's comfort. However, the real challenges may arise when the requirement would be the migration of already existing databases from on-premises to cloud environment, especially in the scenarios of highly transactional databases and huge analytical databases in size. The highly transactional databases are mostly referred to as Online Transactional Processing (OLTP) databases, considering some analytical database batch jobs can generate more archive logs or transaction logs [3].

The OLTP databases could generate a very high count of transactional logs or archive logs. These transactional log sizes sometimes cross over more than the actual database


S. Shukla et al. (eds.), *Data Science and Security*, Lecture Notes in Networks and Systems 290,
https://doi.org/10.1007/978-981-16-4486-3_47

size within one hour. In such cases, migrating OLTP databases with less downtime from on-premises to the cloud environment is very difficult. Because these archive logs or transactional logs should transfer in a faster way to the cloud environment over the network.

Analytical databases usually very large in sizes and usually referred to as Online Analytical Processing (OLAP) databases [4]. Although the OLAP databases hold historical data and most of the operations via direct load, the block changes at the physical level will be very high [4]. So, keeping these changes on track will be difficult for the cloud environment migrations.

In previous research, the analysis was carried out to provide a secure and stable network to transfer such huge data along with subsequential changes of it [5]. From the application perspective, the functionality compatibilities and cost management have been shown as the main migration obstacle [6]. However, there is no research has been carried out on lowering the data quantity for data migration. The proposed model works to unlink the existing model as transferring the data as it is from source to cloud environment during the migration. Rather, the new method will decrease the quantity of data and speeds up the data migration process even on the default network transfer plan. In this paper, the default-free network plan speed was used to move the data from on-premises to the cloud environment and the speed of the network ranged from 750 KiB/s to 850KiB/s.

This paper has been narrated as follows. The second section explains the existing models and their implementation methods along with a depicted explanation. The third section explains the current issues in problem definitions. The fourth section explains the proposed framework and algorithm for implementation. The fifth section explains the function implementation of the proposed model with accurate results. The sixth section helps to understand the analysis of this paper from the perspective of cost-effectiveness and data volumes. The graphs show the benefits of the proposed implementation methods over existing implementations and the findings of the research. The last seventh section concludes this paper with suitable justifications which are derived from the accurate implementation methods.

## 2 Existing Model

The data encryption can be implemented for the secure transfer of data over the internet. From the security perspective, it is mandatory for configuring encryption. There are different types of migration models and methods are available for relational databases. In oracle databases, there some very frequently used methods are available.

### 2.1 Complete Downtime Model

In this approach, the database will be shut down to take the cold backup of it to restore it in the cloud environment [7]. The database will be in a state of complete shutdown till it will get restored to the cloud environment.

As shown in Fig. 1, The database cold backups of the On-Premises database will be stored in object storage via Virtual Private Network (VPN). Later, the database restore will begin on the oracle cloud environment. Here, the Total downtime duration for

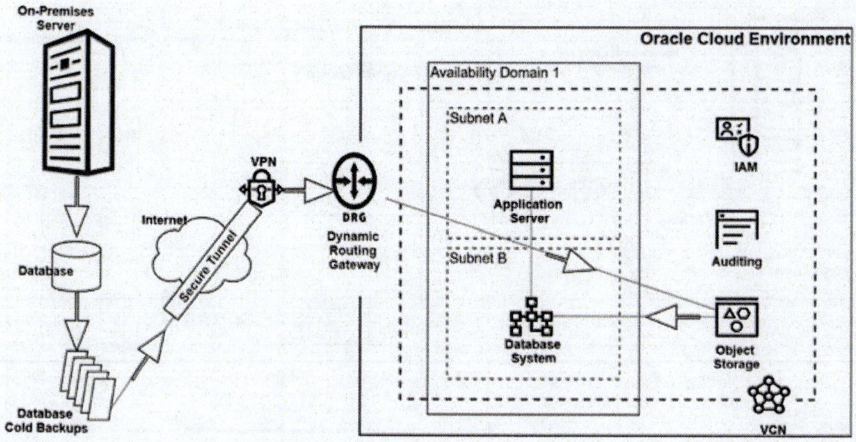


**Fig. 1** Complete downtime model

Database Migration (TDM) can be calculated as the addition of Cold Backup Duration (CBD), Backup Transfer to Object Storage (BTOBS), and Database Restore (DBR).

$$\text{TDM} = \text{CBD} + \text{BTOBS} + \text{DBR}. \tag{1}$$

Here, the Total downtime duration for Database Migration (TDM) can be calculated as the addition of Cold Backup Duration (CBD), Backup Transfer to Object Storage (BTOBS), and Database Restore (DBR).

### 2.2 Primary and Standby Model

In this approach, the primary database need not be shut down for hot backup [8]. The primary database will be in an up and running state while doing the restoration at the standby location.

As shown in Fig. 2, the archive or transaction log's secure shipping is required to get the standby in synchronization with primary [9]. Once they got to sync up, the switchover operation needs to be performed to switch the role of the primary database as standby and vice versa. The application connection route also should be pointed to the new primary database by updating the Connection String (CS). Here, the total downtime duration for database migration (TDM) can be calculated as the addition of the Duration of Switch over (DS) and Time for Application Connection String update (TACS).

$$\text{TDM} = \text{CBA} + \text{TACS}. \tag{2}$$

### 2.3 Bidirectional Data Update Model Using Goldengate.

The Goldengate is a tool from Oracle corporation to configure Active-Active (AA) database setup for bidirectional updates [10].

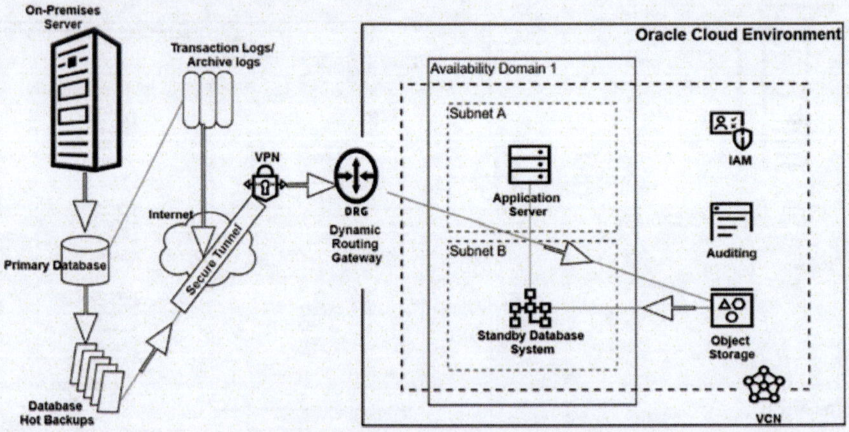


**Fig. 2** Primary and standby model

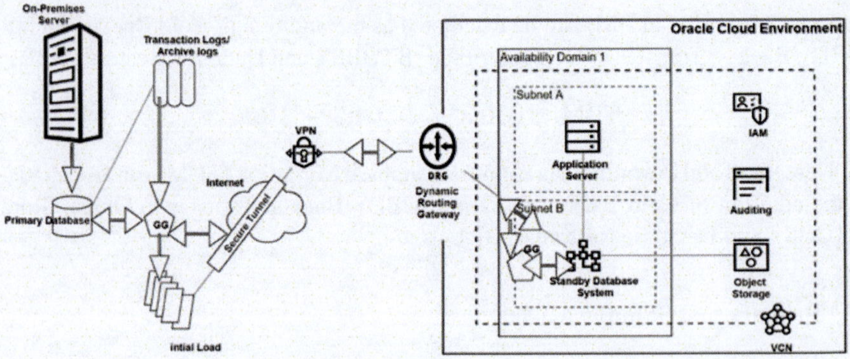


**Fig. 3** Goldengate model

As shown in Fig. 3, The On-Premises and Cloud databases can get the changes of each other via Goldengate (GG) tool. Likewise, two independent databases can get the sync of operations via a secure network. Here, the total downtime duration for database migration (TDM) is almost zero. Because the new connections will be directed to the new cloud database and Once all connections transactions are completed the On-Premises database will be shut down.

## 3 Problem Definition

These existing methods work effectively on small transactional databases. However, obstacles may arise in highly transactional database migrations. The mentioned top migration methods and the rest of them are only concentrating on downtime perspective. In real-time, the generation of archive log size of more than 1 Terabyte (TB) is common. To maintain the sync with primary, the network between should be capable to transfer such huge data over the network to the standby site. The same transfer rate must be maintained even in the AA scenarios as well.

### 3.1 Cost-Effective

The network requirement can cost more and overburden the companies while migrating to the cloud environment for high transactional databases. All most all cloud vendors providing default cost-free of network connection for data porting into the cloud [11, 12]. The proposed model is implemented with a range from 750 to 850 KiB/s speed of network where 150 MiB/s is required for existing migration processes.

### 3.2 The Volume of Data

In the existing model, the generated data must be transferred as it is to the standby site to apply the changes of primary data [7, 9, 10]. The proposed model will merge the techniques to decrease the volume of data movement over the network.

## 4 Proposed Methodology

The data quantity will a play key role to configure an efficient secondary/standby database server. To address these concerns the Incremental Backup Management System (IBMS) has been implemented. In all existing models, the data transfer must be done over the default network or on-demand network. In this paper,

- The research work is done to minimize the quantity of data transfer without compromising the integrity of the data.
- The default network which comes free of cost was used to transfer the data into the cloud. This model can be a significant design from a cost-effective point of view.

The Fig. 4 explains the proposed methodology namely the IBMS algorithm. The hot or online backup is to trigger the database backup when it up and running state. These types of backups will not affect the database availability. However, there will be small server resource consumption. As part of the methodology, A full online backup should be triggered at the primary database. The secondary backups can be restored upon successful completion of the backups in primary via object storage or shared file system.

On the day of cutover, the current state of primary and secondary databases needs to be verified. The least checkpoint SCN should be picked up from the secondary database. If any files are not present on the standby side, those need to be restored via primary service in the secondary database. Upon completion of missed datafiles restoration at the secondary database, the incremental backup should be triggered at the primary side from the least SCN of secondary.

The completed incremental backups can be transferred to standby site or object storage to perform the recovery operation at secondary. The successful recovery will match the least SCN numbers of the primary and secondary.

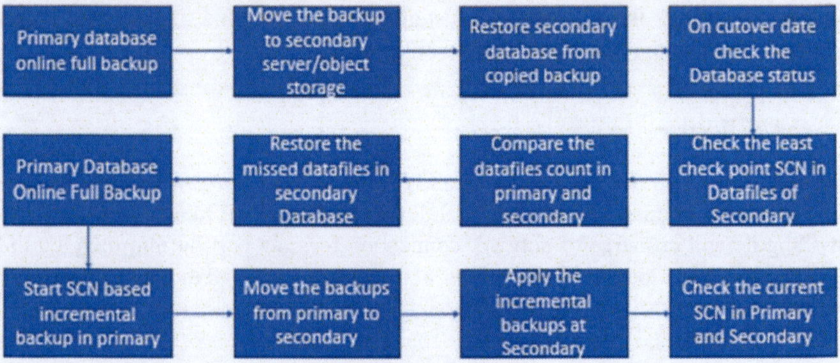


**Fig. 4** IBMS algorithm

The algorithm pseudo-code of the IBMS in the elaborated description as follows.

```
Algorithm:  IBMS
Input: Lowest System Change Number (SCN).
Output: Primary DB and Secondary DB gets in sync.
Step 1:     Start
Step 2:     In Secondary DB: Cancel (MRP);
Step 3:     In Secondary DB: i=Get
            min(checkpoint_Change#) from
            v$datafile_header;
Step 4:     In Primary DB: b=Trigger (RMAN incremental
            backup between (scn=i) and Current SCN;
Step 5:     t=1;
Step 6:     while (t)
Step 7:        if (Primary: total datafiles ==
                   Secondary:total datafiles)
Step 8:                In Secondary DB: Catalog(b);
Step 9:                In Secondary DB: recover (DB)
                       noredo;
Step 10:               t = 0;
Step 11:       else
Step 12:             In Secondary:  Primary service <=
                       Trigger (RMAN (restore missed
                       datafile);
Step 13:    end;
```

## 5 Implementation

The proposed system has been implemented in this section. The full online backup of the 19C database has been taken on the primary side and moved to the secondary database's file system or object storage in cloud.

As shown in Fig. 5, The current lowest SCN of checkpoint change is taken and initiated the backup. As shown in Fig. 6, Upon successful completion of the backup, the backup pieces have been moved to the target secondary database side.

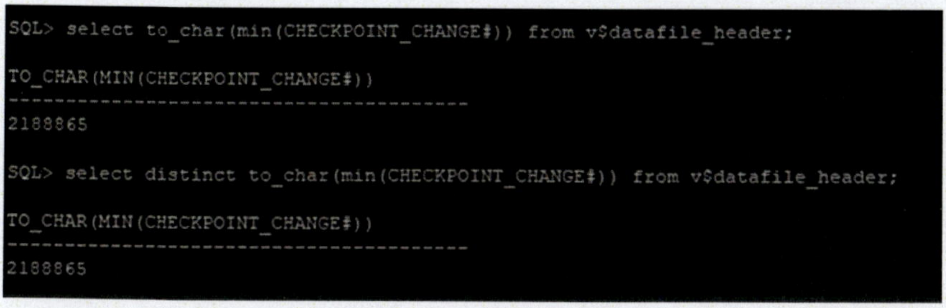


**Fig. 5** Current SCN details

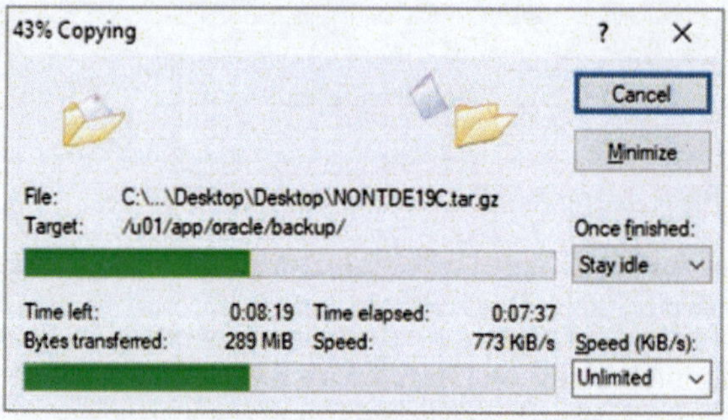


**Fig. 6** File transfer details

**Fig. 7** Recovery at secondary location

As shown in Fig. 7, the backup restore is completed at the target secondary side and the recover database command should fail at SCN 2,188,865. Figure 7 shows the failure of the recover command exactly at the same SCN number as expected.

The Fig. 8. Represents the transaction logs or archive logs for one week along with sizes. The total transaction got generated is 1.25 Terabytes (TB). In the normal process, these transaction logs need to be transferred to the target secondary server. Since the default plan is being used in this paper, the default size that can archive for file transfer is 773 KiB/s as shown in Fig. 6.

```
DAY            THREAD# SIZE IN GB ARCHIVES_GENERATED
--------- ---------- ---------- ------------------
26-DEC-20          1 130.273438                697
27-DEC-20          1 200.498047               1067
28-DEC-20          1 199.615234               1062
29-DEC-20          1 202.975586               1080
30-DEC-20          1  203.53418               1085
31-DEC-20          1 201.577148               1073
01-JAN-21          1 145.964844                777
```

**Fig. 8** Archive logs generation at primary location

```
[oracle@cloud-incremental-server oracle_archivelogs]$ du -sh oracle_incr
3.4G    oracle_incr
```

**Fig. 9** Incremental backup size

## 5.1 Analysis with Existing System

The total time that can be consumed with the existing model for just archive logs transfer to the target side will be 20.09 Days.

$$
\begin{aligned}
\text{Total time for file transfer} &= \text{(Total size of files) / (Current network speed)} \\
&= \text{(1.25 TB) / (773 KiB/s)} \\
&= \text{20.09 Days}
\end{aligned}
$$

This calculation proves that the existing system is not efficient for highly transactional databases with a default-free network plan in cloud computing.

## 5.2 Implementation of Proposed System

The incremental backup needs to be triggered on the Primary side from the SCN 2,188,865 to the current SCN.

As shown in Fig. 9, Upon successful completion of the backup the total backup file size was just 3.4G which contains the total changes of 1.25 TB archive logs actions list.

$$
\begin{aligned}
\text{Total time for file transfer} &= \text{(Total size of files) / (Current network speed)} \\
&= \text{(3.4 GB) / (773 KiB/s)} \\
&= \text{1.28 Hours}
\end{aligned}
$$

The secondary database recovery needs to be started upon moving these incremental backups to the location. These incremental backups can be applied also directly from object storage incase if the backups are transferred to object storage instead of the file system.

The Fig. 10 shows the total elapsed time for the secondary database recovery as 4.35 min.

The Fig. 11 shows the current state of the primary database on server instance-20210″ and current SCN as 10,768,775. The Fig. 12 shows the current state of the secondary database on server "cloud-incremen" and current SCN as 10,768,775.

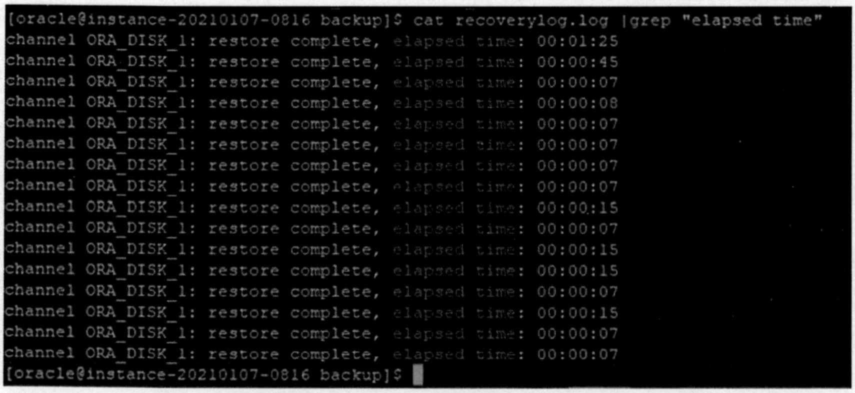


**Fig. 10** Total elapsed time for recovery

**Fig. 11** Primary database status

**Fig. 12** Secondary database status

# 6 Result Analysis and Discussions

The derived results show that database migration can be done efficiently. The total time taken for migration is far better than in synchronization methods. The overall results show that the required goal can be achieved in a better manner.

## 6.1 Cost-Effective Advantage

The default free network plan is enough to move the changes of primary to the secondary site. There is no need to maintain a dedicated network link from 3rd party between primary and secondary sites.

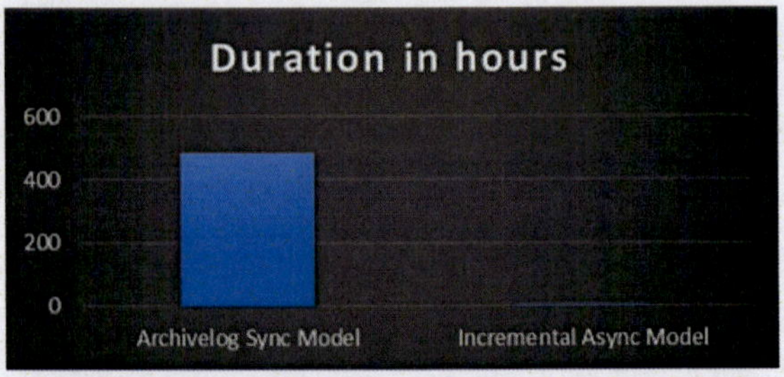


**Fig. 13** Cost-effective comparison with the timeline

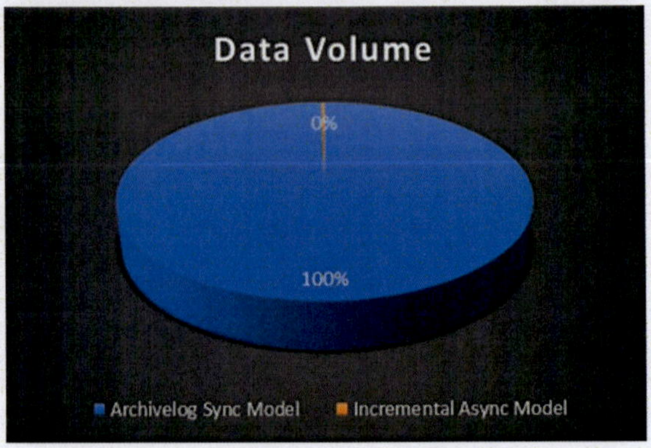


**Fig. 14** Archive logs generated volume comparison

As shown in Fig. 13, the default low-speed network link works very well with a proposed model which not with the existing model. The results clearly show advantages in the proposed model.

## 6.2 The Advantage Over Data Volume

The volume of data that needs to transfer to a secondary site for applying the transaction changes has been decreased tremendously.

As show in Fig. 14, The volume of data of archive logs that must be transferred to secondary location has been decreased from 1.25 TB to 3.4 GB. The migration of databases cannot be done on a single attempt rather it is a phase-wise implementation. In the existing model, after the restoration of consistent complete data, the consequent data changes are streaming continuously to the secondary state. This live streaming method needs to be continued till the date of the cutover. Considering the highly transactional databases in the migration, this continuous streaming method cannot be maintained for dynamic database changes. Because very speed stable dedicated network connection is required between primary and secondary databases.

In the proposed model in the first phase, a consistent data copy or restoration will be done at the secondary site as a previous model. However, the consequent changes are being moved to async mode to the secondary site with frequent intervals.

As per the results, the downtime for the last incremental changes apply at the secondary site will take less time than the live streaming method.

## 7 Conclusion

The Organizations depending upon their business types they might have very highly transactional databases. In those cases, the network cannot keep up with the change's magnitude. In those situations, the secondary will fall behind the primary. So, the proposed model has been intended only for very high transactional databases in async mode. The results also proved that the async mode is suitable and can be applied to the changes at a secondary state within less time when compared with the live-streaming model.

This paper has shown the effective cost management and better data transfer process without compromising the reliability and integrity of the data. In future work, the automatic sync method proposal can extend from this current research work.

## References

1. Fahmideh M, Low G, Beydoun G, Daneshgar F (2020) Cloud migration process A survey evaluation framework and open challenges, arXiv [cs.SE]
2. Ellison M, Calinescu R, Paige RF (2018) Evaluating cloud database migration options using workload models. J Cloud Comput Adv Syst Appl 7(1):1–18
3. What Is an OLTP System? Oracle.com. https://docs.oracle.com/database/121/VLDBG/GUID-0BC75680-5BD4-43A9-826F-CD8837D30EB2.htm. Accessed 25 Jan 2021
4. Surhone LM, Timpledon MT, Marseken SF (eds) (2010) Oracle OLAP. Betascript Publishing, Beau-Bassin
5. Tona AA (2020) DPS-AA: Intranet migration strategy model for clouds. I J Modern Educ Comput Sci 5:55–63
6. Suryawan RB, Ferdiana R, Widyawan (2020) The comparison of cloud migration effort on platform as a service. J Phys Conf Ser. 1577:012056
7. Physical Backup, Oracle.com. https://docs.oracle.com/cd/E57185_01/EPMBK/ch01s02s01s01.html. Accessed 25 Jan 2021
8. Administrator's Reference, Oracle.com. https://docs.oracle.com/en/database/oracle/oracle-database/18/ntqrf/hot-backups.html. Accessed 25 Jan 2021
9. Learn about migrating with Oracle Data Guard
10. Configuring Oracle GoldenGate for active-active high availability, Oracle.com. https://docs.oracle.com/goldengate/1212/gg-winux/GWUAD/wu_bidirectional.htm. Accessed 25 Jan 2021
11. Amazon.com. https://aws.amazon.com/blogs/aws/aws-data-transfer-prices-reduced/. Accessed 25 Jan 2021
12. Pricing - Bandwidth, Microsoft.com. https://azure.microsoft.com/en-in/pricing/details/bandwidth/. Accessed 25 Jan 2021

# Analysis of Different Approach Used in Particle Swarm Optimization Variants: A Comparative Study

Amreen Khan[1](✉), M. M. Raghuwanshi[2,3], and Snehlata. S. Dongre[4]

[1] G.H. Raisoni College of Engineering, Nagpur, India
[2] Department of Computer Engineering, GHRCEM, Pune, India
[3] G.H. Raisoni College of Engineering, Nagpur, India
[4] Department of Computer Science and Engineering, G.H. Raisoni College of Engineering, Nagpur, India

**Abstract.** For every type of system transformation data/information plays a very important role. Everywhere artificial intelligence system is being deployed to reduce human efforts, as well as to improve system performance, accuracy, effectiveness and throughput of system. These artificial intelligent system works only on information feed to the system as an input. Swarm Intelligence is a most effective way of transmitting information between two communication parties. Most of the Artificial Intelligence system is based on Swarm Intelligence technique where the hardware system (Robots) and software is connected to each other for performing task with robots. This paper represents the survey and comparative analysis of various swarm optimization technique based on their working parameter.

**Keywords:** Artificial intelligence · Optimization · Inertia weight · Feasible solution

## 1 Literature survey

The Information (data) plays a very important role in every communication system. Information received by the system should be processed intelligently to obtain a fruitful result. For processing these information artificial system has been used widely [4]. These system used a type of intelligence called as "SWARM INTELLIGENCE" for processing the information (data) received by the system [9]. Swarm Intelligence is an intelligence technique based on animal social behavior. Around 5000 years ago the Egyptian started harvesting honey and they wondered looking at the constructive method of bee hive made by bees in hexagonal shape. It was just a divine power that a bee has while constructing their house in a particular shape and communicates with other bees also. Scientifically it has been found that hexagonal cells can pot much more amount of honey than other type of cells. Like bees every other animal can perform various activities using their way of transmitting data which is nothing but swarm intelligence. In their social activities' like finding food by hunting, searching, playing with other companion, staying calm


S. Shukla et al. (eds.), *Data Science and Security*, Lecture Notes in Networks and Systems 290,
https://doi.org/10.1007/978-981-16-4486-3_48

in their house while protecting themselves from enemy, their overall behavior is totally depended on their computational ability. All these creatures work by finding proper ways or solutions in various situations like while hunting a lion always stalks the target first, then computes the distance for chasing the target, grabbing the target and then killing the target by appropriately selecting the target's body part. All these activities are done by their own intelligence capability where an animal will select, adapt and operate in proper way to achieve their target. Here an animal uses optimization technique to find the best way of hunting for achieving their target. This animal intelligence has been studied by different researchers and then an algorithm was developed called as particle swarm optimization (PSO). It simulates the basic social behavior of FISH, BIRDS, bee, and various types of small insects. PSO is completely based on HUNTING action of Birds. Where every bird is acted as spot with no weight and volume and a solution, the food that the bird wants to achieve. Optimization is nothing but finding the best solution from various available solution under some particular set of rules and observations. Many researchers have developed various solutions for solving linear and non-linear optimization problems. The PSO algorithm is most widely used technique in various artificial intelligence systems. The basic design idea of PSO is mostly related to two researches

1. It's an evolutionary algorithm
2. Artificial Life i.e.it is based on artificial system with life characteristics.

For constructing the co-operative structure with swarm artificial life [10] a system, following five basic principal theories must be used:-

a. Quality: System should be capable of sensing the quality change in environment and according to that change it should a response.
b. Stability: System should not change its working behavior with every change in environment.
c. Diverse Response: System should not be restricted to its limited scope.
d. Adaptability: System should be capable of changing its working as per the basic change is worthy. These five principles are the basic features of any artificial life system.

## 2 Introduction

As a mathematical technique Optimization is concerned with finding the maxima or minima of functions in some feasible region. There's no business or industry that's not involved in solving optimization problems. Particle Swarm Optimization (PSO) is a modern and powerful method of optimization that has been empirically shown to perform well on many optimization problems.

A novel population-based stochastic search algorithm is the Particle Swarm Optimization algorithm (PSO). The algorithm of PSO was first introduced by Dr. Kennedy and Dr. Eberhart in 1995 and its basic idea was originally inspired by simulation of the social behavior of animals such as fish schooling and bird flocking. PSO is based on the nature of group communication, knowledge sharing among individuals, migrating in search space for food. If any particle finds a desirable path then the rest of the members will follow quickly. To solve optimization problems, the PSO algorithm learns from animal's behavior. The popularity of PSO is due to its simplicity of implementation and the ability to swiftly converge to a better solution with a lesser number of parameters to tune. This paper presents a rudimentary review of the PSO method for optimization. As there are more than 2500 plus research variations available on the PSO with algorithm variation and different application domain investigation. Further, this paper elaborates on PSO method in Sect. 3 and the variations are discussed in Sect. 4. Section 5 concludes the writings.

## 3 PSO Algorithm

As stated in the previous section swarm technique is completely based on the social behavior of animals. Literature survey of various swarm-based algorithms is available in various research manuscripts which were conducted [7].

PSO is a random optimized technique based on swarm intelligence. The working principle of PSO is based on hunting and acting of animals specifically birds where every bird is considered as spot/particle with no volume, weight and food or target is considered as a solution in the search space. Every spot has two parameters i.e. velocity and position. It will modify the parameter value by using some set of rules. Getting it set to some optimal value after a mass modification.

$$\left(V^{k+1}\right)i = w * \left(V^{k}\right)i + C1 * r1 * \left(pbest - X^{k}i\right) + C2 * r2 * \left(gbest - X^{k}i\right) \tag{1}$$

$$X^{k} = X^{k} + V^{k+1} \tag{2}$$

Where V is the velocity of the particle ($X^k$) in the search space. W is the weight, C1and C2 are the acceleration factors, r1 and r2 are the random numbers between 0 and 1. Pbest represents the personal best position of the particle in search space and (Gbest) is the global best position of the particle. While $X^k$ represents the current position of the particle [3].

The PSO Methodology:

1. Randomly initializing the position and velocity of each particle.
2. Calculating the fitness value of each particle, and accordingly updating the gbest and pbest values based on a current search condition.
3. Continuously updating the position & velocity value by using above given equation.
4. Update the current pbest value if the fitness value is much better than the current pbest value.
5. Update the current gbest value if the fitness value is much better than the current gbest value. Halting the iterative process if all the conditions are satisfied else continue the execution process by switching back to step3.

# 4 Existing State of the Art PSO Algorithm

## 4.1 Linearly Decreasing Weight Particle Swarm Optimization (LdwPSO):

This algorithm was developed to resolve the static weight's poor effect on searching for best solution during the execution of the algorithm. Generally the mass value of weight displays good results. A larger weight value will give a high rise in a jump for achieving partial value for search while the smallest value will decrease the rise in jump to get a partial value which will increase the execution time [2].

$$W = W_{max} - t * (W_{max} - W_{min})/t_{max} \tag{3}$$

Where $W_{max}$ is the peak value of the weight and $W_{min}$ is the minimum value of the weight, T specifies the no. of iterations. For changing the value of weight the algorithm uses the formula given in the Eq. 3, as shown in Eq. 3 the value of weight totally depends on the value of T. If the value of t will increase then the value of weight will be decreased, and vice-versa.

## 4.2 Linear Decreasing Inertia PSO

The main problem with above discussed algorithm was a linearly decreasing weight value which affects the working of the algorithm. Hence the new advanced linear decreasing inertia-based PSO has been introduced in the paper [8]. This algorithm's performance is totally based on the linear value of inertia in PSO. This inertia parameter was introduced in the original PSO. Researcher Kennedy has observed that by introducing a linearly decreasing inertia into the PSO, the overall performance of the algorithm has drastically changed. By modifying the value of linear inertia will change the ability of search operation in the algorithm. If the value of linear decreasing inertia is large, the algorithm will search a particle globally whereas if the linear decreasing inertia value is low the algorithm will search the particle in the local area range. The inertia value can be set to 0.9 as max and 0.4 as min [1]. This is the reason why the inertia value is called a fluid of the medium where the particle will move. The inertia weight parameter was introduced into the original method of PSO. This addition of linear inertia values enhanced the performance of the algorithm. Here the larger value of inertia weight will facilitate the global search while a smallest value of inertia weight will facilitate the local search.

$$\omega_t = (\omega_{start} - \omega_{end}) * ((T_{max} - t)/T_{max}) + \omega_{end} \tag{4}$$

The inertia weight value depicts the flowing velocity of the medium in which the particle is moving. Hence by setting high initial weight value will decrease the moving speed of viscosity of medium whereas a low value of inertia weight will increase the movement of a particle in a viscous medium.

## 4.3 QPSO (Quantum Based Particle Swarm Optimization)

This algorithm is a result of the deep study and knowledge of physics, modern science, and technology. Also in similar ways, deep interest in exploring the "mechanism of biological evolution" researchers are so much motivated for simulating the idea they studied in various simulators. When two or more ideas joined together can give a much more interesting result. The same happened with the QbPSO algorithm. It's a simulation of Anneal Algorithm (SAA) which was generated earlier by the researcher. After implementation, they had introduced a particle "Q (1)" based on the quantum bit (qbit). After a while they replaced the sigmoid function used and also the best chromosome guidance has been taken to make the algorithm much optimized and its step more optimum. And the algorithm got its new name as discrete PSO or Quantum particle swarm optimization (QPSO). In this quantum study (Theory) the minimum bit (unit) that carries valuable information or data is defined as "QuBIT" which has two states either "0" or "1". It uses the following equation for finding the velocity as below:

$$Q(t) = [q1(t), q2(t) \ldots qn(t)](qi(t) = [q1\,(t), q11\,(t) \ldots 2..q1\,(t)])m \quad (5)$$

Equation 5 represents the length of the particle, q1m (t) represent the probability of particle. Here a highest regression value estimator and lowest trimmed squares value is used to remove the effect-cause by various observations containing outline boundaries values. QPSO behaves like the main optimizer method. Which can give an excellent direction to dynamic regions. Here a simple and easy method named Nelder-Master is used as a local search to finally tune the obtained solution from QPSO [5].

### 4.4 BARE- BONESPSO (BBSPSO)

This is an advanced version of PSO which is a parameter-free PSO known as "BARE BONES PARTICLE SWARM OPTIMIZATION (BBPSO)" which has no additional components such as inertia weight, acceleration coefficients. This algorithm has a very simplified behavioral approach as it doesn't have any velocity-based or related formula. And has much better performance in solving various optimization problems. Instead of velocity function, it uses Gaussian distribution function with the mean sub-function {(Pt)ij + (Pt)qj}/2 and the standard Deviation function as Ptij-Ptgj. That is the reason why the researcher has more interest in investigating this algorithm more precisely and developed much more knowledge. In this algorithm, neighborhood searching is used to make a balance between the exploitation and exploration capability during research process in the neighborhood. It has a jumping technique that is based on Gaussian and Cauchy probability function distribution for tolerating local optimal solutions. This algorithm integrates barebones particle and opposition-based learning techniques to find solutions for constrained non-linear optimization-related problems. Here the velocity is not the main component, it plays as a main component to show the actual recent position of a swarm particle. This velocity improves the capability in exploring the different areas.

### 4.5 CbPSO (Cultural Based Particle Swarm Optimization)

This algorithm is initially introduced by author Reynolds [6] in the year 1996. This algorithm is basically generated and co-related with the human social behavior which is developed to show the model of human social and cultural behavior in a computational system. That accumulates the overall behavior of people in society. The overall processing of this algorithm is divided into two different sub-parts as

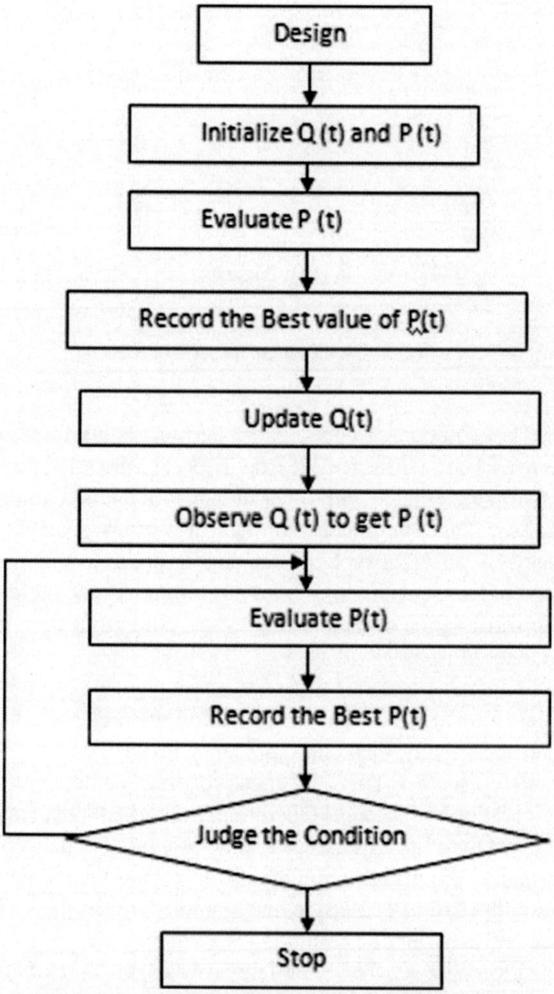


**Fig. 1** Execution flow of QPSO ( Source: QPSO, p. 302)

(1) Belief Space &
(2) Population Space. As shown below in Fig. 2.

These two sub-parts can communicate with other component with the help of some set of rules. Hence we can say that this algorithm provides a very excellent technique for dynamic data knowledge and a very useful technical framework. Which is capable of self-adaptation that can be adaptive in any situation or swarm intelligence optimization system. The lower level main population space (PSO population) contributes group of various individuals to the upper level space (belief population) periodically, and the upper level space continuously evolves these group of individual and then contributes elite individuals to the lower level space. Thus the mechanism of dual promotion and evolution improves the population diversity, and avoids premature convergence.

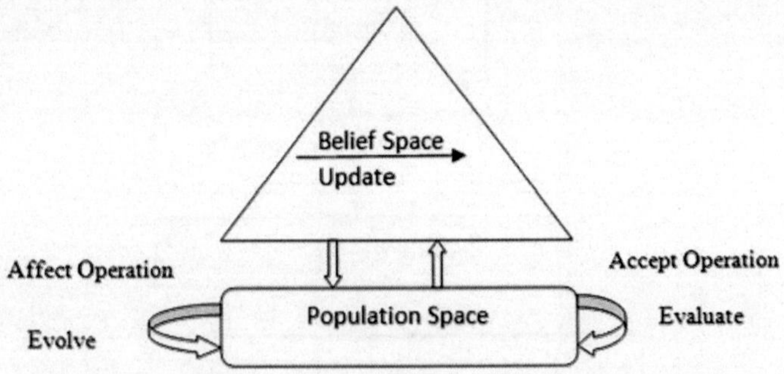


**Fig. 2** Block schematic of CbPSO

1. *Belief Space:* This component contains the various data/information of a cultural activity known as Data (Information Repository). Various types of humans will share their cultural experience with other human when it is required to share.
2. *Population Space:* This component consists of various possible solutions to the problems shared by different types of humans in previous components and hence may be constructed as a population-based algorithm. Above these two components will share their information or data through some communication medium based on some particular set of rules.

Here in this algorithm at the start of the execution process, both the component first initiates themselves receptively, then communicate with each other for exchanging real-time experience with each other. This exchange process considers basic two operational processes affect Operation and Accepts Operation. Individual object in population space component will inherit their life experience to other component name belief space using given operations, hence these both components exchange their knowledge with each other and interact with each in one space and support both of them. The maintenance cost of belief space and experience exchanging is much more but if its knowledge library guides the optimization process. The complete process gets much better and effective. More fruitful experience can generate more excellent optimization solution, with only some specific iterations of search required. Hence this algorithm gives much efficient result than any other evolutionary algorithm. This process of experience exchange is described in detail and given below:

It has the following steps for its execution process:

Algorithm: CbPSO (Cultural based Particle Swarm Optimization).

Step-1] Construct the belief and population space component.

Step-2] Initialize the value of each component, Check the fitness of each Human.

Step-3] End the process if the end step criteria are fulfilled else continue.

Step-4] Update each individual value in both the components.

Step-5] Update the beliefs space initial value.

Step-6] Update the population space initial value randomly.

Step-7] Return to step -2.

Table 1 below shows the comparative analysis of various PSO algorithm variations in context of is performance, technique and efficiency.

**Table 1** Comparative analysis of existing optimization techniques

| Sr. No | Type | Basic Idea used | Pros | Cons |
|---|---|---|---|---|
| 1 | Swarm Intelligence | It uses the shortest path discovery method for finding the source and return back to the communication medium with the use of some natural Phenomena | 1. Better Performance<br>2. Less effected by poor solution. | 1. Difficulty in finding solution for larger no. of nodes.<br>2. Complicated to implement practically. |
| 2 | Evolutionary PSO | Endurance of the best approach from the available approaches | 1. Support multi-objective optimization.<br>2. Very much effective for large no. of nodes. | 1. Difficult to Operate in dynamic data.<br>2. Tradeoff is convergence. |
| 3 | LdwPSO(Linearly Decreasing weight particle swarm algorithm) | Static initial weight effect | 1. Suitable for Dynamic data set.<br>2. Easy to implement. | 1. Static weight causes poor effect.<br>2. Less effective for large weight Value. |
| 4 | LDIPSO(Linear Decreasing Inertia PSO) | Totally based on the linear value of weight in PSO. The addition of linear inertia values enhances the performance of the algorithm | 1. Linear value is static in nature hence easy to maintain.<br>2. Linear value is good Component for search | 1. Extra memory required to store the value.<br>2. Global & local search values is difficult to Maintain. |
| 5 | Non-Linearly (Chaotic) Decreasing Inertia Weight PSO | CHAOTIC is a dynamic AI system which is too much sensitive for its initial value. This system has discrete and dynamic property | 1. Simple to implement.<br>2. Easy to understand.<br>3. It is more effective algorithm. | 1. Progression depends on Specific value. |
| 6 | Evolutionary Based And Random Inertia Weight PSO(EBRIWPSO) | Another type of PSO whose output performance is based on some specific calculation | 1. It is very Simple and efficient. | 1. Sometimes changes in certain parameter will affect the working. |
| 7 | Random And Adaptive PSO (RADPSO) | Random and dynamic technique has been introduced as a new idea to set inertia weight values based on the current swarm diversity and dynamic degree as well as effectiveness of search operation and its performance of the swarm | 1. Convergence problems in already existing system has been solved in this algorithm.<br>2. Very effective algorithm. | 1. Dynamic value needs more attention to be maintained.<br>2. Complex process to implement. |

(*continued*)

Table 1 (*continued*)

| | | | | |
|---|---|---|---|---|
| 8 | AsynLnCPSO | This is another type of particle swarm optimization algorithm which is based on memory management of swarm particle, ability of social learning and self-learning of a particle. | 1. It provides much feasible & optimized solution from various available solutions. | 1. Requires extra memory for storage of its various components value. |
| 9 | CbPSO(Cultural based Particle Swarm Optimization) | This type of PSO is based on basic two components 1. Belief Space 2. Population Space | 1. Easy process to understand. 2. Simple to implement. | 1. Need Extra space for storing two components value. |
| 10 | Discrete PSO (DbPSO) | DbPSO working is completely based on Material acquisition handling problem. | 1. Discrete value processing gives more advantages in this algorithm. | 1. Problems should be related to material only for implementing this algorithm. |

## 5 Conclusion and Future Work

Here at the end of research work, this manuscript concludes the analysis that represents a broader view of the particle swarm optimization algorithm for researchers. This paper describes a brief introduction of PSO and various types of its variants from this algorithm. Since the introduction of PSO has been studied from an early age, the methods branched from this algorithm, and their applications have developed a lot nowadays. Based on detailed survey carried out over many research paper publications of PSO, this study pose various techniques used in these methods. Moreover, various methods derived from this algorithm are briefly introduced with respect to the time limit and other various publication restrictions. Also, a table is presented in this paper expressing the main parameters used in various variants of PSO with their advantage and disadvantage. PSO algorithm plays a very important role in finding the optimal and feasible solutions from various available solutions in every field like economy science, financial services, business strategy, medical science, engineering, etc. This will help to develop novel work in any field of application.

## References

1. Gao Y, An X, Liu J (2008) A particle swarm optimization algorithm with logarithm Decreasing inertia weight and chaos mutation. In: Proceedings of the International Conference on Computational Intelligence and Security (CIS 2008), vol 1, pp 61–65, Suzhou, China, December 2008
2. Li HR, Gao YL (2009) Particle swarm optimization algorithm with exponent decreasing Inertia weight and stochastic mutation. In: Proceedings of the 2nd International Conference on Information and Computing Science (ICIC 2009), pp 66– 69, Manchester, UK, May 2009

3. Liu H, Su R, Gao Y, Xu R (2009) Improved particle swarm optimization using two novel parallel inertia weights. In: Proceedings of the 2nd IEEE International Conference on Intelligent Computing Technology and Automation (ICICTA 2009), pp 185–188, Hunan, China, October 2009
4. Shi Y, Eberhart RC (2001) Fuzzy adaptive particle swarm optimization. In: Proceedings of the Congress on Evolutionary Computation, pp 101–106, Seoul, Republic of Korea, May 2001
5. Xue-Bin S, Zhan-Min L, Cheng-Lin Z, Zheng, Z (2012) CognitiveUWB pulse waveform design based on particle swarm optimization Ad-Hoc Sens Wirel Netw 16(1–3):215–227
6. Evers GI (2009) An automatic regrouping mechanism to deal with stagnation in particle Swarm optimization [M.S. thesis], University of Texas-Pan American, Edinburg, Tex, USA
7. Wang Y, Ai J, Shi Y, Teng H (2007) Cultural-based particle swarm optimization algorithm. J Dalian Univ Technol 47:539–544
8. Shi Y, Eberhart RC (1998) A modified particle swarm optimizer. In: Proceedings of the IEEE World Congress on Computational Intelligence. Anchorage, AK, USA, pp 69–73
9. Alatas B, Akin E, Ozer AB (2009) Chaos embedded particle swarm optimization algorithms. J Chaos Soliton Fract 40:1715–1734
10. Lin CJ, Chen CH, Lin CT (2009) A hybrid of cooperative particle swarm optimization and Cultural algorithm for neural fuzzy networks and its prediction applications. J IEEE Trans Syst Man Cybern C Appl Rev 39:55–68

# Identification of Offensive Content in Memes

Aayush Aman(✉), Gopal Krishna, Tushar Anand, and Anubhaw Lal

Netaji Subhas Institute of Technology, Patna, India

**Abstract.** Social media is an interactive platform for communication that facilitates the creation or sharing of information, ideas, or other forms of expression among people. With the increased usage of such platforms, automatic detection of multimodal content should be mandatory and so the main focus should be on memes. A meme can be briefly described as an image or video depicting thoughts or ideas of people as well as to express culturally-relevant ideas. Although, memes mostly are captioned photos intended to elicit humor. Since memes are multimodal in nature, posting of irrelevant memes that spread hatred, trolling, cyberbullying is increasing day by day. Awful speech, aggression is explored as single modalities such as text or image. However, combining both the modalities text and image to detect whether it is offensive content or not is a difficult task. Though with the use of a multimodal approach this issue can be dealt with up-to a certain extent.

**Keywords:** Multimodal data · Classification · Memes · Offensive content · Modalities

## 1 Introduction

Social media is a dynamic platform facilitating information exchange amongst people and acts as a medium for communication. Social issues like caste system, substance abuse, communalism, cyberbullying, are posted and spreads online through offensive images and texts [4]. Therefore, automatic detection of multi-modal content should be mandatory and so the main focus should be on memes. The pictorial or graphical representation of opinions or perceptions of a particulate group of people is known as a meme. Also, they act as a way for the expression of a culturally-relevant idea. With increasing trends in meme circulation and its abundance in our social life, a proper regulatory method is the need of the hour. Most memes are captioned photos intended to elicit humor. Since memes are multi-categorical in type, posting irrelevant memes that spread hatred, trolling, cyberbullying is showing a significant increase at a rapid rate. Although, hate speeches or memes of violent nature show to have a single fashion, either being text or image. Though considering text and images as a single instance is a strenuous job. The detection and categorization of a meme as offensive or non-offensive is an exigent task due to the implicit depiction of sarcasm and humor through them. This suggests that a meme might not be offensive when considering it as a text-only or image the only instance. Due to this, the combined analysis of memes, combining both, texts and images and considering them as a single modality is required to judge the offensive or non-offensive categorization of the memes [4]. Precise identification and


S. Shukla et al. (eds.), *Data Science and Security*, Lecture Notes in Networks and Systems 290,
https://doi.org/10.1007/978-981-16-4486-3_49

classification of offensive content on social media are very vital for the stable regulation of the platform(s), and this is where the research is headed to. One of the major objectives being the eradication of negativity off such platform(s). To start with, a meme dataset comprising of 749 memes is selected and is split into train, test, and validation datasets. After which, the training dataset is carried forward as an input to the machine learning model(s). The elimination of errors in the datasets is the first step for improved accuracy of the models, to be later on implemented and used. To achieve this, the various algorithm for data pre-processing is taken into consideration for the images in the dataset as well as the text dataset, firstly, identification and elimination of words with non-significant meaning (Stop Words) is carried out, after which strings of equivalent length (Padding) are made off from the textual content in the dataset and finally, tokenization of text for the textual dataset is carried out. Once finishing with the rectification of the text dataset we move on to the image dataset where firstly, normalization of image matrix is done, after which images from the dataset are denoised and resized to a homogenous dimensioning and finally, edges of the images are detected where the images are smoothened by the use of a Gaussian filter for better training results. Once datasets are purified and pre-processing steps are carried out, we move on to model selection and analysis, keeping in mind the outcome precision and process efficiency as the basic measures for the model(s). Model(s) for the text dataset as well as image datasets are selected separately and tests are performed to find out the best-case model in the cases. Later on, outcomes of these experiments would be analyzed to find a possible solution to the prime motive of classification of offensive memes on social media platform(s) and further help in the removal of such content making the internet a safer and better place. The rest of the paper is organized as follows: - Sect. 2 provides a brief review of the earlier established research work. Section 3, provides an insight into the approach taken for the problem statement. Detailed Analysis of the experiment has been marked down under Sect. 4 whereas, Sect. 5, concludes the paper with prospects involving possible space of betterment in the required solution to the problem.

## 2 Related Works

This related section covers the work done in identifying offensive content in text and images. It also describes the research done in the area of meme analysis as well as multi-modality.

Warner and Hirschberg model offensive language by developing a Support Vector Machine (SVM) Classifier which takes in features manually derived from the text and classifies if the given text is abusive or not. Watanabe et al. 2018; Zampieri et al. 2019. Offensive Language Detection [5], it relies on unigrams and pattern of the text for detecting hate speech. These patterns are carefully crafted manually and then provided to machine learning models for further classification. Hate speech identification by Davidson et al. 2017.

Several works proposed convolutional neural networks (CNN) to identify appropriate or in-appropriate images for children, Connie et al. 2018. Identifying Offensive content in an image based on skin detection techniques have been proposed for nudity detection, Arentz and Olstad 2004. [1] The research done by Gandhi et al. 2019 deals with offensive images and noncompliant logos. They developed an offensive and non-compliant image detection algorithm that identifies the offensive content in the image. They have categorized images as offensive if it has nudity, sexually explicit content, objects used to promote violence or racially inappropriate content.

He et al. 2016 proposed a meme extraction algorithm that automatically extracts textual features from data posted during events such as the anti-vaccination movement. The process of extraction is done by identifying independent phrases and by clustering the mutation variant of each phrase associated with the meme. This work studies the convergence and peak times of memes. Drakett et al. 2018, in their research, address online harassment of marginalized groups by abusing memes, using thematic analysis of 240 sample memes. This research studies memes from a psycho-linguistic perspective.

## 3 Proposed Methodology

A meme dataset of 749 memes was picked up which later was split up into train, test, and validation datasets. Different methods of analysis such as the CNN model, LSTM model, and Bi-LSTM models have been implemented on the text, which was extracted from the memes available as CSV datasheets. The datasets were then pre-processed via various algorithms aimed at improving accuracy and removing errors in the datasets. Later on, the VGG-16 model was used on the image datasets after pre-processing the images via pre-processing steps. After which a comparative study of the model outputs was done. Based on the experiments, the models would be fused and concatenated for the formation of multimodal machine learning models that can classify the memes based on both, the textual feed as well as the visual features of the memes. Three multi modals were formulated, by concatenating each of the text-based machine learning models with the available VGG-16 model for the images, namely Stacked LSTM + VGG 16, Bi LSTM + VGG 16, and CNN + VGG16. After which comparative study between the results of these three multi-models would be done to achieve the best-case model.

## 4 Experimental Analysis

### 4.1 Dataset Description

A dataset is a collection of data that are arranged in sequential order. In layman's terms, machine learning is a technique of learning from experiences. It is vital to have self-sufficient data provided to the machine as input. First of all, when we talk about data, the first thing that pops up in one's mind is a sheet comprising of rows and columns. This baseline structure is the traditional structure of data which can be commonly seen when we talk about Machine learning. Although, data also comprises unstructured information like images, etc. which is also used in the field of machine learning for model training. Our theme is the identification of offensive content from a collection of memes. So, in this case, the presence of an image dataset that contains a lot of images is vital. Images from various sources were collected for building our Image dataset, which is to be later used for training the machine learning models by the use of various algorithms to be later discussed in the course of this paper. Firstly, to begin with, we split the dataset into

**Table 1** Summary statistics of meme dataset (avgw: average number of words, avgs: average number of sentences, off: offensive, and n-off: non-offensive)

| Data | Avgw | Avgs | Off | n-off | Total |
|---|---|---|---|---|---|
| Train | 41 | 2 | 187 | 258 | 445 |
| Test | 47 | 2 | 59 | 90 | 149 |
| Val | 45 | 2 | 59 | 90 | 149 |

two parts, one to be used for training of the machine learning model and the other as a testing dataset. Then, the image training dataset was pre-processed using various pre-processing steps like Normalization of the image matrix, denoising of images, resizing of pixels in images, Canny edge detection, and merging of color channels in the image aimed at increasing accuracy and efficiency. For the objective to be accomplished, a text dataset was also required for a better understanding of the Memes. To achieve this, the word contents of the images in the image dataset were extracted and were later put on into a CSV file. As earlier, in the text dataset too various pre-processing steps such as Text Padding Elimination of Stop word(s), Tokenizing of text were done (Table 1).

## 4.2 Model Analysis

**Models for the Text Dataset.** The ultimate aim of the project is to design a classifier that can accurately classify the memes in the following two classes: Offensive and Non-Offensive. So, after the pre-processing of the dataset, we use different classifiers for the text part. So here we are going to use three different classifiers, the first being Stacked LSTM where the sequential arrangement of LSTM layers in a stacked architecture results in the formation of Stacked LSTM. After which, Bi-LSTM is implemented on the same dataset by the use of opensource libraries, such as Keras. And later, the model is trained via a GPU for 10 epochs per batch. Moving further, the third model classifier is used on the dataset, i.e., CNN Model, which falls in the category of deed-forward deep artificial neural networks, i.e., the connection between the nodes does not form a closed cyclic structure [2].

**Models for the Image Dataset.** VGG 16: - VGG comes under the categorization of Convolutional Neural Network, which was presented by A. Zisserman and K. Simonyan from the Oxford University proposed in the paper "Very Deep Convolutional Networks for Large-Scale Image Recognition", 2014 [3]. The implementation of the VGG model is as follows, the conv1 layer's input is fixed to 224 * 224 size for RGB images. The image is passed through a stack of the stacked convolutional layers through which the images were passes, filters with minimal receptive fields, 3 * 3 were used. One of the compositions of CNN uses a filter of 1 * 1, i.e., the spatial padding of convolutional layer input is in a way that the spatial resolution is maintained after convolution; the convolution stride is precontrived at 1 pixel, the padding is 1-pixel for 3 * 3 conv. layers. Spatial pooling is done with help of five max-pooling layers, followed by some of the convolutional layers, though not all of the max-pooling layers are sequenced by convolutional layers. Max-pooling is done in a 2 * 2-pixel window, with stride 2.

**Table 2** Accuracy of Stacked LSTM + VGG-16

| | Precision | Recall | F1-score | Support |
|---|---|---|---|---|
| 0 | 0.85 | 0.19 | 0.31 | 57 |
| 1 | 0.84 | 0.47 | 0.60 | 92 |
| Micro Avg | 0.84 | 0.36 | 0.51 | 149 |
| Macro Avg | 0.84 | 0.33 | 0.46 | 149 |
| Weighted Avg | 0.84 | 0.36 | 0.49 | 149 |
| Samples Avg | 0.36 | 0.36 | 0.36 | 149 |

**Multimodal Approach.** Until now, we were training our text and image parts separately, but for final experimentation, we are going to concatenate the textbased and image-based Machine Learning models for the formulation of a multimodel, which is also known as the early fusion approach. The text and image modalities in their vector form have been fed into the concatenated classifier. In this architecture, both modalities are required to classify the offensive content. Now we combine all of the text models one-by-one with the VGG16 image model for a comparative result study. The final constructed multimodal structures were:

*Stacked LSTM + VGG-16: The same Stacked LSTM approach used in the text baseline has been used in the multimodal experiment. Word embeddings are created using a pre-trained GloVe dataset. The use of pre-trained word embedding leverages the contextual meaning of the word globally. After running the 10 epochs of the multimodal we get the resulting accuracy* (Table 2).

*Bi-LSTM + VGG-16: In this architecture, only one Bi-LSTM has been used. The output of this layer has been connected to the classification layer with a sigmoid activation function which gives out the probability of the offensive class. According to the Bi-LSTM, we use VGG16 to finally concatenate the model. In VGG16, the output of the activation function has been fed to the max-pooling layer which later has been followed by a fully connected layer that also uses "Relu" as an activation function. Instead of a fully connected layer, a Global Average Pooling layer has been used which later is connected to a Dense layer with the Sigmoid activation function to predict class probability* (Table 3).

*CNN + VGG16: Last but not least we are going to build the CNN model concatenated with the VGG16 to check how well the model is performing. CNN approaches are suitable for text as it can be represented in a learnable vector form. As text can be represented in such form, CNN can be relied upon to classify the text data as well. The convolutional layer and max-pooling layer with the output of previously connected to the input of the latter has been used. Three such convolutional blocks have been used before the classification layer. In this experimental setting, image features have been carried out by a pre-trained VGG16 network on the Image Net dataset and textual features have been extracted by using a CNN model. These features are concatenated and fed as input to a stacked LSTM model. The output of the LSTM model is connected to the dense layer which then is combined with the image features to classify the meme. The CNN*

**Table 3** Accuracy of Bi-LSTM + VGG-16

| | Precision | Recall | F1-score | Support |
|---|---|---|---|---|
| 0 | 0.48 | 0.75 | 0.59 | 57 |
| 1 | 0.79 | 0.24 | 0.37 | 92 |
| Micro Avg | 0.55 | 0.44 | 0.49 | 149 |
| Macro Avg | 0.63 | 0.50 | 0.48 | 149 |
| Weighted Avg | 0.67 | 0.44 | 0.45 | 149 |
| Samples Avg | 0.43 | 0.44 | 0.43 | 149 |

**Table 4** Accuracy of CNN + VGG-16

| | Precision | Recall | F1-score | Support |
|---|---|---|---|---|
| 0 | 0.88 | 0.88 | 0.88 | 57 |
| 1 | 1.00 | 0.43 | 0.61 | 92 |
| Micro Avg | 0.93 | 0.60 | 0.73 | 149 |
| Macro Avg | 0.94 | 0.66 | 0.74 | 149 |
| Weighted Avg | 0.95 | 0.60 | 0.71 | 149 |
| Samples Avg | 0.60 | 0.60 | 0.60 | 149 |

*Text + VGG16 approach leverages the CNN architecture text as used in the baseline models. After running the 10 epochs of the multimodal the resulting accuracy is displayed* (Table 4).

### 4.3 Experimental Analysis

On the analysis of the results of the different implemented models, it can be understood that classification of the offensive language within the mems is best done by the CNN model on text, which can be highlighted by its capability of highest recall. It can be inferred that the text classifier models based on CNN show good results in terms of recall when textual data is considered for the memes. Also, VGG16 in the case of the image dataset shows to have a satisfactory result. Later on, three multi modals were formulated, by concatenating each of the text-based machine learning models with the available VGG-16 model for the images. It can be noted that the Stacked LSTM + VGG 16 model has a high precision score of 0.84, whereas the Bi LSTM model shows a decline in precision compared to the first multi-modal classifier being 0.55. The noticeable precision has been noted when using the third CNN (for text) + VGG16 (for image) models as a multi-modal classifier, having the precision of 0.93, being the best precision throughout all of the models. The last multi-modal, the CNN Text + VGG16 multi-modal has the best recall out of all the three multi- modals, being 0.60. Also, when talking about the F-1 scores the best has been the output from the CNN Text + VGG16 multi-modal

**Table 5** Comparative study of the multi-model(s)

| Multi-model(s) | Parameters | Precision | Recall | F-1 score | Support |
|---|---|---|---|---|---|
| Stacked LSTM + VGG-16 | 0 | 0.85 | 0.19 | 0.31 | 57 |
| | 1 | 0.84 | 0.47 | 0.60 | 92 |
| | Micro Avg | 0.84 | 0.36 | 0.51 | 149 |
| BI-LSTM + VGG-16 | 0 | 0.48 | 0.75 | 0.59 | 57 |
| | 1 | 0.79 | 0.24 | 0.37 | 92 |
| | Micro Avg | 0.55 | 0.44 | 0.49 | 149 |
| CNN + VGG-16 | 0 | 0.88 | 0.88 | 0.88 | 57 |
| | 1 | 1.00 | 0.43 | 0.61 | 92 |
| | Micro Avg | 0.93 | 0.60 | 0.73 | 149 |

having a score of 0.73. Thus, from the experiments, it can be inferred that the images in themselves can be labeled as offensive but chances are that the text associated with it be might be non-offensive if considered solely. The text classifier model sometimes might fail to identify the true label. On the flip side, the image classifier models might identify the true label for the same data, when followed by the multi-model classifier (Table 5).

## 5 Conclusion and Future Prospects

In this paper, we use the concepts of machine learning to provide a solution for the classification of offensive contents in memes from the abundant presence of memes on social media. One of the major challenges faced in the area is the multi-modality nature of the memes. When a single instance, i.e., text or images separately are considered the outcome is not necessarily precise and has low efficiency. To better off with the solution, we need to consider both the modalities as a single instance. To help with this, the approach of model concatenation and formulation of multi-model classifiers was our go-to. The best-case text-based model is concatenated with the available image-based classifier model for the formulation of a multi-modal which can combine the textual and visual features of the meme as a single modality and give out satisfactory results. This suggests, as of now models aren't efficient enough to guarantee a 100% accuracy rate but with the increasing scope of artificial intelligence and machine learning if we would be able to make a custom model that would be able to learn at each significant step rather than depending on the dataset available, it would help in sharpening the accuracy which in turn could be used at areas such as social media platform(s) for precise and continuous removal of negativity.

## References

1. Arentz WA, Olstad B (2004) Classifying offensive sites based on image content. Comput Vis Image Understand 94(1–3):295–310. https://doi.org/10.1016/j.cviu.2003.10.007
2. Denny, B.: Convolutional neural networks, http://www.wildml.com/category/neural-networks/convolutional-neural-networks/
3. Simonyan K, Zisserman A (2015) Very deep convolutional networks for large-scale image recognition
4. Suryawanshi S, Chakravarthi BR, Arcan M, Buitelaar P (2020) Multimodal meme dataset (MultiOFF) for identifying offensive content in image and text. In: Proceedings of the second workshop on trolling, aggression and cyberbullying. pp 32–41. European Language Resources Association (ELRA), Marseille, France, May 2020. https://www.aclweb.org/anthology/2020.trac-1.6
5. Watanabe H, Bouazizi M, Ohtsuki T (2018) Hate speech on twitter: a pragmatic approach to collect hateful and offensive expressions and perform hate speech detection. IEEE Access 6:13825–13835

# Prediction of Depression in Young Adults Using Supervised Learning Algorithm

Anushree Chakraborty(✉) and Samiksha Shukla

Christ University, Bangalore, India
anushree.chakraborty@science.christuniversity.in

**Abstract.** Over the years, mental health has achieved an essential role in the pertinent development of a human being, and a large part of the population is affected by it. The most commonly affected community being college-going students, and the most common disorders being Anxiety and Depression. Depression is a leading cause of suicide in individuals, where suicide is the second most prevailing reason for death among 15–29-year-olds. This study aims to identify the different reasons and other factors associated with depression to predict and determine whether an individual faces depressive disorders. For this research purpose, the most appropriate classifier is selected. The absolute accuracy of the proposed model is 91.17%, *i.e.*, the model can correctly predict whether an individual has depression 91.17% of the time.

**Keywords:** Depression · Classification · Oversampling · Recursive Feature Elimination · Random Forest Classifier

## 1 Introduction

Mental Health refers to cognitive, behavioral, and emotional well-being. Despite much progress in various countries, it remains a topic of severe stigma. In recent years, mental health has achieved an essential role in the proper development of a human being. Mental disorders have various types characterized by a combination of thoughts, behaviors, emotions, and relationships with others. Different effective strategies help in identifying as well as treating them. However, due to a considerable gap in availability and understanding of the need for care, it remains something that very few people have access.

Mental disorders have become very common these days, and a high percentage of the people suffering from these disorders are college-going students. Anxiety is the most prevalent mental disorder, followed by depression. Since every person is different, the causes might not be the same for every individual. This research work is an interdisciplinary study that blends healthcare with the computer science field and focuses on identifying individuals' primary reason or fear to provide better counseling and therapy. In this study, various techniques are used to predict the mental health of college-going students.


S. Shukla et al. (eds.), *Data Science and Security*, Lecture Notes in Networks and Systems 290,
https://doi.org/10.1007/978-981-16-4486-3_50

Mental Health is vital and highly sought after these days. Many people suffer from mental disorders from a young age, most of which remain undiagnosed or unidentified. Hence it is a socially relevant issue in the current world. Especially during this pandemic, the world has seen a spike in mental health disorders among youngsters. The research work has been done in consultation with the domain expert. Hence, the proposed system will be useful in identifying whether an individual has depression or not. Various therapists can use the proposed model to identify depression for both online and offline use quickly.

The upcoming section presentation is as per the following: the second section describes the related work for this topic, the third section explains the methodology that has been adopted and contains information related to data used and the preliminary data preparation process, the fourth section gives the details of the experimental results and discussions obtained based on the dataset, and finally, the fifth section presents the conclusion.

## 2 Literature Review

There have been various data-driven studies that try to identify stress factors that relate to the socio-demographic factors, academic factors, relationships with people, and other interpersonal and intrapersonal factors for depression and anxiety among young adults [1–3]. Statistical tests were used to check the causes of depression for both men and women. The results showed that intrapersonal and interpersonal stress could be a reason for Anxiety and Depression for all, where the essential individual-level predictor of the onset of Major Depressive Disorders (MDD) was previous suicidal ideation [4]. For women, performance stress resulted in Anxiety and Depression. Also, women are more prone to receive mental health diagnoses or treatment than men and showed a significantly higher rate of depressive symptoms [1]. Following a similar approach, the classification tree method was used to determine women's risk factors in different postpartum periods, which showed similar results [5]. However, most people use coping strategies and take social support to get out of the state of Depression and Anxiety [6, 7]. Very few men have help-seeking tendencies [1] and hence are more prone to rely upon approach-based coping strategies than avoidant coping strategies. In a study based on the Center for Epidemiological Scale for Depression score, the students were allotted depressive symptoms (either high or low). The Coping Response Inventory-Adult was used to analyze the most prominent coping strategies adopted by them. Data analysis in SPSS showed that the group with high depressive symptoms depended on more avoidant coping methodologies than the group with lower symptoms. Women were seen to utilize avoidant coping strategies in comparison to men [6]. A broad study across 36 counseling centers from different universities showed a general expansion in risk factors related to Depression and Suicidal Ideation among young adults, including Anxiety, eating disorders, fear, substance and alcohol abuse, hostility, and anger [8]. This study helped in identifying the factors and hence gave ways to address those issues. Another study to show the association among various forms of Anxiety and their relationship with depression displayed that in the case of Anxiety, broad neuroticism, Academic Anxiety, and Test Anxiety are the significant reasons for its occurrence in young adults.

The above results also provide an early indication or prediction of emotional distress [9]. Apart from this, a comparative study using Ordinal Logistic Regression, LIME (Locally Interpretable Model-Agnostic Explanations), and Deep Learning analyzed the depressive symptoms based on the features that are either directly or indirectly associated with depression [10]. Studies show that an individual's mental health can be recognized by their daily use of words. For this purpose, tweets were studied to examine the language use and connect it to the behavioral aspects and hence classify it into Depression or Anxiety. Simple Text Mining techniques were used to assess the mental health status of the individuals and provide therapy recommendations for the same [11].

## 3 Methodology

### 3.1 Proposed Model

The proposed system performs the task of identifying the most prominent features or parameters that have a role in determining depressive symptoms in an individual. The next step is to select the most appropriate classification algorithm. The classification algorithm parameters are tuned to give the optimal values, such that the maximum amount of accuracy is achieved.

As shown in Fig. 1, the unbalanced data is dealt with using an oversampling technique. The re-sampled data is divided into separate training and testing set in the ratio of 90:10. The classification models are then trained on the training set, and the testing set is used to check the various performance metrics to identify which classifier gives better performance.

**Model Selection:** Since the problem is a classification type, the algorithm should be chosen to correctly distinguish between the two classes (Yes/No). The algorithms considered are SVM, Logistic Regression, and Random Forest.

**Evaluation of the Model:** The classification models' efficiency was assessed using statistical methods. The final model achieved the highest accuracy of prediction.

### 3.2 Data Description

The dataset used is obtained based on the study of Nguyen et al. [12] entitled "A Dataset of Students' Mental Health and Help-Seeking Behaviors in a Multicultural Environment". It is a comparative dataset that gives insights into the mental health and help-seeking behavior of University students. It has found its use in various other research purposes too.

This data was collected using PHQ-9, the Social Connectedness Scale (SCS), and the General Health Help-Seeking Questionnaire (GHHSQ). One of the main findings from this dataset is how social connectedness is predictive of students' reported depression. It contains 268 observations based on the questionnaires mentioned above expressed by the international students as well as the domestic students at an international university in Japan [12].

The features considered for the study can be divided into three segments: Socio-demographic data, PHQ-9 data, and GHHSQ data.

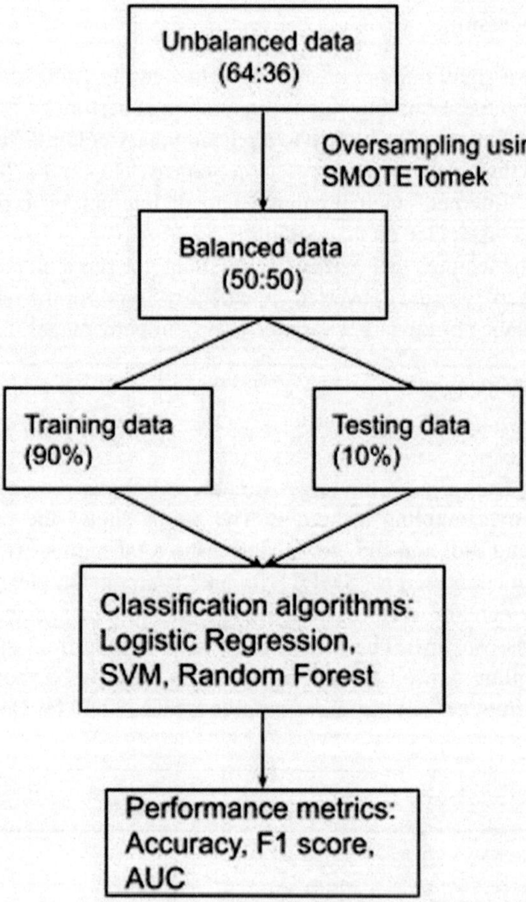


**Fig. 1** Model architecture

a) Socio-demographic data: Gender, Academic status (Graduate/Undergraduate), Age, Age category (1/2/3/4/5), Intimate relationship - boyfriend or girlfriend or spouse (yes/no), and Religious beliefs (yes/no).
b) PHQ-9 data: Suicidal ideation (yes/no), Depression (Target Variable) (yes/no), Depression Type (no/other/major), Total Depression score of PHQ-9, Depression Severity (min/mild/moderate/moderate-severe/severe), and Total Social Connectedness Score.
c) GHHSQ data (Data is present in the form of a Likert scale showing help-seeking behavior - 1 to 7): Intimate Partner, Friends, Parents, Relatives, Mental Health Professional, Phone Helpline, General Physician, Religious Leader, Alone (do not seek help from anyone), Others, and Internet (Website, Social Media, etc.). Also, the binary format (Yes/No) of these GHHSQ features are also present.

### 3.3 Data Pre-processing

The data contained a small number of missing values in the "Intimate" and "Internet" column. To deal with these missing values, imputation is performed based on the other columns' data. The "Intimate" column had a related binary column, "Intimate_bi," and those places where the NA value was present contained "No" in the "Intimate_bi" column. Similarly, for "Internet," another column named "Internet_bi" is present and based on that, imputations take place on that column.

Since most of the features that will be dealt with in this research are categorical, the columns with "object" type are converted to "category" type for the required columns.

The original balance of the target variable (Dep), in percentages, is:

- Depression Yes = 35.82%
- Depression No = 64.18%

As can be seen from Fig. 2, the target variable is unbalanced and needs to be re-sampled using an oversampling technique. The x-axis shows the two categories of Depression - Yes and No, and the y-axis shows the total number of occurrences (in percentage). The function used is "SMOTETomek" from the "imblearn" package. The "imblearn" package contains various re-sampling techniques that generate a dataset with an equal ratio of classes. "SMOTETomek" is one such method from the package which uses both oversampling as well as undersampling to create a balanced dataset. After re-sampling, the current ratio of the target variable is thus, 50:50 for Depression Yes and No.

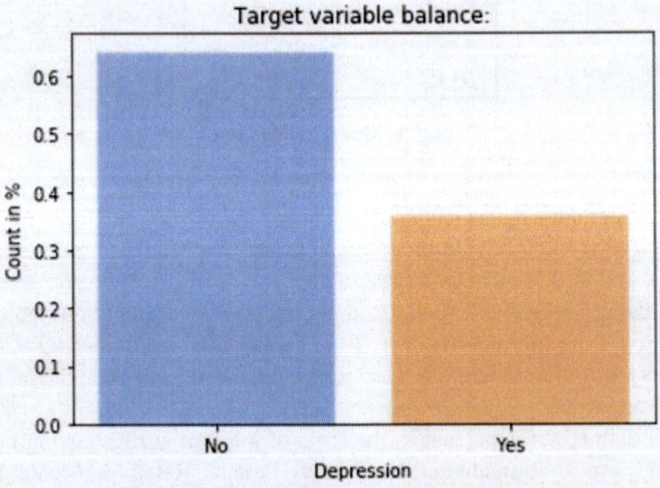


**Fig. 2** Balance ratio of the target variable

### 3.4 Classification Algorithms

Since it is a classification problem, the most crucial task is to predict whether a person might suffer from depression based on various demographical and socio-cultural features. Thus, it means that the problem is to classify depression in an individual in two classes: Yes and No. Various Classification algorithms are available in this regard, out of which the most appropriate algorithms for our purpose are selected.

The models used for this study are - Logistic Regression, SVM (Support Vector Machine), and Random Forest Classifier. Since our predictor variable is binary, Logistic Regression is one of the most preferred classifiers among other supervised learning classifiers; and Support Vector Machine is very useful for high dimensional spaces and is memory-efficient. Moreover, Random Forest helps in reducing overfitting and is more accurate. Hence, out of all the available Supervised Learning Algorithms, these classifiers are chosen for the study. All these algorithms were implemented using the "sklearn" package.

### 3.5 Feature Extraction

The dataset contains many features, so a feature selection technique must be performed to select the most appropriate features. The method used here is the Recursive Feature Elimination (RFE). It removes the weakest features until the number of specified features is reached. RFE is performed for all the models, and the classification report is checked for each one of them. For performing this, RFE was imported from

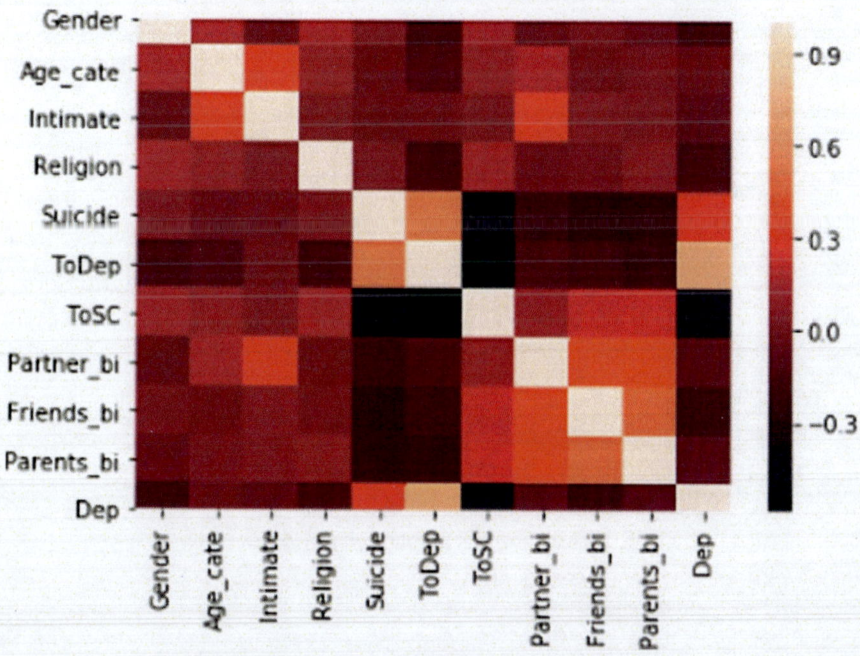


**Fig. 3** Heatmap for the features selected using RFE

the "sklearn.feature_selection" package. Based on the classification report, the most appropriate model and features are selected.

Figure 3 shows the heatmap of the correlation matrix obtained using only features selected by Recursive Feature Elimination.

The distribution plots, as well as the box plots obtained for the selected features, can be shown by the following Figs. 4 and 5:

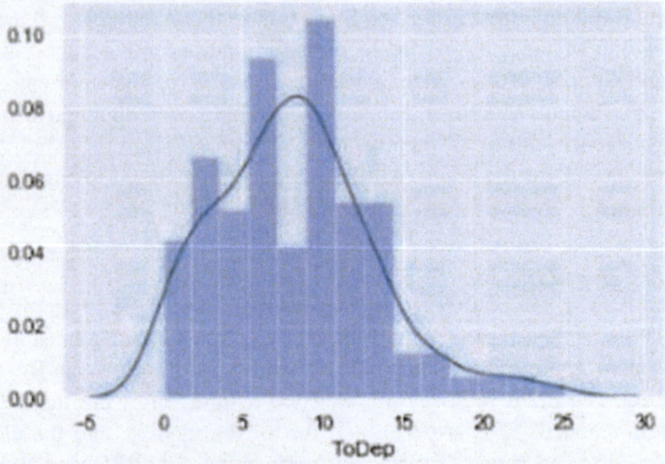


**Fig. 4** Distributed plot of total depression score

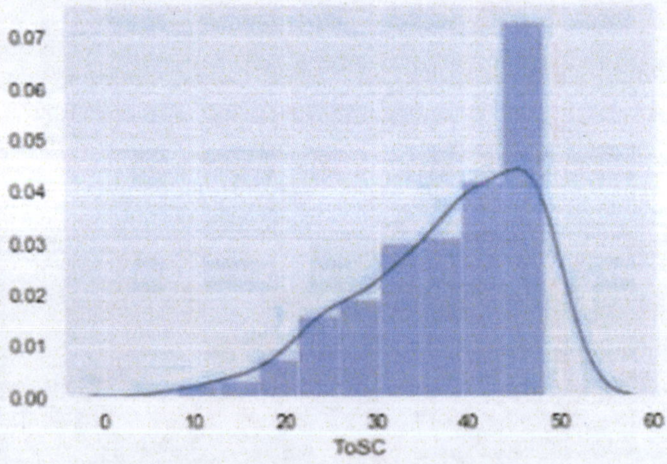


**Fig. 5** Distribution plot pf total connectedness score

After performing RFE with "RepeatedStratifiedKFold" cross-validation, it can be observed that Random Forest provides the highest accuracy of 88.2%. Hence, Random Forest Classifier is selected for building the most optimal classifier for the project. To choose the optimal values of the parameters for the Random Forest classifier, Hyperparameter Tuning is performed. Two methods are used for this purpose: "RandomizedSearchCV" and "GridSearchCV". In "RandomizedSearchCV", only a random sample of a fixed number of hyperparameters is performed, whereas, in "GridSearchCV", all the hyperparameters are tried (Figs. 6, 7, 8, 9, 10, 11, 12, 13).

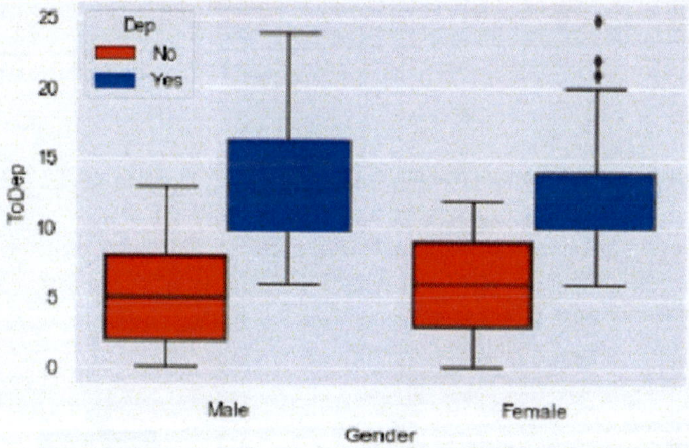


**Fig. 6** Distribution of the genders of the individuals with respect to the total depression score

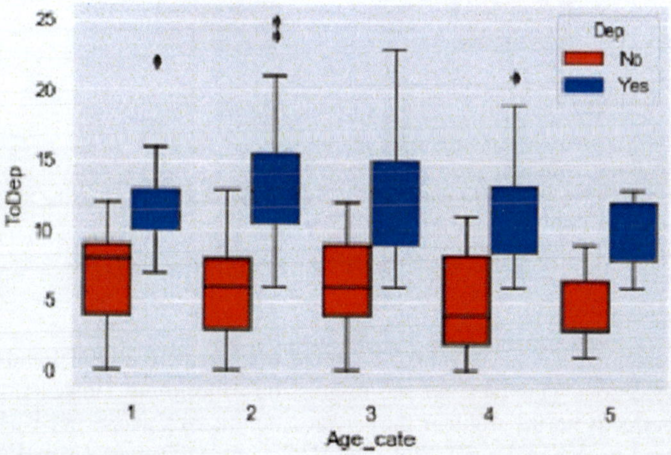


**Fig. 7** Distribution of the various age categories with respect to the total depression score

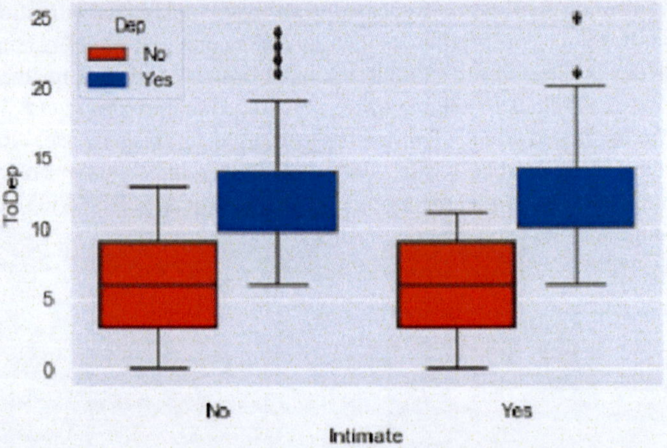


**Fig. 8** Distribution of the intimate relation of the individual with respect to the total depression score

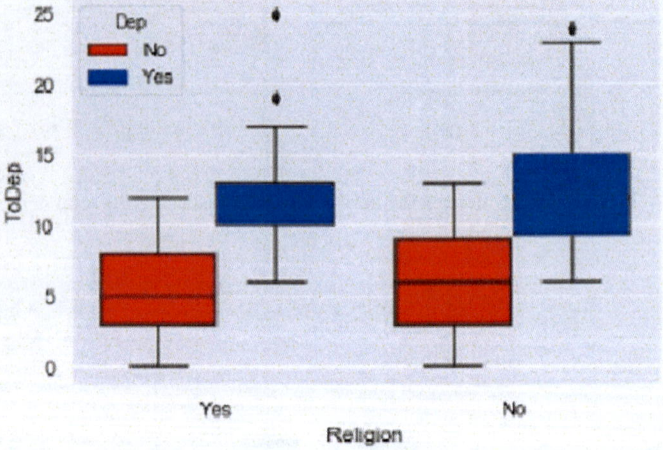


**Fig. 9** Distribution of the religious beliefs of the individuals with respect to the total depression score

## 3.6 Evaluation Metrics

One of the essential tasks in Machine Learning is to create the model and predict the output using various Classification and Regression Algorithms. However, to choose a model that perfectly fits the problem, the performance metrics' scores need to be checked. Here Accuracy, Precision, Recall, F1 score, AUC score, and Binary Cross-Entropy Loss have been used as the evaluation metrics.

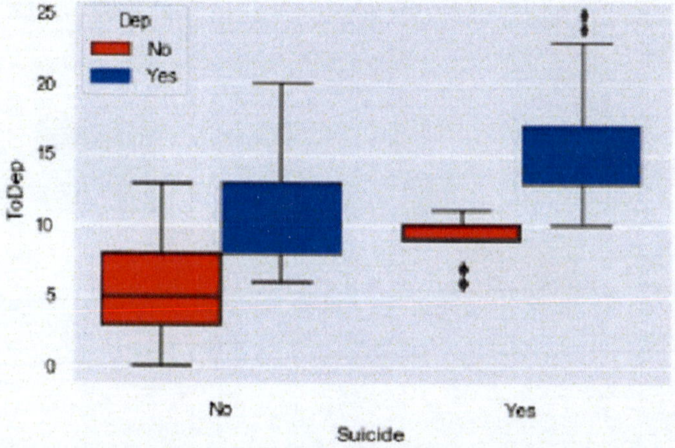


**Fig. 10** Distribution of the suicidal ideation of the individuals with respect to the total depression score

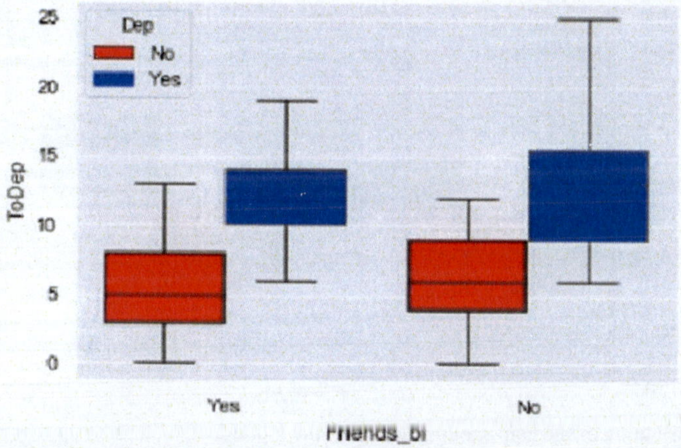


**Fig. 11** Distribution of the help-seeking ability of the individuals with their friends with respect to the total depression score

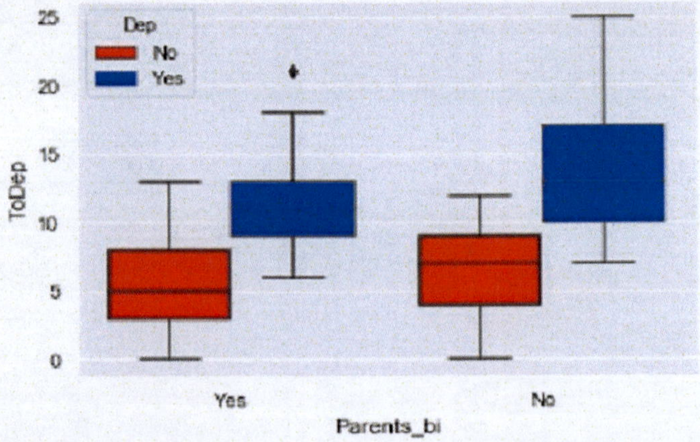


**Fig. 12** Distribution of the help-seeking ability of the individuals with their parents with respect to the total depression score

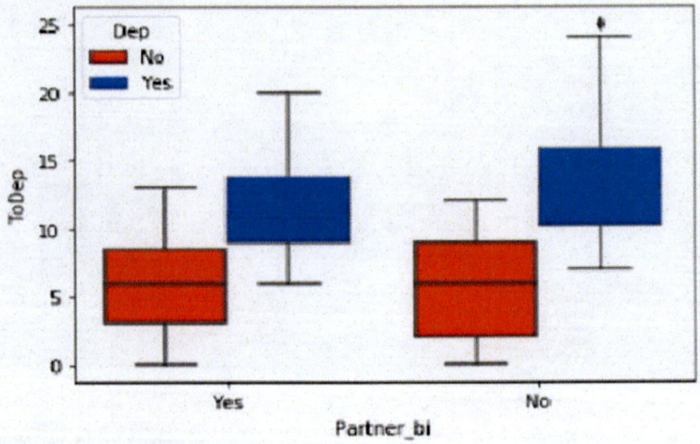


**Fig. 13** Distribution of the help-seeking ability of the individuals with their partners with respect to the total depression score

# 4 Results and Discussions

## 4.1 Precision, Recall, and F1 Score

Precision is the proportion of the True Positives and all the True Positives and the False Positives. A recall is the proportion of the True Positives and all the True Positives and the False Negatives. And the F1 score is the harmonic mean of the Precision and the Recall values.

The various values obtained are:

Table 1 shows the values of the performance metrics obtained for the original data, and Table 2 shows the values obtained after oversampling the data. The performance metrics show better results in the original data than the oversampled data due to the

**Table 1** Precision, Recall, and F1 scores for the original data

| Classification algorithm | Class | Precision | Recall | F1 score |
|---|---|---|---|---|
| Logistic regression | 0 | 1.00 | 0.94 | 0.9523 |
| | 1 | 0.91 | 1.00 | |
| SVM | 0 | 1.00 | 1.00 | 1.0000 |
| | 1 | 1.00 | 1.00 | |
| Random forest | 0 | 1.00 | 0.94 | 0.9523 |
| | 1 | 0.91 | 1.00 | |

**Table 2** Precision, Recall, and F1 scores after oversampling the original data

| Classification algorithm | Class | Precision | Recall | F1 score |
|---|---|---|---|---|
| Logistic regression | 0 | 0.89 | 0.94 | 0.9090 |
| | 1 | 0.94 | 0.88 | |
| SVM | 0 | 0.73 | 0.94 | 0.7586 |
| | 1 | 0.92 | 0.65 | |
| Random forest | 0 | 0.89 | 0.94 | 0.9090 |
| | 1 | 0.94 | 0.88 | |

unbalanced nature of the target variable. After oversampling the values reduce, but the results are not influenced by the unbalanced classes. According to the Table 2, Logistic Regression and Random Forest provide higher precision, recall, and F1 scores than SVM.

### 4.2 Accuracy

Accuracy is the rate of correct predictions/classifications. Since the original data is unbalanced, the models created using it are not reliable. After Oversampling the data to fix the balance of the target variable, the accuracies for Logistic Regression and Random Forest give equal values. Now, the data is balanced, but the most prominent features need to be selected from the entire set of explanatory variables. After applying Recursive Feature Elimination on the balanced data, the Random Forest gives the highest accuracy followed by Logistic Regression. Thus Random Forest shows the highest accuracy scores, so it is selected as the most optimal model for our purpose. The Base Model gives an accuracy of 88.2% and after Hyperparameter Tuning using both Randomized Search and Grid Search Cross-Validation techniques, an exactly equal accuracy of 91.17% is obtained. Thus, from the results it can be seen that, both the techniques give the same accuracy values for their best estimators.

### 4.3 AUC - ROC Curve

Since Random Forest Classifier provides better results, it is selected as the classification model for our purpose. ROC curve is the graph between the True Positive Rate (along the y-axis) and False Positive Rate (along the x-axis). It shows the performance of the classification model at all thresholds.

Figure 14 shows the ROC curve obtained for the Random Forest Classifier model. AUC score tells how much the model can distinguish between the classes. It lies in the range of [0, 1]. The larger the AUC score, the better the classifier is at differentiating within the categories. Here, the AUC score for the Random Forest Classifier is 0.97923. Thus the proposed model can easily distinguish between the classes Yes and No for Depression.

### 4.4 Binary Cross-Entropy Loss

Since the target variable is binary, binary cross-entropy loss is opted. The predicted probability for each class is compared to the actual desired output, and the loss is calculated by penalizing the probability based on how close or far it is from the desired result. It helps in checking the performance of a binary classification model. Lower the classification loss, better is the model. The cross-entropy loss value obtained here is 0.12445, which is considered a good score for a classification model.

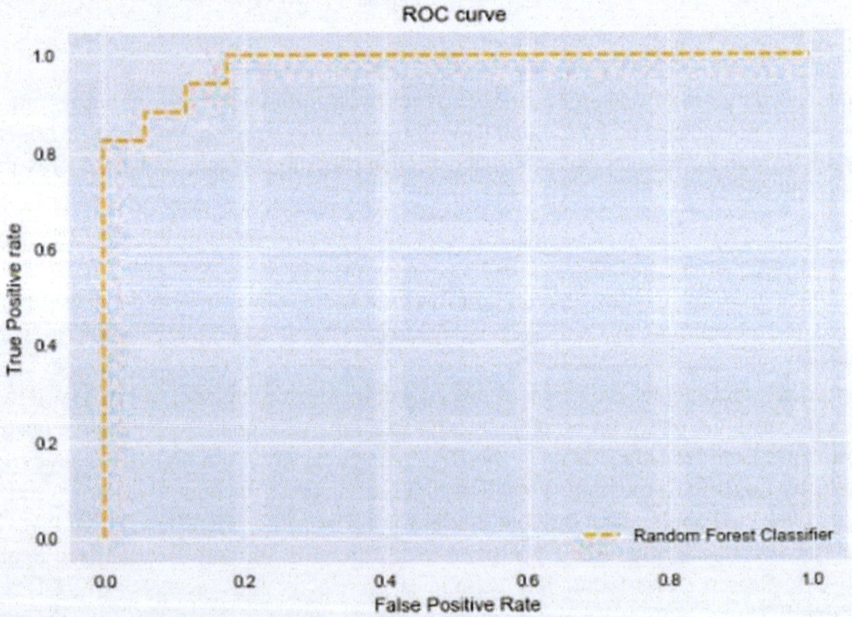


**Fig. 14** ROC curve for Random Forest Classifier

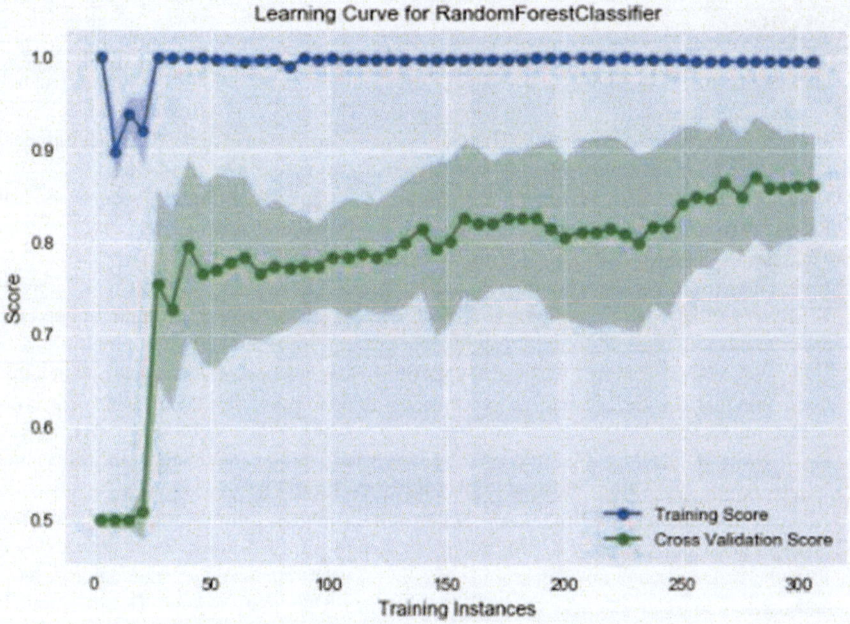


**Fig. 15** The learning curve for Random Forest Classifier

### 4.5 Learning Curve

The Learning Curve shows the relationship between the training scores versus the cross-validation test scores with varying samples. This curve shows the various accuracies for the training and testing samples concerning the cross-validation scores. The x-axis contains the number of training instances considered, and the y-axis has the accuracy scores.

The Learning curve obtained for our Random Forest classifier is shown in Fig. 15. It can be seen that the estimator benefits from an increase in the number of data.

## 5 Conclusion

This research aims to create a classification system that will help predict whether an individual has depression or not based on various features taken into consideration. Thus, after implementing the model, it is observed that the model can correctly identify whether an individual has depression or not with an accuracy of 91.17%. The main challenge faced was to find ways to increase the efficiency of the model. From selecting the technique to choosing the most appropriate features to tune the hyperparameters, all of this required a lot of trial and error and patience. The most significant advantage of this model is that it can quickly identify depression in an individual. However, this model can only be used for identification purposes and not for measuring the severity of depression.

In the future, extensions can be made to classify the levels of depression (mild, moderate, severe) and hence recommend suitable therapy for the same.

## References

1. Hubbard K, Reohr P, Tolcher L, Downs A (2018) Stress, mental health symptoms, and help-seeking in college students. Psi Chi J Psychol Res 23(4):293–305 ISSN 2325-7342
2. Villatte A, Marcotte D, Potvin A (2017) Correlates of depression in first-year college students. Can J High Educ Revue canadienne d'enseignement supérieur 47(1):114–136
3. Mohd N, Yahya Y (2018) A data mining approach for prediction of students' depression using logistic regression and artificial neural network. In: Proceeding of IMCOM 2018, Langkawi, Malaysia, 5–7 January 2018
4. Ebert DD, Buntrock C, Mortier P, Auerbach R, Weisel KK, Kessler RC, Cuijpers P, Green JG, Kiekens G, Nock MK, Demyttenaere K, Bruffaerts R (2018) Prediction of major depressive disorder onset in college students. Depression Anxiety 36(4):294–304
5. Camdeviren HA, Yazici AC, Akkus Z, Bugdayci R, Sungur MA (2007) Comparison of logistic regression model and classification tree: an application to postpartum depression data. Expert Syst Appl 32(4):987–994
6. Sawhney M, Kunen S, Gupta A (2020) Depressive symptoms and coping strategies among Indian university students. Psychol Rep 123(2):266–280
7. Alsubaie MM, Stain HJ, Webster LAD, Wadman R (2019) The role of sources of social support on depression and quality of life for university students. Int J Adolesc Youth 24(4):484–496
8. Roesch JB: Depression and suicidal ideation in undergraduate college students: risk factors and barriers to treatment present within universities. University Honors Theses. Paper 186 (2015)
9. Cassady JC, Pierson EE, Starling JM (2019) Predicting student depression with measures of general and academic anxieties. Front Educ 4:11
10. Choi B, Shim G, Jeong B, Jo S (2020) Data-driven analysis using multiple self-report questionnaires to identify college students at high risk of depressive disorder. Sci Rep 10(1):7867
11. Rath T, Purohit Y, Rajput R, Ghosh A, Shukla S (2021) Therapy recommendation based on level of depression using social media data. In: Jat DS, Shukla S, Unal A, Mishra DK (eds) Data Science and Security, vol 132. Lecture Notes in Networks and Systems. Springer, Singapore, pp 289–297. https://doi.org/10.1007/978-981-15-5309-7_31
12. Nguyen M-H, Ho M-T, Nguyen Q-Y, Vuong Q-H (2019) A dataset of students' mental health and help-seeking behaviors in a multicultural environment. Data 4(3):124

# Analysis of Rule-Based Classifiers for IDS in IoT

Pushparaj Nimbalkar(✉) and Deepak Kshirsagar

Department of Computer Engineering and IT, College of Engineering Pune, Pune, Maharashtra, India

**Abstract.** The Internet of Things (IoT) networks suffered from different types of cyber attacks due to vulnerabilities present in IoT devices. The attacker creates Denial of Service (DoS) and Distributed DoS (DDoS) quickly towards IoT networks. Therefore, to secure IoT networks from such types of cyber attacks intelligent intrusion detection system is needed. This paper proposes the IDS with and without feature selection to detect DoS and DDoS attacks in IoT. The proposed system achieves higher accuracy of 99.9992% with a JRip classifier from the suite of rule-based classifiers using 36 features obtained using pre-processing data phase. The proposed approach brings relevant features using the correlation feature selection method with top-ranked 50% features and achieves higher accuracy of 99.9994% on IoT-BoT dataset compared to 36 features obtained after pre-processing data phase. The proposed system is compared with traditional IDSs in terms of the used feature selection method and dataset.

**Keywords:** Internet of Things · Cyber attacks · Intrusion detection · Feature selection · Rule-based classifiers

## 1 Introduction

The Internet of Things (IoT) is a term used to describe globally connected objects and devices of all types connected to the Internet through wired or wireless connections and other connected devices. The IoT is a massive network of interconnected devices and individuals that store and exchange data on how they are used and their surroundings.

The Internet of Things business intends to increase. Intelligent vehicles, innovative home systems, and connected industrial equipment have more IoT connections than non-IoT connections like smartphones, tablets, and computers. By the end of 2020, 54% of the 21.7 billion operating mobile devices [1] on the planet will be IoT system connections. More than 30 billion IoT connections are anticipated by 2025, with about four IoT devices per person on average.

IoT systems are insecure because they have no built-in protection to protect themselves from attacks. Users, in addition to technological factors, lead to the devices' susceptibility to attacks. Some issues like Hardware shortcomings and computing power ability, Technology for diverse distribution, unaware of compliance issues DoS and DDoS are the most challenging attacks due to many connection requests for a period that forces the machine to slow down and crash. In the security network, it is one of the most popular cyber-attack techniques.


S. Shukla et al. (eds.), *Data Science and Security*, Lecture Notes in Networks and Systems 290,
https://doi.org/10.1007/978-981-16-4486-3_51

The use of artificial intelligence through machine learning and deep learning in IDS provides higher performance. The availability of classifiers and feature selection methods in machine learning plays a significant role in IDS. The work presents rule-based classifiers using feature selection methods to detect DoS and DDoS attacks in IoT networks.

The contributions of the work are as follows

- This paper proposes the analysis of filter-based feature selection techniques in IDS to detect DoS and DDoS attacks.
- The proposed feature selection-based IDS achieves higher accuracy on BoT-IoT dataset.

## 2 Literature Review

The work [2] proposed IDS to detect DDoS attacks using Mahalanobis Distance (MD). The system used principal component analysis (PCA) for feature selection and provided higher precision of 92% using relevant features on KDD Cup 1999 dataset.

The system [3] proposed PCA for dimensionality reduction in IDS. The system obtained a higher False Alarm Rate (FAR) of 1.116% with Softmax regression on KDD Cup 1999 dataset using ten features.

The study [4] presented semantic IDS to detect DoS attacks. The system provides a conceptual ontology model and semantic rules based on network security situation elements.

The work [5] proposed IDS to secure IoT devices from XOR DDoS. The system used the wrapper feature selection method and provided higher accuracy of 97.39% with Support Vector Machine (SVM) on Cowrie dataset.

The study [6] proposed feature selection for IDS with deep learning. The system achieved higher accuracy of 99.2% on UNSW-NB 15 dataset, respectively, with DNN using reduced features.

The system [7] proposed wrapper feature selection to detect malicious traffic in IoT network. The system obtained fewer features using a combination of correlation method and accuracy metrics of the selected classifier. The system achieved lower HTTP DoS sensitivity compared to TCP DDoS and UDP DDoS on BoT-IoT dataset using Random Forest (RF).

The study [8] proposed multi-objective feature selection technique in IDS. The system obtained six reduced features and achieved an accuracy of 99.90% with an extreme learning machine (ELM) classifier on CICIDS 2017 dataset.

The work [9] proposed a multi-layer model to detect DDoS attacks. The model achieved an accuracy of 99.98% with a decision tree on the simulated dataset using manually selected eight features.

## 3 The Proposed IDS in IoT Using Feature Selection with Rule-Based Classifier

The proposed IDS in the IoT network, as shown in Fig. 1, consists of data pre-processing, feature selection using filter methods, and rule-based classifier.

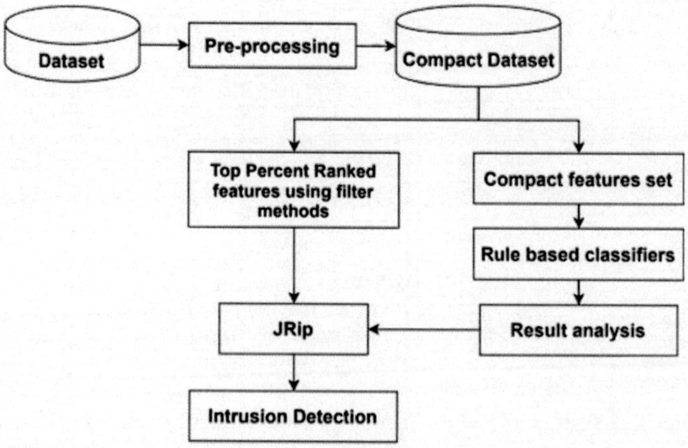


**Fig. 1** Proposed IDS in IoT using feature selection with rule-based classifier

The system captures the network traffic in the IoT network and is used for intrusion detection. The captured network traffic does not operate directly for intrusion detection due to the presence of noise. Therefore, data pre-processing is performed on the captured network traffic. The noisy data consist of missing and NaN values. It also includes some of the network attributes that are represented in different forms. In pre-processing data phase missing and NaN values are replaced with zero. Further, some of the network attributes that create intrusion detection noise are removed from the dataset. The data pre-processing helps to obtain compact data for intrusion detection in IoT.

Firstly, the proposed IDS uses the compact dataset to select the appropriate classifier to detect DoS and DDoS attacks. The system selects different classifiers such as PART, Decision Table, OneR, ZeroR, and JRip from rule-based classifiers. The performance of these rule-based classifiers is measured on the compact dataset. A single classifier that provides higher accuracy with substantial model built-up time is selected for intrusion detection.

Further, IG, GR, CR, CHI, SU, and ReliefF from the suite of filter-based feature selection methods are analyzed on the compact dataset. These feature selection methods produce the score for each feature with rank ranges from zero to one. The system selects top-ranked 25, 50, and 75% features of the total number of features present in the compact dataset. The subsets obtained using top-ranked 25, 50, and 75% for each method are analyzed with the best rule-based classifier. Each feature selection method's performance by selecting top-ranked 25%, 50%, and 75% features is measured and compared with the original feature set using JRip Classifier. The system suggests the relevant feature subset for intrusion detection that provides higher accuracy with minimum build-up time.

## 4 System Implementation and Result Analysis

The proposed IDS is implemented and tested using the Weka tool. The presence of feature selection methods and rule-based classifiers in Weka is used in intrusion detection. The implemented system is tested on IoT-BoT dataset.

**Table 1** Analysis of rule-based classifiers on IoT-BoT dataset

| Classifier | No. of features | Accuracy (%) | Incorrectly classified instances (%) | Model built time |
|---|---|---|---|---|
| PART | 36 | 99.9989 | 0.0011 | 42.43 |
| DT | 36 | 99.9982 | 0.0018 | 135.95 |
| OneR | 36 | 99.9930 | 0.0070 | 6.66 |
| ZeroR | 36 | 99.9334 | 0.0666 | 0.13 |
| JRip | 36 | 99.9992 | 0.0008 | 80.94 |

The proposed system mainly concentrates on detecting DoS and DDoS attacks; therefore, only DDoS and DDoS instances from the dataset [10] are extracted and used for the system implementation. The dataset consists of 715,371 instances of DDoS and DDoS. The dataset consists of 477 normal samples with 42 features excluding labels.

The dataset does not operate directly for intrusion detection due to the presence of noise. Therefore, data pre-processing is performed on the noisy dataset. In pre-processing phase [11] missing and NaN values are replaced with zero. Some network traffic features such as Packet ID, Stime, Ltime, Protocol, Flags, Source IP, and Destination IP that create intrusion detection noise are removed from the dataset. The data pre-processing obtains the compact data with 36 features for intrusion detection in IoT.

The system uses PART, Decision Table (DT), OneR, ZeroR, and JRip [12] from rule-based classifiers, and performance is measured on the compact dataset. The classifiers' performance is compared with each other, and a single classifier is selected for further intrusion detection. Table 1 shows the analysis of Rule-based classifiers on the compact IoT-BoT dataset.

Table 1 shows ZeroR classifier built the model with a minimum of 0.13 s but provides lower accuracy of 99.9334% with maximum Incorrectly classified instances compared to PART, DT, and JRip.

Figure 2 shows JRip offers higher accuracy of 99.9992% compared to other classifiers with 80.94 s model built-up time. Therefore, the system selects a JRip classifier for feature selection-based IDS to detect DoS and DDoS attacks.

IG, GR, CR, CHI, SU, and ReliefF methods [12, 13] present in the Weka are used to feature intrusion detection. These feature selection methods obtain the score for the 36 features present in the compact dataset with rank. IG obtain the score for each feature in descending order ranging from 0.000546 to 0.007627. GR brings the score for each feature in descending order ranging from 0.000487 to 0.322073. CR obtains the score for each feature in descending order ranging from 0.00241 to 0.80615. SU gets the score for each feature in descending order ranging from 0.000967 to 0.474024. ReliefF obtains the score for each feature in descending order ranging from 0.00000104 to 0.83849159. In contrast, CHI brings the score for each feature in descending order ranging from 8324.91664 to 671,607.3644. The system selects top-ranked 25, 50, and 75% features and obtains the feature subsets for each feature method. The subset received using the

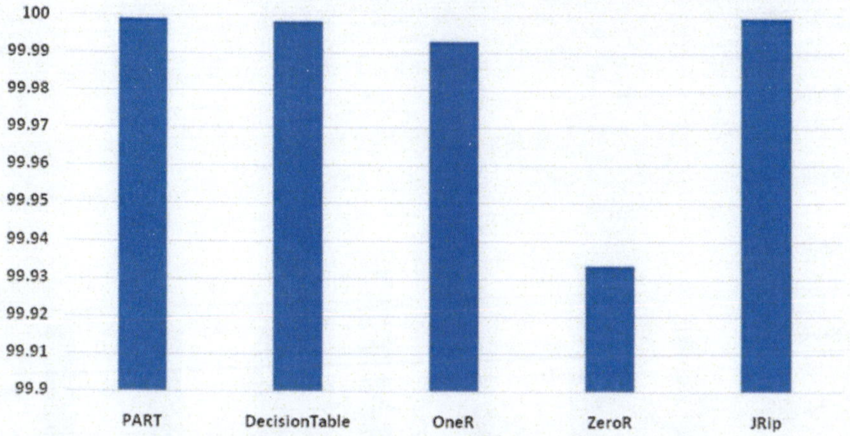


**Fig. 2** Accuracy of rule-based classifiers on IoT-BoT dataset

**Table 2** JRip performance analysis on IoT-BoT dataset using 18 features

| FS method | No. of features | Accuracy (%) | Incorrectly classified instances (%) | Model built time |
|---|---|---|---|---|
| Compact | 36 | 99.9992 | 0.0008 | 80.94 |
| IG | 18 | 99.9993 | 0.0007 | 36.34 |
| GR | 18 | 99.9992 | 0.0008 | 42.45 |
| CR | 18 | 99.9994 | 0.0006 | 49.76 |
| CHI | 18 | 99.9992 | 0.0008 | 36.71 |
| SU | 18 | 99.9992 | 0.0008 | 40.53 |
| ReliefF | 18 | 99.9992 | 0.0008 | 43.38 |

selection of top-ranked 25, 50, and 75% features include 9, 18, and 27 features, respectively. The obtained feature subsets for each feature selection method are tested with a JRip classifier, and performance is measured. The experimentation carried out shows that the subsets obtained using top-ranked 50% features provide similar or improved accuracy with JRip compared to the compact feature set. Table 2 shows the subset's performance analysis using the selection of the top 50% ranked features with JRip for each feature selection method.

Table 2 shows GR, CHI, SU, and ReliefF provides similar accuracy of 99.9992% using 18 features with JRip compared to the compact feature set in a minimum model built-up time.

Figure 3 shows CR achieves a higher accuracy of 99.9994% with JRIp than other feature selection methods and compact feature set. JRip classifier built the model in a minimum of 36.34 s and achieved higher accuracy of 99.9993% compared to GR, CHI, SU, ReliefF, and compact feature set using IG subset of top-ranked 18 features. The

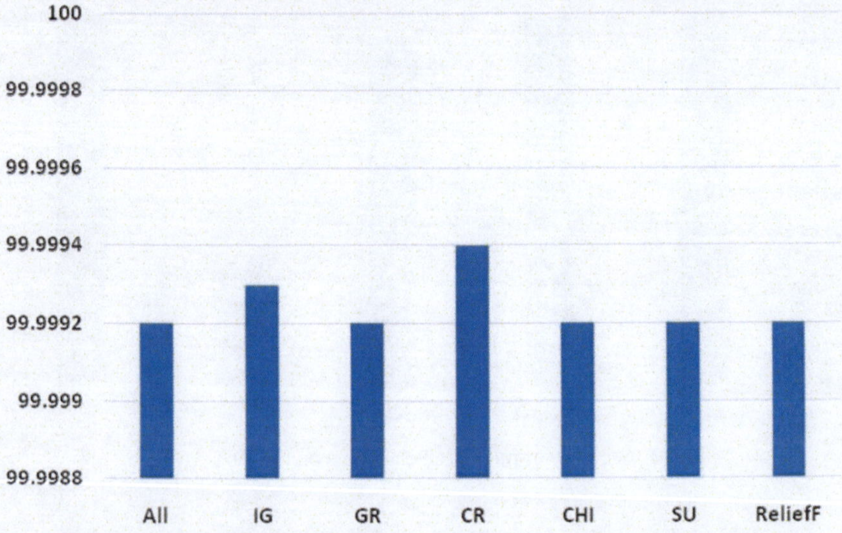


**Fig. 3** Accuracy of JRip classifier using filter-based feature selection methods

performance analysis using top-ranked 50% features suggests that IG and CR provide higher accuracy than the compact feature set.

## 5 Comparison with Traditional IDSs

The implemented IDS to detect DoS and DDoS attacks is compared with the traditional IDSs. Table 3 shows a comparative analysis of the proposed feature selection-based IDS with traditional IDSs.

Table 3 shows the proposed system achieved higher accuracy of 99.9994% with JRip using top-ranked 18 features of CR on the latest IoT-BoT dataset includes new types of DoS and DDoS attacks compared to [2] and [8] in IoT.

**Table 3** Comparison with traditional IDSs in IoT

| Work | FS method | Classifier | Dataset | Accuracy (%) |
|---|---|---|---|---|
| Compact feature set | NA | JRip | IoT-BoT | 99.9992 |
| [2] | PCA | MD | KDD Cup 1999 | 92.0000 |
| [8] | multi objective | ELM | CICIDS 2017 | 99.9000 |
| Proposed | IG | JRip | IoT-BoT | 99.9993 |
| | CR | JRip | IoT-BoT | 99.9994 |

## 6 Conclusion

This paper presented filter-based feature selection methods in IDS to detect DoS and DDoS attacks. The presented system achieved higher accuracy of 99.9992% on IoT-BoT dataset with JRip using 36 features obtained after data pre-processing. The feature selection methods such as CR and IG with top-ranked 50% features in IDS achieved higher accuracy of 99.9994% and 99.9993%, respectively, using JRip compared to the compact feature set.

This paper's presented work will extend to obtain reduced features using a combination of IG and CR in the future.

## References


1. Knud L (2020) State of the IoT 2020: 12 billion IoT connections, surpassing non-IoT for the first time. https://iot-analytics.com/state-of-the-iot-2020-12-billion-iot-connections-surpassing-non-iot-for-the-first-time/
2. Zong Y, Guoyan (2019) A feature dimension reduction technology for predicting DDoS intrusion behavior in multimedia internet of things. Multimed Tools Appl 80:2267122684
3. Zhao S, Wei T, Albert Y (2017) A dimension reduction model and classifier for anomaly-based intrusion detection in internet of things. In: IEEE 15th international conference on dependable, autonomic and secure computing, 15th international conference on pervasive intelligence and computing, 3rd international conference on big data intelligence and computing and cyber science and technology congress (DASC/PiCom/DataCom/CyberSciTech), pp 836–843
4. Xu G, Yan C, Yuanyuan R, Xiaohong L, Zhiyong F (2017) Network security situation awareness based on semantic ontology and user-defined rules for Internet of Things. IEEE Access 5:21046–21056
5. Shrivastava RK, Bashir B, Hota C (2019) Attack detection and forensics using honeypot in IoT environment. In: Fahrnberger G, Gopinathan S, Parida L (eds) ICDCIT 2019, vol 11319. LNCS. Springer, Cham, pp 402–409. https://doi.org/10.1007/978-3-030-05366-6_33
6. Choudhary S, Nishtha K (2020) Analysis of KDD-Cup'99, NSL-KDD and UNSW-NB15 datasets using deep learning in IoT. Procedia Comput Sci 167:1561–1573
7. Shafiq M, Zhihong T, Ali K, Xiaojiang D, Mohsen G (2020) IoT malicious traffic identification using wrapper-based feature selection mechanisms. Comput Secur 94:101863
8. Roopak M, Gui Y, Jonathon C (2020) An intrusion detection system against DDoS attacks in IoT networks. In: 10th annual computing and communication workshop and conference (CCWC). IEEE, pp 562–567
9. Chen Y, Jang-Ping S, Yung-Ching K, Nguyen C (2020) Design and implementation of IoT DDoS attacks detection system based on machine learning. In: European conference on networks and communications (EuCNC). IEEE, pp 122–127
10. Koroniotis N, Nour M, Elena S, Benjamin T (2019) Towards the development of realistic botnet dataset in the internet of things for network forensic analytics: Bot-iot dataset. Futur Gener Comput Syst 100:779–796
11. Kshirsagar D, Sandeep K (2021) A feature reduction based reflected and exploited DDoS attacks detection system. J Ambient Intell Hum Comput 1–13
12. Kshirsagar D, Sandeep K (2020) An ensemble feature reduction method for web-attack detection. J Discrete Math Sci Crypt 23(1):283–291
13. Kshirsagar D, Kumar S (2020) Identifying reduced features based on IG-Threshold for DoS attack detection using PART. In: Hung DV, D'Souza M (eds) ICDCIT 2020, vol 11969. LNCS. Springer, Cham, pp 411–419. https://doi.org/10.1007/978-3-030-36987-3_27

# Towards a Novel Strategic Scheme for Web Crawler Design Using Simulated Annealing and Semantic Techniques

S. Manaswini[1] and Gerard Deepak[2(✉)]

[1] Department of Computer Science and Engineering, SRM Institute of Science and Technology, Ramapuram, Chennai, India

[2] Department of Computer Science and Engineering, National Institute of Technology, Tiruchirappalli, India

**Abstract.** In this era the reflexive response to any question that pops in the mind, is to ask Google. And it is fascinating to witness the speed at which Google is able to give us so many answers with multiple interpretations despite the usage of broken language during type-searching. The module that powers google to provide so many responses is known as a web crawler. Just as the name suggests, it crawls through the web using the hyperlinks or URLs present on each page. The most vital unit of a search engine is the web crawler. When a particular topic is to be searched for, a focused web crawler is put to work. The internet has become the hub of resources of late and the efficiency of web crawlers is extensively put to test. This paper is built over two baseline papers that suggest models for focused web crawling- one for bioinformatics and the other in relevance to building a database for academicians. A multi module hybridized approach which initially classifies the URL database using decision tree and the optimizes with simulated annealing - a meta heuristic optimization algorithm and domain ontology has been devised to lift the efficiency of the focused web crawler. The proposed approach is seen to have an accuracy of 90.12% and an F-measure of 91.62% .

**Keywords:** Decision tree · Domain ontology · Search engine · Simulated annealing · Web crawler

## 1 Introduction

Now-a-days surfing the internet is the immediate solution to any question or information requirement. Be it any area of research or study, the instinctive response to look- out for resources to begin exploration, is to search the web. The internet is an immeasurable source of data. But fishing out what is required and relevant is a real task. If the internet decided to surface anything and everything as the result of a posted query or search, the task of fishing becomes manual and hence tedious. And it only makes sense for the engine to classify and order in accordance to the relevance of the search, at the very time of surfing by filtering the choice of URLs. Crawlers, in simple terms, are bots that index content all over the internet for retrieval at a later point. The crawlers quite literally crawl


S. Shukla et al. (eds.), *Data Science and Security*, Lecture Notes in Networks and Systems 290,
https://doi.org/10.1007/978-981-16-4486-3_52

through the multiple web pages present on the internet. It crawls via relevant references on each webpage and catalogs relevant results. As it is highly improbable to store all the indexed content, focused crawlers work to satisfy particular property or a relevant topic, wherein the index with great detail. The first page that it begins its crawling with is called the seed page and the subsequent pages are child pages. And each child page has relevant results. The relevance of the parent page defines the consequent relevance of the subsequent child page.

*Motivation:* The relevance of results of crawling the web have become extremely important for a number of reasons, beginning at the very identification of obtaining the right dataset for experimentation purposes in order to analyse and identify previously undiscovered facts. There have been various suggested methods to improve the efficiency of focused web crawlers such as the usage of domain ontologies relational methodologies using SQL and so on but yet what could increase the efficiency of a web crawler is an integrated multi module system design that focuses on the increasing the efficiency in the filtering of the indices that in turn define the relevance of a page.

*Contribution:* This paper suggests a model that has multiple crawled web pages from the Internet, storing it in a URL database and allocating it to multiple AI classification fogs using a scheduler and is then classified using a decision tree algorithm [1]. Meanwhile the seed URL is also fed into AI fog [2]. The source of the seed URL is a dataset that consists of URL categories. This is then passed through a classification algorithm to classify the key indices for crawling through referential URLs. The following output is mapped and passed through an aptly designed domain ontology [3] which is further passed through an optimization algorithm [4] and later the level of relevance is calculated using metric measures such as web overlap index and cross entropy to ensure the increased efficiency and throughput in web crawling.

*Organization*: The first section of the paper includes a brief introduction to the concept. The flow of the rest of the paper is as follows-: Sect. 2 consists of related work to the field of study and experimentation. Section 3 illustrates the proposed architecture for the system of anomaly detection. Section 4 consists of the implementation and experimental observation. Section 5 includes conclusion and references.

## 2 Related Works

Manish kumar et al. [5] have made use of the focused crawling methodology for tracing Indian origin academicians working abroad. This aims at providing a connecting platform for those Indian students who are applying for study programs across the globe. The designed focused crawler for Indian origin Academicians (DFCI), has the capability to crawl websites inorder to present them as seed URLs and also focuses on constant information updation. Mani Sekhar et al. [6] in their paper have designed an Optimized focused web crawler (OFCW) and proposed the use of master slave architecture to optimize the design and implementation of the focused web crawlers and have also proposed a solution for predicting the page relevance, which is based on Natural Language Processing (NLP). Their results show that the optimization of the focused crawler is achieved better when the computation is performed in parallel when compared to serial computation. Anish Gupta et al. [7] in their paper have reviewed the different types of approaches of focused crawlers and have proposed focused web crawler design to unmask the hidden techniques of web crawling implementation. Wen-xian Wang et al. [8] in their paper put forward a focused crawler based on Naive Bayes classifier which made use of TF-IDF algorithm to extract the characteristics of page content. They also claimed that their model outperformed the page rank crawler and BFS crawler. Duygu Taylan et al. [9] in their paper have focused on classification of links and have made use of Naive bayes classifier for classification of URLs with simple URL scoring optimization. G Pant et al. [10] in their paper have discussed the rudimentary problems regarding development of an infrastructure for crawlers and have reviewed several topical crawling algorithms, and evaluation measures. [11–27] include references on knowledge centric sematic approaches.

## 3 Proposed Architecture

The proposed system architecture has a selective set of data inputs passed through multiple modules for weeding out the irrelevant URLs. A hybridized approach has been formulated in order to increase the efficiency of the crawler. A URL classification dataset is used for experimentation and has been obtained from DMOZ, a web directory, present in Kaggle. The dataset includes URL categories and the top 4–10% of the categories of URL are taken as the seed URLs for the focused crawler. The seed URLs are the ones that are most relevant to the topic of crawling. A few more categories are then added to further make them more explicit from Wikidata, DBpedia and other real-world notions and the URL base which contains the URLs crawled from the web which are further indexed. These indexed URLs from the URL base is fed into the scheduler. The scheduler then schedules or allocates them to the AI fog.

A fog provides a decentralized structure for resource or storage. Here the URL dataset is placed into multiple fogs for the convenience of retrieval. Fogs allow data storage in easily accessible logical locations closer to the user end, thereby enhancing the performance and reducing the time taken. Crawling URLs are a computationally complex task in terms of filtering, given the large number of data points. Such computationally complex processes when conducted in clouds instead, face issues such as high latency and transmission expenses. Fogs are lightweight and break the large chunk of data and hold it in multiple units of bearable size, that coordinate and cooperate to process in separate units, but all at the same time together.

The data in the proposed framework is automatically sampled, and classification occurs parallelly. In an approximation, about 80% of the data must be trained in a cloud. But whereas the fog reduces this burden. Therefore, multiple fogs in the model for multiple simultaneous processing are installed and further integration of AI to fogs enhance the filtering process. Eventually the process then moves into further classification of the URLs using the Decision tree algorithm. As the input URLs are indexed and the less relevant URLs are required to be discarded at each level, this algorithm proves to be efficient enough due to its simplistic nature. URL topic mapping is then used to map the key attributes obtained from the tokenization of preceding URL in the focused web crawler to procure a more relevant succeeding URL, mapping to the thesaurus of domain relevant terms, making each relation more distinct.

The next stage consists of a domain ontology in order to increase the specificity and information density for achieving better results and to interpret the underlying domain with greater ease. Ontologies have proven to enhance the efficiency of semantic text analysis by a great fold due to the specificity of classes included in the domain. A topic set containing enriched indexes and words defining the relevance of the URLs are obtained and then passed into a metaheuristic optimization algorithm, namely Simulated Annealing [28] which is probabilistic in nature. The concept of the algorithm has been adopted from a metallurgical process in which an object is heated up to a desirable temperature and then cooled gradually to scale down defects. It is used for optimizing both bound constrained cases where an objective is defined with flexible variables. In the case of a focused crawler, there are constraints of specific indexes and classes.

The formerly mentioned, randomly generates a new point in every iteration. During every iteration this optimization algorithm sees the neighbouring states 'X*' of the current state 'X'. The degree of search which is defined by the distance of the new point from the old point, is based on a probability distribution. The transition probability from one state to another is defined by the acceptance of the probability function (1), which in turn is defined by the listed Eqs. (2) and (3), where 'e' is the energy of X and e new is the energy of X* and T being temperature which varies with time and is a global variable.

$$P(e, e_{new}, T) \tag{1}$$

$$e = E(T) \tag{2}$$

$$e_{new} = E(X^*) \tag{3}$$

The average outcome of the probability function is the local optimum, but the best solution would be the global optimum. The output is in the range [0,1]. The function tends to 0 when T tends to zero and e > enew, else it tends to increase, that is positive. Now to determine the actual contextual similarity of the optimized resultant URLs, a metric measure called web overlap index is used to calculate the co-occurrence to recognize the similarity or relevance, on top of which cross entropy is used to enhance the same process. Web overlap index indicated by Eq. (4) is a kind of page count-based co-occurrence measure formulated for web search engines to weigh the semantic similarity of the enriched key indices, extracted from the URLs in this case. It is used to measure the co-occurrence of two terms based on the page count. Assuming the two words are 'w1' and 'w2', the page count is returned when these words are filled as input.

$$\text{WebOverlabIndex} = log\frac{p(w1, w2)}{\min(p(w1, w2))} \tag{4}$$

The output is generally supported by an empirical value wherein the words that are more closely related have an index > 0.5. Cross entropy is the measure of divergence or difference between two probability distributions for a mentioned set of random variables. Assuming the existence of two variables, here the key indices, 'p' and 'q', the cross entropy of a distribution q relative to a distribution p over a given set is defined by the Eq. (5), Where 'Ep' is the expected value operator of the distribution 'p'. Once the semantic similarity is drawn using both the metrics, the system outputs the refined set of URLs at the click of search.

$$H(p, q) = -Ep[\log q] \tag{5}$$

## 4 Implementation

The proposed architecture has been built on Python 3.9.0 using Google Collaboratory, which has been stacked on multiple GPUs and hosted by Jupyter notebook. The system has been designed using tools of NLP. Few of the main library files that have been included for the implementation of the proposed design includes NLTK, Sci-kit learn, Flash Text, and Scrapy. The model is run on Windows 10 Operating System installed over intel core i7 8th Gen Processor supported by 16 GB RAM. The data URLs are preprocessed, tokenized and made ready for utilization by the classifier. The input-controlled hybrid measure utilizes various methods to refine the data at the very time of input in order to save time and define a concrete path for crawling to support the motive of the focused crawler. Two baseline models have been adopted for performance comparison. One of them has utilized a master slave method, where multiple slaves work together for one master crawler to reduce the processing time. The other, aims at building a database of academicians using a focused crawler experimenting upon the depth of the hyperlinks, selection of keywords and so on using Crawler4j and open source, using a java-based program.

The proposed approach had been compared with two other variations of its own frame- work and the two adopted baseline models for experimentation, namely, Proposed Approach without AI Classification Fog, Proposed Approach without Simulated Annealing, OFCW, DFCI. Precision(%), Recall(%), Accuracy(%), F-measure(%) and Harvest Rate have been calculated for the mentioned, depicted by Eqs. (6), (7), (8), (9) and (10) respectively. From Fig. 2 we can see that the accuracy of performance of the methodology proposed is 90.12%, the harvest rate is 95.82% and an F-measure is 89.89% whereas the accuracy of the baseline papers OFCW and DFCI are 78.21 and 81.32% respectively. The proposed approach without the simulated annealing displays a decreased accuracy of 85.12% because of the comparatively slow performance of the model due to the presence of a hardware and multiple index enrichment modules, as the metaheuristic optimization algorithm plays an important role in reducing the processing time (Fig. 1).

$$\text{Precision} = \frac{\text{True number of Positives}}{\text{True number of Positives} + \text{False number of Positives}} \tag{6}$$

$$\text{Recall} = \frac{\text{True number of Positives}}{\text{True number of Positives} + \text{False number of Negatives}} \tag{7}$$

$$\text{Accuracy} = \frac{TP + TN}{TP + FP + FN + TN} \tag{8}$$

$$\text{F} - \text{Measure} = \frac{2(\text{Precision} \times \text{Recall})}{\text{Precision} + \text{Recall}} \tag{9}$$

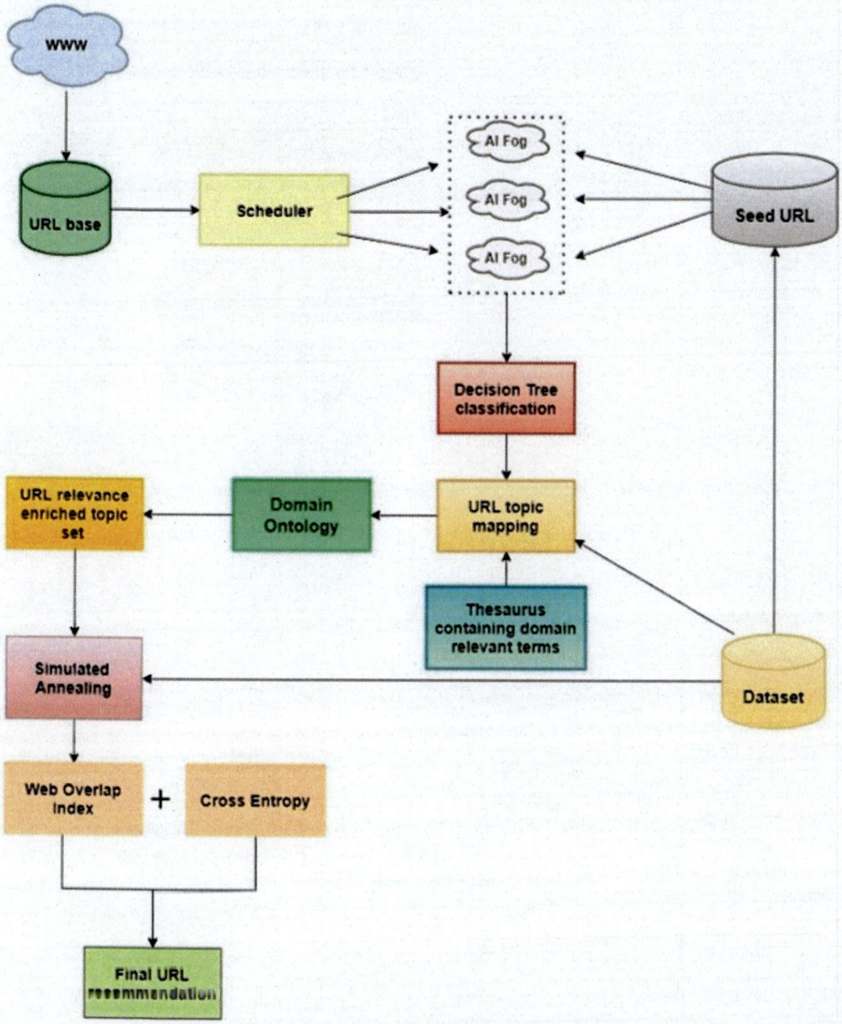


**Fig. 1** Proposed system architecture

$$\text{Harvest Rate} = \frac{\textit{number of relevant web pages}}{\textit{total number of web pages}} \tag{10}$$

The approach without the AI classification fog displays an accuracy of 84.63% due to the reduced efficiency as the fog plays the important role of segmenting the load of URLs and acts as an interface between the classifier and the scheduler. Figure 3 shows a line graph plotted against the number of pages crawled to the actual number of relevant web pages fetched, also called harvest rate. The harvest rate can be defined as the ratio of the relevant web pages to the total number of web pages retrieved.

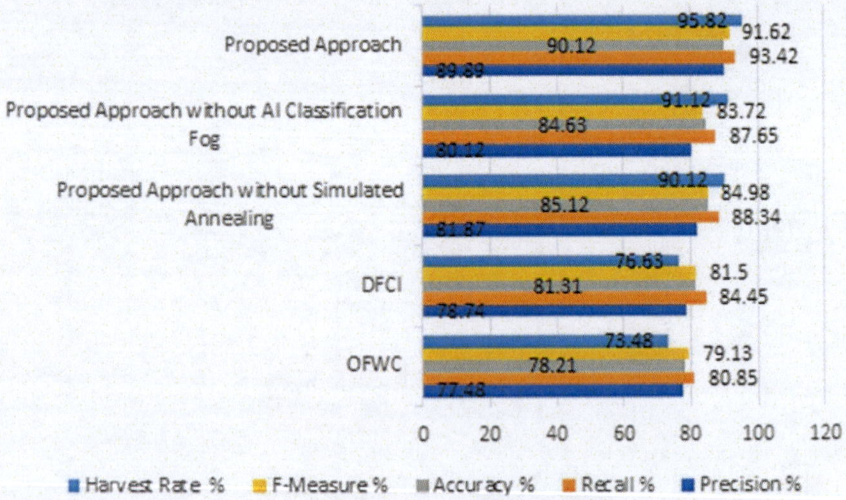


**Fig. 2** Performance comparison graph experimented methods

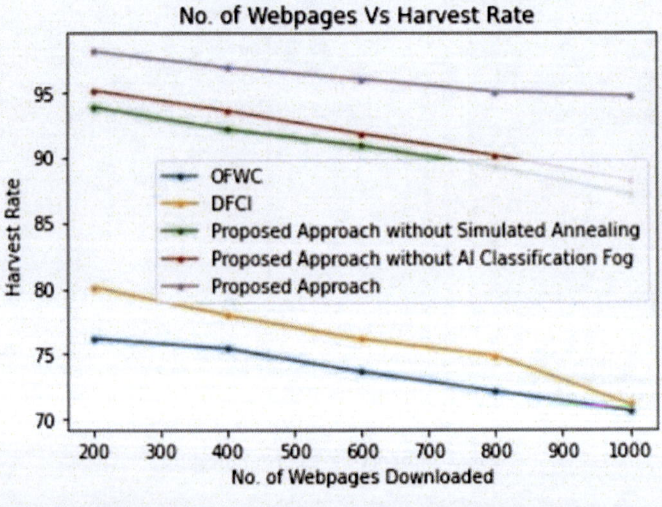


**Fig. 3** Harvest rate per number of web pages crawled

The percentage of harvest rate for downloading 200–1000 webpages in total, in increments of 200 for the experimented approaches are 76.18, 75.42, 73.62, 72.12, 70.63 for OFCW, 80.12, 77.95, 76.12, 74.85, 71.12 for DFCI, 93.87, 92.12, 90.87, 89.23, 87.12 for Proposed Approach without Simulated Annealing, 95.12, 93.58, 91.79, 90.12, 88.2 for the proposed Approach without AI Classification Fog, 98.12, 96.87, 95.99, 95.01, 94.71 for Proposed approach, respectively. This depicts the correctness of the crawler. Figure 4 depicts the processing time for each of the experimented methods and the proposed approaches seen to be much more efficient than the others with an execution

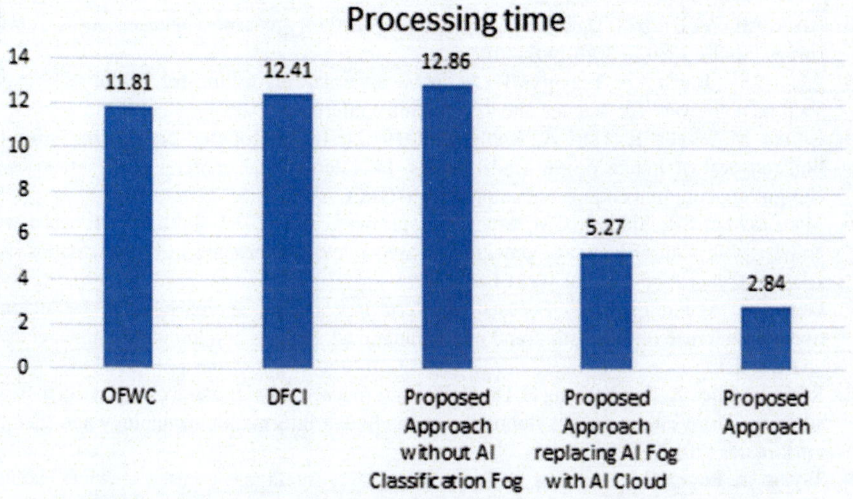


**Fig. 4** Processing time comparison for proposed approach measure in ms

time of 2.84 ms, therefore displaying a clean success in the objective of improving the efficiency of a focused crawler. The proposed approach has proved to be better than the adopted baseline models OFCW and DFCI, due to the inclusion of the optimization algorithm to hold a high and steady efficiency and the AI fog, overcoming the drawbacks of multi master slave processing systems. The speed of simultaneous processing with AI fog is balanced and validated by the Simulated Annealing algorithm, hence catering to the requirement of a reliable improvement.

## 5 Conclusion

The requirement of the web is undeniable in the present worldly scenario. Be it a matter of research, development, casual surfing, studying, people are highly dependent on search engines and web crawlers. Overcoming the disadvantages of the adopted baseline models, which was to use multiple slaves in the master slave model, the proposed approach utilizes an AI classified fog for reducing the time taken for traversal between scheduler and classifier and a meta heuristic optimization algorithm to compensate for the queuing time in the OFWC baseline model. Various factors that affect crawling were also adopted from the DFCI baseline model. The proposed approach is seen to have an accuracy of 90.12% showcasing a 11 and 9% increase from the baseline models respectively and a highly optimized processing time of 2.84 ms.

## References

1. Quinlan JR (1986) Induction of decision trees. Mach Learn 1(1):81–106
2. Sobecki A, Szyman´ski J, Gil D, Mora H (2019) Deep learning in the fog. Int J Distrib Sensor Netw 15(8):1–17

3. Ali A, Alfayez F (2018) Hani Alquhayz semantic similarity measures between words: A brief survey. Sci Int Lahore 30(6):907–914
4. Ahmad SR, Bakar AA, Yaakub MR (2015) Metaheuristic algorithms for feature selection in sentiment analysis. In: Science and information conference (2015)
5. Kumar M, Bhatia R, Ohri A, Kohli A (2016) Design of focused crawler for information retrieval of Indian origin academicians. In: International conference on advances in computing, communication, and automation (ICACCA)
6. Mani Sekhar SR, Siddesh GM, Manvi SS, Srinivasa KG (2019) Optimized focused web crawler with natural language processing based relevance measure in bioinformatics web sources. Cybern Inf Technol 19(2):146–158
7. Gupta A, Anand P (2015) Focused web crawlers and its approaches. In: 1st international conference on futuristic trend in computational analysis and knowledge management (ABLAZE)
8. Wang W, Chen X, Zou Y, Wang H, Dai Z (2010) A focused crawler based on Naive Bayes classifier. In: Third international symposium on intelligent information technology and security informatics (2010)
9. Taylan D, Poyraz M, Akyokus S, Ganiz MC (2011) Intelligent focused crawler: learning which links to crawl. In: International symposium on innovations in intelligent systems and applications (2011)
10. Pant G, Srinivasan P, Menczer F (2004) Crawling the web. In: Web dynamics, pp 153–177. Springer, Heidelberg https://doi.org/10.1007/978-3-662-10874-1_7
11. Deepak G, Teja V, Santhanavijayan A (2020) A novel firefly driven scheme for resume parsing and matching based on entity linking paradigm. J Discrete Math Sci Crypt 23(1):157–165
12. Deepak G, Santhanavijayan A (2020) OntoBestFit: A best-fit ocurrence estimation strategy for RDF driven faceted semantic search. Comput Commun 160:284–298
13. Kumar N, Deepak G, Santhanavijayan A (2020) A novel semantic approach for intelligent response generation using emotion detection incorporating NPMI measure. Procedia Comput Sci 167:571–579
14. Deepak G, Kumar N, Santhanavijayan A (2020) A semantic approach for entity linking by diverse knowledge integration incorporating role-based chunking. Procedia Comput Sci 167:737–746
15. Haribabu S, Kumar PSS, Padhy S, Deepak G, Santhanavijayan A, Kumar N (2019) A novel approach for ontology focused inter-domain personalized search based on semantic set expansion. In: Fifteenth international conference on information processing (ICINPRO), pp 1–5. IEEE, December 2019
16. Deepak G, Kumar N, Bharadwaj GVSY, Santhanavijayan A (2019). OntoQuest: an ontological strategy for automatic question generation for e-assessment using static and dynamic knowledge. In 2019 fifteenth international conference on information processing (ICINPRO), pp 1–6. IEEE, December 2019
17. Kaushik IS, Deepak G, Santhanavijayan A (2020) QuantQueryEXP: A novel strategic approach for query expansion based on quantum computing principles. J Discrete Math Sci Crypt 23(2):573–584
18. Varghese L, Deepak G, Santhanavijayan A (2019) An IoT analytics approach for weather forecasting using raspberry Pi 3 Model B+. In: Fifteenth international conference on information processing (ICINPRO), pp 1–5. IEEE, December 2019
19. Deepak G, Priyadarshini S (2016) A hybrid framework for social tag recommendation using context driven social information. Int J Soc Comput Cyber-Phys Syst 1(4):312–325
20. Deepak G, Priyadarshini JS (2018) A hybrid semantic algorithm for web image retrieval incorporating ontology classification and user-driven query expansion. In: Rajsingh E, Veerasamy J, Alavi A, Peter J (eds) Advances in Big Data and Cloud Computing, vol 645. Springer, Singapore, pp 41–49. https://doi.org/10.1007/978-981-10-7200-0_4

21. Deepak G, Gulzar Z (2017) OntoEPDS: Enhanced and personalized differential semantic algorithm incorporating ontology driven query enrichment. J Adv Res Dyn Control Syst, 9(Specia):567–582
22. Shreyas K, Deepak G, Santhanavijayan A (2020) GenMOnto: A strategic domain ontology modelling approach for conceptualisation and evaluation of collective knowledge for mapping genomes. J Stat Manag Syst 23(2):445–452
23. Deepak G, Kumar AA, Santhanavijayan A, Prakash N (2019) Design and evaluation of conceptual ontologies for electrochemistry as a domain. In: 2019 IEEE international WIE conference on electrical and computer engineering (WIECON-ECE), pp 1–4. IEEE
24. Deepak G, Priyadarshini JS (2018) Personalized and enhanced hybridized semantic algorithm for web image retrieval incorporating ontology classification, strategic query expansion, and content-based analysis. Comput Electr Eng 72:14–25
25. Deepak G, Ahmed A, Skanda B (2019) An intelligent inventive system for personalised webpage recommendation based on ontology semantics. Int J Intell Syst Technol Appl 18(1/2):115–132
26. Deepak G, Kasaraneni D (2019) OntoCommerce: an ontology focused semantic framework for personalised product recommendation for user targeted e-commerce. Int J Comput Aided Eng Technol 11(4/5):449–466
27. Santhanavijayan A, Naresh Kumar D, Deepak G (2020) A novel hybridized strategy for machine translation of Indian languages. In: Reddy V, Prasad V, Wang J, Reddy K (eds) Soft computing and signal processing, ICSCSP 2019. Advances in intelligent systems and computing, vol 1118, p 363. Springer, Singapore. https://doi.org/10.1007/978-981-15-2475-2_34
28. Van Laarhoven PJM, Aarts EHL (1987) Simulated annealing: theory and applications MAIA, vol 37. Springer, Dordrecht, pp 7–15

# Implementation Pathways of Smart Home by Exploiting Internet of Things (IoT) and Tensorflow

Rahul Sarawale[1(✉)], Anupama Deshpande[1], and Parul Arora[2]

[1] J.J.T. University, Rajasthan, India
[2] JSPM's ICOER, Pune, India

**Abstract.** Today's world is getting influenced by cutting edge technologies. We have experienced how technology played a key role during COVID-19 pandemic period in all over the world including fields like health sector, information technology, education, finance etc. With the rapid and huge development in electronic hardware technologies in the last decade resulted into Artificial Intelligence (AI) is now driving numerous fields. The homes have been becoming smart since the last decade and now AI is putting intelligence into the smart home forming a smart intelligent home. The things or devices of the smart home are controlled and networked by the means of the Internet of Things (IoT) technology. The computer vision is an emerging field which is embedding a smart vision to IoT based systems including smart home as well. The paper work presented here is based on experimentation with smart home using IoT and Computer Vision technologies.

**Keywords:** Artificial intelligence · Internet of Things (IoT) · Machine Learning · Smart home

## 1 Introduction

The basic ignition of this work is paved in automation. Then the automation could be from the endless list including office, home, factory, school/college campus, robotics, forest caring, agriculture, space vehicle etc. This automation brings more merits like reliability, easiness, correctness, less time and efforts in operation etc.

After doing rigorous literature survey, we decided to work on home automation utilizing promising technologies. The smart intelligent home implementation is becoming the essentiality of our homes from various perspectives like security, caring elderly persons and children, home automation, smart city as well.

The main motivation of this work is to exploit the essence of cutting edge technologies including Artificial Intelligence, Computer Vision and IoT for betterment of human life and societal benefits. This paper is logically organized in sections as – It first introduces Technological know-how as a part of Introduction in connection with the building a smart intelligent home, then it opens the research doors by presenting State-of-the-art of smart home realm, then it covers the Methodology, later it illustrates the Result and Discussion with the proof of concepts of the pilot project work and at the end technically it wraps up with the Conclusion and Future scope.


S. Shukla et al. (eds.), *Data Science and Security*, Lecture Notes in Networks and Systems 290,
https://doi.org/10.1007/978-981-16-4486-3_53

## 1.1 Technological Know-How

This work focuses on pathways to implement smart home by utilizing machine learning and IoT technologies. This section briefly introduces the technical aspects of the proposed work.

### 1.1.1 Hardware

For demonstration purpose, the hardware utilized for this work includes Raspberry Pi module, pi camera, screen monitor, motion sensor, light bulb, smart phone etc. The Raspberry pi is the most powerful computer for prototyping the embedded and IoT applications.

The pi module is used to control the devices. Here laptop or desktop can be used as a monitor. The PIR sensor can be used for presence detection of human. Pi camera is used to capture images. We can use any dedicated voice assistant like Google assistant, Amazon Echo or we can install ALEXA app on a smart phone.

### 1.1.2 Software

The software part comprises of Node-RED tool, Raspberry pi OS, Node-RED, Tensorflow Node-RED nodes, VNC viewer, PuTTY, ALEXA App etc.

The VNC viewer is installed on the pi and the laptop for the purpose of desktop sharing. The Node-RED is the programming tool to wire up the hardware and APIs. We can install ALEXA app on smart phone as a voice assistant. The PuTTY is used to run session remotely. We can conduct a remote session using this software. With SSH connection type and providing "ip address" of Raspberry pi module in PuTTY, we can connect with pi module. Thus we can operate pi module as a computer. We can install operating systems like Windows, Linux, Android or Raspberry Pi OS (formerly known as Raspbian OS) on the pi module.

The SSD is a single shot detector. It means that this detector requires a single image for detecting multiple objects [12]. TensorFlow is growing framework in the field of machine learning. It comes with features like high availability and flexibility. The object detection can be implemented with the help of TensorFlow [13].

Pose estimation is utilized to estimate the human pose from a video as well as an image. Basically pose estimation is a computer vision technique. In videos or images, we can identify the human figure like elbow position. This technique cannot recognize the human means that who is in video or image. The PoseNet receives input in the form of processed camera image and then it gives the keypoints information. The keypoints are marked as part ID. These points are given with a confidence score and the score ranges from 0.0 to 1.0 [14].

# 2 State-of-the-Art

The state-of-the-art of the smart home realm depicts challenging practical approaches presented by authors, researchers and different companies/firms. Some of the influencing review notes are summarized below.

The paper [3] demonstrated Rudas system. It provides energy efficient and low cost system. The system uses a single board computer, microcontroller, camera, sensors, web technology, server system, database system. The proposed system uses a fuzzy logic artificial intelligence to control the light intensity and air condition of a room. In this system, the home owner can monitor and manage all smart home devices. Ultimately, the proposed work includes security, monitoring and energy saving.

The paper [4] presented a Computer Vision based in an IoT platform. The platform is called as Midgar. The computer vision is used to analyze the picture and returns the result to IoT platform. The proposed system can provide automation and improve the security related to homes, industries, towns, cities etc.

The paper [5] proposed a people detection system which combines computer vision and IoT technology. The system uses a Raspberry PI 3 and PIR sensor that are used for monitoring and getting alerts when human movement is detected.

The paper [6] demonstrated an intelligent security system based on visual surveillance. The system provides a home security with a high level of security. The developed system can be used in home automation and company premises. The developed system is designed using Beagle Bone Black (BBB) board. The system includes GSM module for the purpose of communication and OpenCV for video surveillance purpose.

The smart home system [1] utilizes a Raspberry Pi with a camera and a PIR as a motion sensor. Here, Pi module captures images with the help of camera and then captured images can be sent through Email via TCP/IP. When the system identifies an intruder then voice alert commands can be sent by authorized person.

The smart surveillance monitoring system [2] uses mobile technology in connection with home security and control applications. It provides smoke detection and human detection. It uses Raspberry Pi with OpenCV software and PIR sensor and as per the algorithm the captured images by camera can be sent to user recipient's mail via Wi-Fi.

The daily activities can be done with the help of IoT devices and this is very helpful for elderly people. The elderly people can schedule their daily activities, get required information, dial emergency call etc. Such activities can be done with their voice. The IoT technology can be used to make easeness in their life without taking any help. The voice recognition softwares are available in the market including Siri, Alexa, Cortana, Google assistant etc. These recognition systems provide a reminder or a voice message related to any activity in absence of family person or nurses or caretaker. The problems associated with elderly people can include falling down that may cause serious injury, memory loss that may cause missing of the medical doses, social isolation may cause depression etc. The voice recognition system can provide assistance in different applications including doing daily activities, budgeting, billing, emergency actions, security, weather updates etc. [7].

## 3 Methodology

This paper focuses on work carried out in making home to be a smart intelligent home. We have incorporated this work with experimentation. We found that, more experimentation yet to be done in view of making home to be smart intelligent home. There is a huge scope towards the aforementioned research field.

We carried out experimentation towards smart intelligent home. It includes hardware namely- Raspberry Pi, Laptop, Pi camera, sensor and software namely- Raspbian OS (Raspberry pi OS), Node-RED, Tensorflow. For the experimentation, Raspberry Pi 3 Model B+ is used. The Node-RED is installed with "npm" command on Raspberry pi [11].

We can use Docker installation for the same. The Node-RED can be started with a command "node-red-start". After successful starting the node-red, just go to the browser and type http://localhost:1880 or you can use "ip address" to open node-red editor with the help of http://<ip-address>:1880. Then on workspace you can drag a node from the palette. As per the application, you can drag the nodes and then can wire up the nodes as per the flow. Finally you can deploy the flow and so that the selected flow will exist in node-red editor.

The aforementioned steps are followed and deployed the flow. The PIR sensor connected to pin number 32 that is GPIO 12 pin of the Raspberry pi. The Raspberry pi node is dragged in the node-red flow. Then pi camera node, Tensorflow nodes, email node, image preview node and debug node are dragged in the flow. The flow is successfully deployed and executed.

Whenever a person comes near to PIR sensor, it senses the movement of human, animals and some other objects. Basically it senses the infrared (IR) radiation and this radiation changes based on the temperature [10].

As PIR detects human movement, pi camera captures the image and the captured image is stored in a specified drive location. The same image is picked up and sent via email node to the specified email id. The tensorflow nodes can be used for person detection. The "tf coco ssd", "cocossd" and "posenet" tensorflow nodes are used in the flow. The coco-ssd detector is a single shot object detector used for jpeg image. It is delivered via an msg.payload in a Node-RED flow. This coco model is loaded locally and so it works locally. The coco-ssd is trained to recognize the objects like- person, car, bus, traffic light, bicycle etc. The posenet basically estimates the human poses [8, 9].

The work also focused to help elderly person in doing activities. In this work, the demonstration of voice controlled device is presented. The light bulb is turned ON and OFF with the help of voice. The elderly persons are not able to wake up or repeatedly go and turn ON and OFF the switches. So they can use voice to control the home devices. To make this in action, we utilized Alexa APP, Redmi Note 8 Pro mobile handset, Raspberry Pi.

## 3.1 Algorithm Steps

This section enlists the algorithm steps of implementing the proposed smart home.

1] Detect motion using PIR sensor.
2] Capture an image using camera once motion is detected.
3] Save the captured image in the specified directory.
4] Obtain the saved image file from the given path.
5] Configure Email node with details and send email with the captured image attachment to the recipient
6] Configure tf coco ssd, coco ssd and posenet.

7] Configure debug and image preview nodes.
8] Deploy Node-RED flow and observe the received mail, image preview and debug message.
9] Configure ALEXA App for the devices to be controlled using voice commands.

# 4 Result and Discussion

This section illustrates experimentation results with the proof of concepts and distributed into three subsections as – Node-RED flow execution, Node-RED debug message and Voice assistance operation.

## 4.1 Node-RED Flow Execution

The Fig. 1 shows the Node-Red flow. The debug message is shown in Fig. 2. The separate debug message is also mentioned below in Table 1.

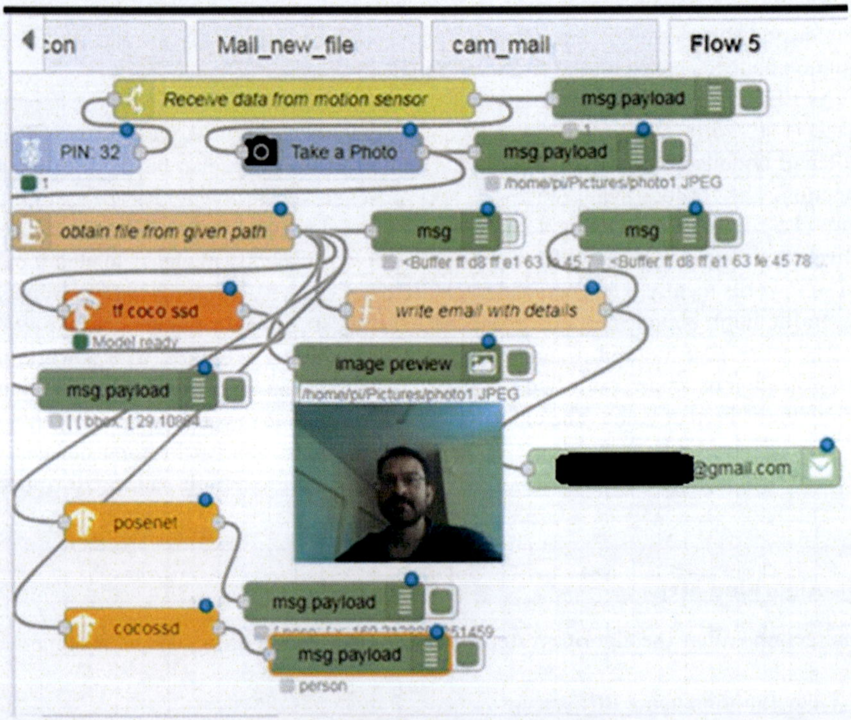


**Fig. 1** Nod-RED flow for the smart home

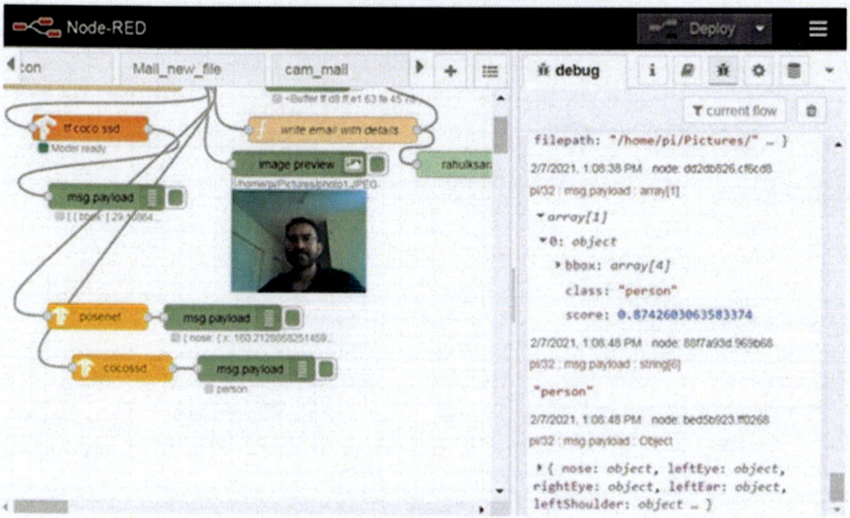


**Fig. 2** Node-RED Debug message

The "tf coco ssd" and "cocossd" tensorflow nodes are utilized for person identification. The "tf coco ssd" identified the person with a score of 0.8742603063583374.

The "posenet" tensorflow node is used to give the coordinate details. The coordinate statistics for nose- x: 160.21288682514592 and y: 130.8382098127432, for leftEye-x: 180.81808958535993 and y: 114.01294990272373, for rightEye-x: 145.26695874878405 and y:117.32782861138134, for leftEar-x: 206.35998677650778 and y:138.40425127675098, for leftShoulder-x: 255.74199750729574 and y:213.7329006566148, for rightShoulder-x: 100.92439050340467 and y:232.67580024927042 and for center-x: 178.3331940053502 and y: 223.2043504529426.

The mail is triggered when person is detected. The mail is sent with image attachment to the recipient. The sent mail and received mail are shown in Fig. 3 and Fig. 4 respectively.

## 4.2 Node-RED Debug Message

The Node-RED displays a debug message. The below tabular result is a part of debug message which is displayed while executing the flow.

The first half of the below Table 1 points out the statistics in connection with messages generated by the debug nodes and they are depicted as msg.payload in the Node-RED flow as shown in Fig. 1 above. Also this part shows the path of the image wherein the captured image is saved automatically as per file path specified in photo node of the Node-RED flow. The first half also mentions the tensorflow results as a part of human identification and tells the result as "person" with the amount of percentage by which the tensorflow model identified as the person.

**Table 1** The Debug message generated in Node-RED flow

| | |
|---|---|
| 2/7/2021, 1:08:32 PMnode: 437deadc.f6f884<br>pi/32 : msg.payload : number 1<br>2/7/2021, 1:08:35 PMnode: cc22af79.f2b18<br>pi/32 : msg.payload : string[29]<br>"/home/pi/Pictures/photo1.JPEG"<br>2/7/2021, 1:08:38 PMnode: 690bcf0a.feb45<br>Your Attachment File for Sun 13:08 : msg : Object<br>{ topic: "Your Attachment File for Sun …", payload: buffer[64377], _msgid: "5a2c98e4.02dda8", filename: "/home/pi/Pictures/photo1.JPEG", filepath: "/home/pi/Pictures/" … }<br>2/7/2021, 1:08:38 PMnode: dd2db826.cf6cd8<br>pi/32 : msg.payload : array[1]<br>array[1]<br>0: object<br>bbox: array[4]<br>class: "person"<br>score: 0.8742603063583374<br>2/7/2021, 1:08:48 PMnode: 88f7a93d.969b68<br>pi/32 : msg.payload : string[6]<br>"person" | 2/7/2021, 1:08:48 PMnode: bed5b923.ff0268<br>pi/32 : msg.payload : Object<br>object<br>nose: object<br>x: 160.21288682514592<br>y: 130.8382098127432<br>leftEye: object<br>x: 180.81808958535993<br>y: 114.01294990272373<br>rightEye: object<br>x: 145.26695874878405<br>y: 117.32782861138134<br>leftEar: object<br>x: 206.35998677650778<br>y: 138.40425127675098<br>leftShoulder: object<br>x: 255.74199750729574<br>y: 213.7329006566148<br>rightShoulder: object<br>x: 100.92439050340467<br>y: 232.67580024927042<br>center: object<br>x: 178.3331940053502<br>y: 223.2043504529426 |

**Fig. 3** Sent mail

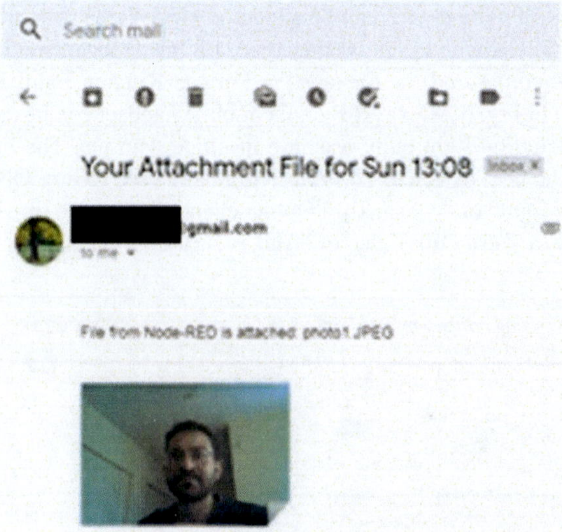


**Fig. 4.** Received mail

On the other half of the below table, the results of "posenet" node are depicted with the co-ordinate values for nose, leftEye, rightEye, leftEar, leftShoulder, rightShoulder and center as mentioned in the right half of the Table 1.

## 4.3 Voice Assistance Operation

The voice controlled demonstration is carried out with the help of Alexa APP. For the demonstration, the mobile handset Redmi Note 8 Pro is used. The smart home skill is

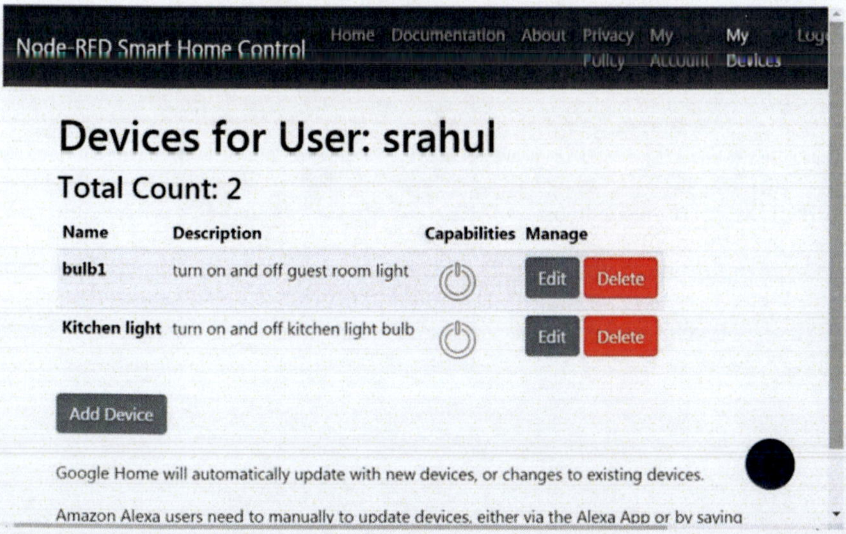


**Fig. 5** Adding device on Control App

added on Alexa APP. The device can be added on Node-RED Smart Home Control as shown in Fig. 5. The Kitchen light is added here for the experimentation.

The group is configured on Alexa APP. For the Kitchen light, Kitchen group is created as shown in Fig. 6. Alexa APP on the mobile handset can be used to control the home devices including light bulb, washing machine, Fan etc. The light bulbs can be controlled with the help of voice. The voice command used to turn ON the light bulb is "Alexa, Turn ON light" or "Alexa, Turn ON Kitchen light" and to turn OFF the Kitchen light bulb is "Alexa, Turn OFF light" or "Alexa, Turn OFF Kitchen light".

**Fig. 6** Alexa App Dashboard

The Kitchen light is turned ON and OFF with the help of voice commands as shown in Fig. 7 and Fig. 8 respectively.

**Fig. 7** Light ON on Alexa App

**Fig. 8** Light OFF on Alexa App

## 5 Conclusion and Future Work

In this work, we demonstrated the pathways towards intelligent smart home by exploiting the technologies like IoT and Computer Vision technique. With the help of Raspberry Pi module and Smart phone we can control and manage our home locally as well as remotely.

For the demonstration purpose, we used Node-RED platform. As the motion is detected by motion sensor, the camera captures the image of the person then mail node sends the captured photo over mail to recipient mail id from the given sender's mail id. We can monitor our home remotely and can provide access to authorized person. The "tf coco ssd" and "cocossd" tensorflow nodes are utilized for the purpose of human detection. The voice assistant operation also incorporated in this work for elderly people.

The future scope of the work can focus on making energy efficient smart intelligent home. The security concerns and issues may come into picture as we are dealing with internet and hence to overcome such threats we need to focus on security aspect of the smart home. The high performing AI algorithms can also be incorporated to improve the performance of existing smart home models.

**Acknowledgements.** We would like to thank J.J.T. University, Rajasthan for a great support. We also would like to thank the researchers and authors who contributed directly and indirectly in the field of Machine Learning, AI and Smart home technologies.

## References

1. Anwar S, Kishore D (2016) IOT based smart home security system with alert and door access control using smart phone. IJERT 5(12):504–509
2. Chuimurkar R, Bagdi V (2016) Smart surveillance security & monitoring system using Raspberry PI and PIR sensor. IJIRAE 2(1):1–6
3. Crisnapati PN, Wardana INK, Aryanto IKAA (2016) Rudas: energy and sensor devices management system in home automation. In: IEEE Region 10 Symposium, pp 184–187
4. Garcia CG, Meana-Llorian D, Pelayo G-Bustelo BC, Cueva Lovelle JM, Garcia-Fernandez N (2017) Midgar: detection of people through computer vision in the Internet of Things scenarios to improve the security in Smart Cities, Smart Towns, and Smart Homes. Future Gener Comput Syst 76:301–313
5. Othman NA, Aydin I (2017) A new IoT combined body detection of people by using computer vision for security application. In: IEEE 9th international conference on computational intelligence and communication networks (CICN), pp 108–112
6. Sefat MS, Khan AAM, Shahjahan M (2014) Implementation of vision based intelligent home automation and security system. In: IEEE international conference on informatics, electronics & vision (ICIEV)
7. Aging in place. https://aginginplace.org/voice-recognition-innovation-and-the-implications-for-seniors/
8. Node-RED tensorflow node. https://flows.nodered.org/node/node-red-contrib-tensorflow
9. Node-RED for object detection using tfjs coco-ssd. https://flows.nodered.org/node/node-red-contrib-tfjs-coco-ssd
10. Passive infrared sensor. https://en.wikipedia.org/wiki/Passive_infrared_sensor
11. Node-RED on Raspberry pi. https://nodered.org/docs/getting-started/raspberrypi
12. Pienaar SW, Malekian R (2019) Human activity recognition using visual object detection. In: IEEE 2nd wireless Africa conference (WAC), pp. 1–5
13. Xie L, Guo X (2019) Object detection and analysis of human body postures based on TensorFlow. In: IEEE international conference on Smart Internet of Things (SmartIoT), pp 397–401
14. Pose estimation. https://www.tensorflow.org/lite/examples/pse_estimation/overview